# THE OIL FRATERNITY IN TEXAS

# THE OIL FRATERNITY IN TEXAS

## MORAL ECONOMY AND PETROLEUM ENGINEERING SCIENCE

## EDWARD W. CONSTANT II

TEXAS TECH UNIVERSITY PRESS

This book is typeset in EB Garamond. The paper used in this book meets the minimum requirements of ANSI/NISO Z39.48-1992 (R1997). ♾

Designed by Hannah Gaskamp
Cover designed by Hannah Gaskamp

Library of Congress Cataloging-in-Publication Data

Names: Constant, Edward W., II, 1943– author. Title: The Oil Fraternity in Texas: Moral Economy and Petroleum Engineering Science / Edward W. Constant II. Description: Lubbock, Texas: Texas Tech University Press, [2024] | Includes bibliographical references and index. | Summary: "Fierce competition among the many deals in an oil play advantaged intra-deal altruistic cooperation among its participants, leading to the emergence of a robust, cooperative moral economy in the oil fraternity in Texas. This fraternity coevolved with, and was transformed by, dynamic petroleum engineering science."—Provided by publisher.
Identifiers: LCCN 2023052500 (print) | LCCN 2023052501 (ebook) |
ISBN 978-1-68283-220-2 (paperback) | ISBN 978-1-68283-221-9 (ebook)
Subjects: LCSH: Petroleum industry and trade—Texas. | Petroleum industry and trade—Moral and ethical aspects—Texas. | Petroleum—Economic aspects—Texas.
Classification: LCC HD9567.T3 C66 2024 (print) | LCC HD9567.T3 (ebook) |
DDC 338.2/728209764—dc23/eng/20231207
LC record available at https://lccn.loc.gov/2023052500
LC ebook record available at https://lccn.loc.gov/2023052501

24 25 26 27 28 29 30 31 32 / 9 8 7 6 5 4 3 2 1

Texas Tech University Press
Box 41037
Lubbock, Texas 79409-1037 USA
800.832.4042
ttup@ttu.edu
www.ttupress.org

# THE OIL FRATERNITY IN TEXAS

## MORAL ECONOMY AND PETROLEUM ENGINEERING SCIENCE

### EDWARD W. CONSTANT II

TEXAS TECH UNIVERSITY PRESS

This book is typeset in EB Garamond. The paper used in this book meets the minimum requirements of ANSI/NISO Z39.48-1992 (R1997). ⊗

Designed by Hannah Gaskamp
Cover designed by Hannah Gaskamp

Library of Congress Cataloging-in-Publication Data

Names: Constant, Edward W., II, 1943– author. Title: The Oil Fraternity in Texas: Moral Economy and Petroleum Engineering Science / Edward W. Constant II. Description: Lubbock, Texas: Texas Tech University Press, [2024] | Includes bibliographical references and index. | Summary: "Fierce competition among the many deals in an oil play advantaged intra-deal altruistic cooperation among its participants, leading to the emergence of a robust, cooperative moral economy in the oil fraternity in Texas. This fraternity coevolved with, and was transformed by, dynamic petroleum engineering science."—Provided by publisher.
Identifiers: LCCN 2023052500 (print) | LCCN 2023052501 (ebook) |
ISBN 978-1-68283-220-2 (paperback) | ISBN 978-1-68283-221-9 (ebook)
Subjects: LCSH: Petroleum industry and trade—Texas. | Petroleum industry and trade—Moral and ethical aspects—Texas. | Petroleum—Economic aspects—Texas.
Classification: LCC HD9567.T3 C66 2024 (print) | LCC HD9567.T3 (ebook) |
DDC 338.2/728209764—dc23/eng/20231207
LC record available at https://lccn.loc.gov/2023052500
LC ebook record available at https://lccn.loc.gov/2023052501

24 25 26 27 28 29 30 31 32 / 9 8 7 6 5 4 3 2 1

Texas Tech University Press
Box 41037
Lubbock, Texas 79409-1037 USA
800.832.4042
ttup@ttu.edu
www.ttupress.org

*For*
Edward W. Constant
For richer or for poorer,
In sickness or in health,
An oilman.


Donald Thomas Campbell
Who asked all the right questions,
And knew almost all the right answers.


Thomas Parke Hughes
In loco parentis.


*And for*
Shadow
The great, beautiful, old wild gray cat,
A Russian Blue it seems,
Whose soul held secrets only his eyes could tell,
For he hadn't the tongue to speak.

In all scientific explanation, we explain by explanators which will themselves need explanation.

# CONTENTS

ILLUSTRATIONS                                              xi

                    INTRODUCTION                            3

CHAPTER 1:    THE TIME OF THE PREACHER                      7

CHAPTER 2:    THE MORAL ECONOMY OF THE OIL
              FRATERNITY IN TEXAS                          33

CHAPTER 3:    ESCAPE INTO THE ABSTRACT                     57

CHAPTER 4:    EVOLUTION'S EXPERIMENT:
              BORN OF MONOPOLY AND
              SIRED BY ARBITRARY POWER                     97

CHAPTER 5:    MERE THEORY AND
              SPECULATION                                 117

CHAPTER 6:    REASONABLE MINDS                            145

CHAPTER 7:    USEFUL RELATIONS                            171

CHAPTER 8:    FRUITFUL EMPIRICISM                         199

CHAPTER 9:    JUSTICE, SWEET JUSTICE                      221

CHAPTER 10:   MY HEROES HAVE ALWAYS
              BEEN COWBOYS                                241

ACKNOWLEDGMENTS                                           261
NOTES                                                     263
BIBLIOGRAPHY                                              315
INDEX                                                     355

# ILLUSTRATIONS

## FIGURES

30     Figure 1.1. Oil Production in Texas, and US percent Texas, 1900–1955

30     Figure 1.2. US and Texas Oil Production, and percent Texas, 1935–2019

62     Figure 3.1. Average Investments over Time in Public Good Games with Stable Groups

62     Figure 3.2. Average Investments over Time in Public Good Games with Random Groups

64     Figure 3.3. Effect of Culture upon Public Good Contributions and Antisocial Punishment

72     Figure 3.4. Evolution of the Proportion of Altruistic Cooperators, P, in the Global Population

73     Figure 3.5. Effects of Within-Group (WG) and Between-Group (BG) Selection in Each Generation

74     Figure 3.6. Net Conditional Defection in Each Generation

81     Figure 3.7. Sangren Dialectics of Alienation

82     Figure 3.8. Dynamic Alienated Representation Model

93     Figure 3.9. Instantaneous Present Value of Exponentially Declining Oil Production

93     Figure 3.10. Cumulative PV of Exponentially Declining Oil Production

118     Figure 5.1. The East Texas Oil Field

139     Figure 5.2. Evolution of Within-Deal Altruistic Cooperation with More Frequent Invasions (Colonization Events), Every Third Iteration

175     Figure 7.1. Relation between Gas and Liquid Permeabilities and Liquid Saturations

179     Figure 7.2. Permeability-Saturation Curves for Consolidated and Unconsolidated Sands

187     Figure 7.3. Permeability-Saturation Curves for Oil and Gas at 15 to 25 Percent Connate Water

242     Figure 10.1. Oil Prices, 1870–2020, in Nominal and Real US Dollars

242     Figure 10.2. US Oil Consumption and Source, Millions of Barrels per Day, 1950–2019

243     Figure 10.3. Annual World, US, and Texas Crude Oil Production, 1971–2019, in 1,000s of Barrels

246     Figure 10.4. Vertical, Deviated, and Horizontal Wells

247     Figure 10.5. Mudmotor Power Section, Longitudinal View

## TABLES

75     Table 3.1. Cohorts' Representation in the Global Population

94     Table 3.2. Production History for Uncontrolled versus Controlled Fields

106    Table 4.1. Papers in the Annual Volumes of the Petroleum Division of AIME

114    Table 4.2. Oil and Gas Laws, Texas and United States, 1876–1935

140    Table 5.1. Significant Case Law, 1866–1943

197    Table 7.1. Employment by Sector, Members of the Society of Petroleum Engineers (N=180)

## PHOTOS

8      Charlie, Red, Virgil, and Louis Wimberly and Edward W. Constant, El Dorado, Arkansas, 1925

200    Claude K. McCan, James S. Abercrombie, Ernest O. Thompson, H. G. Nelms, Herman Brown, circa 1959

259    Edward W. Constant, Matagorda County, Texas, 1975

# THE OIL FRATERNITY IN TEXAS

# INTRODUCTION

Oilmen rank right up there with Cowboys, Indians, and Gangsters as cultural icons. Texas independent oilmen, usually not carefully distinguished from those in the rest of the mid-continent or California, are legendary, good and bad: the "Greatest Gamblers," the fiercest competitors, the bitter-end defenders of free enterprise, ever ready to brawl, bet, or bed. And since the late nineteenth century, popular economists, politicians, and scriptwriters have gleefully explored and exploited the "oil cartel," although whose cartel it is and whom it serves are still a little fuzzy. These stereotypes or caricatures (take your pick) might contain a glimmer of fact, but overall, they're about as historically reliable as docudramas or *Dallas*.

Because of their importance to economic life, the world oil industry, that in the United States, and that in Texas also have spawned substantial, serious historical, economic, and political inquiries. The field is immense and may rival that of the industry or even the State of Texas itself. This study is littler. It is tightly focused on three facets of the oil business in Texas: the distinctive oil fraternity that comprised and constituted Texas oilmen; the emergence, development, and articulation of petroleum engineering science and practice, which transformed what oilmen did, how they did it, and what they believed about the world and themselves; and the tortuous path by which regulation of oil and gas production in Texas accommodated, exploited, negotiated, neglected, and sometimes promoted or resisted dynamic changes in scientific understanding of subsurface hydrocarbon resources.

"The oil fraternity in Texas" is their own term and bespeaks the common history and strong communal bonds that came to connect oilmen in Texas, especially those engaged in exploration and production. The oil fraternity in Texas evolved and in turn was defined by its nuanced moral economy, a concept E. P. Thompson articulated and applied to the eighteenth-century English crowd and its enforcement of communal, pre-capitalist economic norms. The primary tenets of the moral economy of the oil fraternity in Texas may be taken to include opportunity for those who belong (in lieu of a basic right to subsistence in Thompson's English communities), altruistic cooperation within the fraternity, privileged access to the resources of Texas, fraternal autonomy and self-regulation, equality and liberty within the fraternity, and useful knowledge.

The first chapter of *The Oil Fraternity in Texas* recounts the early history of oil in Texas, the origins of the population from which the oil fraternity emerged, and the cultural baggage they brought along. The second chapter depicts the moral economy of the oil fraternity in Texas, mostly in their own words. The problem with this, or any, attribution of moral economy and the within-group altruistic cooperation it entails is that it violates the core dogma of orthodox economics: Economic behavior is always and only self-interested. Agents' actions may be strategically "nice" short term, but at some level they are invariably motivated by egoistic incentives. The logic of selfishness is compelling. In any population—of oilmen, business firms, turtles, or amoebas—altruists are disadvantaged relative to opportunists: opportunists benefit from others' altruism but incur no costs. Thus, the reproductive or economic success of opportunists is inevitably greater than that of altruists, and the altruistic dolts are driven to extinction. The intellectual hegemony of egoistic incentives in economics infected political science, anthropology, history—virtually every academic discipline concerned with human behavior. Moral economy is an overt challenge to the primacy of egoistic incentives.

This hegemony of selfishness has been especially deleterious for understanding the oil business. While oilmen *are* ferocious competitors, and there are real and substantial material differences between majors (firms with interests in hydrocarbon exploration, production, refining, distribution, and marketing) and independents (those with interests primarily in exploration and production), this competition and these differences are set in the context of a shared moral economy. Similarly, many orthodox interpretations view the Texas Railroad Commission as a cartel. Although the accusation is leveled often enough in intrafamily squabbles, no consensus exists among historical actors in real time, nor among economists and historians later, about whose selfish interests this alleged cartel serves. Instead, from its origins in railroad regulation in the late nineteenth century until the mid-1970s, the Texas Railroad Commission unproblematically saw its role as that of protector of Texas producers and Texas resources from the depredations and exploitation of outsiders and as a guarantor of opportunity for Texans—including the oil fraternity.

Happily, the last thirty years have seen resurrection in evolutionary theory of an alternative to individual self-interest: group selection. As Darwin himself speculated, a tribe rich in altruists who cooperate, even if it is hazardous or costly to do so, will most likely prevail in conflict with a tribe of mostly opportunists, who individually defect when they sense an advantage in doing so: Altruism can increase in the global population. Contra egoistic incentive presumptions, group selection can rescue altruism. Chapter 3 contains a quantitative simulation model, Group Selection Deal (GSD), adapted from biology and behavioral economics, for how altruistic cooperation could have evolved under the circumstances of the early oil business in Texas. The mathematical mechanics of relevant models are contained in boxes inserted at the end of the chapter; the arguments are verbally summarized in the text. Those not much amused by maths can skim this chapter, so long as they are aware of the takeaway message: Far from being motivated solely by egoistic incentives, group or deal selection in the early years of the oil business in Texas produced a well-winnowed, close-knit oil fraternity possessed and defined by a nuanced moral economy. It ain't *only* about the money.

These first three chapters set the stage for the oil fraternity's waltz across Texas, although it more often resembled a honky-tonk dance floor crowded with folks who didn't know how to dance, peppered with some who were indecently inebriated by sudden great wealth. What brought order to this sometimes-unseemly promenade was petroleum engineering science.

Petroleum engineering, here construed as what is more properly called reservoir engineering, coevolved with the oil fraternity in Texas: the two interacted so intensively that each became the predominant element in the other's selection environment. While an engineering science (as any science) has its own internal logic and dynamic, it doesn't evolve in a social or cultural vacuum. The ongoing adventures and misadventures of the oil fraternity, and its legal travails, channeled the path-dependent evolution of petroleum engineering. Conversely, the emergence and development of petroleum engineering radically redefined settled rules of property and transformed the oil fraternity's world. Because of prorationing (state regulation of oil and gas production), the coevolution of petroleum engineering and the oil fraternity is perhaps more visible than might ordinarily be the case: Petroleum engineering was repeatedly, and literally, put on trial. Administrative and legal processes introduce their own distortions, but they also publicly lay bare internal disputes and the contingent, contested character of engineering knowledge and its use, both within the profession and among those most affected by, and influential in, its evolution.

From the emergence of a scientific petroleum engineering in the mid-1920s into the mid-1960s, the oil fraternity was vexed by a convoluted and frequently uncivil epistemological and legal struggle simultaneously to understand subsurface oil and gas reservoirs and their responses to production, and to contrive a legally tenable and effective system of regulation that conserved the resources yet still assured opportunity for the oil fraternity in Texas. Solving these problems, iteratively, piecemeal, sometimes acrimoniously, and with lots of trial and error (usually in court), is what this book is about. For all its analytical complexity, this remains a very limited study. As important as they are in their own right, refining (and its basis in an equally dynamic chemical engineering science), transportation, distribution, or marketing, as well as national oil, energy, or antitrust policies, receive scant attention other than as they directly affected exploration and production in Texas. The same goes for issues of labor, gender, and ethnicity. In a tome of finite length, broadening scope would necessarily degrade focus.

Despite its limitations, this narrow and particular history has broader implications. Much of the best work on collective cooperation, self-organization, and self-regulation—arguably recognizable as moral economies—has concerned Indigenous communities, from rice farmers in Bali or Sri Lanka, to fishermen in northwestern Mexico or Thailand, to subsistence agriculturalists in Guyana. These moral economies, too, often are transformed by dynamic scientific knowledge, rarely without conflict, and not always for the better. The suspicion is that moral economies emerge to constrain egoistic behavior in most economic communities, even within and across modern organizational boundaries. Likewise, the convoluted, trial and error course of petroleum engineering science and practice implies that science and engineering are, to borrow the title of Richard Westfall's monumental biography of Isaac Newton, "never at rest." Any application of science to policy is likely to be fraught, contested, and acrimonious—and corrigible and hypothetical like scientific knowledge itself.

Historians try to impose order on all this chaos. They tidy. Tidy it wasn't in the time of the preacher in the year of '01.

# CHAPTER 1

# THE TIME OF
# THE PREACHER

It was the time of the preacher
In the year of '01
An' just when you think it's all over
It's only begun.

—WILLIE NELSON[1]

Once upon a time, not so very long ago, in a place not too far and not too near, was Texas.[2] There Everyman, if he just walked upright and had a little bit of luck, could be Rich. Everyman who came there was a pilgrim, but the grail he sought was more likely to be filled with potable water, or black crude oil, than the sweet red wine of eternal salvation.[3] Many in fact probably weren't real sure about the difference, and didn't much care. The oil fraternity in Texas, the fraternity of Everyman, the men who would become the first citizens of that place, were the last of those generations of European descent who looked out on America with the eyes of the Conquistadores and the moral certitude of the Pilgrims. They were the last of those legions who saw in the ground unimaginable riches and an unbounded future, both theirs just for the making, and could honestly call it El Dorado.[4]

Like Cowboys, and Pilgrims, and Conquistadores, oilmen are legendary, even mythical now: bigger than life, boomers, wildcatters, by their own description "the greatest gamblers." Their individual identities were for the most part not much known outside the oil fraternity itself, and few are remembered now. The more successful have their names plastered on every other university hall or library, church education building, or hospital annex in Texas. The less fortunate now are only the pale images typical of a transitory life, men without surnames, mostly even without given names, just "Slim," "High Pockets," "Mexican Joe," "Shorty," "Curly," "Old Man Jim," "One-Eyed McCleary," "Yellow Young," "Buckskin Joe," "Red," "Barefoot Ed." And like ancient, lost battlefields, the places of their lives echo and are still: Desdemona, Daisetta, Sour Lake, Batson's Prairie, South Liberty. But the oilmen's places are

The oil fraternity in Texas: (L–R) Charlie Wimberly, Red Wimberly, Virgil Wimberly (tool pusher), Louis Wimberly, and Edward W. Constant (foreground), El Dorado, Arkansas, 1925. (Photo courtesy of Edward W. Constant)

those more of rusting pipe and rotting storage tanks and the eternal, sweet smell of crude than of grandiose monuments, elaborate commemorative rituals, or marble museums.

The world that birthed the oil fraternity in Texas was not the frontier of legend but the (literally) grubby frontier of cowboys and dirt farmers, of poverty, drought, crop liens, debt, and, as historians so politely phrase it, "agrarian discontent." Theirs was a time in Texas when the county courthouse was still "the capitol of all the world that mattered."[5] Away from the larger towns and the railroads, distance expanded and time slowed down: roads and bridges, for the most part, were still in the future. To cross the upper end of Galveston Bay east of Houston you hired a rowboat;[6] or, further down the bay, at low tide, a wagon to take you, hub deep, right across Red Fish Reef from San Leon to Smith Point. To buy pipeline right-of-way from Oklahoma to Port Arthur before automobiles and roads were common meant depending "on trains that ran seldom, freight trains, section hand cars, saddle horses, buggies, and hacks."[7] Even in the early 1920s, to go from Wichita Falls to Ranger or Desdemona, or virtually anywhere in South or West Texas, meant following dirt tracks cross-country in a Model T Ford.[8] Identity was personal and local and face-to-face. For an outsider to go oil prospecting along the Sabine River required local escort: the only strangers ordinarily seen back in those parts by themselves were revenuers, and they were likely to be shot.[9]

Rare indeed is the historical epoch that can be said unequivocally to have begun in a given place, in a given year, on a given day, at a given hour. But Texas's, and the world's, oil century began on one of those cold, raw, drizzly Gulf Coast days, exactly at 10:30 a.m., on January 10, in the year 1901, on the "Big Hill" four miles south of Beaumont known forever after as Spindletop. The oil century began with an eruption the likes of which American eyes had never seen: the Lucas "gusher," the first oil well to wear, or deserve, the name, "came in," blowing six tons of four-inch drill pipe followed by maybe 75,000, maybe 100,000 barrels of oil a day a hundred feet over the top of the derrick. It blew for ten days until finally capped, while the crew frantically built widening rings of dirt levees to try to contain the oil.

What made oilmen, and the oil fraternity in Texas, exceptional was opportunity, opportunity on a scale and of a magnitude provided virtually no other generation in no other place in no other time. Spindletop came in toward the end of the long deflation that followed the War Between the States and the sad years of agrarian discontent. The average salary for a schoolteacher in Texas was $300 a year; the average annual per capita personal income $138.[10] A single oil well producing 20,000 barrels a day, which was just middling in those years, could pay out more in that single day than an honest man could ever hope to earn in a whole lifetime of honest toil. But unlike the remote New World of the Conquistadores or the city upon a hill the Puritans imagined, the oil fraternity in Texas, and the opportunity it offered, was open, free, and accessible to Everyman, to Everyman who would belong to the fraternity, adhere to its folk ways, speak its language, and most of all, believe in its opportunity.[11]

## GENESIS

When Spindletop came in, oil was not entirely new to Texas. Commercial oil production first developed at Corsicana in 1895, following the accidental discovery of oil in 1893 while drilling a water well. Local businessmen contacted James M. Guffey and John H. Galey (Guffey and Galey), Pittsburgh wildcatters, who sent cable tool or standard drilling rigs and crews south. All wells in the old oil areas of the Northeast (Pennsylvania, Ohio, West Virginia) were drilled using cable tools until well into this century, and some few are still used there. Cable tools were ideal for shallow wells in hard rock formations but did poorly in soft ground or shale. A cable tool rig chisels. A "bit," or big chisel, is hung on the end of a rope (hence the term "drill string," still applied to drill pipe) and repeatedly picked up and dropped, thus breaking up rock. Periodically, the drill string is removed from the well, the bit may be sharpened, or "dressed," and a clamshell bucket used to bail the chips out of the hole. More casing or pipe may be set, and the process repeated.

Apparently the cable tool rigs performed poorly at Corsicana, and most of the drilling was done by newly introduced rotary rigs. Best evidence suggests that rotary rigs had been used to drill water wells on the Great Plains beginning in the 1880s, and they seem to have been used extensively for that purpose in North Texas. Rotary rigs work on a different principle. A rotary rig has a horizontal turn table, or rotary table, and a vertical "kelly joint"—think of an old-fashioned record player with a square hole and a long, square vertical spindle that can slide up and down as the disk is rotated. Kelly joints are typically thirty to sixty feet long. A bit is attached to a "joint" of heavy pipe, which is attached below the kelly joint: as the rotary table turns, it turns the kelly joint and hence the pipe and the bit. As the hole deepens, the drill string is lengthened by adding joints of pipe below the kelly joint. Periodically, the whole string has to be pulled to change the bit. A drilling fluid, originally just water, then, literally, mud (and now sophisticated chemical mixtures, still called "mud") is circulated at high pressure down the inside of the drill pipe, out through holes in the bit, and back up the outside of the pipe, between it and the wellbore wall. The mud flushes out the cuttings and lifts them up the hole (in place of bailing), stabilizes the hole wall, which is critical in soft formations, and cools the bit. Rotaries dominated drilling in soft Gulf coastal formations from the beginning and are generally much faster at "making hole." But their general adoption was slow: they were so fast that they had a tendency to punch right through productive formations without anyone noticing. As late as 1917–1918, an estimated 100,000 of the 109,000 wells completed in the United States were drilled with cable-tool rigs.[12] Not until the development of reliable, efficient rotary core bits in the mid-1920s did rotaries become more popular, and cable-tool rigs remained competitive for hard formations, especially for shallow wells, such as some of those in West Texas, until development of hard rock cone bits for rotaries, also in the 1920s.[13]

As is to be expected when both pride and money are at stake, some animosity arose at Corsicana among the outsiders, the cable-tool drillers, and the indigenous rotary drillers. The rotaries "were drilling wells in days where it took them [the cable tool rigs] months, and they didn't like that competition."[14] Moreover, the Yankee standard rig drillers, the true aristocrats of a craft technology largely bereft of formal engineering, did not always get the respect they thought their due: said one local recruit of the outlanders, "Some of 'em couldn't fish a bird cage out of a washtub if you gave them a pitchfork. That's the God's truth."[15] Many cable tool drillers ultimately made the transition to rotaries, others did not. As one said on taking his leave of a rotary crew, "You can have this rotatin', vibratin' chain breakin' sonofabitch; I'm goin' home." None of the Corsicana wells were particularly prolific, averaging less than ten barrels a day, but ultimately the field did produce over two million barrels of oil.[16] But what became the pattern for the evolution of the oil fraternity in Texas was set. Many of the people who "made" Spindletop started at Corsicana.

But Spindletop really made oil in Texas, and the oil fraternity in Texas. Patillo Higgins, a Beaumont real estate broker, booster, and the town's only gunsmith, had been trying for nearly ten years to get a well drilled on the Big Hill, which towered a magnificent twelve feet above the surrounding coastal plain (pushed up by a subsurface salt dome). He had persevered despite repeated failure and loss of hole. No preacher ever fought sin harder or longer, or with less apparent success, than Higgins fought doubt, failure, and derision. Frustrated alternately by lack of money to drill and repeated failure when there was money, Higgins and his backers in Beaumont cut a deal with Captain Anthony F. Lucas. Lucas was a graduate of the University of Graz (in mining engineering) and of the Austrian Naval Academy who had worked for the previous six years core drilling salt dome formations in South Louisiana, including Avery Island. His primary purpose had been exploring potential locations for salt and sulfur mines, not oil. But, "At that time no man knew as much about the salt domes of the coastal area and the topographic mounds which mark their occurrence."[17]

The wisdom of the sages offered little encouragement. John D. Archbold, effectively Standard Oil of New Jersey's chief operating officer under John D. Rockefeller, had famously sworn in 1885 to drink every gallon of oil produced west of the Mississippi. Calvin Payne, Standard Oil's production expert, examined the Big Hill prospect in mid-1899 and concluded "that there was no indication whatever to warrant expectation of an oil field on the prairies of Southeastern Texas," while C. Willard Hayes, soon to be the United States Geological Survey's chief geologist, who visited in the fall of 1899, warned that he could see "no precedents for expecting to find oil in the great unconsolidated sands and clays of the Coastal Plain." But Lucas, the engineer, too had fallen under the spell of the preacher: He later wrote to Everette Lee DeGolyer, "The plain fact of the matter is that I am not a trained geologist, hence do not see my way to give the proper or necessary interpretation to my—well—visions."[18]

Lucas also lost one hole to sand. He then made a deal with Guffey and Galey, who sent the Hamill brothers, Curt, Al, and Jim, with their rotary rig down from Corsicana to try to complete a well.[19] The Hamills succeeded, in large part because they used "mud" as a drilling fluid to stabilize the hole and prevent sand intrusion, as well as hold down gas pressure. The sources are ambiguous on whether it was the first time mud had been so used. What is clear is that mud was exactly that, dirt and water mixed: Early Deane, one of the legion who "made Spindletop," recalled Curt Hamill "telling about running cattle through the slush pit to mix up the mud, chew it up for him, and put it so they could use it."[20] But mud solved the problem of drilling in soft formations with a rotary rig. The Lucas gusher, the "geyser of oil," completed January 10, 1901, at 1,160 feet, was faith's, determination's, and innovation's reward.

Only the epic nineteenth-century gold rushes, and maybe Pithole in Pennsylvania, offered any parallel to the frenzy the Lucas gusher set off, but Spindletop was a whole lot more accessible. It was the first great bonanza that could be easily and directly reported on by the new media of nearly instantaneous mass communication, the telegraph and the mass-circulation newspaper, and directly accessed (except for the last four miles) by an integrated system of mass transportation, the railroad. Of course, some attention to detail was still required: one of the Heywood family of California oilmen telegraphed frantically by his brother to "come immediately to Beaumont" went to Beaumont, California.[21]

El Dorado itself could never have created the unimaginable wealth for Everyman, or the illusion of unimaginable wealth, that Spindletop did. A commissary clerk who had bought four acres of land with his life's savings of $60 (so he wouldn't spend it frivolously) sold out for $100,000. A Black farmer who had been trying to sell his farm for $150 sold out for $20,000, then got home to find another man willing to pay $50,000. The first purchaser took the $30,000 profit; the second buyer in turn sold out a few weeks later for $100,000.[22] At one point at the height of the boom, a promoter climbed up on a chair in the lobby of the Crosby Hotel in Beaumont waving a hundred $1,000 bills, which he said he would pay for a single acre in proven territory. He was laughed down, because that was the price for a single twenty-foot by twenty-foot lease next to a producing well.[23]

Spindletop spawned the tradition of Texas big rich: big deals, big dollars, big talk, and big broke. "Nouveau riche" never would be much of a pejorative in Texas, if only because it was so likely to be temporary. But rich or poor, Everyman with ingenuity had opportunity. Some of the deals were ingenious indeed. Edward M. House, soon to be Woodrow Wilson's éminence grise, leased the dedicated streets and alleys of Gladys City from the town's developers, which gave him a "perfect gridiron on two-thirds of the hill" at Spindletop. He never had to drill to prove acreage: anybody who hit told him where to drill.[24]

Ironically, the deal that paved the way for some of the sleaziest of the later promotions was the work of none other than James Stephen "Big Jim" Hogg, the trust-busting former Governor of the State of Texas and father of the Texas Railroad Commission. Hogg and a group of friends had created the Hogg-Swayne Syndicate to buy a 15-acre lease for $310,000, of which they had only $40,000 in cash. Pressed to come up with the other $270,000, Hogg got the bright idea of dividing and subdividing the tract into ever smaller leases.[25] The result was that ultimately some "leases" were only eleven feet wide by twenty-two feet long, and wells were drilled so densely that the legs of the derricks intertwined.[26] When the big fire, to become a rite of passage for Texas oil booms, came in 1902, most of the densely packed wells were destroyed. But the Hogg-Swayne Syndicate made out like bandits, although allegedly the more egregious promotions were not perpetrated by them but by the promoters to whom they had sold. Spindletop was on its way to the moniker by which many inside, and many more outside, Texas would know it: "Swindletop." Hogg himself had so much the Midas touch that when he "picked up a sulphur pyrite that had blown out of one of his wells" and "facetiously announced that it was gold" he very nearly did set off a gold rush. Indeed, "The story got so good that Hogg, momentarily taken in by his own imagination, seriously considered organizing a gold mining company."[27]

The euphoria did have a liquid basis in fact. Based on projections of initial output, half the original six wells at Spindletop could have equaled the 68,000,000 barrels a year, or 185,000 barrels a day, which Russia, then the world's largest oil producer, had produced in 1900. Those three wells easily could have topped the 58,000,000 barrels the United States had produced that same year. The Lucas gusher alone could produce twice as much oil as all the wells in the State of Pennsylvania, then the leading oil producing state; put differently, the Lucas well alone could have produced half the United States' total output.[28] The Standard Oil monopoly, which controlled (by ownership or contract) 48,000,000 of those 58,000,000 barrels of oil produced in 1900, died that raw winter morning in 1901, a good ten years before the United States Supreme Court would officially inter it.

From the time oil had been discovered at Titusville, Pennsylvania, in 1859, right through the years of Standard Oil domination, right through the radical transformations that followed the beginning of the oil century at Spindletop in 1901, the oil fraternity's freedom "to produce and prosper" had been essentially untrammeled by any government or any man, not even John Davison Rockefeller.[29] Oilmen came from a culture where you didn't mess with a man's dogs, hogs, kids, whiskey, womenfolk, or stuff, pretty much in that order.[30] When the time came, they didn't much cotton to anybody, especially not the state, telling them how to produce "their" oil. As far as they were concerned, Standard Oil of New Jersey and revenuers were cut from the same cloth: their experiences with neither had been altogether satisfactory.

The "rule of capture" governed who owned oil, who was rich and who was poor. The rule of capture was originally an English common-law doctrine applied to wild game: if you shot it on your land, it was legally yours.[31] Given the fugacious nature of oil underground—the "vagrant character of the mineral," as one annoyed jurist later termed it—American common law by the 1880s uniformly recognized capture as the governing principle of oil ownership: if it came out of a hole in the ground on your land, it was yours.[32] Since capture determined ownership, discovery of a new oil pool invariably resulted in a chaotic rush to drill and produce as much oil as possible. The self-interest of each owner and lessee, as defined by the rule of capture, compelled him to get his oil (or, practically, as much as possible) before his neighbor drained it away. Hence the prototypical "oil boom."

The situation in Texas was even more complicated. The Constitution of the State of Texas recognized a landowner's fee ownership of minerals in place under his land, including oil and natural gas.[33] The first reported American case concerning ownership of subsurface oil, in 1854, acknowledged the surface owner's claim, even if oil were "a peculiar liquid not necessary nor indeed suitable for the common use of man."[34] Moreover, in Texas, mineral rights are property rights separately alienable from surface rights: they can be sold or inherited separately.[35] Nevertheless, practical reason compelled Texas courts, as it had those in other states, to make exercise of the constitutional property right in subsurface oil and gas contingent upon capture. (In Texas, the same doctrine applies to subsurface water rights.)[36] This juxtaposition of absolute property right and immutable contingency resulted in an even stronger compulsion to drill and produce. Everyman had a right to the oil and gas in place under his land. Everyman had a right to drill a well on his land and produce as much as possible. But given capture, no man had a right to sue his neighbor for drainage, nor could he be held liable for drainage from his neighbor's property. All he could do was try diligently to recover his property, and maybe just a lagniappe more, first.

Oil is conventionally found in subsurface reservoirs or pools, where it has accumulated over eons in porous, sedimentary rock formations capped by impervious strata.[37] Surface ownership, however, is almost never congruent with a subsurface pool, so ownership of oil in place is typically split among several, or in densely settled parts of the state, among several hundred or even several thousand surface tracts. Landowners rarely had either the capital or the technical and organizational capabilities to drill oil wells. Mineral rights therefore were leased to oil "operators" who thereby had the opportunity, and the obligation, to drill and complete wells. Lessors typically got a "bonus," an initial payment (often nominal), plus annual "rentals" for a fixed term, unless production were established. Once oil started to flow, lessors were entitled to an "override," a specified share of oil production, usually one-eighth, but sometimes more, without any costs deducted; the lease would remain in effect so long as production continued.

Leases impose both explicit and implicit obligations on lessees, the most critical of which are an obligation to drill "off-setting" wells to prevent drainage, full development of the sub-surface resource, or, implicitly, "conscientious development," and restoration or compensation for surface damage.[38] In a case originating in Spindletop in 1902 (concerning a lease executed in 1900) against J. M. Guffey Petroleum Company, "a partnership composed of J. M. Guffey, John H. Galey, and Anthony F. Lucas," the Court of Appeals of Texas held that Guffey Petroleum had forfeited its lease on land belonging to one Clara Chaison and was liable for damages due to its failure to honor its "implied obligation . . . to use reasonable diligence and care to develop the property and protect the property, and this obligation required it to sink as many wells as the exercise of such diligence and care would suggest under the circumstances."[39] Implied covenants were reaffirmed and broadened in repeated Texas cases.[40] Ultimately, implied covenants included at least eight separately mandated but overlapping obligations.[41] From the beginning, the force of law was laid on top of brute self-interest to compel the most frantic possible development and production.[42]

Petroleum geology (at least as it applied to Texas) and petroleum engineering were still in their infancy through the first decades of the twentieth century and provided little useful guidance. Absent coherent grounds upon which to evaluate leases, wells, or prospects, credit institutions were sorely lacking. The men who made the deals that made the oil fraternity in Texas were forever long on hope and short on cash: even if (usually temporarily) rich, their prospects invariably outran their means. Shortage of capital was compounded by risk. Over the entire span from 1859 to 1939, the eminent petroleum geophysicist Everette Lee DeGolyer Sr. estimated the odds against a "wildcat" well—one drilled some distance from established pro-duction, often in places only a wildcat could find, hence the name—at 30 or 40 to 1, perhaps worse than 100 to 1 in the beginning. By the late 1930s, wildcat odds against were still 9 to 1.[43] And even drilling in an already discovered oil field was uncertain: ultimately a field was defined empirically by "step-out" wells, its extent delimited by the ring of dry holes surrounding it.

With benefit of present-day petroleum engineering science, the consequences of early production practices are apparent.[44] Uncontrolled flush production ordinarily resulted in a net recovery of total oil in place of less than 10 percent: 90 percent of the oil was wasted without ever being brought to the surface, left underground, unrecoverable by ordinary means. A good part of the 10 percent that was produced also was wasted through evaporation, leakage, and the inevitable fire. Not until the mid-1920s had geologists and nascent petroleum engineers established that recovery or production from a subsurface reservoir (or pool) of liquid hydrocarbons depended upon the viscosity of the oil in place (which is reduced by dissolved natural gas, under pressure, in the reservoir), the porosity of the reservoir rock, and the available "drive," or production mechanism, free gas cap or dissolved gas, which serves to force oil through the formation toward and up and out of the wells. ("Water drive" as a primary production mechanism was widely recognized only after the chaotic natural experiment in the giant East Texas field in the 1930s.) What was clear, even early on, however, was that uncontrolled flush production (letting each new well in a pool flow wide open, maybe even "blow in") resulted in rapid exhaustion of whatever reservoir pressure was available, water coning, and large amounts of oil left underground.[45] In contrast, conscientious application of conventional, scientifically informed recovery techniques by the mid-1930s could recover about 70 percent of oil in place. Application of present-day secondary and tertiary recovery methods, such as carbon-dioxide injection or the use of surfactants, claim to recover upward of 96 percent of oil in place.[46] Enhanced recovery techniques (including rational management) have in effect multiplied traditional oil resources by nearly an order of magnitude.[47]

The economic effects of unrestrained flush production were equally unpleasant. Oil is an unusual commodity: in the short run, both its supply and its demand are inelastic. Sudden additions to or removals from the market cause radical short-run price movements. Abrupt discovery and development of new fields, which the rule of capture assured would be exploited in the most rapid and inefficient manner possible, alternated with equally sudden depletion.[48] The effect was very much like that caused more recently by alternation of politically induced shortage and glut: large price fluctuations, consumer anger and mistrust, and an inability of producers to coherently plan exploration or intelligently manage production.[49] As Standard Oil managers had learned in the 1870s, producing oil and transporting, refining, and marketing oil are two very different problems. The major economic effect of Spindletop, as it would be for later booms, was to collapse the price of oil: from a Standard-set $1.13 a barrel at the end of 1900 to, in southeast Texas, $0.03 per barrel by mid-1901. Potable water went for more, $6.00 a barrel, another pattern that would be repeated in oil booms time and time again. Markets were undeveloped, transportation limited, and production controls unheard of.

All told, in 1901 alone, an estimated $235,000,000 was "invested" in Texas oil, most of it in bogus promotions. By early 1902, only about $80,000,000 had actually been sunk in land, leases, wells, or production equipment. James A. Clark and Michel T. Halbouty estimate ultimate production revenue for the entire first (shallow) Spindletop boom at not more than $25,000,000.[50] Hence "Swindletop." Beaumont boomed anyway, exploding from a country town of fewer than 10,000 people to 50,000-plus. Order and sanitation suffered: the oil fraternity came to refer to diarrhea as "the Beaumonts."[51] Where the oilmen went, of course, the four-flushers and vultures followed. By 1903, however, the party was dying. The rule of capture, and greed untempered by anything but sheer ignorance, had taken its toll.

But the oil fraternity was born. Most were local boys "born to pick cotton."[52] It was at Spindletop that the language of the oil fraternity took form, "that a well borer became a driller, a skilled helper . . . a roughneck, a semi-skilled helper . . . a roustabout, and a beginner . . . a boll weevil."[53] There the folkways of opportunity took shape: "It was on this hill that a 'shoe-stringer' would 'poor boy' a hole down by splitting his interest with his crew, the landowner, the boarding house, the supply house, the saloon keeper, and sometimes the madame at his favorite bawdy house."[54] At Spindletop the oil fraternity learned not to trust the world outside and developed their habits of xenophobia, self-reliance, and communal solidarity. There they formed their representations of themselves and their history: there they learned to believe that "real" oilmen had nothing to do with promoters and charlatans, with men only in the "money racket."[55] It was there that they learned never to trust people who, in their parlance, "didn' unerstan' the oil bidness." Who they did learn to trust were their friends, the sources of rumor and information, of cash and credit and help, a pattern that would characterize the oil fraternity from then on. Like every succeeding oil boom in Texas, Spindletop birthed an extraordinary number of firms, partnerships, and continuing business relationships. The oil fraternity learned to trust themselves and their hunches, and not to believe what the "experts" said. Rich or poor, who among them could but puzzle when a well drilled fifty feet southeast of the Lucas gusher came up dry, or that the Lucas well itself, ever cantankerous, never made any money.[56] The oil fraternity learned to trust itself, and the prima facie evidence of a drill bit, and that was about all.

As had their fathers in cattlemen's associations and vigilance committees, they organized themselves to order their world: In response to the catastrophic fire in Spindletop in the fall of 1902, a field safety committee, with no legal standing, assumed authority to require removal of wooden derricks once a well was completed and the erection of a brick housing around each well. In view of the millions of barrels of oil already in storage, the same committee took the extraordinary step of imposing field rules requiring those without storage to shut in their wells.[57]

The oil fraternity was a first-name community of itinerants, poised to waltz across Texas, gathering converts, recruits, camp followers, and fortune as it went.[58] Not every man knew every other, but few were those who didn't know somebody who knew somebody who did know just about anybody else.[59] And they had something to offer no revival, no religion, no political crusade ever could. Even thirty years later in East Texas, mineral rights under land that might sell for $5 or $10 an acre in fee simple could be leased for $1,000 an acre and a one-eighth overriding royalty interest, plus a guarantee that a well would be drilled. And surely the Texas across which the oil fraternity waltzed to the siren song of opportunity was most of all a place of toil and tribulation, and, for most, poverty.

Few in the oil fraternity, even the multimillionaires, did not have living memory of poverty on the land and the lifelong bondage promised by the crop-lien system. E. P. (Matt) Matteson recalled:

Now, Mr. Higginbotham, that owned the famous Higginbotham lease north of Breckenridge up there, he had sixteen of those gushers on the place. And he told me that before the time, he raised a big family of girls without a mother.... But he said, never had $10 in my life that I could hold up and call my own, until this. If I took in a dollar, I already owed it. I made a cotton crop, it was eaten up before I took it to the gin. Lord, I just never had $10 to call my own.[60]

What an oil boom meant to people dirt poor and deep in debt is apparent in Walter Cline's description of those that flocked to Burkburnett in 1919 and 1920:

Of course, there was a tremendous influx of people. Crop conditions had been very poor over most of Southern Oklahoma and Western Texas and there were thousands of families who were suffering for enough food and clothing and shelter to carry them through the winter months.... None of them had ever been out of debt in their life. Didn't think there was enough to get out of debt with, you know.[61]

## DEMOGRAPHICS: SOUTH CAROLINA'S CHILDREN

As a population, a social collectivity, and a culture, the oil fraternity in Texas emerged and evolved. David Hull, in his analysis of selection theory (the more general process which includes evolution) argues that entities that evolve by definition change; they therefore cannot have immutable Aristotelian "essences." They are spatiotemporal particulars, that is, historical.[62] What matters in the evolution of such entities, including cultures, is ancestor-descendant lineages. As Robert Boyd and Peter Richerson showed, and, now, a host of investigators have demonstrated, the preeminent mechanism in the evolution of cultural lineages is density-dependent bias, which may be instantiated by vertical transmission from parent to offspring, within kin-groups, or within any well-bounded deme.[63] Symbolic (including linguistic) cultures enhance the importance of horizontal transmission among non-kin who are associated in social groups, and may be instantiated by imitation (of those perceived to be successful or prestigious, for example) or intentional acculturation. Biological descent is a marker for places to look for indicators of conceptual descent: artifacts, language usage, ritual forms, or ideological representations. Conversely, material culture—artifacts and their names, techniques, or technologies—serve as tracers for the movement of populations and their descendants.[64] Understanding the oil fraternity in Texas and its evolution begins with who they were, where they came from, and what they brought with them.

Usage of "the oil fraternity in Texas" thus far has been pretty fast and loose. Put simply, if you live in Texas, work in the oil business, drill oil wells, truck drill pipe, gauge tanks, mix mud, run a pumping station, lease your land, sign affidavits or division orders, appear in a Railroad Commission hearing, sue your lessee or your neighbor over oil, belong to the Texas Independent Producers and Royalty Owners Association, the Society of Petroleum Engineers, the American Society of Petroleum Geologists, or go to any of their meetings, search titles, take leases, work for the Oil and Gas Division of the Railroad Commission, make oil and gas loans, bitch about government regulation of the oil business, spend oil money, or ever want to, you're part of the oil fraternity in Texas. So conceived, the oil fraternity in Texas is extraordinarily large and extraordinarily inclusive, which gives it some of its historically unique characteristics.

There are no direct measures of the oil fraternity, its membership, or its growth: it is not exactly a formal organization. There are, nevertheless, several indirect indicators of its scale, scope, and growth over time. By 1940, for instance, 133 of Texas's 254 counties had oil production; over 195,000 wells had been drilled, of which some 54,000 were dry; of the 134,236 successful oil wells, 39,022 had been abandoned, leaving a total of 95,214 producing oil wells.[65] By 1948, there were 1,248 different oil fields in Texas with 106,738 producing wells located in 189 of the 254 counties. There were roughly 28,000 individual producing leases with approximately 147,000 more awaiting drilling and development.[66] Ten years later, in 1958, 210 of the 254 counties had oil or natural gas production; 43 percent of the land area of the State of Texas was under lease or held by production—114,955 square miles out of a total of 267,388, an area some 20 percent larger than the United Kingdom. There were 1,700 different oil or gas fields containing nearly 180,000 individual producing wells.[67]

The oil fraternity in Texas as well as in the rest of the midcontinent was socially and ideologically inclusive: it recognized few distinctions of wealth and virtually none of class. The myth of the fifth-grade dropout working as a driller or tool pusher one day and making his millions the next was true just often enough to sustain the faith. The men who made exactly that transition were in fact legion, although few were widely known beyond the confines of the fraternity. But despite this inclusivity, real and imagined, the oil fraternity in Texas remained in critical ways a closed society. From the discovery of Spindletop in 1901 into the 1960s, the oil fraternity in Texas was ethnically, religiously, and culturally homogeneous and, in Texas, politically and socially dominant. With its populist ideology (the uncharitable might call it demonology), and southern, rural lineage, the oil fraternity was and largely remained lily-white, male, and Protestant. And despite the crucial roles played by immigrants from the older Pennsylvania and Ohio oil fields (such as J. S. Cullinan), the fraternity in Texas was drawn mostly from Texas and adjacent states. Within Texas, a few Black or Mexican landowners participated from the sidelines. A few women, usually by marriage or inheritance, became substantial players; ironically, those few that did sit at the table were reputed to be among the toughest traders in Texas. But in general, the oil fraternity was just that.[68] And ideologically, if not always factually, the oil fraternity was built from the bottom up. As one of their number, E. I Thompson, put it:

> He's come up the hard way, 90% of 'em. There are rare exceptions, but 90% of them went through, they learned their work in the field, and they worked for drilling contractors, they worked on rigs, and then they got a rig together possibly, and then they got out and got hold of a lease, and they, they did their own field work, and they came up the hard way, and some of them, you can name them by the dozens, some of the most successful and really wealthy men of the state today, a few short years back were living on chili and struggling and going around with a patch in their pants.[69]

They were men rooted in the soil of Texas, and in its agrarian, evangelical, and cooperative past.

No demographic history of the oil fraternity exists.[70] But, since the oil fraternity was so ethnically and geographically homogeneous, sources providing a basis for inference do exist. Those sources are of two types: specific geographic studies of immigration to Texas, and general demographic or census data on the population of Texas. The former accounts, particularly those of cultural geographer Terry G. Jordan, are especially helpful. Jordan's work explicitly addresses two concerns: the differences between German and Anglo-American farmers in Texas in the nineteenth century, and the origin of western, open-range cattle ranching.[71] At their maximum in 1870, German immigrants composed only about 7 percent of the white population of Texas, and the counties most heavily settled by German immigrants were, with a couple of exceptions, not those most important in the early development of oil production in Texas. Thus, the influence of German traditions on the oil fraternity in Texas was minor.

But in his study of nineteenth-century farming styles, Jordan did carefully consider the origins and the routes various groups took to Texas. From 1850 on, the white population of Texas was overwhelmingly composed of native Texans and southern immigrants. In 1850, 32 percent of white Texans were born in Texas; 49 percent were southern, which together accounted for 81 percent of the population. In 1880, 49 percent of white Texans were native-born, while 33 percent were southern, together 82 percent of the state's population. In both 1850 and 1880, immigrants from the Gulf and Upper South predominated. In both periods, the bulk of the newcomers were yeoman farmers and mostly non-slaveholders before 1860.[72]

Southern source areas all had high cattle to people, and even higher hog to people, ratios. Both cattle and hogs for the most part ran free in the pine barrens and swamps. By long-standing custom, crop fields were fenced to keep animals out, while animals enjoyed the open range: before cheap barbed wire, large-scale fencing was not economically feasible. Open-range land typically was owned by someone, but if he had not fenced and planted it, it was considered free, to be exploited in common. As Jordan argues, this pattern of extensive agriculture was rational: land was abundant and cheap, and more was there for the taking just over the western horizon; labor and capital in contrast were dear. Often, Southerners, and Texans, moved on just for the sake of moving.[73] This pattern of extensive resource utilization and high mobility characterized American economic development from the beginning until well into the twentieth century.[74] What differed in the southern agrarian experience was communal exploitation of a common, presumably infinite, resource—the land—for benefits both individual and collective: the two were not seen as mutually contradictory.

Jordan offers a convincing portrayal of the way local communities from older regions in the South were replicated in Texas: immigrants in a given county in Texas tended to come from a specific county or locale in another southern state, often traipsing after an initial pioneer.[75] Robert C. McMath Jr. found a similar process on even finer scale in the counties of the Texas Cross Timbers that he examined.[76] Populations from extended families to larger tribes to ethnic groups have followed this pattern since before time.

Jordan's work on the origins of western open-range cattle ranching exhibits, in richer detail, the same community replication. Moreover, since Jordan is looking not just at migration but also at cultural diffusion, his study links biological descent and cultural propagation. "Culture," as used here, is inclusive: it is our species' "extrasomatic means of adaptation," in Leslie White's memorable phrase. Jordan portrays the way cattle-raising culture spread from seventeenth-century South Carolina to the nineteenth-century Great Plains, via migration, adaptation, and "creolization." Jordan's argument comprises three primary contentions. "Open range cattle herding" developed indigenously among Anglo- and African Americans in eastern South Carolina during the seventeenth century, then diffused westward via two routes. The major route ran along the Gulf Coast and inland pine barren strip from Georgia and Florida through Alabama, Mississippi, Acadian Louisiana, and into East Texas and the Gulf Coast of Texas. Some "creolization"—the admixture of Spanish and French names especially— apparently occurred in Florida and, even more so, in South Louisiana, but that admixture did not include characteristically Hispanic techniques or customs. The other diffusion route passed through western South Carolina into Tennessee, Kentucky, southern Illinois, Missouri, Arkansas, and into Northeast Texas. Distinctive Anglo-style cattle ranching dominated the three principal early cattle (and hog) raising areas of Texas (Southeast Texas and the middle coast, East Texas, and Northeast Texas), which were settled by immigrants from the United States. True hybridization with Hispanic techniques and customs (and vocabulary) did not occur until the 1870s in the Cross Timbers and Heart of Texas areas of Central Texas, from whence what is recognized as "western" open-range cattle ranching passed first to North Texas and the Panhandle, and then across the Great Plains west of the 100th meridian. As Jordan concludes, the origins of western cattle raising "genealogically and culturally" belong to South Carolina's children.[77]

Jordan traced cultural diffusion using two sets of markers: language, and techniques, practices, and customs. He demonstrates that Anglo-American names for things, originating in South Carolina, predominated even in Texas (outside South Texas) well into the 1870s: "cowpen" rather than "corral," "drover" or "cowboy" rather than "vaquero," "stockman" rather than "rancher." The South Carolina tradition also produced distinctive techniques and practices. Preeminent were the use of open (unfenced) common range and what amounted to near total neglect of livestock: "cowhunts," only much later called "roundups" (as opposed to weekly "rodeos" in the Hispanic tradition), occurred usually only once, at most twice, a year. Owners identified their stock by unique (registered) ear markings or brands, but the "brands" were simple alphanumeric designs very different from the highly embellished, often Moorish inspired designs of the Hispanic ranchers. At roundup, calves would still be with their mothers, so their proper ownership could be established and the newborns branded; stock to be sold could be separated. Cattle and hogs shared the range, with hogs usually outnumbering cattle. Sheep or goats were as rare among Anglos as hogs were among Hispanics.

Anglos did not work cattle from horseback: they lacked both the lasso or lariat and the horned saddle; but "drovers" did use horses for transportation. Stock was actually controlled not by men on horseback but by "cow dogs" trained to locate, round up, and even "cut" cattle. Anglo cattle were of English-colonial descent; archetypical "longhorn" traits from South Texas herds did not begin to appear in North Texas until the 1860s.[78] The dogs themselves were lineal descendants of breeds developed in South Carolina, and the best of them were trained to bring down a running steer by jumping up and grabbing him by "the nose, lip or ear."[79] Quite apart from aesthetics, the advantages of using dogs to run stock in pine barrens or brush where man on horseback couldn't go is obvious, and dogs continue to be so used. Eventually, when Texas stockmen became ranchers, they started to use Quarter Horses (apparently because Southerners bred them to race a quarter mile) as "cutting horses" to work cattle. As Jordan observes, "A quarter horse trained as a cutter is well named: it is three quarters dog."[80]

Jordan's studies are paradigmatic exemplars of lineal biological descent (man, dog, hog, and cow) and of lineal cultural descent.[81] Two specific properties of Southern cattle culture matter. First, the Southern culture of cattle raising, associated as it was with yeoman farmers, was the product of a frontier of apparently limitless resources. It was a non-zero-sum world, where one man's gain was not another man's loss. Second, the "traditional communalism of the Anglo frontier"[82] expressed a historical experience in which competition and cooperation were not only not antithetical or mutually exclusive, but rather were equally essential to individual and collective survival. As Jordan concludes: "It is more logical to seek the origin of such associations [Stockmen's Associations] in the traditional communalism of the Anglo frontier, the same communalism that produced claim clubs, house raisings, wagon trains, and home-guard militia."[83]

These migration and population patterns persisted until the middle of the twentieth century, by which time the social and cultural character of the oil fraternity in Texas was formed. The population of Texas, from which the oil fraternity was drawn, continued to be dominated by native-born Texans and Southern immigrants. In 1900, nearly 71 percent of Texans born in the United States were born in Texas; almost 26 percent had come from the South, defined as bordering states and the Southern states plus Missouri. According to the same conventions, in 1940, nearly 82 percent of Texans were natives of Texas; another 14 percent were Southern immigrants. Thus in both 1900 and in 1940, at least 96 percent of the relevant population of Texas was either born in Texas or had Southern origins. More than a few had living memory of the letters "G. T. T." scrawled in family Bibles or on signs nailed to the doors of abandoned tenant farms: "Gone to Texas."[84]

What matters is the ethnic and geographic stability of the population from which the oil fraternity in Texas was drawn. Despite the ambiguities in the numbers introduced by the way Mexican and Mexican American residents of Texas were enumerated, for a hundred years, from 1850 to 1950, the majority of the Anglo population of Texas was either born in Texas or had immigrated from a Southern state. Minimally 80 to 90 percent of the population from which the oil fraternity descended was homogeneous, ethnically, geographically, and culturally. More likely, for the critical period between 1900 and 1940, which were the oil fraternity's formative years, upwards of 95 percent of the fraternity's parent stock shared those common roots.

The major demographic change that did occur for this population was the shift from rural to urban. Yet the cultural effects of this shift can easily be exaggerated. As late as 1940, Texas was nearly 60 percent rural. Even by 1960, only fifteen of Texas's 254 counties had urban populations greater than 100,000, although urban places did contain over half the state's total population. The five biggest cities, Houston, Dallas-Fort Worth, San Antonio, and Austin, held just over a quarter of the state's people. More importantly, prior to 1960, it is doubtful that any town in Texas would be considered by somebody outside of Texas as being truly urban or cosmopolitan or as having anything other than a rural, southwestern culture (Dallasites might argue, but they pretend a lot).[85]

Both geographic studies of "trails to Texas" and census data show the biological lineage of the population that birthed the oil fraternity in Texas: they were sons of the rural South. More important, as Jordan demonstrates, genealogical lineage and cultural lineage go together, as expected on theoretical grounds. This genealogically and culturally homogeneous population provided the cultural resources out of which the oil fraternity in Texas would produce its world.

## CULTURAL FORMS

The oil fraternity in Texas did not spring fully formed from Mammon's brow, nor was its culture contrived or constructed de novo on the occasion of its birth. Neither did the oil fraternity in Texas simply replicate or reproduce ancestral culture or inherited ideology. Rather, the oil fraternity in Texas created its own culture from inherited symbolic and behavioral resources. Key elements of its inherited culture included the unproblematic melding of competition and cooperation in the context of an egalitarian republic of small producers, local autonomy and self-regulation, and resistance to external authority, exploitation, or coercion. The oil fraternity was heir to the agrarian poverty of the late nineteenth century and its political analyses and programs, and to its modes of political action and expression.

### Communal Life

Four generations ago, Mody Boatright angrily argued that American historiography, still living in the shadow of Frederick Jackson Turner, had badly overstated the purported "rugged individualism" and essential asociality of American frontiersmen, and that the "principle of mutuality" played a far greater and more central role in rural life than commonly acknowledged.[86] More recently, a new generation of social historians has focused on the complexity and the "habits of mutuality" characteristic of rural communal life in the eighteenth and nineteenth centuries, while other scholars interested in popular agrarian protest movements have rewritten the history of the Farmers' Alliance and the People's Party.[87] Relevant here are the culture of Southern agrarians, their particular collective ideological representation of themselves and their world, and their specific ritual forms and modes of social organization. In each respect, the world of the Southern yeoman farmer maps onto features of the oil fraternity in Texas.

Perhaps the hardest thing to imagine about the world we have lost is a society for which competition and cooperation are not only not mutually exclusive but also not even antithetical. As Boatright argues, from the very beginning of frontier settlement, opportunity and survival depended upon cooperation. Sources agree, as Jordan, McMath, and Steven Hahn demonstrate, that migrations west typically involved subsets of well-defined communities. More importantly, migrants reproduced their cultural institutions and forms from South Carolina and Georgia right across the South and into Texas.[88] Indeed, Gary Kulik argues that such traditions reach at least back to England and the "Country thought" of eighteenth-century Commonwealthmen.[89] But culture and representation were not just replicated: they were modified by descent, adapted to new environments and novel situations. As Hahn observes:

> Popular ideologies and cultural forms also have a dynamic quality even if their modes of expression remain unaltered. . . . Customs and institutions organizing and governing various aspects of social life not only prove readily adaptable to new conditions but provide powerful means for mobilizing and transforming collective behavior.[90]

Once folks settled, cooperation continued to be the leitmotif of communal life. The community collectively cleared land and built houses and churches and barns and fences. Neighbors cooperated with each other in felling trees and collecting timber for buildings and firewood. The ancestral term for such cooperative endeavor persists, albeit now with pejorative connotation: "log rolling." The community imposed its own local order and regulated the behavior of both its own members and outsiders passing through. Closed communities have their dark side, a provincialism that turns easily into xenophobia or even paranoia. Anti–horse thief associations, cattlemen's associations, and vigilance committees could be harsh in their justice, but their justice was likely better than none at all. As Boatright observes, "This deliberate and considered mass justice is not to be confused with mob violence. It is yet to be proved that it erred more frequently than the courts presided over by learned justices in robes."[91] Or as one Texas jurist answered when asked why a man could go free for murder in Texas but be hung for stealing a horse, "Well, we got a fair number of people that need killing, but we don't have any horses that need stealing."

To the extent that an extra-local state was thought to have a role, it was only to guarantee this "republic of small producers," or essentially the power of the community to regulate itself. As Boatright argued, those who saw frontiersmen as impatient of any control misread him. Rather, "He resented regimentation by the governing class in the East, but he did not hesitate to regiment himself, and, in so far as he was able, others whose interests conflicted with his."[92] This social control more often than not took the form of "a set of folkways" enforced by "public opinion" rather than by the formal apparatus of the legal system. Local order brooked little outside interference in communal affairs. Overt intrusion could be met with force of arms, as it was in the Darlington, South Carolina, insurrection in 1894 against Governor Ben Tillman's liquor dispensary system.[93] As William Link put it, in general, what "outsider" Progressive reformers who tried to work in the rural South "lamented as a decline in 'community' was a contentious, intensely localistic, rural participatory democracy."[94]

None of these communitarian impulses implied a lack of self-reliance, individualism, acquisitiveness, or competition. The right and opportunity to do well, and the rights and prerogatives of private property were unquestioned, but, paradoxically, only in communal context. As Steven Hahn maintains, "the independence to which Upcountry yeomen aspired hinged on social ties, on "habits of mutuality" among producers, that imported to their culture a communal, prebourgeois quality whose egalitarian proclivities sharply distinguished it from that of the planters."[95]

As the oil fraternity did a century later, cash-poor upcountry farmers traded and swapped and bartered face-to-face and personal, and this "configuration helped govern the terms of trade and shaped understandings of the proper working of the marketplace."[96] Obligations were mutual: "The better off helped the worse off in times of trouble, for the bonds of mutual aid were all the social security that anyone had."[97] And, "When things had a money price, it was more often than not a 'fair price' rather than a market price, and when the market came calling, the community prevailed: public auctions of estates or, more rarely, of foreclosed property, were almost always 'closed,' with buyer and price designated by communal consensus well in advance."[98]

Neither wealth nor station automatically implied deference. Those to whom deference was due understood that it was theirs precisely because they need not assert it. Legendary cattleman Charles Goodnight, born three days after the Republic of Texas,[99] blazed the Goodnight-Loving Trail across the High Plains after the Civil War to move Texas cattle, and open-range cattle culture, to Wyoming.[100] Yet in 1881 when Goodnight organized the Panhandle Stock Association of Texas to combat rustlers and the transit of diseased herds across the Panhandle range, Everyman, or at least every man judged to be decent, sat at the table:

> In emphasizing the point, Goodnight said that "we organized for the purpose of mutual benefit, cooperation, and protection, taking in any settler who would join us, whether he had one cow or ten, guaranteeing that our attorneys would take care of his legal battles and our inspectors would take care of his cattle interests." Throughout the life of the Association the man with one cow or saddle horse had equal rights with the man who counted his cattle by the tens of thousands.[101]

When representatives of newly arrived British and eastern American cattle corporations made the mistake of attacking this frontier equality and tried to get the by-laws of the Association amended to allocate "extra votes on the basis of cattle owned," Goodnight opposed them, unequivocally and successfully, even though the change would have given him and his part- ner the largest number of votes in the Association. He told his biographer, J. Evetts Haley: "I knew nothing about oratory, but I got up and told them plainly that such a move would defeat the purpose of the Association, which was to give the little man equal rights with the big man, and before I'd see such a rule passed, I'd disband the whole organization."[102] There is more than a little irony in Goodnight's pledge to assure democracy and equality or else he'd by God personally disband the whole outfit. But Goodnight's attitude expresses the ambiguity southern agrarians, Texas cattlemen, and, later, the oil fraternity in Texas, felt about money, power, and status, and their place in the egalitarian republic of small producers. Yet underneath the surface there is the profound sense of obligation the great felt for, truly, their peers. In a world of opportunity, where resources were to anyone who could see, infinite, and where, with hard work and little good fortune, any man could be as good as any other, equality, mutual obligation, and reciprocity only made good and moral sense.

Even tenant farming, which became the curse of the South after the Civil War, originally was construed as a means to opportunity and mutual benefit. "Land poor" owners would "farm out" a place to a family, which ostensibly would give it a start toward fee ownership and the owner some return on an investment he could not otherwise afford to exploit. The practice, and the terms "land poor," "lease poor," and "farm out" persist in oil-fraternity usage, still with the ancestral connotation of opportunity and mutual benefit. Before the war, tenants typically were subject to little supervision, and were obligated to share a percentage of their crop in lieu of rent. Only later, as commodity farming replaced subsistence farming, were tenants entitled only to receive a percentage of the market value of what they produced.[103] Yet as the country filled up in the 1870s and 1880s, and as small farmers began their century of travail, old customs of land tenure and use came under increasing attack. From the Georgia Uplands to the Texas Cross Timbers, bitter quarrels over "hunting seasons," trespassing, and "stock laws" foreshadowed the firestorm of discontent that would fuel the Farmers' Alliance, the Grange, the People's Party, and William Jennings Bryan's Free Silver Crusade.

For two hundred years, a hungry man could hunt the woods or fish the streams to feed himself and his family; surely hunger knew no season. Now, in the name of conservation, wealthy landowners wanted to impose hunting seasons and limit access to unimproved land. For two hundred years, custom, then law, had held each farmer responsible for fencing his crops, not his livestock. No man could hold his neighbor liable for damages his neighbors' cattle did to his crops, nor was any man responsible for where his cattle went or what they did. Truly, good fences made good neighbors. Any farmer who wanted to, and could afford to, could put stock on the land. Any farmer who wanted to go to the trouble could claim wild hogs or cattle, or horses, as his own. The commons was opportunity open to Everyman. But Everyman was obliged to participate in the annual cowhunt, and Lord help the man found taking a calf from a branded mother that wasn't his. Now, in the name of conservation and sound management, "stock laws" would turn the world inside out. Livestock would have to be fenced up, rather than allowed to run free. Now crops would be unfenced instead.[104] Among the most active advocates of stock laws were the hated railroads, frustrated by claims when their trains ran down livestock on their rights of way. Liability now would lie heavily on dumb animals trying to eat and poor farmers trying to just get along. When prorationing of oil and engineering rationality came to Texas clothed coyly in the fig leaf of "conservation," it would wear a familiar and altogether unpleasant visage.[105]

## THE POLITICS OF DISCONTENT

The agrarian communities of the Uplands, the wiregrass, the Piney Woods, and the Cross Timbers had always harbored a certain amount of what Robert McMath, following Eric Hobsbawm, calls "social banditry." In McMath's North Texas counties, the most notable example of the type is the legendary train robber Sam Bass, the "Robin Hood of the Cross Timbers," who apparently had the plain good sense to steal gold from the Union Pacific and Texas and Pacific Railroads and not cattle from his neighbors, although, unlike Robin

Hood, he apparently also had the good sense to rob the rich and keep it. More typical were the stockmen's associations and the night riders who protected the rights of "squatters" against eviction or foreclosure and cut the fences of those who would block trails, cut off water, or fence in the historic commons, even if it was legally theirs. But more and more, as yeoman farmers became entangled in the cash nexus of the market, dreams of opportunity literally "came a cropper." Tenant farming, or sharecropping, spread, and even the share of the crop a tenant farmer had a right to receive was eaten away by the crop-lien system, which generally specified what crop—cotton—a debtor or tenant had to plant. Farmers, borrowing to live this year, ate tomorrow's hope.[106]

The Farmers' Alliance, which came roiling out of North Central Texas in the mid-1880s, gave concrete expression to the ideology of cooperation and the "republic of small producers," as well as to the corresponding demonology of "middlemen," bankers, insurance companies, railroads, and the Eastern "money power." The Alliance grew directly out of traditions of communal action and custom: "The Alliance fit comfortably into the interlocking collection of voluntary associations and neighborhood settlements that bound together frontier society. In that setting the ritualistic affirmation of community, the protection of livestock, and the organizing power of the voluntary associations were very important."[107] Faithful to their belief in individual initiative, responsibility, and self-reliance, farmers determined to solve their problems of poverty, debt, and tenancy themselves. Alliance chapters all across Texas tried to "bulk" their crops at harvest time: to sell collectively in order to have some bargaining power with merchants and purchasers. Similarly, Alliance chapters tried to collectively develop lines of credit for, and, in some cases, to jointly purchase, supplies and manufactured goods.

When these strategies miscarried, the Alliancemen, like their corporate cousins, sought market power in scale. Their cooperative movement culminated in 1887, when the Texas Farmers' Alliance created the Texas Exchange, financed by the "joint-note plan." All Alliancemen would pledge their individual property to jointly secure a note, the proceeds from which would be used by the Exchange to collectively purchase supplies for the year. At harvest time, all their cotton would be collectively marketed through the Exchange and the note paid off.[108] But the great Texas Exchange too came a cropper. Banks were unwilling to advance substantial sums against the farmers' joint notes, which is understandable in view of the odd legal status of the notes: foreclosure would have been curious. The Farmers' Alliance also made fatal managerial errors. They smoked their own dope and consistently overestimated both their number and their assets. In the absence of long-term credit sources, they operated on short-term credits provided by manufacturers and suppliers. Given their ideological commitment to brotherhood, they made no provision for bad or uncollectible debts. And given their conviction that merchants and middlemen were leeches on the body economic, they grossly underestimated handling costs, and made no allowance for overhead costs whatsoever.[109]

In the 1880s, the Farmers' Alliance had worked within the Democratic Party in Texas, and many of its members had fervently supported the reform Democrat and stem-winding stump orator, James Stephen Hogg, first for attorney general of Texas and then in 1890 for governor. Like so many Texans of his time (and later), Hogg saw the problems of Texas as largely inflicted upon her by "foreign," that is, non-Texan, corporations: insurance companies, banks, speculative land companies, and, most of all, railroads. As attorney general, he had attacked the insurance companies and had secured passage of the second state antitrust statute in the country (a month later than Kansas), and one of the strongest. The Texas antitrust law not only provided for triple damages for those injured by anti-competitive practices, but also required that those found guilty in Texas courts immediately post in escrow the entire amount of the judgment—in cash. The cash bond provision was written into Texas law specifically to protect the "little man" from the legal and financial resources of large corporations, which could drag cases out on appeal all but forever.[110]

When Hogg ran for governor in 1890, the centerpiece of his campaign was support of an amendment to the Constitution of the State of Texas providing for a Railroad Commission. Hogg won the election, and the amendment passed, leading to the creation of the Railroad Commission of the State of Texas in 1891. But although he shared much of the farmers' anger, and all their mistrust of "foreign" corporations, and though he gave booming voice to their rage, Hogg, like so many of the farmers themselves, was at heart what T. R. Fehrenbach so nicely phrased "folk conservative."[111]

The population that fathered the oil fraternity had little reluctance to appropriate the power of the state to secure its economic and social interests—simple justice, in its eyes—provided that whatever was done was under communal control and safe from external subversion or corruption. The oil fraternity's search for solutions to the problems of overproduction, price instability, chronic waste, and the continual depredations of "the money power" would recapitulate the path their forebears had trod: local (field) communal attempts to control production and waste, voluntaristic programs to collectively coordinate broader production restraint, and eventual, and reluctant, acquiescence in use of the power of the state (Texas) to secure their goals.

## RHETORIC AND RITUAL FORMS

What persisted, from the pine barrens of South Carolina to the Texas Cross Timbers, from the Farmers' Alliance to the oil fraternity, was a common sense of communal life, of fraternity, equality, cooperation, and, most of all, opportunity. What persisted was a representation of the man who tilled the soil or produced the oil as the true producer of wealth, the true foundation for a good society. What mattered was that Everyman have opportunity. Not that all would succeed: Everyman after all was the protagonist of *The Pilgrim's Progress*, and he was subjected to many and varied temptations of flesh and spirit.[112] But faith, at its roots biblical and evangelical, promised that despite the timeless rhythm of sin and redemption, of backsliding and being saved, most decent men, if given a fair chance, would succeed. With them the community, and the republic of small producers, would prosper.

The forms of representation to which the oil fraternity was heir, passed down to them through the Farmers' Alliance, Populism, and reform Democracy was, and remained, evangelical.[113] More importantly, for Alliancemen, as for oilmen later, the vocabulary and idiom of apocalypse and redemption "provided the original categories for perceiving the world, categories of understanding and social articulation that could be readily converted to secular and political ends."[114] When William Jennings Bryan decried the "Cross of Gold," he spoke for many and for Everyman and for Texas.

If the idiom of the progenitors of the oil fraternity in Texas was evangelical, so too were their ritual forms. According to Robert McMath, the organization of the Farmers' Alliance directly appropriated that of "the Methodist church, which, a century before and on another frontier, had developed a system whereby itinerant preachers gathered converts from camp meetings into local 'classes,' suppled them with literature, and visited them regularly for services."[115] More spectacularly, the Alliance organized periodic mass encampments that often lasted several days and drew well over 10,000 people. Such encampments, with their carefully arranged rituals of hymn singing, speeches, prayers, lectures, and testimonials, replicated evangelical camp meetings and revivals.[116]

The world that birthed the oil fraternity, if poor and downtrodden, was also rich and complex. It was a world of cooperative rugged individualists, of poor people who wanted to get rich. It was a world rich in idiom and form, ritual and representation. It was a world of insular, interdependent communities who saw themselves bathed in a universal light of salvation and blessed opportunity.

## WALTZ ACROSS TEXAS

As Spindletop played out, the fraternity of oilmen that had been forged there spread out across Texas. They went first to the next obvious places, other Gulf Coast salt domes that could be easily identified on the surface: Saratoga and Sour Lake, Batson's Prairie, Humble, Barbers Hill. Each boom pretty much recapitulated the ones that had gone before. Sour Lake, discovered in 1902, like Spindletop, had its "Shoestring Strip" of leases "twenty-seven and four-tenths feet wide, and twenty feet deep."[117] Only after the obligatory great fire, which also replicated the Spindletop experience, did the whole thing crater because of subsidence. Oil production in Texas actually declined between 1905 and 1912, the result of quick drilling and even quicker depletion of obvious Gulf Coast locations.

But beginning in 1912, as the oil fraternity accumulated experience and honed its exploration and drilling skills, Texas oil production began a nearly monotonic rise. The more naïve notions of the origins of oil and where to find it were slowly replaced: a Houston newspaper, for example, had solemnly explained when Spindletop was discovered that a "river of Oil flowed underground from Corsicana to Beaumont to Spindletop."[118] But even useful theories were still country-boy simple: "trendology," and "creekology." Trendology was little more than a game of connect the dots: wildcatters just drew a line between places of known production and went looking for leases on the assumption that oil fields followed the "trend," or direction of gross geological features, which sometimes they did.[119] Creekology rested upon an equally simple hypothesis: everybody knew that Texas was basically tilted from northwest to southeast, and that water runs downhill. Therefore, anytime a river, or even a creek (hence "creekology"), "comes down and winding," and "goes out two or three miles, you ought to be suspicious." Some subsurface structure, likely to be oil bearing, caused the watercourse's deviation.[120]

By the early 1920s, more academically respectable theories of oil-bearing formations gained currency: faults (cracks in the earth's strata, accompanied usually by vertical displacement on either side of the crack, which tends to trap oil in porous strata which slide up against impermeable formations) and anticlines (upfolds or arches of stratified rock which trap oil), and better understanding of salt dome formations and their characteristics. But oil "prospecting" was still very much an empirical art. Even a trained geologist like Bill Wrather, who, with Landon Cullum, discovered the Desdemona field, could only go out and locate a likely limestone strata in a water well, trace it cross-country through other water wells, find its ultimate outcroppings in creeks or along hillsides, and get some idea what its slope was and where faults or anticlines might be present.[121] More scientific forms of exploration, such as rigorous application of micropaleontology, or the use of torsion balances, did not come until the late 1920s.

Nevertheless, the oil fraternity was learning, and exploration, drilling, and production all grew rapidly after 1918 (see Figures 1.1. and 1.2.: Oil Production in Texas). Expanded exploration and oil production could not be obtained by more intensive activity in situ, as was the case with manufactured industrial output after the age of water power. Even exploitation of the other fossil fuel, coal, was more geographically concentrated. The nature of oil as a resource is such that new production by definition entailed geographic expansion to new fields, to new regions of Texas, and, therefore, the necessary, and enthusiastic, recruitment and inclusion of new members of the oil fraternity: The social structure of the oil fraternity in Texas was, by the nature of the resource and its distribution, open ended, geographically and socially.

The evolution of the oil fraternity in Texas is not without its paradoxes. In their ideological representation of themselves, they were the last and greatest defenders of the bastions of free enterprise, the fiercest opponents of government regulation and oppression in any of its many and pernicious guises. Michel T. Halbouty, Houston independent oil operator and soon to be Ronald Reagan's official "energy advisor," spoke (1978) in rhetorical terms more akin to those of a nineteenth century revivalist, or Farmers' Alliance circuit rider, than a "statesman of industry":

> The tyranny of government is everywhere! . . . We no longer live in a nation controlled by the principles upon which it was founded. Instead, we live in a land where federal politics dominates, controls, tyrannizes. The tentacles of a monstrous bureaucracy are enveloping and crushing our incentives and enthusiasm to produce and prosper—which breaks the morale and spirit of man.
>
> I would as soon be governed with a rifle at my head as to be bound hand and foot and gagged with the red tape of regulation.[124]

This is no mealymouthed organization man sitting in front of his computer carefully calculating his hoard of utilities. This is a real oilman, lifeblood of free enterprise and the great white hope of a free society. This is capitalism red in tooth and claw, the way God meant it to be.

The irony is that from the day in 1931, in the depths of the Great Depression, that Michel Halbouty graduated from the Agricultural and Mechanical College of Texas (Texas A&M)

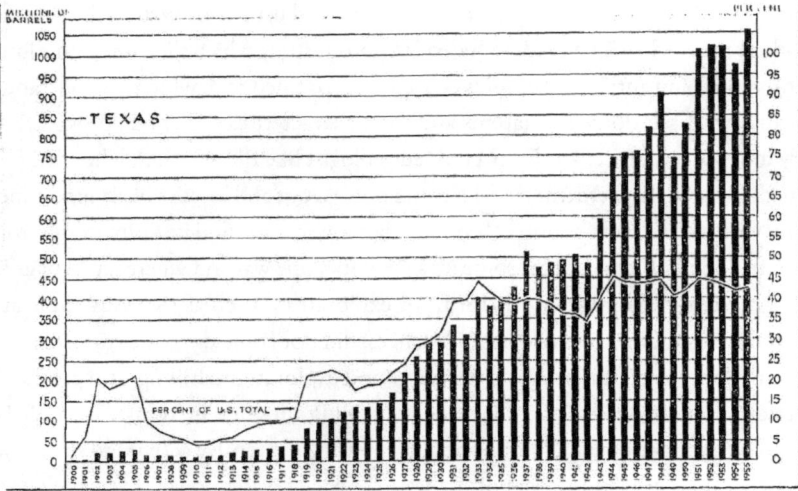

Figure 1.1. Oil Production in Texas, and US percent Texas, 1900–1955[122]

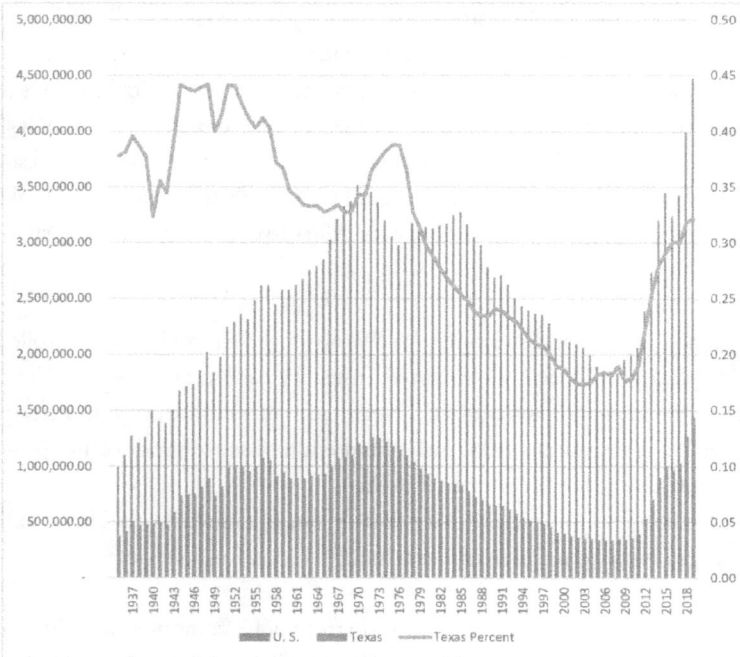

Figure 1.2. US and Texas Oil Production, and percent Texas, 1935–2019[123]

with a degree in geology and petroleum engineering and went to work as a chainpuller for the Yount-Lee Oil Company, he lived, and prospered mightily, in what was arguably the most stringently regulated industry in the United States. Colonel Ernest O. Thompson, long-time chairman of the Texas Railroad Commission, in 1950, on the Commission's role in oil regulation, commonly called in Texas and the other mid-continent states, prorationing:

> The regulation starts with providing rules for proper spacing of wells before drilling, testing of casing, protection against blowouts, protection of fresh water encountered in drilling, and then when oil or gas is found we test the wells and set a proper allowable for the wells and the fields so that the reservoir energy will be utilized in the highest degree and then, above all, when the ability to produce oil or gas is greater than the market demand, we allocate the market fairly to each of the 1700 oil fields and then distribute the fields' allowables among the 117,000 producing wells of the State so that every field and every well in each field has the opportunity to share ratably in the market demand for Texas oil.
>
> In the beginning, we lost cases in the trial and error process, but today our orders stand sustained by all of the courts. As equally important, these orders have also been sustained by an enlightened public opinion.[125]

Halbouty himself had as much to do with establishing petroleum engineering as the final arbiter of subsurface property rights and as the basis for regulation as did any other single individual. Some might dismiss the disjuncture between rhetoric and reality as mere mendacity, or as the ideological ravings typical of those who confuse being rich with being wise. But even in a culture in which telling tall tales and pulling legs are considered highly reputable art forms, the contrast is extreme.[126]

# CHAPTER 2

# THE MORAL ECONOMY OF THE OIL FRATERNITY IN TEXAS

This [popular consensus] in turn was grounded upon a consistent traditional view of social norms and obligations, of the proper economic functions of several parties within the community, which taken together, can be said to constitute the moral economy of the poor. An outrage to these moral assumptions, quite as much as actual deprivation, was the usual occasion for direct action.

—E. P. Thompson, "The Moral Economy of the English Crowd in the Eighteenth Century"[1]

By the time the oil fraternity in Texas entered its second generation, its social and cultural character—its moral economy—as well as its patterns of growth and development, were well defined. The "moral economy of the oil fraternity in Texas" is an analytical abstraction and a narrative convenience; the elements into which it is decomposed are intended to capture, however incompletely, a complex and elusive social reality, which is a fancy way of saying it's all made up.[2] The spatiotemporally particular moral economy of the oil fraternity in Texas may be taken to comprise six tenets: opportunity, altruistic cooperation within the fraternity, privileged access to the resources of Texas, fraternal autonomy and self-regulation, equality and liberty within the fraternity, and useful knowledge, the more direct and empirical the better. "Tenets" is perhaps a poor word choice: despite their evangelical heritage, most oilmen never have been much on creeds, doctrine, or anything else it took a preacher to say or a lawyer to understand. But there are distinctive patterns in the behavior of oilmen and in the way they recount that behavior. Those patterns reveal the moral economy of the oil fraternity in Texas.

## MORAL ECONOMY

"Moral economy" remains controversial. One difficulty has been that many of those who have used the concept have treated it, whether tacitly or explicitly, as spatiotemporally unrestricted, that is, as a natural kind with some defining essence, which it is not. Rather, what is called "moral economy" is, first, a set of spatiotemporal particulars, "moral economies" plural, which, second, evolve in time. Moral economy, then, is a diverse collection, from different societies and different periods, of social arrangements regarding property, usufruct, income, social and material prerogatives, mutual obligations, subsistence rights, and economic activities, which are thought to have broad structural similarities.

The paradigmatic explication of moral economy is English social historian Edward Palmer Thompson's reinterpretation of eighteenth-century English food riots:

> The food riot in eighteenth-century England was a highly complex form of direct popular action, disciplined and with clear objectives. It is of course true that riots were triggered off by soaring prices, by malpractices among dealers, or by hunger. But these grievances operated within a popular consensus as to what were legitimate and what were illegitimate practices in marketing, milling, baking, etc. This in turn was grounded upon a consistent traditional view of social norms and obligations, of the proper economic functions of several parties within the community, which, taken together, can be said to constitute the moral economy of the poor. An outrage to these moral assumptions, quite as much as actual deprivation, was the usual occasion for direct action.[3]

Food "riots" in eighteenth-century England were not simple banditry, rebellion, or social disorder but rather local and disciplined enforcement of communal norms regarding the "just" (customary) price of grain, the privileged access of the local community to locally produced foodstuffs, and the fundamental right to subsistence of even the meekest members of the community. Direct collective action such as food riots rarely entailed outright theft or expropriation—merchants or those hoarding grain (more often than not depicted as outsiders) were duly paid the customary price. Thompson was a devout Marxist and deliberately contrasts this pre-industrial "moral economy" with later "political economy," or the market economy of capitalist industrialism. For his eighteenth-century English crowd, erstwhile "economic" issues were settled within a traditional framework of communal value and action.

Thompson's concept of moral economy has been widely exploited and extended, most notably by James C. Scott in his work on Southeast Asian peasant communities.[4] For Scott, the moral economy of the village is a "local system of action, status, influence, and authority." It entails local autonomy and self-regulation, and mutual support within the community (especially an obligation of the better-off to assist those in need). Most of all, as in the moral economy of the English crowd, the moral economy of the village mandates privileged access: "the village and its inhabitants have a right to the resources (e.g., land, forests, water, pasturage) they have traditionally used and [external] claims on local resources are only admissible after the customary subsistence needs of the villagers have been met."[5]

Moral economy is also central to the pursuit of science, including engineering sciences such as petroleum engineering. The most fully developed application of moral economy to scientific inquiry is that of Robert E. Kohler in his examination of *Drosophila* research in evolutionary biology during the 1930s, although the behaviors he describes appear in virtually all modern accounts of the emergence and development of scientific disciplines or specialties.[6] How scientists go about being scientists is governed by a delicate balance between cooperation and competition, by mutual support and sharing, by inclusion and exclusion, and by privileged access to laboratories, research materials, and experimental equipment and methods, unpublished results, and speculations and ruminations among those acknowledged by their peers to belong to the community. Disciplinary boundary definition and maintenance, internal governance, research evaluation and authentication, and professional autonomy and authority are analogs to the elements of the moral economies of the English crowd, a Vietnamese village, or the oil fraternity in Texas.

Except perhaps in Thompson's almost too-clever rhetorical sense, moral economy carries no connotation of conventional morality. For the oil fraternity in Texas, anything but. Nor does moral economy imply egalitarianism or any particular hostility to inequality of wealth or status. What is entailed is a moral system which governs both the internal relations among community members—their obligations to each other—and relations between the collective and the external world. Thompson's and Scott's characterizations of moral economy, together with what is easily recognizable as the moral economy of agrarian Southern communities to which the oil fraternity in Texas was heir, are the starting points for constructing the moral economy of the oil fraternity.

Usage of the term "moral economy" has diverged, and the original concept is still contested.[7] Nevertheless, altruistic cooperative behaviors characteristic of a moral economy have been noted across a broad spectrum of social collectivities: diamond merchants, furniture manufacturers, aeronautical engineers in Southern California in the 1930s, Balinese agriculturalists, and the pirates of the Caribbean.[8] Moral economy likely applies also in unexpected places: among used car dealers, Hollywood producers and agents, and venture capitalists and other petty thieves. Thompson himself was remarkably generous and gracious about how his idea has been used and extended, perhaps an example of intellectual moral economy.[9]

The oil fraternity in Texas, however remote in clock time and distance, was close in lifetime and experience to the moral economy of the English crowd in the eighteenth century, and, as then, the shackles of the cash nexus still rankled. C. V. Terrell was the chairman of the Railroad Commission of Texas in 1932. He had made his reputation prosecuting cattle rustlers in Wise County, northwest of Fort Worth, in the 1890s, and had served on the Railroad Commission since 1924. Testifying before the Texas Senate in November 1932:

> I want to call your attention to one thing I remember when I was a boy. . . . We had four flour mills in our section of the country and we boys would hitch up the team or get on the old gray mule with a sack of wheat and go to that mill where they would grind it and we would bring back part of the flour and pay them part of the flour as the toll to the miller and the people were happy and prosperous people then. Where are they now? I am condemning nobody, but I want to say this. When the

Interstate Commerce Commission forced the Railroad Commission to give stoppage to mills and transferred all wheat to those particular mills with money, they broke every independent mill, and you cannot find one today hardly in this country and that is the result.[10]

## THE MORAL ECONOMY OF THE OIL FRATERNITY IN TEXAS

Except maybe to New Yorkers, the oil fraternity in Texas aren't exactly peasants. But a local and fraternal system of action and meaning, of representation and value, which is homologous to other moral economies, does define the oil fraternity: the fraternity's own, unique, spatiotemporally particular moral economy. It is not identical in content to any other moral economy. But if "subsistence" in Thompson's or Scott's moral economy is replaced by "opportunity," opportunity specifically for members of the oil fraternity in Texas, a similar structure applies. Other ideals gave concrete behavioral expression to, and assured and preserved, opportunity: altruistic cooperation, privileged access, fraternal autonomy, equality, and useful knowledge.

In life, the moral economy of the oil fraternity is holistic, not neatly categorized and codified; it has no "constitution." Moral economy has to be inferred, from behavior, and from what actors say, from how people themselves ascribe meaning to their actions and to their world. The problem is that among folks not given to abstraction, as the oil fraternity is not, formal rhetoric is likely to be stilted and stylized, rhetorical in its pejorative sense. For this reason, the best historical source for the moral economy of the oil fraternity in Texas, for understanding the referents of its representations of itself, is its "little tradition," the things oilmen talk about among themselves, the meanings they attach within their own community. The oil fraternity is after all largely an oral culture. Luckily, the Oral History of the Texas Oil Industry Collection at the Dolph Briscoe Center for American History, which contains interviews covering mostly the period before 1950, and which has already been used to portray who the oil fraternity was, offers helpful materials. In order to let folks speak for themselves, these oral histories are the primary source for depiction of the oil fraternity's moral economy. Oral history is controversial: it is demonstrably lousy at accurate factual recall. But for now, facts are secondary. The goal is to invoke a culture, its representations, and its mythology, and for these purposes, especially for an oral culture, oral history is unsurpassed.

What oilmen mostly talk about are deals and plays and booms and personalities. Deals particularly are recounted in exhaustive, even excruciating, detail. Deals are the exemplars for the practice of the oil business: Like paradigmatic experiments in science, they serve not only as models and as templates for behavior, but also as guides to moral judgment and authority, and to wisdom and to success as a practitioner.[11] Stories convey the subtle nuances of moral economy: "Salting" a well core, telling a prospective investor a good, oil-bearing core came from a well when it didn't, like salting a gold mine, is a big no-no. But carelessly leaving a bogus dry core in a bucket near a well, as if discarded, is delightfully clever: a wily "scout," snooping about, sees what he will, while the successful wildcatter goes about quietly buying up more leases. For the oil fraternity, these "war stories" parallel accounts in other traditions of battles, migrations, and natural disasters, of deeds of great heroism or contemptible cowardice, of divine or profane intervention. Oral histories, plus occasional use of biographies in the hagiographic tradition, which are themselves almost invariably cast as oral narratives, open the inner world of the oil fraternity and its moral economy.

## Opportunity and Cooperation

For the oil fraternity in Texas, opportunity first and foremost and always was why they came: why they left their farms, and often their families, why they lived in tents and worked twelve-hour days seven days a week, why they came to parts of Texas only arachnids could love. Their exuberant belief in opportunity was written into the way they customarily named wells. Each well on any lease was always going to be the first, or next, of many more, as in Lou Della Crim No. 1, or Santa Rita No. 1, or Bear Ranch No. 1, or Yates No. 30-A, or Daisy Bradford No. 3 (the third try on the same lease, which finally produced the discovery well for the East Texas oil field). But if opportunity, opportunity for men with nothing but a handshake and a smile, was why they got into the oil business, cooperation was how they stayed. No more for them than for their Alliancemen forebears were competition and cooperation antithetical. Walter Cline, one of the founders of the Texoma Company, and later mayor of Burkburnett, and then Wichita Falls:

> . . . but I had an opportunity to get some rotary work, and no money. So I went down to the National Supply Company where old Ode Chatwell had charge and told them very frankly my troubles as honestly as I could. Had some work and no money, but I thought I could make some if I could get some iron to dig with and they sold me a complete drilling rig from crown block to boilers with all the equipment and furnishings and tools and the fittings. And I didn't have any money so I couldn't pay them anything. Then my old friend W. W. Graham sold me a derrick pattern, a wooden derrick, of course in those days, on credit.
>
> I was planning to do most of the drilling myself, except I didn't have any cash to pay my roughnecks. I went down to the old former State Bank at Burkburnett . . . and again told my story of one wife and two children and no money, a contract and a desire to work. And Mr. Moore and Mr. Daniels loaned me a thousand dollars.[12]

In a cash-poor land, leases were paid for with hope and often nothing more than a promise to try to get a well dug. Landon Cullum, on putting together the deal that discovered the Desdemona field in North Texas:

> I got most of 'em [the farmers] to agree, and we had a meeting at the school house one night and I told 'em I wanted to treat 'em all alike. That even though some of 'em had more land than the other one that they would all get benefit out of a well, and that I'd give 'em a dollar a farm, and we'd put it in the bank [the leases would be

placed in trust in the local bank, their execution contingent on drilling a well] and we'd start a well within six months. So that night I had a notary public with me, and that night before we'd left there we'd signed up three thousand acres of leases.[13]

For men for whom tent revivals and evangelical camp meetings were central features of cultural life, for men who remembered the Farmers' Alliance, with its itinerant lecturers and its chapters at every other crossroads schoolhouse in North Texas, neither the scene nor the means were unfamiliar. Yet the next move was pure wheeling and dealing:

> However, I kept on until I got 6000 acres. Then I reported back to Wichita Falls that we had this acreage, and we sat down and worked out this kind of a deal that Mr. Harvey and Mr. Kell and Mr. Perkins would underwrite the cost of a well, and Wrather and I would have a quarter interest in the deal, carried free in the first well; and we would try and make a deal on some of the acreage if we could and relieve them of the drilling obligation, and if that was the case then we would have to revamp our deal so they'd have an interest in it the same as we would. They agreed to that; so Wrather got on the train and went to Pittsburgh to see his old friend F. B. Parriott, who was with Benedum-Trees.
>
>   They [Benedum-Trees] had sold out so they were ready for another deal. They said they would take this field and for three thousand acres of our leases they'd carry us for one-fourth interest and the first three wells. After that we'd have to pay our way. They would guarantee to market the oil, which at that time was rather hard to market.[14]

The custom, indeed the obligation, of the better-off to "carry" the less well-off to assure them opportunity became characteristic of how deals were done and was so ubiquitous in the oil fraternity that it spawned its own legal lexicon and, because the terms of such deals were often ambiguous and complicated, its own case law.[15]

Cullum's next big play was Electra, and how he got in, and got out, is typical of how opportunity worked in the oil fraternity:

> Well, I happened to know Johnny Suman. He was head man for Rio Bravo [the oil-and-gas division of the Southern Pacific Railroad] then. He had been out in West Texas where I was with the Humble, and he and Bill Wrather were great friends; so I called him up I went down and made a deal with him on 160 acres of that section [640 acres, a square mile] that Rio Bravo owned there so we could drill the well. Didn't pay him anything for it; they just gave it to me for the well. So I came back, and we drilled a well on it, and we got a three or four hundred barrel well right off the bat, and so we started in on that 160 acres, drilled a string of wells, and we got a well in every durn well we drilled.
>
>   Magnolia [the Texas wholly owned subsidiary of Socony Vacuum, later Mobil]

came along and wanted to know if we wouldn't sell that and sell the property we had down at Desdemona and the Gillis lease out at Burkburnett. Well, we jockeyed around with them for a while, and finally they made us an offer of three million dollars.[16]

Cullum's odyssey was replicated over and over again in the oil fraternity, sometimes with the happy ending, more often not. John H. Wynne:

So we got the idea we'd like to get this block [of leases] together. And we were going to get this block out there, and the easiest way we could figure to do it was to throw a big barbecue.

Of course . . . we paid all the expenses. First we got us a couple of steers and had them barbecued, some barbecued goat, got a few barrels of beer and invited all these German and Bohemian people to come to the big barbecue. And we let them know the reason we were having this barbecue is that we were goin' to get up a lease block to drill 'em a well out in this particular section Some three or four hundred people showed up, as I remember. Of course, directly we had the big barbecue and all of them got full of beer, why it became my part to get up and announce our intentions and what we were going to do. We gave them a synopsis and what our plans and everything were, and we had these leases already made out. We were pretty sure we were going to get them, you know. We had two or three big landowners we had interviewed, you understand, to set the thing off. We'd says, "Well, Mr. Smithers here (or whatever his name was), would you be willing to give your lease in to the Fayetteville State Bank for a period of one year, with a contract attached to it, that there must be a well drilled within a mile or two miles to a certain depth? And we think it is a good prospect or we wouldn't be out here spending our money. Would you give your lease?" And of course, he had already been interviewed and he'd say, "Yes." And he would come up and sign the lease, and then the next guy. We would have another one the same way. Oh, all was perfectly legitimate, you understand. And all the little fellows would crowd around. So now we would have ninety percent of our block signed up.

So we got that block together, and we got our well drilled—the usual dry hole.[17]

No revivalist ever painted the golden streets of heaven any shinier, or the sawdust trail any rosier, or made redemption any easier, or stage-managed the whole show any better, except maybe for the ending.

True to the tradition of Spindletop, even the most spectacular wells in Texas were apt to be "poor boyed." For example, E. C. Laster, the driller on Columbus Marion "Dad" Joiner's discovery well for the giant East Texas field in 1930, Daisy Bradford No. 3, recalled:

So the rig was repaired and rebabbited and fixed up generally and started to make additional hole. We continued the drilling when we could secure help and fuel. We were frequently without oil or repairs for the rig. And the checks were frequently unpaid due to lack of funds on the part of Mr. Joiner. And then, most of the crew were local men who had farms nearby. And when the funds were completely gone they would frequently go back in the fields. On some occasions I appeared for work and there wasn't anyone, maybe one and the next day there may be two. I would have to wait until they got laid by or some particular job done before they could return to the well.[18]

In addition to poor boying wells, or what usually was called running a "bean job" in West Texas because the sole operating capital for the well consisted of a hundred-pound sack of beans,[19] opportunity in the oil fraternity was provided by any number of other arrangements: anything men long on hope and ambition and short on cash and credit could devise. Most common were "farm outs," deals in which someone who had a block of acreage under lease would in effect sublease a portion of it in return for getting a well drilled. The term, of course, was carried over from pre–Civil War usage, when "farm out" meant to give tenant farm families a chance to get started on otherwise unused acreage. The deal Cullum cut with Suman and Rio Bravo at Electra was a farm out. Usually, the outfit offering the farm out was a "lease poor" larger operator or had lots of property (like Southern Pacific). What the people offering the farm out got was a well, which they probably could not afford to drill, in the middle of a block of leases they had already paid for, and which might be close to expiring. If the well came in, all or a good part of their acreage would be "proved up"; if the well were dry, they were not out a dime—and had the only certain information in the oil business for free. What the man taking the farm out got was opportunity: to promote a well, to get rich, and all for little or no upfront cash.

Most deals got complicated fast as each party tried to fit short means to long ends. Other common provisions were for dry-hole money or bottom-hole money. Either the outfit making the farm out, or people with adjacent or nearby leases or land, would agree to indemnify the driller if the well were dry, or to pay a certain amount if the well were drilled to a specified depth. Again, risk was spread out, information shared, and opportunity created. One of Michel Halbouty's deals exemplifies the sometimes-byzantine nature of opportunity in the oil fraternity. Just as World War II was ending, Halbouty got a farm out from Bud Lund and Ken Crandall of the California Company (a Socal subsidiary) on acreage in Louisiana on which they had already drilled seven dry holes. To put together a large enough block to promote a well, Halbouty next acquired farm out acreage from Shell, and leased additional land himself. Both California and Shell retained a one-sixteenth interest in their farmouts, which, net of the landowners one-eighth override, left Halbouty with a thirteen-sixteenths working interest in the leases. He tried to sell three-quarters of his thirteen-sixteenths to raise capital. But, given the dry holes, Halbouty couldn't raise any money. He was broke. A friend kicked in $10,000, about half what he needed to get a well down. Finally, desperate, Halbouty went back to Lund and Crandall and offered to sell them an interest in their own farm out for enough money to drill the well. After some unpleasant words, Lund finally concluded,

"You know any son of a bitch who's got the nerve to ask a company to join in drilling acreage it farmed out in the first place ought to be helped. By God, I'm going to do it!" The farm out became the Ashland oil field.[20]

A host of social adaptations, inventions, and customs assured Everyman the opportunity to try. Custom dictated that poor-boy oilmen had a virtual right to expect better-off partners in a deal to "carry" their working interests, interest free, until production could cover costs.[21] After World War I, less well-off promoters and brokers typically could trade a "working interest" for a smaller "overriding" royalty interest, which, like the landowners' royalty, paid a fraction of production with no costs deducted.[22] Such royalty provided instantaneous cash flow once a well came in and increased the small operator's payout rate, and thus his opportunity to participate in new deals. Given the near certainty of hitting oil anywhere in the vast East Texas field, by the early 1930s promoters and operators had morphed the hated crop-lien system into the oil payment, which L. L. James, Dad Joiner's lawyer, described:

> A owns a lease on B's land. A wants to drill a well and he can't, so he hires C to drill the well and gives him $30,000 to be paid out of the first oil, out of a certain fraction of the first oil to be produced from that well. It may be an eighth, it may be a quarter, it may be a half, any amount agreed upon. . . . An oil payment which would pay itself, pay its money back, which would pay the money back in six months would sell for about 50 or 75 cents on a dollar.[23]

These transactions too could get complicated, involving multiple parties, carve outs (provisions for payments to be made from a specified share of production until an obligation was satisfied), and trade-outs, as well as convoluted sequences of contingencies.[24] However untidy, in the depths of the Great Depression, the oil fraternity evolved an indigenous, flexible system of discounting, a futures market, liquidity, and opportunity all rolled into one.[25]

Structurally, opportunity in the oil fraternity is similar to subsistence claims in a peasant moral economy or the shared obligation to participate in a cowhunt. W. B. Hamilton, one of the principals in the Texoma company, gave voice to this altruistic obligation in the oil fraternity:

> . . . rarely would anybody drill a wildcat well that we wouldn't support them by buying some of their acreage. We didn't make any money out of it. One year . . . [we] spent over $400,000 in assisting other people drill their wells, and not a tract of that land which was purchased for the Texoma Company ever produced a barrel of oil as long as we owned the company. But it did give a lot of support, economic support to the community and to the small operators.[26]

Even the biggest and toughest oilmen were little once and needed somebody's help, and they remembered it. Clint Wood, who worked in a Beaumont lumberyard when Spindletop came in, recalled the Texas Company's early days:

. . . the Texas Company they got him, J. S. Cullinan and turned the thing over to him. And he took the Sharp boys there; he had known them up at Corsicana . . . and I used to sell 'em lumber, wait ninety days and then take a ninety day note from Texas Company. Because they was puttin' all the money they could git into those earthen tanks, storin' out oil, Cullinan was.[27]

But the oil business, like "hogs in the timber," was perceived as a non-zero-sum game. E. E. Townes, who started in the oil business at Spindletop and was for many years the Humble Company's chief counsel, portrayed the "game against nature" the oil fraternity played, and the inference it drew from it:

There's one thing that distinguishes the profit made in the oil business from profits made in a great many other businesses. A concern may make a million dollars by the discovery of oil on a wildcat lease and it and its employees and the stockholders and the public, the landowner, the boarding house keeper, the merchant, everybody shares in the benefit of that million dollars and nobody loses. Now, you go out and trade horses. If one man—if the trade turns out to be a good trade for one man, he makes, and the other fellow has lost that. Or you buy a piece of land and then something happens and you sell it. The fellow that sold it, lost it. But in the oil business, if anybody makes money, it's a new, it's the development of a new product that was worthless to anybody before and nobody loses.[28]

What opportunity meant for the oil fraternity was the chance, with emphasis on chance, for Everyman to become rich. Johnny Wynne, again: "Of course, you've got to have an incentive. And the incentive was . . . that the ten million dollars was just around the corner, as sometimes it was. In my case it wasn't but I have no kicks to register, I did all right."[29]

Binding on all who would belong to the oil fraternity in Texas was its moral economy's primary ethos, opportunity. As Wynne explained it:

. . . it got to be a very common thing as the oil business went along here in Texas to get what we call a "checkerboard" block. People would rush in and there would be five or six operators and maybe six or eight big companies. Wouldn't any of them have enough leases to justify drilling a well. And then some guy in the middle of that, some independent or big company would call in John Doe, the independent, and say, "We want you to work out a deal, drill a well, test this acreage out . . . and we'll give you a drill site and say, ten thousand dollars to start you with."[30]

Mr. Doe would then cajole (promote) the other leaseholders in the block into contributing their fair share of drilling costs to get the well down.

Of course, not everyone cooperated. Opportunistic free-riding was always a compelling option: why pay for what you can get for free? But even if free-riding didn't kill the whole deal, non-cooperators were likely to be excluded from future deals:

> Then some of them were heels and wouldn't donate anything. They were the guys that would get a free ride. . . . And the way they would get back at them, well that guy being in the same business he'd want to do that same thing some time, and when he'd call on them, why he found out he couldn't get any money. So it was a cooperative enterprise.[31]

Within the broad yet stringent normative constraints of moral economy, every individual was free to promote any lease he could, drill anywhere he could get the money to, believe in any prospect he chose to for whatever reasons he might adduce, cut any deal he could get his confreres to go along with, spend his money however he saw fit. It is Lawrence Goodwyn's point once again. Cooperation and competition are antithetical only in contemporary "economic" eyes: the oil fraternity could and did order its world differently.

## Privileged Access

The oil fraternity in Texas, like southern farmers in the nineteenth century, like peasant societies in almost any century, like the English crowd in the eighteenth century, saw the usufruct of the land as theirs by right. It is a sentiment rarely overtly stated, so obvious to those who hold it as normally not to need explicit articulation. Just about the only time rights of privileged access are explicitly asserted is when someone external to the community threatens, or is perceived to threaten, to mess with them. The men who became the first generation of the oil fraternity in Texas were heirs of Southern agrarian protest and of all that implied: hatred of the "trusts," the "money power," the "East," and a determination to protect what was rightfully theirs. The attorney general of Texas expressed the sentiment nicely in 1907, when he vowed to use the State's strengthened antitrust law, "the most drastic antitrust law enacted by any state," to "drive every trust and unlawful combination out of Texas," for the simple and obvious reason that their "well known purpose was to appropriate the territory of Texas for their greed and exploitation."[32] Privileged access for the oil fraternity in Texas was expressed most often not as a direct positive claim but as a vehement, sometimes violent, denial of external claims against communal resources, or as an apocalyptic vision of external threat.

From the very beginning, the oil fraternity sought to secure its rights of privileged access. The fraternity's sentiments and fears—and their highly stylized representation—is exemplified in the row over control of the Texas Company. The Texas Company had been formed in 1902 from the Spindletop holdings of Joseph S. Cullinan, the Hogg-Swayne Syndicate, and the Sharp brothers' Producers Oil Company. Financial backing came from Chicagoan John W. Gates and New Yorker Arnold Schlaet, acting for the Lapham brothers (the leather trust). Miss Ima Hogg (Governor Hogg's daughter)[33] recalled the concerns of those founding of the company:

The men all felt at that time, the Texas men, rather—I don't know how men from outside felt—but Texas men felt they needed protection. They knew that there was a great deal at stake in Beaumont, and they felt they needed some good council as to strengthening their position against outside invasion. There were larger companies, I'm sure, at that time, who could have crowded out all the Texas interests.[34]

Even the Chicago financier John W. Gates, in bed with the leather trust or not, shared Cullinan's and Hogg's determination to keep "control of oil in Texas."[35]

Unfortunately for Cullinan and the Texas Company, Gates, who had steadfastly supported Cullinan's ambitious but risky expansionary polices, died in the summer of 1911. He was replaced on the Texas Company's board of directors, at Cullinan's urging, by his son and sole heir, Charles G. Gates, who in fact took little interest in company affairs. The Texas Company itself saw net income decline by two-thirds in 1911 and 1912, compared to 1910, due largely to Cullinan's rapid expansion in the Electra field in North Texas and in the Caddo Lake field on the Texas-Louisiana border. Worse, Charles G. Gates too died, from a stroke at age thirty-two, in the fall of 1913. The younger Gates's death brought to a boil simmering disagreements between Cullinan and the Texans and the more conservative Schlaet and other Eastern investors.[36]

One of the country's nastier proxy fights ensued, pitting the Texans against the "Easterners." What is of interest is the way Cullinan and the Texans represented the conflict to themselves, to the Texas Company's stockholders, and to the oil fraternity in Texas. From a Statement to the Stockholders of the Texas Company, from J. S. Cullinan, November 17, 1913 (the eve of the crucial Stockholders' meeting):

(1) From a small and modest enterprise, the stock of which was held entirely in Texas, your company has grown in financial strength and resources until now the majority of the stock is held in the North and East.

(2) Its original management, its corporate attitude and activities were branded with the name Texas and Texas ideals. . . .

(4) The Texas Company was organized in 1902 under Texas laws; being a Texas corporation, it is to be conclusively presumed that it was the intent that Texas laws as to management should be observed, and, accordingly, that actually as well as nominally the headquarters and governing authorities should be kept and maintained in Texas.[37]

Just cause or no, the Texans were outvoted, with the result that Cullinan, William C. Hogg (Governor Hogg's son), and James L. Autry (the Texas Company's chief counsel), as well as a number of others, all left the Texas Company. The attitudes they expressed in their statement to stockholders were not mere public-relations flapdoodle. They said the same thing privately, even after the issue was decided; as James L. Autry wrote in December 1913:

The main issue presented was as to whether the business of the Company, from the executive and legal standpoint, was to be operated from New York, and according to New York ideas, or from Texas, and according to Texas ideas. . . . [T]he alternative of getting out was the only possible one, from our point of view, upon it appearing that our ideas were not to prevail.[38]

F. C. Proctor, Gulf's general counsel and the person who ran Gulf's operations in Texas, and who himself had come out of the Beaumont law firm of Carlton, Proctor, and Townes, sent Autry a handwritten note:

Nov. 26, 1913
"Personal"
Dear Autry:
I am awfully sorry you have quit the Texas Company. I know from our personal talks you have no regrets and are really glad to lay down the burden of this work. The regret is personal with me and I had hoped you would remain as long as I shall. The situation is that for now nearly nine years we have occupied similar positions toward two commercial rivals . . . at times it has probably tested the ability of both of us to keep such rivalry within its proper limits and yet I feel we have succeeded with result that we have established a firm relation toward each other of mutual respect and confidence.

Of course, your successor Beaty is also my personal friend and I anticipate nothing but pleasant relations with him—but still I am very, very sorry you have quit the game.[39]

Texans had no doubt as to whom their patrimony of oil rightfully belonged, nor about by whom and how it should be controlled.

The same attitude was shared by E. E. Townes, also of Carlton, Proctor, and Townes, who became chief counsel for the Humble Company. When the Humble Company was being formed, Townes had already agreed to become General Counsel for the Sun Company and all associated Pew interests.[40] With the exception of Ross Sterling and Walter Fondren, all the original Humble founders were former clients of his firm. Townes asked the Pews to release him from his commitment, which they did; he went to Houston rather than Philadelphia. What the founders wanted Humble to be was clearly defined:

We decided that we were not going into the stock selling business, as we didn't intend to take advantage of legal technicalities, that we did not want to organize under the laws of the state of New Jersey or Delaware or Arizona, but that we would organize under the corporation laws of the state of Texas and comply with those laws . . . a corporation that expected to deal on the top of the table.[41]

That the Texas founders of what would grow to be among the world's largest integrated oil companies should so clearly assert notions of privileged access to Texas's oil resources for

Texans or, specifically, for the oil fraternity in Texas, is in part an indication that by virtue of their size, and their sources of capital, they were in some sense suspect. Yet there is no evidence that either Cullinan, Pennsylvanian and former Standard Oil ally or no, or Townes, or any of the other Texas men associated with major oil companies, were disingenuous in their protestations. What is significant is that only they felt the need to voice what the rest of the oil fraternity knew as existential fact: Texas oil was for Texans.

## Fraternal Autonomy and Self-Regulation

Communities with a moral economy impose their own order on economic behavior within the community, and attempt, with greater or lesser success, to impose that order on relations with outsiders. The core of the moral economy of the oil fraternity in Texas was opportunity and cooperation: the one depended upon the other. The harshest sanction in any community governed by moral economy is ostracism. De facto ostracism was imposed by the oil fraternity on those who transgressed its mores, or on strangers, not so much overtly as by simply not dealing with them. If someone were an outsider, somebody who, in that most damning of phrases, "didn' unerstan' the oil bidness," or if he was known as a "four flusher," a "high binder," a "tin horn," or any one of a variety of other wonderfully rich and inventive (and likely libelous) descriptions, he was apt to be left out. And to be left out was to be denied access to information, to help, to participation: to be denied opportunity.

The oil fraternity in Texas also exhibited a talent for more formal self-organization and self-regulation. As descendants of the Farmers' Alliance and the People's Party, as well as of stockmen's associations and vigilance committees, the oil fraternity inherited not only an antitrust, xenophobic ideology, but also specific forms of social action and organization. In virtually every field in every region, from Spindletop to the Permian Basin, from the Gulf Coast to North Texas, producers repeatedly tried to organize producers' cooperatives to store, transport, and, most importantly, collectively market Texas crude oil. They tried to do with oil what the Farmers' Alliance had tried to do with cotton. In each case, the oil fraternity's efforts foundered in the face of the rule of capture (and a lessee's legal obligation to produce flat-out that it entailed), simple greed, the complexities of the world oil market, and the ever-present fear of antitrust prosecution. Yet what the oil fraternity tried to do, and how they interpreted their plight and their failure, tells a lot about their moral economy and their ideology, and presages their later cooperative creation of Texas Railroad Commission regulation.

One the most ambitious attempts at collective marketing was led, in 1916, by none other than the men who would within a year create the Humble Oil and Refining Company, William Farish, Ross Sterling, and Harry Wiess. In words he would have handed back to him in the 1920s, Farish voiced a lament with which any dirt farmer could harmonize and any victim of John D. Rockefeller empathize:

> The producer has never in the history of the industry been paid a fair price continuously. The refining industry has always bought up its flush production at a large discount. I know that supply and demand determine prices, but it is manifestly unfair

for the refineries to say to us that they are paying to us the market price. They make the market. Commerce in our product, except through them, has been destroyed. Perhaps I might say that they had destroyed it. Whether it be right or wrong, lawful or unlawful, they have, through allied companies, come almost wholly to control the production, handling, piping, and refining of oil in the State of Texas. What the little fellow gets for his product is through their grace.[42]

William Jennings Bryan might have said it better, but he would have meant the same thing. While attempts by producers to collectively market oil invariably failed prior to 1925, modes of social action, and the location of evil outside the fraternity and Texas, persisted.

Even though the oil fraternity's more grandiose projects for collective marketing miscarried, as had those of the Farmers' Alliance, more restricted forms of oil fraternity self-organization and regulation flourished and provided a repertoire of responses which likewise persisted. As noted, following a series of catastrophic fires in the Spindletop field in 1902, operators organized a field safety committee that not only controlled how wells were completed and the kinds of structures that could be retained, but also, to prevent further fires, required those without storage to hold their oil in the ground, an extraordinary limitation on the right to produce.[43] Producer-organized field-safety committees became a fixture of succeeding major oil plays but had no legal authority. Texas's first legislation regulating oil exploration and production, enacted in 1899 in response to problems in the Corsicana field, required operators to seal off water or oil bearing strata above the main "pay" (producing strata) with casing, and to "plug" abandoned wells, but provided no monitoring or enforcement: compliance depended upon individual civil suits.[44] The exercise of field safety committee authority went well beyond the law and rested upon a moral consensus among operators backed by the implicit threat of more forceful action. Like western miners and Panhandle stockmen before them, the oil fraternity evolved forms of collective self-regulation, which would carry forward into voluntaristic pooling, field unitization, and joint operation.[45]

The oil fraternity also tried to expel the charlatans and frauds, at least the ones from outside, from its midst. When the mail-order swindlers descended on Burkburnett in 1918, oilmen in Wichita Falls called a mass meeting, which led to formation of an Oil Investors Association. F. G. Swanson, Wichita Falls attorney, and the Association's chairman:

> . . . a bunch of promoters, some of them crooks, would come in and organize, put on a fast sell, high pressuring suckers. . . . And they would be gone before any sucker would find out that he had been hooked. . . . No blue sky laws moved fast enough. . . . We had several discussions on the organization of the Oil Investors Association, and I suggested that nothing would be effective except an open forum. . . . And lean up against the slander and libel laws as heavily as the mail order swindlers were leaning against misuse of the mails and swindling statutes. . . . We gambled on the idea that in slander and libel suits we might have the sympathy of the juries, whereas the swindlers probably would not have.[46]

No anti-stock law night-rider or cattlemen's association vigilante ever figured it any better, or understood his community, or moral economy, more completely.

One of the major consequences of these deeply rooted traditions of self-regulation, in the oil fraternity as in other social collectivities, was axiomatic opposition to external control or manipulation. Prior to the 1920s, the most fearsome external threat the oil fraternity perceived was "the trusts." During World War I and the early 1920s, "the trusts" were joined by "the government," which soon came to dominate the demonology. Texas Railroad Commission regulation, specifically prorationing, ultimately was accepted only because it was negotiated and constructed, in a prolonged and sometimes acrimonious process, to be *not* alien government interference but rather fraternal self-regulation.

## Liberty, Fraternity, and Equality

The oil fraternity in Texas was an egalitarian republic of highly unequal "small producers." Its moral economy stipulated and depended upon egalitarianism without the expectation of equal outcomes: it was a system that could exist only when opportunity was, or was thought to be, Everyman's to grasp. Oil in Texas from the beginning was discovered without the permission or the blessing of "experts." Expertise again and again was simply wrong; and while the "experts" looked asinine, some good ol' boy who had just played the right hunch and got lucky went laughing all the way to the bank.[47] That right to be wrong—in defiance of any authority, in contempt of any expertise—is central to the way the oil fraternity in Texas would construct prorationing and, more importantly, its scientific foundations.

Money, of course, was what opportunity in the oil fraternity was all about: equality was a means, not an end. Yet given the high probability, and observed frequency, of reversals of fortune, good or bad, that characterized the oil business, the oil fraternity by and large was spared the sort of moral prissiness that afflicted, and afflicts, European landed aristocrats, American industrial capitalists, professional corporate managers, and politicians. Money and virtue were rarely conflated. E. P. "Matt" Matteson on the effects of new money on old poor: "Most of the farmers and ranchmen were worse off after they got oil than they were before . . . they'd made renegades out of their boys and made monkeys out of theirselves . . . there was very few that the money actually ever did them any good."[48] The predominant view, consistent with the oil fraternity's cultural heritage and moral economy, was that money, however desirable, rarely changed a man's basic character. Walter Cline:

> The truth is that, in my judgement, they run pretty well true to form. The good, honest, well-intentioned, hardworking, we might even say church-going and Christian, man, almost without exception will run true to form. He'll take his money and build a Sunday School room onto the local church or maybe tear it down and build a new church as Uncle Gash Hardin [John G. Hardin] did at Burkburnett. . . .
> There is, however, always the fellow who has always been a stinker. He's at heart a loafer and a cheat and due, usually, to the thrift and determination of his wife who works side by side with him, he finally acquires a plot of ground and you go out and

drill a well and get oil on that stinker's property and he runs just as true to form as the decent fellow does. He shows up down in the lobby of some big hotel in a few weeks all dressed up like a sore thumb and grinning like a jackass eating briars, with a blonde headed girl on his arm and leaving the old lady home to continue to throw water down to the calves and feed the horses and look after the farm and keep the children in school . . . when they get the money, brother, they put their tail over the dashboard and they start.[49]

The stories that became legend are of people who changed not at all. Old Lady McClesky, whose family owned the discovery well for the Ranger field, and built the first hotel there, used her first oil money to buy a new axe, because she was tired of the old one with slivers on the handle.[50] Daisy Bradford, on whose farm the discovery well in East Texas was located, continued to run her boardinghouse despite her good fortune. The stories are told with mixed emotions: part respect, part amazement.

There is a great deal of "there but for the grace of God go I" in the oil fraternity, and for good reason. Too many made theirs and then lost it, sometimes several times over; too many came too close too often not to inspire the sense of social symmetry that underlies true egalitarianism. Such emotions are most often expressed as a kind of gentle self-deprecation, or even fatalism. Joseph M. Weaver:

I remember, to show how things jumped in the Ranger boom, having 285 acres leased from one of the good farmers of this country and when the boom broke, came to me, for not his own money but money that he had been offered, and offered me $285,000 for this 285 acres lease that not more than eight months before he had leased to me at $125 an acre. I did not make the trade; I was too intelligent. I kept the lease and then drilled two dry holes on it.[51]

Walter Cline, who had sold his house in Burkburnett to one of his drillers just before the boom hit:

He leased it. I think he got 20 or 40,000 dollars for a lease on those two corner lots. They went out and moved my little tin garage off and drilled a well right in my backyard, right where my garage was and got—the cussedest flowing well you ever saw. We're smart, ain't we?[52]

The egalitarianism that imbued the moral economy of the oil fraternity in Texas pretty much precluded invidious distinctions based purely on money or class. Such did not imply that all were equal: clearly, some were more equal than others. But true to the traditions of South Carolina's children, deference was earned, not due. And nobody, but nobody, was too rich or too big to be taken down. Clint Wood, who was working as a sawmill superintendent when Spindletop came in, recalled an early encounter with J. Edgar Pew, one of the Philadelphia

Pews, who had been sent south to manage Sun Oil operations at Spindletop. Seems that Pew grew a little impatient at delayed delivery of the custom shaped and cured lumber Sun used to line oil storage pits:

> Well, he says, "I'll tell you. I've been down here in the South now for three weeks and I haven't seen a son of a bitch in this South that can do anything." Well, when he opened his mouth, I just climbed on his back. And I gave him an awful lickin'.
>
> And I made him take his damned orders and git out of there with 'em. Well, he went to two, three other sawmills there and they wouldn't touch it. . . . And so Mr. Gilbert come down to see me; he was the president of the mill. Asked me to take it just to help these oil men out. . . . Mr. Pew come down, and boy we were the best of friends from then on that ever was. Because they never was a better man lived than Ed Pew, but he just got on the wrong foot.[53]

The image of Ross Sterling, as president of the Humble Company, laughing and slapping a driller on the back and dealing with him on a first name basis is an enduring one, and still influenced employer-employee relations even in larger oil firms in Texas into the second half of the twentieth century.[54] It is something one cannot imagine a Carnegie or a Frick or a Rockefeller, much less an Alfred P. Sloan or a Roger Smith, doing. Liberty to go your own way, the equality of all who were deemed worthy to sit at the table, were part of the moral economy of the oil fraternity in Texas.

## Useful Knowledge and Wise Use

Prior to the 1930s, and in many cases even later, members of the oil fraternity in Texas by and large harbored ambiguous feelings about science, engineering, and conservation. Most probably would have endorsed some variant of broader American notions of good science: the Baconian view that "true" scientific knowledge was empirical and experimental, and, most of all, useful, as opposed to speculative or, synonymously, "theoretical" and therefore useless. Engineering similarly was cognized as something practical and useful. Resources, including oil, were surely put on this earth to be used wisely, yes, but first and foremost, to be used. Petroleum engineering as a body of recognizable disciplinary knowledge, and as a profession, was just emerging in the mid-1920s. Prior to that time, to the extent that most members of the oil fraternity were concerned with scientific knowledge at all, it was with geology, for it alone might offer guidance to the riches all sought. Yet no authority, certainly not anything that purported to be scientific knowledge, could stand between any man and his convictions, his right to be wrong.

Geology sometimes had proved useful, both the broad-brush surveys of academic geologists and the practical knowledge of the academically trained. W. B. Hamilton, who founded the Texoma Company, credited the geological survey of North Texas, conducted by Dr. J. A. Udden of the Bureau of Economic Geology at the University of Texas, with convincing him that oil was likely to be found in the vicinity of Burkburnett.[55] Crude folk geological notions such as "trendology" and only slightly more sophisticated "creekology" served as fairly effective guides for oil men. More importantly, those who found success using them, like Roy Cullen, swore by them. Academically trained geologists, such as Bill Wrather, could use more theoretically informed knowledge of stratigraphy to identify various strata and follow them cross-country, in outcroppings or even in water wells, to locate anticlines that might contain oil pools.

Only in the second half of the 1920s did torsion balance and seismic surveys, and micropaleontological analyses, arrive. Torsion balance surveys plotted small changes in the earth's gravitational field to detect underground formations. In refraction and later reflection seismology, or "shooting," a series of small explosive charges were set off at the bottom of a number of shallow holes drilled for the purpose. Sensitive motion detecting instruments, strung out in lines and attached to recording devices, picked up the shock waves from the charges propagated through the earth or reflected by underlying strata. By recording the patterns of the shocks, and timing how long they took to reach different instruments, geophysicists could make inferences about the depth, composition, and contours of subsurface strata, and with considerable practice and even more luck, identify formations likely to be oil bearing. Micropaleontological analysis entailed microscopic examination of core samples taken from a number of wells to identify (using characteristic microscopic fossils) specific strata as they appeared in different wells. Sedimentary strata were normally laid down in the same vertical order. By using drilling logs of core depths, geologists could infer contours of subsurface formations. Moreover, differences in the depths at which the same stratum appeared in closely spaced wells indicated the likely presence of faulting, or anticlines, which frequently also signaled oil bearing traps. Yet both geophysical exploration and stratigraphic mapping using micropaleontological analysis suffered serious limitations, especially early on. Geologists were wrong so often that one oil company executive got to calling them "the a'oh boys" after their typical response when told they had been wrong again.[56]

Experience told the oil fraternity in Texas, especially during its first two generations, that experts were often egregiously wrong. As noted, Spindletop itself defied the experts. So did the Scurry reef. So did East Texas. Even in the mid-1970s, so did the Austin Chalk, as did the Barnett Shale in the 1980s. Yet scientific knowledge, and scientific reasoning, could pay. Michel Halbouty deserves credit for discovering the High Island field, finally, in 1931. The obvious salt dome east of Galveston Bay had resisted years of exploration. It took Halbouty, the petrologist, to figure out that the salt column forming the dome did not rise vertically, "like a barber's pole," but rather mushroomed out like a toad stool. Everyone else had either drilled off to the side of the "top" and found nothing or stopped drilling when they hit salt. Halbouty realized that the trick was to drill through the overhanging salt to reach the oil trapped against the lower shaft in up-tilted sedimentary layers. He was right.[57]

Of course, anywhere lots of money and lots of uncertainty mix, the charlatans come to feed, and the oil business was no exception. The most flamboyant of the species were the "doodlebuggers," a name also applied, derisively but not always inaccurately, to geophysical shooting crews. Landon Cullum:

... he [the doodlebugger] usually had some kind of a little box with some chemicals in it, and he'd hang that on his shoulder, and then he had two or three rods he'd tie together in the shape of a "Y" or a "U" or maybe a "V" and he'd go out here and he'd walk around and find an oil well or some oil. "Yeah, it's right here! Right here!" Then, of course, he wanted to promote. . . . One of them had him a great big station wagon all fixed up with all kinds of gadgets in it. Why, if you got around the thing, it'd shock you.

... there was another one had a jeep rigged up. He had a big silver ball in there. . . . He would get up in the front seat and turn the motor on. Then he'd turn him on some other stuff. He wouldn't let you see what it was, and that ball would whirl around, and he had him a gadget fixed up so he could throw static electricity through the air. That made it jump back and forth; and brother, when he got over an oil pool, that static electricity would really have a fit.[58]

The oil fraternity learned to mistrust self-proclaimed "scientists" about the same way their nineteenth-century forebears had learned to mistrust patent medicine drummers.

Oilmen had to see to believe, and they were recalcitrant even then. Michel Halbouty's experience in introducing chemical mud mixtures to drilling is instructive. Gulf coastal wells from Spindletop on were frequently plagued by what drillers and tool pushers called "heaving shales," which, containing a high proportion of clays, "seemed to suck the fluid out of drilling mud," with the result that the shale would expand and plug the well whenever the drill string was pulled and mud circulation stopped. Supported by Frank Yount, Halbouty finally concocted a mud using hexametaphosphate. Halbouty's mud, which was extraordinarily thick when still, and could therefore hold back the shale, also had the necessary property of immediately reliquefying when agitated, as by a drill bit. But its composition had to be exact. It failed its first field test at High Island in 1932. In a profanity-laden exchange Jürgen Habermas would recognize as a social negotiation of situation, Halbouty got out of the drilling crew that they had watered, and thereby ruined, his mud. The conversation concluded with the Yount-Lee tool pusher telling the crew, "All right, I'm passing the word, starting right here, that anybody that tries to fuck with this mud program is going to get run off."[59] One well at a time, one rig at a time, one crew at a time was how petroleum engineers would produce the social authority and the epistemological warrant of scientific knowledge.

## Small World Social Networks

An oil boom demanded quick decisions and decisive action, whether to buy or sell a lease, cut a deal, participate in a well, help somebody out, extend credit, or hire a driller and rig crew. All information was uncertain, but success depended upon timely gossip and fast-moving rumor—and knowing whom to believe and whom to trust. The dynamics of oil deals and life in the boomtowns quickly formed enduring small world social networks in which personal knowledge could flourish. Such individual recognition is usually associated with localite communities of relatively limited spatial extension. Other than science, most of the other collectivities to which a "moral economy" has been ascribed share this characteristic, however much their other particulars might differ. The oil fraternity in Texas differs on this dimension, too. The oil fraternity developed historically as an itinerant community, growing and spreading out as it waltzed across Texas from oil play to oil play, from boom to boom. At the same time, improved transportation and communication loosened spatial constraints on community. Thus, the oil fraternity became a spatially extended "community without propinquity," knitted together by increasingly dense and complex small world social networks.[60]

The oil fraternity's emergent small world networks fostered one of the fraternity's more unusual characteristics: its dependence on "handshake" deals and purely oral commitments. Landon Cullum, again:

> It used to be a man's word was good, and a shake of hand was all there was to it, and a man lived right up to it and his word was good. It was up to you whether or not to consider it good, but 95 out of 100 were just as good as gold.

Automobiles, roads, telephones, and airplanes, rather than disrupting these small world networks, enhanced their importance.

The oil fraternity in Texas, like other social collectivities, sustained its core self-representations by locating the sources of evil and malfeasance, as well as the origins of transgressors, beyond the pale. John H. Wynne:

> . . . of course, it's been true in the oil lease business, there were a bunch of hangers-on, their word wasn't good for anything. But they were just outsiders as far as being into the real integral part of the business, in other words, they would do anything to get a dollar. . . . Of course, we had those by the hundreds . . . but they didn't last long . . . if you were not on the up and up, and you didn't play a straight forward game, you didn't last.[61]

E. E. Townes, on the origin of written lease assignments:

> As a matter of fact, the writing of an assignment or conveyance from one man to another was the exception instead of the rule until after oil was brought in in Oklahoma about 1906 or '07, and a number of Beaumont oil men went up to Oklahoma to traffic and trade in leases and undertook to deal in the word-of-mouth manner, in which they had done in Beaumont, and got badly swindled and cheated. After Oklahoma experiences, then we begun [sic] to have more written instruments in the oil business.[62]

Or on the unhappy need for written contracts to drill wells, rather than just verbal agreements, Carl Angstadt observed, " . . . well, I think it started in Kansas, to where the blue-sky operators and promoters, crooked promoters showed up."[63]

## MÉNAGE À TROIS: MAJORS, INDEPENDENTS, AND THE RAILROAD COMMISSION OF TEXAS

The characterization of the oil fraternity as a collective whole governed by a moral economy contrasts with customary academic (and popular) accounts of the oil business in Texas. Those accounts typically tell one of two stories: either they stress the antagonistic relationship between majors and independents, or, obversely, they treat the oil business as a unitary, rapacious "cartel" which "captured" its purported regulator, the Railroad Commission of the State of Texas.[64]

That majors and independents were part and parcel of the same organic whole hardly precludes some considerable disharmony, mostly about who was the dog and who was the tail and who was going to wag what. Conflicts of interest were real enough. Just within the State of Texas, majors and independents often found themselves on opposite sides of arguments about prorationing itself, about access to pipeline and storage facilities, about well-spacing and allowables, and about the various incentives provided for discovery wells and for stripper well operation. Yet variance of opinion among majors and among independents most often was greater than any identifiable variance between the two groups. And majors and independents held a variety of positions in common: opposition to any sort of federal control of the oil business, support for percentage depletion (although differences did arise over tax treatment of foreign production), and fealty to the system of State regulation and Interstate Oil Compact Commission (IOCC) coordination that was in place by 1940.

The men who worked for the majors (especially within Texas) and the independents were born of the same soil and the same culture. For the seven sisters, Anthony Sampson called it simply "the Texas pipeline," the stream of mostly operating people who joined the majors, often part-time in high school or college, in Texas, or Oklahoma, or Louisiana, and then moved up through the corporate ranks.[65] Managers, no less than engineers, leave their identities at home. Moreover, because of Texas's stringent antitrust laws, the integrated majors who operated within Texas most often did so through wholly or partially owned subsidiaries: Roxana (Shell), Stanolind and Pan American (both Standard of Indiana), Pure Oil (Standard of California), Magnolia (Socony Vacuum, then Mobil), Rio Bravo (Southern Pacific Railroad), and of course Humble Oil and Refining (Standard Oil of New Jersey, after 1919). Such arrangements established a Texas identity for such firms. Humble particularly maintained a remarkably strong differentiation from Standard of New Jersey.[66] Even when wholly separate corporate forms were not used, as in Gulf's case, Texas people typically ran Texas operations (Underwood Nazro and F. C. Proctor, for example). To see majors and independents as different animals requires not seeing people at all.

Majors and independents were knitted together in a complex web of interdependence. That interdependence in turn reflected the functional differentiation of majors and independents within the oil business, as they each pursued or exploited their respective comparative advantages. Based upon their study of West Texas wildcatters, Roger and Diana Olien argue that independents had a significant edge in two specific oil industry functions: exploration and discovery, and stripper or marginal well operation.[67] The standard pattern was for wildcatters to discover new fields, do initial development, then sell out, often to majors or other, larger independents, and use the money either for riskier "step-out" wells or for new wildcat wells somewhere else. While majors did drill wildcat wells, independents predominated in discovering new production. Even late in the oil fraternity's evolution, during 1950–1956, independents completed 76 percent of new wells in the United States while controlling only 42 percent of oil production. Independents also drilled 87 percent of the dry holes.[68] But this apparent differentiation also entailed extraordinary interdependence. Independent operators normally sold their oil to forward-integrated refiners, for whom such purchases were essential. Exploration itself often involved complex cooperative arrangements, frequently involving one or several majors and one or more independents: shared geophysical or geological information, farm outs, bottom-hole or dry-hole money, lease swaps, and so on.

The other domain in which independents were advantaged was in operation of stripper or marginal wells and in some secondary recovery projects. Independents typically operated on relatively modest scales, with a minimum of overhead or staff support. Their management style was direct, hands-on. Stripper wells, those capable of producing less than twenty barrels of oil a day, are normally found in older, long-developed, and therefore relatively well understood, fields, and usually are pumped rather than free-flowing. Because of their low overhead and intimate local knowledge, independents could operate such wells, or modest secondary recovery projects developed in such fields, more cheaply and efficiently than highly bureaucratized, overhead-laden integrated producers.

Conversely, the majors enjoyed offsetting comparative advantages in other stages of the oil business. As Alfred Chandler argues, large, bureaucratic, multi-unit business enterprises persist if and only if the coordination provided by the "visible hand" of their managerial hierarchies can generate economic efficiencies greater than those obtainable through market coordination.[69] High rates of throughput or velocity are necessary for expensive managerial coordination to pay off. In mass-production industries, such velocity was typically generated by sector-specific mass production, continuous-flow technologies combined with exploitation of transportation, communications, and organizational innovations to expedite distribution, which is exactly how integrated oil producers-refiners-marketers operated.

The functional division of labor between independents and majors, and their complex pattern of competition, interdependence, and cooperation, is replicated across multiple contemporary sectors: chemical products and pharmaceuticals, electronics, computer programs, and video games. Inventors and entrepreneurs funded by "venture capital" do the innovation and early development; later development, manufacture or production, and distribution typically is taken over by a much smaller number of established firms who buy out the successful pioneers.[70] Perhaps the way functional differentiation and interdependence went together for majors and independents in the oil fraternity was best expressed, with the easy familiarity that comes with years of cooperation and friendship, by an independent oil operator when trying to talk Morgan Davis, then president of the Humble Company, into letting the independent swap his working interest in a deal for a (lesser) overriding royalty. Humble, as was its custom, was developing the new discovery with great care and what looked to be all deliberate speed: the independent wasn't going to see any money that did not go right back into development wells for a long time to come. As the independent put it, "You know, a mule and a jackrabbit can't plow together."

The Railroad Commission of the State of Texas has a central, and unusual, place in this drama: it is both actor and arena. As noted, from its origin in railroad regulation in the late nineteenth century, the Railroad Commission saw its role as that of protector of Texas producers and resources, and as a guarantor of opportunity for Texans. This Populist-Progressive orientation was not hostile to *Texas* enterprise, but rather looked to its protection and encouragement as the pathway to prosperity for the whole state. By 1934, the Railroad Commission was, by its own insistent proclamation, integral to the oil fraternity. In its hearings, formal and informal, in its testimony, in its arguments and decisions, in its framing of its appellate cases, the Railroad Commission, together with specific courts designated by Texas Statute Law, became co-constructor with the rest of the oil fraternity, including the community of petroleum engineers, of a common social and economic reality.

As arena, as a formal, legal, institutionalized stage, the Railroad Commission of Texas is the critical site for the self-production and reproduction of the oil fraternity, as well as petroleum engineers. The early statewide prorationing hearings resemble nothing so much as mass Farmers' Alliance encampments, rich in emotion and finely textured ideological expression. But even in later formal, full Commission hearings, in informal "hearings" with a single commissioner, or in field hearings sometimes conducted by a senior female clerk, the processes of ritual production and collective construction, mutual accommodation, and amalgamation went on. And the Commission was the locus in which both the fraternity and the engineering community wove themselves into the larger political, economic, and social fabric of Texas, and, sometimes resentfully, of the nation. To understand the Railroad Commission of Texas and its place in the oil fraternity is to understand government of, by, and for people who didn't much care for any government at all.

For oilmen, all that mattered, all that needed to matter was the deal and the deal and the deal. In the oil fraternity in Texas, Everyman was the equal of any other and had the sovereign right and opportunity to make any deal he wanted on the best terms he could get with anybody he wanted. From the eulogy for H. L. Hunt, the great-granddaddy of them all,[71] delivered by his long-time attorney, Sidney Latham in W. A. Criswell's First Baptist Church in Dallas, December 2, 1974:

> I have seen him turn the stroke of a pen into a million dollars and to enjoy the full harvest of great wealth. But in all candor, I think the greatest and most genuine happiness I ever saw or heard him exhibit was that Sunday night that he called me to tell me than he had just joined the Church within the hour just passed. He seemed to feel that he had made the best and biggest trade he had ever made in all of his career of big trades. He had traded the here for the hereafter, and he was pleased.[72]

It is the ultimate moral economy of equals. And anybody who thinks God got the better of that deal doesn't know much about H. L. Hunt or the oil fraternity in Texas.

# CHAPTER 3

# ESCAPE INTO
# THE ABSTRACT

The difficulty of living backwards and thinking forwards is that you become confused
about the present. It is also the reason one prefers to escape into the abstract.

—MERLYN, IN T. H. WHITE, *THE BOOK OF MERLYN*[1]

The previous two chapters purport to offer a passable reconstruction of the early history of the oil fraternity in Texas: its historical, demographic, and cultural origins, and the emergence of its spatiotemporally particular moral economy. But neither the spatiotemporally particular moral economy of the oil fraternity in Texas nor any other "moral economy" for that matter had ought to exist. Moral economy and the altruistic cooperation at its core directly challenge the egoistic incentive assumptions at the foundation of orthodox microeconomics as well as the genetic self-interest axiom of individualistic biological selection. Devotees of egoistic incentives have labored mightily, and with some success, to accommodate empirically observed economic cooperation. But, like the "selfish gene" view of biological evolution a few years back, the account of economic behavior provided by self-interest alone is at best incomplete, if not misleading. In biology, the selfish gene has been supplanted by a broader view of evolution and development ("evo-devo") in which epigenesis and environmental interactions play major roles, while theories of coevolution and group selection among organisms and species supplement more individually focused accounts.[2] Nevertheless, among mainstream economists and their fellow travelers in behavioral and experimental economics and in political science and history, egoistic incentive interpretations remain rampant.

Understanding the oil fraternity's plight and the solutions it evolved decomposes into four intertwined problems. First, because egoistic incentives are both an analytical assumption and an ideological commitment, it takes some effort to tease out precisely what self-interest theories claim, how those claims have been modeled and tested, and where they have succeeded and where they have proven inadequate. Egoistic incentive-based accounts alone arguably fail to account for the emergence of large-scale cooperation among anonymous agents and therefore for moral economies. Second, an alternative theoretical approach, group selection, can account for large scale cooperation and for moral economy. Group Selection Deal (GSD), a quantitative, multilevel selection model tailored to the circumstances of the oil fraternity in Texas, offers an abstract explanation for how the emergence and evolution of the oil fraternity's moral economy could have occurred. Third, abstract models of whatever sort leave "cooperate" largely undefined. In theoretical models, cooperate is simply depicted by terms in equations. In behavioral economics experiments, cooperate is narrowly reified, and precisely defined for experimental subjects: "provide a specified benefit $b$ to others at a specified cost $c$." These conceptions of cooperate radically underdetermine real-world behavior. Agents must somehow "know" what cooperation is in a specific social and cultural context. A qualitative model, alienated representation, purloined from P. Steven Sangren, resolves this difficulty and depicts how cultural subjects produce both themselves and social collectivity, and thereby learn what it is to cooperate. Finally, these analytical issues are wrapped in a fundamental economic conundrum peculiar to the oil fraternity. The value of oil and gas doubly declines: Oil and gas are produced over time, and the revenue stream they may generate is subject to the usual Present Value calculations. But hydrocarbons also are depleting resources: They are used up, their ultimate recovery a function of how they are produced. These four analytical problems, which lie behind reconstruction of the oil fraternity's historical experience, are examined in detail in the four sections of this chapter.

Egoistic incentive models and GSD, which is a variant of group selection models, are all based upon fitness functions or equations (as they are known in evolutionary biology), or payoff functions (as they are called in economics). Their notation can be fearsome looking, but the equations really amount to nothing more demanding than high school algebra. These functions are critical in constructing replicator dynamics, the process by which, for example, the proportion of altruists (cooperators) and opportunists (non-cooperators) in a population change over time. These replicator dynamics are the causal mechanisms underlying George Price's mathematical parsing of selection into within-group and between-group effects, which allows characterization of how group or multilevel selection works.

Use of group selection concepts to explain the moral economy of the oil fraternity in Texas is an overt affront to the hegemony of egoistic incentive explanations across the social sciences and humanities. Orthodox economics is notorious for its missteps and erroneous prognostications. A good part of its difficulties may derive from its foundational assumption that only pure self-interests animate all economic behavior. The suspicion is that moral economies and the selection dynamics that give rise to them are ubiquitous in virtually all communities, from research scientists to technological entrepreneurs to subsistence fishermen to the oil fraternity in Texas. Except in the most impersonal of market-mediated transactions, economic action blends self-interested competition and altruistic, within-group cooperation, situated in spatially and temporally specific communal or organizational cultures.

## WHY MORAL ECONOMY SHOULDN'T EXIST

Moral economy is a species of altruistic cooperation among non-kin. It runs afoul of the "paradox of altruism," which afflicts a broad spectrum of disciplines, including genetics, epigenetics, evolutionary biology, game theory, economics, behavioral economics, political science, anthropology—and history. The paradox is easily stated if not so easily resolved. In any population of any sort whatsoever—alleles, genes, cells, schools of fish, flocks of birds, primates, hominids, or tribes—that includes both altruists, defined as those who cooperate and benefit others at expense to self, and opportunists, who don't, selection should drive altruism to extinction because opportunists gain from others' altruism but incur no costs. Yet altruism, or at least altruistic behavior, seems to persist not only among the cells within our bodies but also among groups of nonhumans and humans alike.

Established evolutionary theory recognizes two forms of altruism as being consistent with self-interest: kin selection, or inclusive fitness, and reciprocity. According to "Hamilton's rule" for inclusive fitness, altruism is genetically self-interested if the benefit provided by the putative altruist multiplied by her degree of relatedness to the recipient is greater than the cost to the altruist; hence, somatic cells don't ordinarily produce new critters, and aunts help nephews.[3] Or, in J. B. S. Haldane's wonderfully precise and personalized calculus, "I am prepared to lay down my life for more than 2 brothers or more than 8 first cousins."[4] Similarly, as Robert Trivers demonstrated, altruistic behavior is self-interested if the probability of its being reciprocated by the recipient, or other observers, times the expected future benefit, is greater than its cost.[5] Critics of James Scott's use of moral economy offer reciprocity as the alternative, egoistic-incentive explanation for observed altruistic behavior.[6]

Inclusive fitness requires kin recognition or kin-group isolation, while reciprocity demands recognition of likely reciprocators as well as a high probability of repeated encounters.[7] Neither approach is very good at accounting for the emergence of altruistic cooperation within large populations of anonymous, unrelated individuals. Therefore, on this basis, and the axioms of rational choice consistent with it, both orthodox economic theory and game theory, as well as evolutionary biology, denigrate the possibility of widespread, truly altruistic cooperation among non-kin. As Michael Ghiselin famously summarized it, "Scratch an 'altruist' and watch a 'hypocrite' bleed."[8]

## Game Theory: The Prisoner's Dilemma and Its Heirs

The iconic game-theoretic explanation for why moral economies, or truly altruistic cooperation in general, should not exist in nature or in society is the Prisoner's Dilemma (PD). This "game," first articulated by theorists at the RAND Corporation in 1950, serves as the linchpin for adamant arguments in evolutionary biology and even more so in economics. If a player (agent or actant) in a PD is self-interested, is rational, and believes the other player is also self-interested and rational, he will play his best response to his opponent's possible strategic choices (Box 3.1. The Prisoner's Dilemma [PD]). These simultaneous choices define the Nash equilibrium of the game—that is, neither player can improve his score by unilaterally altering his strategy. Given the payoff matrix of the PD game, each player is individually better-off, come what may, if he defects. Therefore, mutual defection is the robust game theoretic prediction for rational agents who can't coordinate their choices, even though both would be better-off if both cooperated. Precisely the same logic applies to an $n$-person Prisoner's Dilemma, which is more easily depicted mathematically as an $n$-person Public Goods game with fixed contributions (Box 3.2. $n$-person Prisoner's Dilemma [PD]). All the games recounted here are games with complete information. All players are assumed to be "rational," that is, they seek to maximize their subjective utilities (operationalized as payoffs or fitnesses)—they respond only to egoistic incentives—and they are assumed to know that all other players also are rational in the same sense.

In an Iterated Prisoner's Dilemma (IPD), the game is repeated among the same players. Although game theorists' "folk theorem" suggests that in an infinitely repeated (iterated) Prisoner's Dilemma, rational agents eventually would learn to cooperate,[9] in a finite game, any cooperative sentiment falls victim to backward induction: a rational agent, believing that other agents are also rational, will reason that even if he were to cooperate, the other would certainly defect on the last round when it is in no one's interest to cooperate; thus, he should seek advantage by defecting in the next-to-last iteration. But he also reasons that another rational agent will reason the same way, therefore he should defect a round earlier, and so on, until all rational agents defect in the first round and ever after.[10] Again, the same logic afflicts an Iterated $n$-person Prisoner's Dilemma: everybody defects all the time. The evolutionary implication of PD games for altruistic cooperation is devastating: If fitness depends upon payoffs, selection should drive altruism to extinction.

The provision of "public goods," those goods and services which all share but at least some must contribute to provide, have the same problem, as does the joint exploitation of "common-pool" resources. A Public Goods (PG) game is an $n$-person Prisoner's Dilemma which allows variable individual contributions (Box 3.3. Public Goods (PG) Game). The Nash equilibrium for a PG game, as for an $n$-person Prisoner's Dilemma (PD), as for a two-agent PD, is all defect; in an iterated PG game, all defect remains the unique equilibrium for the entire game.

Subsurface oil pools or reservoirs underlying multiple surface tracts with diverse ownership are prototypical common-pool resources, which have been of considerable theoretical and political interest since Garrett Hardin's seminal 1968 article, "The Tragedy of the Commons," the prediction written into the title.[11] The structure of Common-Pool Resource (CPR) exploitation is that of a Public Goods game with a more complex payoff function (Box 3.4. Common-Pool Resource (CPR) Exploitation as a Non-cooperative Game). An oil reservoir has many of the same features as communally held and exploited pastures (the original commons), open-water or open-access fisheries, subsurface aquifers or geothermal formations, flowing water, air, global climate, or public spaces or domains generally (such as a town square or the internet). The defining characteristic of a common-pool resource is that some set of actants (individuals, groups, corporations) collectively has unrestricted access to the resource, but their individual exploitation of the resource has the potential, usually realized, to impose negative appropriation externalities upon other users. The aggregate payoff function is concave, rising with total investment up to a point, then declining as the resource is overexploited. For common parameter values, the highest average (Pareto optimal) returns depend upon considerable restraint: average investment typically must remain well below even Nash equilibrium values, which in turn are well below break-even levels. The consensus of

CPR research, both theoretical and experimental, is that a CPR exploited non-cooperatively, that is, by rational utility maximizers heedless of the negative externalities they impose, will be used inefficiently, degraded, or destroyed.[12]

These game-theoretic predictions foment "the paradox of altruism," the conflict between what is theoretically predicted and the seemingly altruistic behaviors observed in our species and in others. At least four modifications to the basic PD/PG game structure have been proposed to accommodate this inconvenience without sacrificing the core egoistic incentive assumption: the ascription of "$\alpha$-normative" dispositions to actants (which provides them an ad hoc, non-material "psychological" payoff for cooperation), obedience to an imaginary "choreographer" which can coordinate self-interested cooperative play, systems of reciprocity based upon observation or reputation, and the emergence of Sanctioning Institutions (SI), which provide egoistic incentives for cooperation.[13]

The most well-developed of these extensions are $n$-person IPD/IPG games that add "sanctioning institutions" in which actants, at some cost to themselves, can punish ostensible non-cooperators (Box 3.5. Public Goods (PG) Games with Sanctioning Institutions [SI]). If information were adequate, and sanctioning effective, self-interested cooperation could be stabilized even among opportunistic agents animated by egoistic incentives. Unhappily, just as PG games are susceptible to first-order free-riding (opportunists benefit from others' contributions but don't contribute themselves), sanctioning institutions are subject to second- or third-order free-riding: Even if an actant contributes, he may not punish; if second-order free-riding (not punishing) is punished, an actant may contribute and punish first-order transgressors but not sanction second-order free-riding, and so on in an infinite regress. In theory, self-interested cooperation eludes stabilization by sanctioning.[14]

## Empirical Results

Game theoretic results inspired and coevolved with robust traditions of experiment in social psychology and behavioral economics, which have often produced surprisingly contrarian results. In a variety of laboratory contexts, many experimental subjects, albeit mostly university undergraduates from Western, Educated, Industrialized, Rich, and Democratic backgrounds (WEIRDs),[15] cooperate initially in PD, PG, or CPR games in defiance of what they should "rationally" do: typically 30 to 60 percent of subjects cooperate in the first round of anonymous games, although, in iterated games, cooperation declines rapidly in the face of others' defections.[16] More elaborate experiments with communication (reducing the anonymity constraint) or with sanctioning institutions yield mixed results.

The paradigmatic public goods experiments with and without sanctioning were reported by Ernst Fehr and Simon Gächter (Box 3.5. Public Goods (PG) games with Sanctioning Institutions [SI]).[17] Fehr and Gächter found that even in anonymous games, sanctioning opportunities stabilized cooperation in all conditions, but much more so in the Partner treatment, where the same four subjects stayed together for all 10 rounds of the game (Figure 3.1. Average Investments over Time in Public Good Games with Stable Groups (10 groups), and Figure 3.2. Average Investments over Time in Public Good Games with Random Groups (18 groups)).[18]

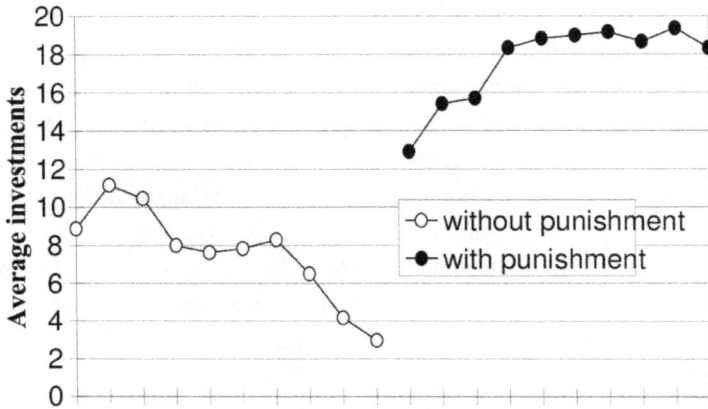

Figure 3.1. Average Investments over Time in Public Good Games with Stable Groups (10 groups)

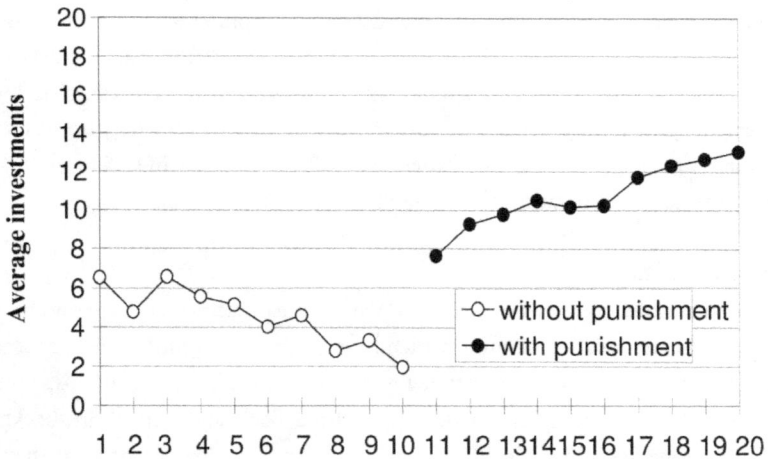

Figure 3.2. Average Investments over Time in Public Good Games with Random Groups (18 groups)

Contributions with sanctioning far outstripped those without sanctioning, which converged towards zero by the end of 10 rounds, as in virtually all other generic PG game experiments. Fehr and Gächter speculated that the observed superior performance in the Partner treatment was due to the emergence of "implicit coordination" among the anonymous subjects in repeated encounters; implicit community might be more apt. Fehr and Gächter's original experiments spawned a lively and innovative mini tradition within behavioral economics. Later work demonstrated that total payoffs, net of sanctioning costs, to all players in PG games with sanctioning institutions on average were less than those without such arrangements, while still further experiments suggested that much longer time horizons might advantage sanctioning.[19] Up to now, sanctioning institutions in PG-like games have not been shown to reliably produce cooperation.

As did public goods experiments, Common-Pool Resource (CPR) experiments produced ambiguities. Elinor Ostrom and her colleagues ran a series of laboratory experiments with her original CPR model (Box 3.4. Common-Pool Resource (CPR) Exploitation as a Non-cooperative Game). Because the results were uniformly disappointing, the research group designed a new series of experiments with sanctioning opportunities (institutions) and limited communication.[20] Sanctioning alone actually reduced net yields (including costs of fining and punishment), while sanctioning with communication could increase them substantially, although results varied considerably from group to group.[21] "Blind revenge" sanctioning also appeared. While communication and sanctioning in some groups ameliorated damage to the common-pool resource, in others it did not.

## The Primacy of Culture

These experiments all used WEIRDs as experimental subjects.[22] This practice, however necessary and convenient, naturally raises questions about how culture-dependent responses in public goods games and in behavioral economics experiments generally are. To address this issue, Herrmann, Thöni, and Gächter conducted public goods experiments across different cultures.[23] Their sample populations were still undergraduate students from mostly upper middle-class backgrounds in complex societies with high market integration and were economically and demographically matched. Results were sobering: in five of sixteen societies, contributions with punishment did not exceed those without punishment. Even though subjects were anonymous, in some cultures, they used punishment as retaliation: low contributors punished anonymous high contributors, whom they apparently suspected of having punished them, and any positive effect of punishment on contributions largely vanished (Figure 3.3. Effect of Culture upon PG Contributions and Anti-social Punishment). Herrmann, Thöni, and Gächter concluded that the efficacy of opportunities to punish in stabilizing cooperation in public goods games depended upon subjects' intrinsic cultural commitments, likely acquired by conformist transmission.[24] Later theoretical investigation showed that antisocial punishment can prevent the evolution of prosocial (altruistic) punishment altogether.[25]

In contrast to Herrmann et al., Henrich et al. pursued a broader investigation of notions of fairness, equity, and cooperation across a diverse set of fifteen small-scale societies.[27] They used three different experimental games, the Dictator Game (DG), the Ultimatum Game (UG), and the Third-Party Punishment Game (TPG) as measures of social norms of fairness. For example, in an Ultimatum Game (UG), one subject, the proposer, receives an endowment, which he can propose to share, in any proportion or not at all, with the other subject, the recipient. If the recipient agrees, the endowment is split accordingly. If the recipient does not agree, neither subject gets anything. A purely rational (utility maximizing) recipient would accept any offer greater than zero; thus a rational proposer would always make a minimal offer. Rejection of unequal offers is taken to measure the recipient's normative sense of fairness and equity, and his willingness to altruistically incur costs to enforce it. A proposer's greater than

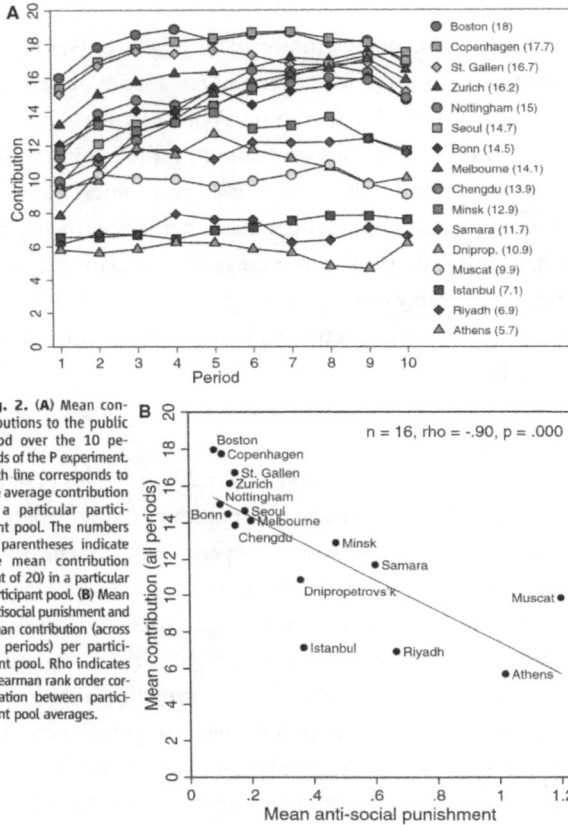

**A**

**B**

n = 16, rho = -.90, p = .000

**Fig. 2. (A)** Mean contributions to the public good over the 10 periods of the P experiment. Each line corresponds to the average contribution of a particular participant pool. The numbers in parentheses indicate the mean contribution (out of 20) in a particular participant pool. **(B)** Mean antisocial punishment and mean contribution (across all periods) per participant pool. Rho indicates Spearman rank order correlation between participant pool averages.

**Figure 3.3. Effect of Culture Upon PG Contributions and Anti-Social Punishment[26]**

minimal offer indicates both his own sense of equity and his judgment about what offers the recipient is likely to reject. (In Western societies, offers hover around 50 percent; offers much lower than that are typically rejected.)[28]

Henrich et al. discovered wide variations in the behavior of different populations in these games, and therefore big differences in attributed norms. After normalizing socioeconomic factors across populations, Henrich et al. explored several macro-cultural features that might account for the variances they observed among their fifteen societies: reported participation in a "world religion" (Christianity, Islam) versus local sects, community size, residence patterns, sources of livelihood, and market integration. Only market integration, operationalized as the proportion of family caloric intake purchased, strongly and consistently correlated with "prosocial" behavior (making and demanding relatively fair offers, sanctioning those who make low offers). Henrich et al. interpret these results as supporting the hypothesis that other-regarding preferences and prosocial norms coevolved with markets, or market integration, and were necessary for the evolution of market economies.[29] But recall that E. P. Thompson saw "prosocial" moral economy as a local and disciplined communal reaction *against* the exploitation inherent in an impersonal market economy.

Further research suggests other hypotheses. For example, in a radical revision and extension of Max Weber's *The Protestant Ethic and the Spirit of Capitalism*, Jonathan F. Schulz, Duman Bahrami-Rad, Jonathan P. Beauchamp, and Joseph Henrich define the "Western Church" as the Roman Catholic Church from circa AD 400, and, after AD 1500, as it and its Protestant successors. They demonstrate that these religions, especially the Catholic Church during the Middle Ages, eroded the intensive kin-based institutions that appear to be the hominin default, and that continue to predominate in many cultures.[30] Uniquely, the Western Church rigorously suppressed incest and related family practices, such as marriage between cousins or among in-laws and step-relatives and actively discouraged polygamy, "co-residence of extended families, clan organization, and community endogamy." Schulz et al. hypothesize that the longer a population was exposed to this intervention, the less conforming, and the more individualistic, independent, analytically minded, and impersonally prosocial it became: the hallmarks of WEIRD societies. Historical religious suppression of extended-family institutions, going back to the early Middle Ages, produced the societies and the attitudes which birthed market capitalism.

In a broader study of the relationship between egoistic incentives and "prosocial" dispositions, Samuel Bowles provides a comprehensive summary and perceptive critique of prior theoretical and empirical investigations. His primary interest is in public policy, how a wise "Aristotelian legislator" might use material incentives, or not, without "driving out" subjects' intrinsic prosocial preferences. It seems that financial incentives designed for *Homo economicus* "knaves" often have the opposite of intended effects and render subjects permanently less other-regarding. In distinction to previous findings, Bowles rejects market participation as the cause of prosocial orientations and looks instead to "nonmarket aspects of liberal social orders."[31] The causal arrow runs from cultural change to market, not the other way round.

The apparent ubiquity of moral economies suggests an alternative historical conjecture about their relationship to the evolution of capitalist markets and of prosocial dispositions. Moral economies seem to evolve within preexisting or emergent communities, with or without propinquity, if those communities are exposed to impersonal, competitive, selectionist environments (such as capitalist markets or scientific practice), and if groups with more within-group cooperators are advantaged compared to groups with fewer. Moral economies exist at different scales, in various degrees of articulation, in peasant villages, on job sites, and in fishing camps, and in firms, bureaucracies, schools, "invisible colleges," universities, and across organizational and institutional boundaries.[32] These moral economies are spatiotemporal particulars, and their domains and prescriptions and proscriptions are restricted. But if multigenerational cultural experiences of diverse, cooperative moral economies (and the group selection processes that produce them) become more common as societies become more differentiated, then the more generalized but conditional "prosocial" preferences observed in behavioral economics experiments might arise.[33] Since functional economic and social differentiation, scientific and technological change, and the emergence of markets are often colinear, it may appear as though market participation encourages prosocial dispositions, although totalitarian environments might inhibit these changes. The origin of broader prosocial sentiments is likely multivariate, with complex, intertwined, and idiosyncratic path dependencies.[34]

Nevertheless, clever game theorists, behavioral economists, and evolutionary biologists, unwilling to budge much on their foundational egoistic-incentive presumptions, have suggested a number of other game theoretic scenarios that might offer a better account for observed cooperative behaviors, and, failing that, have advanced qualifications to "strict rationality" itself, usually some form of "bounded rationality."[35] Prominent among the alternative games are the Hawk-Dove or "chicken" game,[36] the Snowdrift game,[37] the Stag Hunt game,[38] and the Centipede game. What distinguishes the Hawk-Dove, Snowdrift, and Stag Hunt from the Prisoner's Dilemma (PD) is a transposed payoff matrix which shuffles the rank ordering among actants' strategic options. The Centipede game is a different type.[39] While these alternatives to the Prisoner's Dilemma (PD) might depict aspects of an oil deal, none captures the idealized structure of an oil play as well as does an iterated $n$-person PD or PG game.[40] Even though game theoretic models' ability to predict or analyze behavior is challenged by empirical results, these models do define critical parameters and the likely relationships among them in real-world interactions.[41] But so far, neither theoretical explorations nor empirical results offer cheery prospects for altruistic cooperation among large numbers of anonymous, purely egoistic-incentive-animated actors in Prisoner's Dilemma–like situations: Neither theory nor empirical results save the phenomena or resolve the paradox of altruism.

The apparent conflict between what rational, self-interested agents in theory ought to have done and moral economy—how folks actually behaved within the cultural matrix of the oil fraternity—leads to the familiar Mexican stand-off between the "two cultures." Cold-eyed rationalists, in the thrall of egoistic incentives, see "moral economy," alleged "altruism," and cultural explanations generally as at best paradoxical, or, more likely, as romanticized shills for behavior that is really at some level self-interested. Erstwhile humanists accuse rationalists, especially orthodox economists, of being pseudo-scientists who willfully ignore empirical refutation of their foundational assumptions.[42] Well, yes and no.

## THE ORIGINS OF MORAL ECONOMY BY MEANS OF GROUP SELECTION: THE GROUP SELECTION DEAL (GSD)

As hypercompetitive as the oil business was, the Hobbesian *bellum omnium contra omnes* predicted by microeconomics or game theory never quite materialized. Moral economy evolved instead. Group selection, appropriated from evolutionary biology, can provide a disciplined and robust *selectionist* account for the emergence of an altruistic moral economy, even beginning with a mostly self-regarding mob like that at Spindletop. There's nothing romantic or sentimental about it. Given appropriate population structure and replicator dynamics, altruism evolves in the global population because altruists do better.[43] Depending upon specific parameter values in the model, GSD selects which among possible equilibria—all altruists, all opportunists, or an unstable mix—evolves in the global population.[44] The GSD model developed here is one variant in a population of game-theoretic group or multilevel selection models now on offer, and is tailored specifically to the circumstances of the oil fraternity in Texas.[45]

The core claim in multilevel or group selection is that behavior that reduces individual fitness but benefits other group members can be selected for if the average fitness of individuals in groups with relatively more benefactors is sufficiently greater than the average fitness of individuals in groups with fewer or none. Even though opportunists *always* do better than altruists within all groups in which both appear (since opportunists benefit from others' altruism but incur no costs), altruism can still increase in the global population.[46] The necessary conditions for group selection for altruism are that the population be divided into groups; that the groups vary in their proportion of altruists; that behavior be in some sense heritable or otherwise replicated; that group "output," its constituents' reproductive success however measured, be at least a partial function of the benefits to group members conferred by its altruists; and that the groups be isolated, but also at some point in the life history of their constituent entities, "merge and re-form or otherwise compete in the formation of new groups."[47] To be sufficient, the differential fitness advantages of those in altruist-rich groups must be large enough to overcome universal within-group selection against altruists.

In the early 1970s, George R. Price, then an American research fellow in genetics and biology at University College London, showed that selection in these circumstances could be mathematically parsed into within-group effects, which favor opportunists, and between-group effects, which may favor altruists.[48] His achievement is summarized in what is eponymously referred to among non-economists as the "Price equation." The equation is not itself causal: the causal muscle in any group selection model is provided by the underlying replicator dynamics or fitness equations.[49] But Price's great accomplishment was to precisely characterize "the mean change of *any* property of a population of things, genes, mental representations, or the hydrogen atoms in the Andromeda galaxy, [that] can be decomposed into two parts, the extent to which the properties of things covary with the effect of selection, and the rate at which the things themselves change with time."[50] Indeed, "The most general form of the Price Equation . . . is exact for anything one can measure about the constituent parts of any evolutionary system."[51]

This spatiotemporally unrestricted causal mechanism—group selection—animates the evolution of the spatiotemporally particular moral economy of the oil fraternity in Texas. "Deals" are the groups upon which selection acts. "Reproductive success" is a deal's economic success, which more often than not simply allows those in the deal to continue or expand participation in the oil fraternity: that is, to produce offspring, or interests in more deals. Cooperation within deals or groups—adherence to the oil fraternity's moral economy—varies among deals, as does the groups' success. The origin of the moral economy of the oil fraternity in Texas therefore is by means of deal, or group, selection.[52]

In the vernacular of the oil fraternity, deals are composed of "interests," virtually always denominated as a fraction, 1/2, 1/4, 3/16, 1/32, 1/256, and so on, the arithmetic of those more familiar with rulers and yardsticks than decimal accounts. Consider, then, a large population of such interests (our actants or replicators or strategies) who get into oil deals: how large this population is doesn't matter much so long as it's very large compared to group or deal size. Following David Hull, interests are both replicators and interactors. As interactors, they cooperate (or not) with other interests in the same deal, but, in the model, have little intercourse with those in other deals. As replicators, interests reproduce: a deal's economic success, if any, allows the interests it comprises to differentially replicate, or become better-off and participate in new deals. Deals themselves are higher-level interactors. Deals compete with other deals in an oil play to secure leases, drill wells, and extract oil or gas. Or, for a lower prospective payoff, a lease may be sold or traded before a well is drilled. Either way, deals are the locus of selection and of interests' fates: they are the unit of selection upon which (interdeal) competition acts. Together, an iterated sequence of multilevel selection and replication constitutes the replicator dynamics of the population of interests and defines an ancestor-descendant lineage of interests.[53]

When word of a new oil discovery got out, it set off a frenzy that would make feeding sharks look like the churched at a Sunday school picnic. Whoever drilled the discovery wildcat well tried to lock up as many nearby leases as possible, as did, of course, everybody else in the oil fraternity and everybody who wanted to be. Such circumstances placed a tremendous premium on the ability to "get there fustest with the mostest"[54]—to acquire leases, get wells drilled and completed, and get oil out of the ground. More importantly, a developing oil boom enormously magnified opportunity costs. Together, capital scarcity, intrinsic riskiness, uncertain information, and the need to pursue as many opportunities as possible as quickly as possible motivated multi-participant deals, while transaction costs and the necessity of acting fast limited deal size, typically to two to five substantial "working interests" (those who were actually supposed to pay for leases and for drilling and completing wells).[55] Urgency largely mooted legal enforcement of agreements, often only oral anyway, and meeting obligations was mostly voluntary: Given the contingencies and uncertainties of oil exploration and development, "contracts" were classically incomplete.[56]

By and large, the men who became the oil fraternity in Texas were strangers when they met, and initially had little expectation of ever meeting again. Because nearly all oil operators worked as closely as they could to the margins of the possible, even small (or a small number of) defections in a deal—opportunists who reneged on their obligations—could spoil the deal, while the liquidity that would allow even the willing to fulfill their commitments was an endemic problem. Like their agrarian forebears who had been "land poor," oilmen were often "lease poor," and, in remote new fields without transportation or marketing infrastructure, just as often "oil poor."[57] Thus, being in a deal with altruists who would not only reliably honor their own commitments but also "carry" a man over the rough spots could be the difference between holding a deal together and being an oilman, or being, once again, a roughneck or sharecropper.

Yet whatever the collective incentive to cooperate in a deal, the countervailing individual incentive to defect was always greater. In the land of unlimited opportunity, opportunity costs were equally unbounded. Resources devoted to honoring one's own commitments or to carrying others could not be invested in other deals. The more deals a man could participate in, the greater his chances of striking it rich. Thus, it was in every agent's interest to whine, wheedle, "poor mouth," stall, obfuscate, and welch. Although some of his deals might "crater" as a result, he would still be better-off trusting to the fickle luck of the drill-bit in ever more deals.[58] Each deal, then, has the structure of a classic, $n$-person Prisoner's Dilemma, the unique Nash equilibrium for which, despite the greater collective payoff if everyone were to cooperate, is universal defection.[59]

The many deals in an oil boom constitute a population of deals or groups into which individual actants or replicators or strategies, that is, our interests, sort. The interaction of two ideal types of replicators—altruistic cooperators and self-regarding opportunists—can be modeled as an evolutionary game, the dynamics of which ensure that more successful strategies replace less successful strategies.[60] Multiple generations or iterations of deals can depict the secular evolution of a global population—a lineage—of these replicators or interests.[61]

Consider, then, a large population of such interests who get into oil deals. Suppose this population initially comprises a minority, say one-third, of conditional altruistic cooperators who will hew to the tenets of moral economy, and a majority, two-thirds, of opportunists or defectors who will not. This specification is conservative and is warranted by the empirical evidence in behavioral economics and is consistent with the specific cultural heritage of the population from which the oil fraternity in Texas was drawn.[62] Sort this population into deals with eight equal-sized interests.[63] In each deal, there will be an initial proportion of altruists, and an initial proportion of opportunists. Absent expectations of repeated interactions beyond the present deal, the best response of each player to the strategies of the other players in the deal is defect, that is, not only fail to honor his explicit commitments in the deal, but also ignore the altruistic, cooperative injunctions of the oil fraternity's moral economy.[64]

Despite the need for haste, deals do have duration in time: they typically last longer than transactions but perhaps not so long as at least some marriages. Participants have enough time to observe one another and behave accordingly. Assume then that conditional altruistic cooperators cooperate if at least half of the participants in the deal (including self) do, always defect if they are the only would-be cooperator in the deal, or defect with high probability if they are in the minority. Conversely, assume opportunists always defect unless they are in the minority, in which case they may conditionally cooperate with a modest probability—presumably a conformity effect. Altruists who conditionally defect are pertinacious and opportunists who conditionally cooperate incorrigible: 90 percent of either will revert to type in the next iteration of the model.

Assume each altruist in a deal provides benefits, such as timely honoring of commitments or "carrying" others, to all other participants in the same deal, altruists and opportunists alike, excluding himself (which eliminates the possibility of self-interested altruism). An altruist's costs to provide these benefits include both actual out-of-pocket costs and opportunity costs. Oil deals also entail information or transaction costs; assume these costs are also disproportionately incurred by the altruists in the deal.[65] Opportunists provide no benefits to others, incur no costs to cooperate, and bear a smaller proportion of transaction costs, but do benefit from others' altruism.

Formally, GSD is a normal form game in which players move simultaneously. But in distinction to games with complete information, or to Bayesian games with incomplete information,

GSD is sparse, and doesn't require many of the usual game-theoretic assumptions. Despite GSD's structural similarity to an $n$-person Prisoner's Dilemma, its entities, altruists or opportunists, are simply determinate strategies subject to selection, and need not be rational utility maximizers, nor have uniform marginal utilities, nor possess common knowledge of rationality, nor know the exact structure of payoffs, nor be Bayesian reasoners.[66] All that is required is that they behave either cooperatively or opportunistically (as defined), with the specified probabilities, conditional on the observed behavior of other participants in their deal.[67] (The model does assume that these observations are accurate.) GSD's big-time departure from orthodoxy is that its altruistic cooperators are not motivated solely by egoistic incentives—they don't necessarily seek to maximize their individual payoffs—but rather knowingly cooperate, under the specified conditions, even though it's not in their material self-interests to do so.[68]

In GSD, after selection, or the completion of a deal, each altruistic and each opportunistic interest will have a new fitness or reproductive success, that is, new economic circumstances, as a consequence of the deal.[69] Deals may benefit from "horse tradin'," that is, selling or swapping leases, or, best case, from drilling and completing a producing oil or gas well, although dry holes are never an unlikely result. This model captures only successful deals, those that have positive fitness payoffs. All wells are assumed to be equal, although in reality, both the size of fields and the quality of wells are variable.[70] There is no reason to believe, however, that well quality is systematically related to the proportion of altruists or opportunists within deals. In the fitness equations for altruists and opportunists, group size, baseline fitness, the benefit provided by each altruist to each other group member, and the costs to provide those benefits to each other member of the group can take any value so long as fitness is a linear function of these variables.[71] In our GSD implementation, these values are constants for computational simplicity. (Parameter definitions and basic equations are presented in Box 3.6. The GSD model.)

The reproductive success of the group, the economic success of each deal, is simply the sum of its participants' ex-post fitnesses. The proportion of altruistic interests in each deal declines after selection if there are any opportunists in the deal, since opportunists invariably do better. Only the superior reproductive success of *groups* with more altruists might allow altruism to evolve in the global population. After selection in each iteration, the global population will comprise a different proportion of altruistic cooperators and opportunistic defectors, who will then again randomly sort into new deals. This basic model can be recursively iterated for as many generations of deals as desired.[72] As is widely recognized, for these model specifications, random assortment typically does not preserve sufficient between-group variation to permit altruism to evolve, and it does not for the parameter values in our model. The proportion of conditional altruistic cooperators in the global population monotonically declines.

There are two standard modifications to these replicator dynamics that help: truncation and assortative group formation. Our implementation of truncation selection is distributed, unlike Michael Wade's classic experiments with flour beetles (*Tribolium castaneum*), in which all groups with fitnesses below a certain threshold were terminated.[73] The intuition here is that "opportunists foul the nest," that is, they ruin, with some probability, any deal in which they participate, as a linear function of their presence: The probability a deal will be truncated or extinguished is the product of a baseline probability and the proportion of opportunists in the deal. More opportunists yield a higher truncation probability. In our model, only half of even all-opportunist deals fail outright and are truncated, which means fully half of the deals in which everybody defects still get lucky—but then that's the oil business. Altruistic interests who happen to be in truncated deals also are extinguished. Truncation selection applies to entire deals (groups) rather than to individuals.[74]

Modest foresightedness alters the effects of truncation. Biological models tacitly assume that nonhuman subjects are blissfully unaware of truncation. Self-conscious hominins with a modicum of experience likely are not: they can estimate the expected values of cooperating or defecting, that is, discount payoffs by the probability their group will be truncated, and act strategically. For sufficiently high truncation probabilities, the expected value of cooperating is greater than that of defecting, even though the undiscounted return to cooperation is less. In this circumstance, when expected values are considered, individual incentives align with group interests, the unique Nash equilibrium is all-cooperate, and purely self-regarding actants will behave cooperatively. In our model, however, with its modest truncation probabilities, the expected value of cooperation invariably is less than the expected value of defection, and cooperation is always altruistic. Thus, conditional on specific truncation probabilities, this model demarcates self-regarding cooperative equilibria from altruistic cooperation, even though observed behaviors may be the same. Counterintuitively, *more* benign environments (with lower truncation probabilities) are more conducive to selection for true altruism. Determining whether observed cooperation is self-interested or altruistic requires knowing the specific selection dynamics.

Assortative group formation permits cooperators who recognize other cooperators within deals to (probabilistically) associate with them in future deals.[75] The intimacy of the deal provides those in it accurate and reliable information about who is cooperating and who is welshing. Cooperators' mutual recognition depends upon direct observation within the deal, and is not a priori, as in the "greenbeard" scenario, nor is it a product of reputation.[76] Since cooperators' mutual recognition is affected by the "noise" associated with the presence of opportunists, altruists' learning depends linearly upon the proportion of altruistic cooperators in the deal. Once ensconced in an all-altruist deal, altruists normally stay there and only randomly reassort with a small probability at the initiation of each new boom.[77]

Assortative group formation occurs after truncation and selection—after deals are done—and applies to the next iteration of the model. The excess transaction costs incurred by conditional altruistic cooperators, four times that of opportunists, may be taken to represent the information costs of identifying other cooperators and of putting together and sustaining all-altruist deals: the cost of assortative group formation. Assortative group formation has the indirect effect of ostracizing opportunists but entails no additional costs to do so. Thus, neither sanctioning institutions nor "strong reciprocators" who are willing to incur costs to

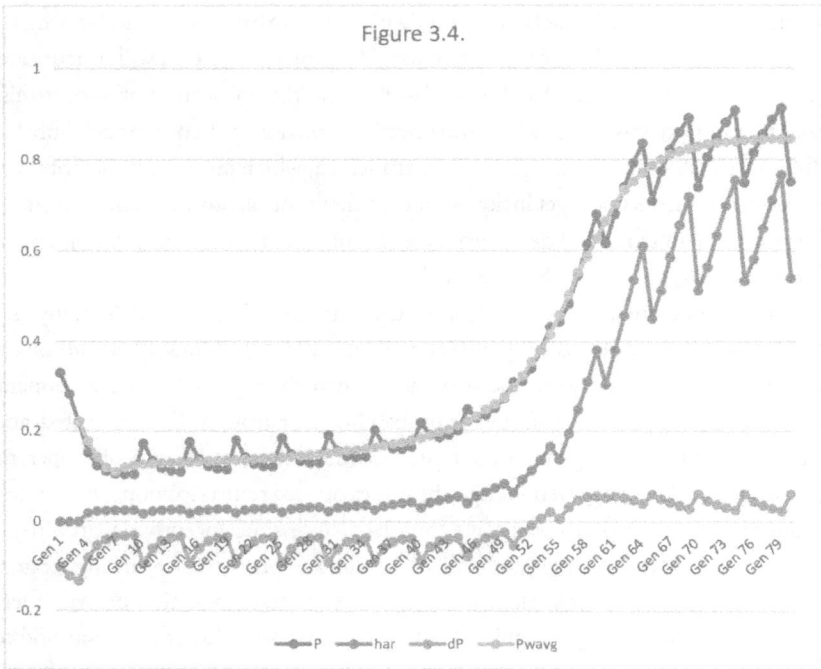

**Figure 3.4. Evolution of the Proportion of Altruistic Cooperators, P, in the Global Population**
*har* = proportion of population who are conditional altruistic cooperators and associate in deals only with other cooperators as a result of assortative group formation, after reversion. dP = $\Delta P$ = net change in the proportion of conditional cooperators in each generation (iteration). Pwavg = 5 iteration moving average P. Independent variable values: $L = 7$, $b = 4$, $n = 8$, $c = 1$, $ct = 1$.

punish transgressors are necessary.[78] Moreover, while assortative group formation is purely a (probabilistic) consequence of direct observation within deals, it leads naturally to the emergence among altruistic interests of small-world social networks in which few "links" or intermediaries connect the elements (nodes) of the population.[79] Assortative group formation and emergent small-world social networks obviate any need for "global" reputational information, but do encourage continual exchange of "truthful gossip" within segments of these selected or exapted subpopulations.[80] Truncation selection and assortative group formation combined have the happy effect of ameliorating the tendency of random group formation to erode the between-group variation necessary for group selection.

Certainly, Texas's oil century provided ample stage for all this. Like the great, recurrent waves of nineteenth-century revivalism, the itinerant oil fraternity collected new adherents as it spread out from play to play, from boom to boom, across Texas and then Louisiana and Oklahoma.[81] This "Waltz Across Texas" approximates a colonization process in which colonists (members of the oil fraternity) mix with local populations, and is formally equivalent to periodic invasions of the original population and its descendants (the oil fraternity) by external entrants (from the indigenous population). In the GSD model, each boom runs

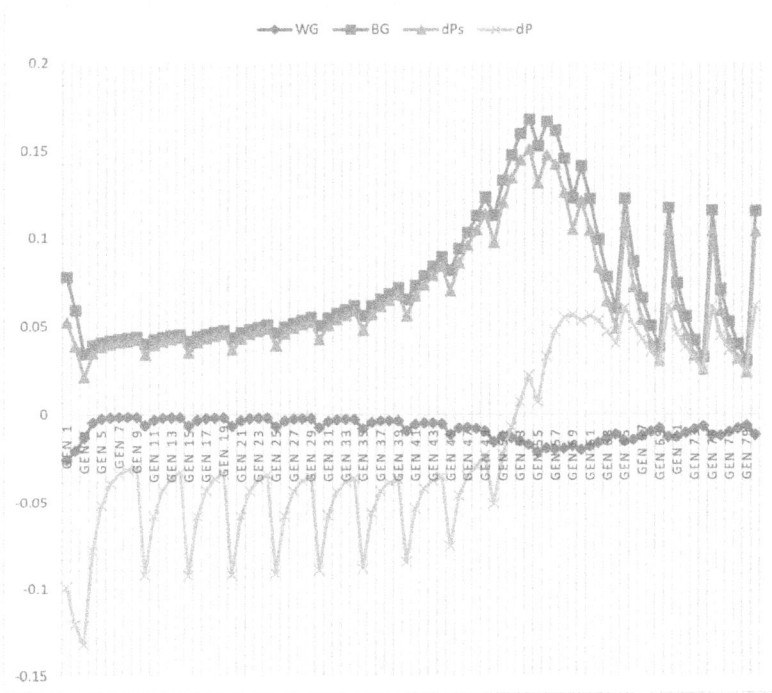

**Figure 3.5. Effects of Within-Group (WG) and Between-Group (BG) Selection in Each Generation**
WG = effect of within-group selection. BG = effect of between-group selection.
dPs = $\Delta$Ps = GPdP ("George Price $\Delta$P") = net change due to within-group and between-group selection, including distributed truncation selection. dP = $\Delta P$ = total net change in the proportion of conditional altruistic cooperators in each generation (iteration) as a result of selection (including truncation) *and* conditional defection and conditional cooperation. Independent variable values: $L = 7, b = 4, \text{n} = 8, c = 1, ct = 1$

for five iterations or generations (ten initially), at which point the fraternity decamps for the next boom. The intra-boom generations portray both the absolute growth of each oil field and the incessant "horse tradin'" characteristic of a developing oil play. Each new boom (colonization event) introduces a new cohort of entrants, 30 percent of the population, who displace an equal number of existing interests. The entrant cohort comprises the same proportion of conditional altruistic cooperators as the original founder population, one-third.[82]

For the values of the independent variables specified, Figure 3.4. shows the evolution of the proportion of altruistic cooperators in the global population, the five-generation moving average of the proportion of altruistic cooperators, the change in the proportion of those cooperators in each generation or recursion, and the emergence of all-altruist groups. In the earlier generations, cooperation sharply declines—think "Swindletop" and its immediate successors—before slowly recovering. Group selection for altruistic cooperation within deals, despite within-deal selection favoring opportunism, ultimately increases the proportion of altruists in the global population.[83]

As noted, the Price equation parses selection into within-group effects, which invariably

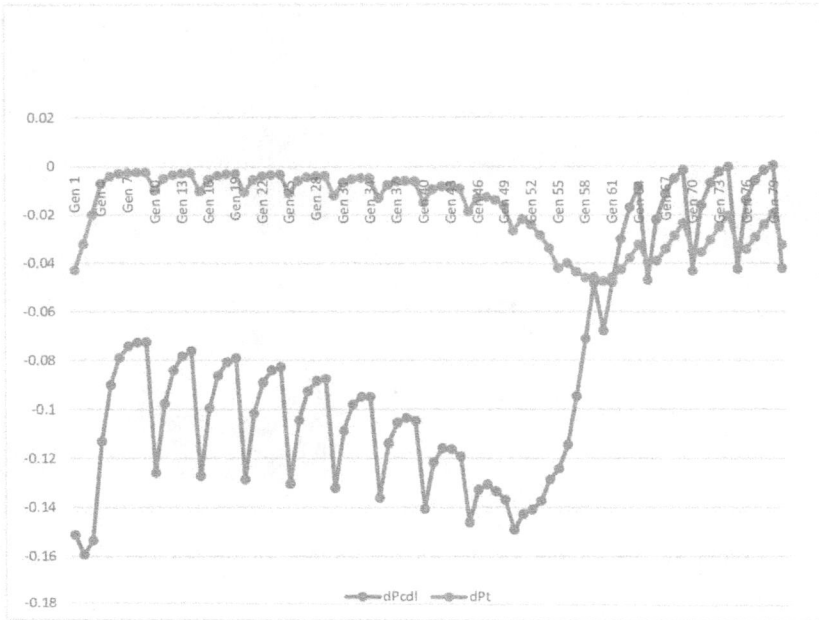

**Figure 3.6. Net conditional defection in each generation**, $\Delta$ PCDL = dPcdl, and effect of distributed group truncation alone, $\Delta$ PT = dPt. Independent variable values: $L = 7$, $b = 4$, n = 8, $c = 1$, $ct = 1$.

favor opportunism, and between-group effects, which reflect the covariance between group fitness and the proportion of altruists in the group and may favor the evolution of altruism (Box 3.6. The GSD model). The net change in the proportion of altruists in the global population as a result of selection is the sum of these two terms.[84] The magnitudes of within-group and between-group selection in each iteration are depicted in Figure 3.5. together with their net effect on selection. The Price equation captures only the change in the proportion of altruists resulting from selection (including truncation selection). The total net change in the proportion of altruists in the global population in each iteration of the model results from the combination of conditional defection and conditional cooperation within groups, and selection (including truncation) within and among groups.

Given the preponderance of opportunist-heavy deals in earlier generations, the replicator dynamics are dominated initially by altruistic interests' conditional defection (Figure 3.6. Net conditional defection in each generation), which enhances the likelihood that their deals (and they) will be truncated. And although distributed truncation reduces the number of altruists in the global population in every generation, it disproportionately eliminates deals rich in opportunists, thereby strengthening the group selection of which it is a part. In later generations, the proportion of altruistic cooperators in the global population rapidly increases, reflecting the growing presence of all-altruist deals (Figure 3.4. Evolution of the proportion of conditional altruistic cooperators in the global population), which amplifies the among-group

## Table 3.1. Cohorts' Representation in the Global Population

Proportion of the global population who are conditional altruistic cooperators *and* descended from a given entrant cohort, and each cohort's representation in the global population, in Generation 74 (after selection, in the fourth iteration after the thirteenth colonization event).

| Cohort | Proportion Who Are Altruists | Proportion of Global Population | Cohort | Proportion Who Are Altruists | Proportion of Global Population |
|---|---|---|---|---|---|
| 0 | 0.0002 | 0.0002 | 8 | 0.0993 | 0.1070 |
| 1 | 0.0005 | 0.0005 | 9 | 0.1183 | 0.1275 |
| 2 | 0.0015 | 0.0016 | 10 | 0.1292 | 0.1393 |
| 3 | 0.0044 | 0.0047 | 11 | 0.1280 | 0.1380 |
| 4 | 0.0111 | 0.0120 | 12 | 0.1357 | 0.1463 |
| 5 | 0.0244 | 0.0263 | 13 | 0.1555 | 0.1677 |
| 6 | 0.0460 | 0.0496 | | | |
| 7 | 0.0735 | 0.0793 | Total | 0.9276 | 1.0000 |

variation upon which the rate of evolution depends.[85] These patterns are consistent with historical accounts of the evolution of oil booms in Texas. The replicator dynamics in this model have substantial resistance to perturbation, to repeated and regular colonization events, and altruistic cooperation comes to dominate the global population.[86] Even though defection is always the better individual option, and cooperation is never self-interested, group selection still can rescue even strong altruism.

Collective claims to privileged access to resources, and the xenophobia associated with them, which canonically define moral economy, are best seen as products of this group selection, not its cause. The clue is in the "sawtooth" pattern evident in Figure 3.4. In later generations, colonization results in a sharp but temporary decline in the proportion of cooperators in the population. Even new entrants didn't have to be very bright to figure out that getting into deals with itinerants from within the oil fraternity yielded a higher probability of encountering trustworthy cooperators, given no other information, than dealing with outsiders.[87] This preferential attachment moves the oil fraternity's small-world networks toward a scale-free topology, although the high level of social integration would remain, indeed, likely would increase.[88]

Limiting "outsiders" access to communal resources—privileged access—is widely regarded as a requisite for collective self-regulation of a common-pool resource's exploitation.[89] The replicator dynamics of the GSD model, the emergence of small-world networks among altruistic cooperators, and, especially, the evolutionary preeminence within the itinerant oil fraternity of all-cooperator deals, have the de facto effect of limiting access in newly discovered oil pools as well as in developed fields. Yet an opportunity ethos was the core of the moral economy of the oil fraternity in Texas—but opportunity only for those who belonged to the fraternity and hewed hard and fast to the tenets of its moral economy. Subject to its assumptions, our model portrays this opportunity ethos in action. The proportion of the global population who are conditional altruistic cooperators and descended from a given entrant cohort, and each cohort's representation in the global population, are shown in Table 3.1. As these results indicate, the model's replicator dynamics provide substantial opportunity for later altruistic entrants: Because of repeated colonization events, "first movers," the founders, do not dominate. Thus, this model secures an apparent contradiction: both privileged access and opportunity.

Conditional altruistic cooperation and conditional opportunism are probabilistic phenotypic traits or strategies that are subject to selection; they are "program-based behaviors" embodied in interests.[90] Altruistic cooperation prevails via group or deal selection, and the consequent differential propagation of altruistically cooperative interests in the population. The GSD model, based as it is upon an $n$-person Prisoner's Dilemma (PD) and group selection, offers a better portrayal of the evolution of the oil fraternity in Texas and its moral economy than individualistic, egoistic-incentive based models.

The GSD model, simple as it is, captures the multilevel dynamic of the moral economy of the oil fraternity in Texas. What is true for the evolution of an original lineage or population of interests in the GSD model is true also for any derivative lineage of interests in a metapopulation of lineages that might be produced by parallel colonization events—by multiple, synchronous oil plays that recruit interests randomly from existing lineages, plus new entrants. The same applies to a suprapopulation of "agents" who "own" the interests that constitute the lineages. These agents might be of any size or character—individuals, partnerships, firms, or corporations—and can own single or multiple interests in a single or in multiple lineages. Thus, the same analytic schema accommodates everything from the poor-boy driller who takes an interest in a single well in lieu of wages to Standard Oil of New Jersey. "Agents" are simply the arithmetic sum of their interests. Because both altruistic and opportunistic behaviors are conditional on the behavior of other interests in a deal, a single agent can comprise some altruistic and some opportunistic interests in the same or in different lineages. Hence the proportion of altruists in the suprapopulation of agents represents agents' aggregate phenotypic behavior (that of their interests). But given the determinate assignment of behavior to conditional altruistic cooperators and to conditional opportunists and the replicator dynamics of the model, expunging opportunistic behavior or replicators in lineages expunges opportunistic types, of interests, of strategies, and of agents.[91]

Moral economy and altruistic cooperation, however well-established within deals and however definitive of the culture of the oil fraternity in Texas, do not logically entail cooperation among deals or groups of lessees in an oil field tapping a common subsurface pool or cooperation among operators in different oil fields to restrain total production. The evolution of altruistic cooperation in our GSD model depends upon variation among the population of deals in an oil play, or, for a lineage, in a sequence of plays. There is insufficient variation among oil fields for the same group selection dynamics to work at a higher level. But, as Samir Okasha argues, understanding how altruistic cooperation might evolve in deals, or among entities at any lower level in a multilevel system, "could help explain . . . their tendency to behave altruistically toward [others], or to aggregate with them in collectives, or to police their selfish tendencies."[92]

Moreover, based upon their game-theoretic investigations, Jonathan Bendor and Dilip Mookherjee maintain that within large groups (their focus is collective political action of the sort that exercised Mancur Olson), a "federated structure" that combines local, small-$n$, altruistic cooperation, and a centralized agency that monitors and possibly sanctions lower-level behaviors, provides more stable collective cooperation than either voluntaristic local or centralized authoritarian organization alone.[93] In addition, Rajiv Sethi and E. Somanathan argue that if defectors and free riders were not too common, such a hybrid structure could emerge through "a continuum of stable states in which resource exploitation is effectively supervised by a centralized monitoring authority funded by voluntary contributions."[94] The evolution of the moral economy of the oil fraternity in Texas via group selection among deals, and the willy-nilly, ad hoc emergence of multilevel monitoring and regulatory institutions— field-safety committees, field-advisory committees, field engineers and umpires, individual field regulations, Railroad Commission field and district offices, the voluntaristic Central Proration Committee, ultimately statewide prorationing under Railroad Commission supervision—corresponds to the processes these theorists envision. It starts with the deal and the necessary altruistic cooperation it requires. Larger-scale cooperation is an emergent property of this exapted population: The well-winnowed altruistic cooperators who had evolved in the oil fraternity in Texas by 1920 would be more likely to cooperate on a larger scale and to fabricate higher-level regulatory structures, just as their "rugged individualist" communitarian forebears had done, or tried to do.[95]

In GSD-world, altruism and moral economy prevail universally. There is one restriction. The suprapopulation of agents could not have consisted of only a few large firms that owned all interests: Group selection would not have occurred. Moral economy would not have evolved. The folk wisdom that despised "monopoly power" feared more wisely than it knew.

## Agent-Based Models

The principal alternative to a game-theoretic or replicator model like GSD, with or without complete information, is an agent-based model, or ABM. These models offer a richer and potentially more realistic portrayal of the evolution of cooperation among large groups of non-kin than do sparse game-theoretic accounts. An ABM program endows (usually) a large number of individual agents with specified, multiple social and behavioral dimensions. These agents are capable of a much greater variety of responses and initiatives than the simple determinate cooperator or defector strategies in GSD. These ABM agents can be from multiple species, and can be simulated as individual actants, groups, corporate or institutional entities, or natural resources, such as fish species. ABM entities may interact in a variety of ways: as random individuals, within partitioned groups, on a spatial matrix, between groups as groups, or with a specified but variable environment. ABMs ostensibly can capture more of the complexity, contingency, and path dependency by which self-organization might emerge.[96] Explicitly programmed ABMs themselves likely can be supplemented or supplanted by more diverse and open-ended artificial intelligence (AI) social simulacra based upon chatbots, which autonomously learn and may innovate.[97] In contrast, even with its convolutions, GSD is a relatively simple tabletop (or, now, desktop) "toy" model with minimal computational requirements. ABMs get big and complex fast, and demand not only substantial programming support but also significant computer resources: to get robust results, an ABM can require multiple runs of 100 different parameter permutations for a hundred-thousand iterations each.[98]

Thus far, patterns of the emergence of cooperation in large-scale, agent-based models have not departed significantly from those produced by corresponding game-theoretic models, although ABMs do offer much more detailed portrayals of social dynamics and hold out the promise of discovering novel properties or processes.[99] For now, GSD adequately illustrates how within-deal altruism could have emerged in the oil fraternity in Texas, and how selection could have produced a fraternity disposed to altruistic cooperation.

## FLESH ON THE BONES

Abstract models, including GSD, radically underdetermine behavior. As spatiotemporally unrestricted mathematical models, the Prisoner's Dilemma and Public Goods games, as well as GSD and its underlying replicator equations, leave "cooperate" and "defect" undefined; they are simply terms in equations with some surreptitious, commonsensical meanings. In experimental reifications, these terms are operationalized very narrowly and unambiguously: contribute $x$ or not. The GSD model characterizes "cooperate" or "defect" as "conform to the tenets of the moral economy of the oil fraternity in Texas" or not (hardly a precise specification), which is reified as "provide benefit $b$ or not." The correspondence relations between spatiotemporally unrestricted theories (for example, selection theory expressed in fitness equations) and spatiotemporally particular models and their specific parameter values (such as our GSD implementation) can be tedious but usually not so obscure: particular models are typically more or less straightforward mathematical or logical derivations. It is in applying models to real-world instantiations that things get dicey. Beyond the most constrained and well-specified experiments (Public Goods games in the laboratory, for example), how is it that subjects know what cooperation *is*, or, in coordination games, what behavior to coordinate on, even if they want to? Does one drive on the left or the right? Since guessing is likely to be counterproductive, some sort of "acculturation" or "learning" is usually invoked to explain how folks just "know." That they all know the same thing, at least locally, and pretty much stay in their pews, remains problematic. As Daniel Nettle notes in his discussion of the application of the Price equation to cultural evolution, "What is missing is the dialogic activity underlying the (constant, ongoing, interpersonally negotiated) transformation of cultural content."[100]

With Merlyn, the happiest solution to such knotty problems is more abstraction. For our purposes, helpful would be a conceptual model that connects individual behavior and representation with collective "culture" and demonstrates how each is produced out of the other. Genuflecting toward culture may be politically correct and analytically expedient, but it doesn't exactly solve the problem. More specification is needed. P. Steven Sangren, a cultural anthropologist, offers a better-specified model, albeit tailored to a culture far removed from that of the oil fraternity in Texas. But suitably modified, Sangren's conceptual schema is promising.

## ALIENATED REPRESENTATION AND THE PRODUCTION OF SELVES AND COLLECTIVITIES

Sangren's primary interest has been religious cults in contemporary Chinese culture, specifically regional cults in the vicinity of the town of Ta-ch'i in northwestern Taiwan.[101] A good part of his work focused on a secondary deity, Ma Tsu, which is derived from a related cult in mainland Fukien Province.[102] Although he has so far focused his attention mostly on explication of Chinese religious practices, Sangren's conceptual model for their social production is clearly intended to have broader application.

The cult deities with which Sangren dealt are local and multiple. Different deities predominate in the ritual and spiritual lives of different geographic areas. No deity is exclusive: an individual subject holds a pantheon of deities sacred and participates in private and public observances devoted to each. A temple primarily dedicated to one deity often also has side altars for other deities. Social and economic class have no discernible effect on which deities are worshipped by which individuals, although class may affect the form and rhetorical rendition of personal observances. Most of the cult deities in Chinese culture are historical in that they are presumed to have taken on human form in specific times and places in the past. Sangren, of course, dismisses any realism to his informants' assertions about the existence or propensities of supernatural agencies: Neither Ma Tsu nor any other deity exists outside the social and psychological processes of their naturalistic production.

Sangren argues that the individual social self as well as collective consciousness are simultaneously produced by a set of "mutually authenticating" social processes:

> Chinese images of supernatural power operate as alienating fetishes in two mutually authenticating ways. Chinese deities are alienated representations of the self-productive power on the one hand of social collectivities, and on the other hand of individual subjects. Moreover, the processes in which these alienated representations are produced are, at one and the same time, the processes in which both cultural subjects and social collectivities are also produced.[103]

Chinese cult deities, including Ma Tsu, are the dialectically alienated representation of a social group's collective historical experience, especially those episodes in that experience in which the group as a collectivity acted efficaciously.[104] Alienated representation of community in a cult deity neatly locates the power and reality of the group outside itself. It makes that power and identity transcendent and, therefore, in principle at least, beyond egoistic tampering, appropriation, or corruption. This process effectively masks the sociogenesis of collective solidarity, which depends, paradoxically, on location of group order and legitimacy outside the social group.

The second part of Sangren's model ties this collective alienated representation of the social group to production of individual selves in collective idiom by means of individual action in social context: in individual worship, ritual participation, pilgrimage, and testimony.[105] Individuals conduct private religious observances and go to temple altars to ask intervention or to give thanks by burning incense. In Ma Tsu's case, the "blackness" of the icon's face, presumably coated with the soot of much devotional incense, is taken to be simultaneously testimony to the power of the deity and of the devoutness of her worshippers. Individual subjects also participate in more public ritual observances, such as pilgrimages and processions. In these ritual contexts, subjects both learn to produce themselves in collective idiom and simultaneously secure communal authentication of their self-productions, which in turn authenticates the community's alienated representation (the deity) itself.[106] (Figure 3.7. Sangren dialectics of alienation.)

While Sangren drew his inspiration from Karl Marx's notion of commodity fetishism, his account resonates also with Émile Durkheim's *The Elementary Forms of the Religious Life* and with aspects of Jürgen Habermas's *Theory of Communicative Action*.[107] Habermas, indeed, extends the notions of self-production and communal authentication beyond formal ritual occasions to mundane conversation: his classic example is negotiation of beer-fetching at a construction site.[108] Nevertheless, Sangren's dialectical conception of the relationship between individual self-productions and collective representations is more explicit and better defined than either Durkheim or Habermas, and lacks the teleological tinge of either.

There is an inherent variability in both individual self-productions and in the collective representations individuals hold, but all are subject to communal authentication—that is, selection. Yet there are no essences that all members of the relevant set unequivocally share. As David Hull noted of species, conceptual systems, and even language communities, "More often than not, more variation exists within a species than between closely related species. Sometimes a single set of genes exists that all, or nearly all, the organisms belonging to a particular species possesses; sometimes not. . . . This same sort of variation characterizes social groups and conceptual systems."[109]

Wherever there is variation, there is the potential for selection, and therefore, for evolution. Sangren's model is relatively static. This stasis doesn't reflect a necessary condition but that the territorial cults Sangren investigated were relatively stable during his field work. Such need not be the case: given variability, the processes of self-construction and alienated representation can produce change, both in self-productions themselves and in communally authenticated representations of the collective.[110]

## A Dynamic Model for Self-Production and Collective Representation

A little reflection suggests several conjectures. First, in the processes by which groups dialectically produce alienated representations of themselves, it matters not one whit what form that alienated representation takes: animism or pantheism, Alexander the Great made living god by his victorious army, Jehovah, Ma Tsu, the Father of Our Country, or whatever.[111] Second, there is no reason why such representation must take supernatural form. Mythologized history

Level:          Individual                                    Collectivity

```
┌─────────────────────────┐   collective   ┌─────────────────────────┐
│ Subject produces self in │ ◄──────────────│ Alienated representation │
│ collective idiom         │     idiom      │        (Ma Tsu)          │
└─────────────────────────┘                └─────────────────────────┘
             │                                            ▲
             │              (re) produces                 │
             └ ─ ─ ─ ─ ─ ─ ─ ─ ─ ─ ─ ─ ─ ─ ─ ─ ─ ─ ─ ─ ─ ┘
                        alienated representation
```

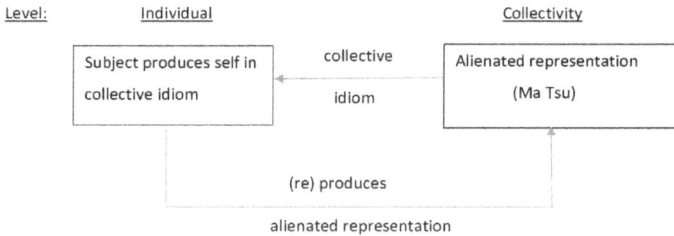

**Figure 3.7. Sangren dialectics of alienation**

serves just as well. Whatever their theological or historiographical importance, the distance between Ma Tsu and Jesus of Nazareth, or Beowulf and Ulysses, or Herodotus, Thucydides, and Thomas Woodrow Wilson, for that matter, is just not that great.[112] The myth, or the encoded ideology, can function as collective alienated representation whether or not it invokes the supernatural.[113] Any representation that serves to promote communal solidarity and to mask its sociogenesis, and which provides the idiomatic basis for individual self-production or reproduction, will do, even the historical rights of small producers to the usufruct of the land, or the sanctity of due process of law. Secular ideology or mythologized history work as well as gods, and better than neoclassical economics. The functional value of alienated representations lies not in the veracity with which they portray the world, but in the coordination and altruistic within-group cooperation they can elicit.[114]

Second, dialectically defined historical collectivities are like species. They emerge, evolve, change, adapt, migrate, merge, and often, and almost always unwillingly, go extinct. Like biological species, such social groups can form relatively isolated demes, in which new characters, that is, novel or differentiated alienated representations, can become fixed more easily. Just as individual members of biological species exchange genes intraspecifically and, on occasion, interspecifically, so individual members of social groups exchange self-productions and, consequently, alienated representations of the group itself, both intra-group and occasionally with outsiders. But no more than species share some set of essential alleles, nor language communities some set of essential meanings, do social groups share some set of essential representations of themselves or of the group. They share just enough.

In a dynamic model, what connects individual self-productions in collective idiom and the concomitant reproduction of the group's representation through time is, in David Hull's terminology, conceptual ancestor-descendant lineages. Each individual enactment—testimony, in both its religious and secular senses, "publication" in its original seventeenth-century meaning,[115] participation in ritual in any of its many guises—reproduces with greater or lesser fidelity the idiom of group representation, itself the cultural product of historical individual enactments. Individual self-production produces variability, even idiosyncrasy, which produces flexibility and adaptability in group interpretation. Such variability permits alienated

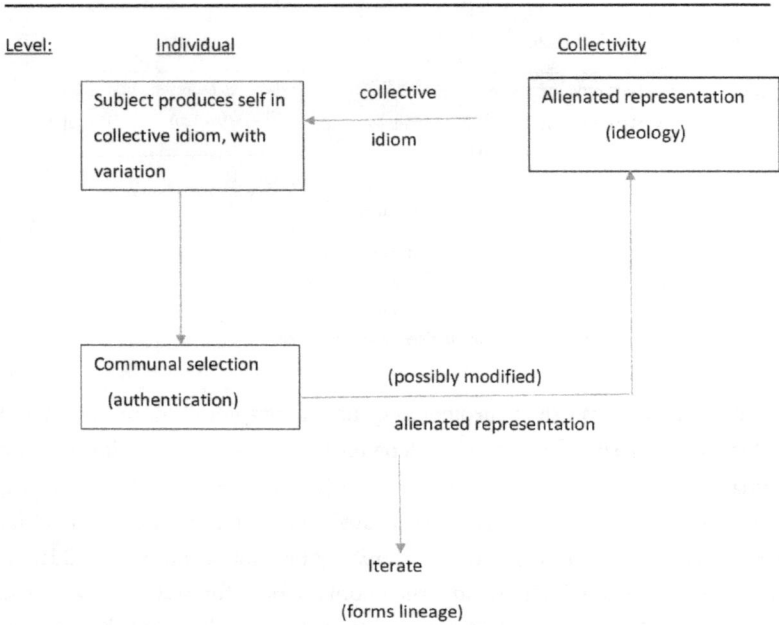

Figure 3.8. Dynamic alienated representation model

collective representation and dialectical self-reproductions to change or evolve, and to reflect new environmental contingencies or historical experiences, as well as simple fad or drift: "Conservation" can mean many things to different collectives, and to the same collective at different times (Figure 3.8. Dynamic alienated representation model). Behavioral referents of a relatively stable ideological representation can shift to more nearly fine track the exigencies of ongoing communal experience.

Southern agrarian communities, the oil fraternity in Texas, and the first generations of petroleum engineers are connected by coterminous biological and cultural ancestor-descendant lineages. Yet there is no implication that the representations or the habits of mind or the behaviors that were characteristic of these groups were genetic or sociobiological in origin. It is cultural lineal descent with modification that matters. Nevertheless, in relatively isolated cultural or conceptual demes, biological descent substantially increases the likelihood of cultural transmission or replication: conformist bias and vertical transmission. Our adaptation of Sangren's alienated representation model is evolutionary: it depends upon variation and selective retention in a population, as all evolutionary models do. As noted, Price's equation, used to explain the evolution of altruistic cooperation in the oil fraternity in Texas, characterizes *any* hierarchically decomposable multilevel system which evolves, including alienated representations.[116]

For our history, a dynamic alienated or ideological representation model has several virtues. It provides a mechanism by which novitiates learn to produce themselves in collective idiom—to "talk the talk and walk the walk"—thereby simultaneously reproducing the collectivity and its representation. It is how terms in abstract equations, "cooperation" or "defection," are reified in mundane behavior—how oilmen learn what to *do* in the myriad business and social situations they might face. "Testimony"—the endless "war stories" about deals that are the forte of the oil fraternity—provides the correspondence rules that teach actants how to behave. It is how they learn how. For a historian, this model warrants the use of a wide variety of evidence, from oral history interviews to court testimony to legal opinions to articles in scientific or professional journals, to infer collectivity, representation, and moral economy and its tenets.[117]

## THE PRESENT VALUE OF A NON-RENEWABLE COMMON-POOL RESOURCE (CPR)

Despite a robust and thriving moral economy, the oil fraternity in Texas was plagued by overproduction, surface and subsurface waste, and price instability. Historical actants—oilmen and lawyers and regulators at the time—and historians and legal scholars since, blame the rule of capture. In addition, antitrust statutes precluded coordination or cooperation to restrain either drilling or production, especially if they came with any hint of coercion or infringement upon property rights. This mantra is indeed the conventional wisdom we have recited so far and on which we will continue to rely. But this historically situated account, whatever its validity, elides a more fundamental problem.

Oil is a depleting, non-renewable resource for which time matters doubly. The Present Value of future oil production depends upon two things: the rate of production decline from a well or lease, which is a function of pressure-decline across the entire field, and therefore the production rate of all other wells drawing from the same pool, and an imputed discount rate. Although clouded by uncertainty, experienced oilmen early on could glean a pretty fair sense of a well's likely rate of decline, even soon after discovery. But new income tax laws in 1913 and 1918 required precisely defining reserves or predicting future production in order to claim depletion deductions from income derived from current production. Geologists and nascent petroleum engineers had first plotted production decline curves and suggested they might be extrapolated into the future in 1909. By the end of World War I, hyperbolic production decline, and the exponential decline to which it most conveniently reduces, became established, and IRS sanctioned, dogma.[118] These logarithmic decline curves were empirically, not theoretically, derived, and were inferred mostly from data on what turned out to be "depletion-type" or dissolved-gas drive reservoirs; suitably modified, however, the basic technique of decline curve analysis can be applied to other types of fields.[119]

Discount rates are more problematic. Virtually all the hundreds of millions of dollars that flowed into "Swindletop" and the other early oil booms in Texas was speculative equity investment, and a goodly proportion of later exploration and development was funded by income from production (retained earnings, in corporate-speak). Debt secured by oil reserves played almost no role for the simple reason that there was not enough known about the new kinds of reservoirs being discovered in Texas for a prudent lender, especially one with fiduciary responsibilities for other peoples' money, to extend long-term credit.[120] Thus, while prevailing interest rates can be fairly used as a yardstick for comparison, formal discount rates used by oilmen are unrecorded, and probably nonexistent: their discount rates, therefore, are imputed (made up), and represent personal discount rates on a future largely animated by the extraordinary perceived opportunity costs of operating in an environment of unlimited opportunity. Together, these factors put the oil fraternity in an unusual, if not unique, bind (Box 3.7. The present value (PV) of an exponentially declining production stream).

The thought experiment in Box 3.7 supports several inferences.[121] First, under reasonable assumptions, flush or unrestrained production extracts far more oil in the first few years of a field's life, despite egregiously lower ultimate recovery, than restrained or prorated production. Second, even under favorable assumptions about restrained or prorated production and discount rates, prorated production does not financially outperform flush production for some time (five years, in our hypothetical example). Third, absent favorable discount rates, restrained production is *never* advantageous: Being good don't pay. Even a Nash equilibrium would have been a welcome respite.

None of the equations presented here, for game theory, for GSD, or for Present Value, is in itself either true or false. All are mathematical tautologies. The abstract models represented with these equations are spatiotemporally unrestricted. They are ideal typical in Max Weber's sense, intended to capture essential features of spatiotemporally particular empirical situations.[122] Likewise, results from experiments thought to instantiate these models are valid only for a specific set of subjects in a specific experimental setting. Qualitative models inspired by observation, no matter how meticulous, such as that adapted from Sangren, are similarly ideal typical. No verisimilitude between idealized model or their experimental reification and the "real" oil fraternity in Texas and its moral economy—themselves historical constructs—can be claimed. But Weberian similitude is more plausible: To the extent the models or their experimental reifications display population structure and replicator dynamics, game theoretic imperatives, social behavior, or financial realities that mimic those that characterized the early years of the oil fraternity, they suggest answers to otherwise puzzling historical questions about how and why the oil fraternity in Texas and its moral economy evolved as they did.

There is no accessible, reliable quantitative historical data to rigorously test these conjectures. In the absence of data to support statistical analyses of such models, Poteete, Janssen, and Ostrom suggest they be evaluated by their intrinsic plausibility (given background knowledge of their domains), by whether their behavior is sufficiently transparent and by what additional understanding they offer.[123] On these criteria, GSD and alienated representation models, as well as PV calculations, are revealing. As with all models derived from abstract theories, they are accepted as provisionally, hypothetically, and corrigibly true (if idealized) representations of a "real" situation.[124] Recursively, they serve as both historical explanation and as selectors for what to attend. The models are recited here in detail because they are *causal*.[125] They abstractly depict how the moral economy of the oil fraternity in Texas evolved, the problems the oil fraternity confronted, the constraints it suffered, and the options it had—the austere Platonic reality behind the sound and the fury of the rambunctious, meandering, occasionally absurd, historical cavalcade.

By and large, real oilmen wouldn't have much truck with any of this. These are not the terms in which they would likely think about themselves or their lives. It is no small irony that the very social identity and solidarity the oil fraternity's ideological representation of itself produced, and the seemingly natural, everyday immutability of the moral economy that representation dialectically sustained, permitted to oilmen an unreflective, almost guileless view of their world and their natural place in it. Merlyn notwithstanding, for oilmen, abstraction was tantamount to obfuscation, and ideology was a dirty word, something concocted by Bolsheviks and unwashed malcontents in the dark basements of "foreign" urban-industrial ghettos. Abstraction and ideology were almost as bad as evolution its own self.

---

## Boxes
### Box 3.1. The Prisoner's Dilemma (PD)

Suppose two miscreants are arrested for a robbery. The constabulary holds them separately, with no communication between them, and offers each a deal: rat on the other and go free immediately, while the other serves 10 years. If they rat on each other, each gets 7 years. If they both keep their yaps shut, each will spend a year in jail pending trial, then go free. This situation is ordinarily presented as a symmetric, normal form game (players move simultaneously) with the payoff matrix or contingency table:

|  |  | Player II (column) | |
|---|---|---|---|
|  |  | C | D |
| Player I (row) | C | -1,-1 | -10,0 |
|  | D | 0,-10 | -7,-7 |

$C \equiv$ cooperate (with each other); $D \equiv$ defect (rat). The first number in each pair is the payoff to the row player (I), the second the payoff to the column player (II). (Often in games with symmetrical payoffs, such as this one, only the row player's payoffs are displayed.)

If these agents are rational, each will play his best response to the other player's strategies (possible choices or moves). If Player I cooperates (doesn't blab), but his partner in crime (Player II) defects (blabs), Player I serves 10 years while his partner goes free. Likewise for Player II. While both would be better-off (spending a year in jail) if both cooperated (with each other), each player is better-off no matter what the other does if he defects. Thus, the unique Nash equilibrium of the game—no player can unilaterally improve his payoff by changing his choice—is both defect: $0 > -1 > -7 > -10$. Hence the "Prisoner's Dilemma."[126] The same structure applies to a PD with positive payoffs.

## Box 3.2. *n*-person Prisoner's Dilemma

Payoffs in an *n*-person Prisoner's Dilemma to a focal actant *i* may be expressed as a Public-Goods (PG) game with a fixed "cost" to cooperate:

$$\pi_i = -x_i + m\sum_{j=1}^{n} x_j \quad \text{with } x_i \in \sum\left(x_j\right)$$

$\pi_i \equiv$ payoff to focal actant *i*

$x_j = x_i \equiv$ fixed cost to cooperate (contribute)

$m \equiv$ amount by which contributions to the public good are multiplied, the positive effect of cooperation; $0 < m < 1.0 < mn$

$n \equiv$ group size

A defector benefits if others cooperate, but incurs no cost ($x_i = 0$); thus the Nash equilibrium in an *n*-person Prisoner's Dilemma, as in a two-person PD, is all defect and none contribute.

## Box 3.3. Public-Goods (PG) Game

The payoff to a focal actant *i* in a Public Goods (PG) game is:

$$\pi_i = e - x_i + m\sum_{j=1}^{n} x_j \quad \text{with } x_i \in \sum\left(x_j\right)$$

$\pi_i \equiv$ payoff to focal actant *i*

$e \equiv$ endowment

$x_i \equiv$ *i*'s contribution to the public good

$x_j \equiv$ contribution of actant *j* to the public good

$m \equiv$ amount by which contributions to the public good are multiplied, the positive effect of cooperation; $0 < m < 1.0 < mn$

$n \equiv$ group size

A Public Goods game (PG) is an *n*-person Prisoner's Dilemma (PD) in which each player may vary his degree of cooperation, that is his contribution, $x_i = [0, e]$, and therefore, his cost. In an *n*-person PD or PG, a contributor shares in and cannot be excluded from collective benefits (and thus is only weakly altruistic). But a free-rider (opportunist) who does not contribute or cooperate ($x_i = 0$) not only keeps all his own endowment and incurs no cost, but also shares in collective benefits and is *always* better-off than any contributor. Hence the Nash equilibrium is all defect in a one-shot PG, and for all rounds of an Iterated PG (IPG) game. Aggregate collective welfare would be maximized if all actants contributed their entire endowment in each round, which is multiplied by *m*, but it's not something rational agents would do.

**Box 3.4. Common-Pool Resource (CPR) Exploitation as a Non-Cooperative Game**

Consider a common-pool resource (CPR) with a concave production function, $f\sum(x_j) = a\sum(x_j) - b(\sum x_j)^2$, where $x_j$ denotes an individual agent's investment in the CPR. Assume a symmetric, noncooperative CPR game in which all $n$ actants have the same endowment $e$ to invest, have equal and unrestricted access to the CPR and to an alternative, risk-free investment with a rate of return $w$, share complete information on the character of the game and its payoff function, are risk-neutral between investing in the CPR or in the alternative investment, and make simultaneous and anonymous investment decisions. In this circumstance, the payoff, the revenue earned, by a focal actant $i$ who invests $x_i$ in the CPR is:

$$\pi_i = w(e - x_i) + \left[\frac{x_i}{\sum(x_j)}\right]\left[a\sum(x_j) - b(\sum x_j)^2\right] \text{ with } x_i \in \sum(x_j)$$

$a$ and $b$ are arbitrary weighting parameters

The maximum collective payoff occurs with an aggregate investment in the CPR of $\sum x_j = \frac{(a-w)}{2b}$. This total investment is Pareto optimal: at this level, no actant can increase his investment, and his return, without making another worse off. The Nash equilibrium in the game is $\sum x_j = \left(\frac{n}{n+1}\right)\left(\frac{a-w}{b}\right)$. At this equilibrium, each actant plays his best response to the choices of others, and cannot increase his return by further investment in the CPR. In conventional economic terms, the Nash equilibrium occurs where marginal revenue from further investment in the CPR equals the marginal return from the alternative fixed-return investment (fixed opportunity cost), or MR = MC. "0 net return" on investment in the CPR is realized when the return equals $w$, which occurs when $\sum x_j = \frac{(a-w)}{b}$; in conventional economic terms, average revenue from investment in the CPR equals marginal cost, the return on the fixed alternative, AR = MC.

Any symmetric game has a symmetric equilibrium, in this case the Nash equilibrium in which all actants' investments are equal, or $x* = \frac{(a-w)}{(n+1)b}$. If this game were repeated or iterated a finite but very large number of times among a large number of actants (who know their own history and that of aggregate investments), Harsanyi-Selten selection theory holds that this symmetric equilibrium would be selected, via either a replicator dynamics in which more successful strategies out-reproduced the less successful or a mutation-learning-imitation process in which successful strategies differentially replaced the less successful. Since the Nash equilibrium constitutes subgame perfection in each iteration, it is also the (selected) symmetric equilibrium for a finite series, the larger game. The theoretically robust prediction for the entire game, and for each subgame, is therefore the Nash equilibrium, which is, for this game specification, invariably suboptimal:

Pareto optimal return > Nash equilibrium return > 0 net return.

**Box 3.5. Public-Goods (PG) games with Sanctioning Institutions (SI)**

In Fehr and Gächter's experiments on sanctioning in public-goods games, the payoff function for actant $i$ without sanctioning opportunity is as for the generic public goods game above (Box 3.3. Public-Goods (PG) Games). With sanctioning, the payoff function for actant $i$ becomes:

$$\pi_i = [e - x_i + m \sum_{j=1}^{n} x_j][1 - (0.10)(\sum_{j \neq i} p_i^j)] - \sum_{j \neq i} c(p_i^j)$$

$\pi_i \equiv$ payoff to focal actant $i$

$e \equiv$ endowment

$x_i \equiv i$'s contribution to the public good

$m \equiv$ amount by which contributions to public good are multiplied, the positive effect of cooperation; $0 < m < 1.0 < nm$

$n \equiv$ group size

$p_j^i \equiv$ assessment against $i$ by subjects $j$

$c(p_i^j) \equiv$ cost to $i$ to punish subject $j$

In these experiments, subject $i$'s payoff in each round is reduced by 10 percent for every "punishment" token assessed against him, but not below 0, which prevents retaliation for behavior in prior rounds. The cost to punish $c(p_i^j)$ is strictly increasing in $p_j^i$, which inhibits draconian punishment. $m$ is sufficiently small that an actant does not benefit enough from his own contribution to make it self-interested, while $mn$ is sufficiently large that all actants contributing is Pareto improving.[127] Parameter values in these experiments: $e = 20$; $m = 0.40$; $n = 4$.

## Box 3.6. The GSD Model

Parameter Definitions

| | |
|---|---|
| N | number of interests in the global population; initially = 3600. In each iteration of the model, N is reset to 3600. As is common in such models, "we assume that density-dependent regulation maintains a constant population size of $MN$ [$M$ groups of $N$ each = N in GSD] at each census."[128] This density-dependent regulation affects altruists and opportunists alike. |
| P | proportion of conditional altruistic cooperators in the global population; initially one-third = 0.3333 |
| $n$ | number of equal-sized interests in each deal = 8 |
| $p_i$ | proportion of conditional altruistic cooperators $a$ in each deal: = {0/n, 1/ n, 2/ n, . . . 8/n } |
| $a_i$ | number of conditional altruistic cooperators $a$ in each deal: $i$ = {0, 1, 2, . . . 8} |
| $1 - p_i$ | proportion of conditional defectors or opportunists $o$ in each deal |
| $o_i$ | number of conditional defectors or opportunists $o$ in each deal = $n - ai$ ; $i$ = {0, 1, 2, . . . 8} |
| L | largesse, the baseline fitness payoff to each interest in a consummated deal = 7 |
| $b$ | benefit conferred on other deal participants by each altruistic cooperator in a deal = 4 |
| $c$ | cost to cooperator to provide benefit $b$ = 1 |
| $c_t$ | transaction costs = 1 |
| Pe | proportion of conditional altruistic cooperators in an entrant cohort |

$= 0.3333$

## Basic Equations

Hypergeometric distribution:

$h_i \equiv$ proportion of the population in groups of type $i$, with $i = \{0, 8\}$ cooperators.

$$h_i = \frac{\dfrac{A_e!}{a_i!(A_e - a_i)!} \cdot \dfrac{(N_e - A_e)!}{(n - a_i)![(N_e - A_e) - (n - a_i)]!}}{\dfrac{N_e!}{n!(N_e - n)!}}$$

$N \equiv$ Global population

$A \equiv$ Number of altruists in global population

$a_i \equiv$ Number of altruists in each group of type $i$

$n \equiv$ Group size

Fitness within groups: $wa_i, wo_i, n_i', p_i'$ after each iteration. Following Elliott Sober and David Sloan Wilson,[129] fitness $w$ for an altruist in a group of type $i$, $i = \{1, 2, \ldots 8\}$, indicating $a_i = \{1, 2, \ldots 8\}$ altruists in each group type, is,

$$wa_i = L + \frac{(b)(np_i - 1)}{n - 1} - (c + 0.60c_t)$$

Similarly, fitness for an opportunist in a group of type $i$, $i = \{0, 1, 2, \ldots 7\}$, indicating $a_i = \{0, 1, 2, \ldots 7\}$ altruists in each group type, is,

$$wo_i = L + \frac{bnp_i}{n - 1} - 0.15c_t$$

The last term in each equation, transaction costs, is added to Sober and Wilson's formulation. "Offspring" or reproductive success ($n_i'$) for a group of type $i$ is,

$$n_i' = n[(p_i)(wa_i) + (1 - p_i)(wo_i)]$$

The proportion of altruists in each offspring group of type $i$ is,

$$p_i' = \frac{(n)(p_i)(wa_i)}{n_i'}$$

Conditional defection and conditional cooperation: $hc_t$. Conditional defection and cooperation occur after hypergeometric distribution but before truncation and selection. The transmogrification of conditional cooperators ($a$) into opportunists ($o$) in groups with $a = \{1, 2, 3\}$ depends upon a baseline probability $d$ and the proportion of opportunists in the deal, $(1 - p_i)$. Thus $d_i = (d)(1 + 1/n - p_i)$. Conditional cooperation, the transmutation of opportunists into conditional cooperators in groups with $a = \{5, 6, 7\}$, depends upon a baseline probability $dc$ and the proportion of altruists in the deal, $p_i$. Thus, $dci = (dc)(pi)$.

Distributed truncation selection: $ht_t$. Distributed truncation selection, following the "opportunists foul the nest" intuition is:

$$ht_i = [1 - (1 - p_i)(truncp)][hc_i]$$

where $ht_i$ is the proportion of the population in groups with $i = \{0, 1, 2, \ldots 8\}$ cooperators in each group after truncation. $hc_i$ is the proportion of the population in groups with $i = \{0, 1, 2, \ldots 8\}$ cooperators in each group after conditional defection and conditional cooperation; $truncp = 0.50$ is the baseline truncation probability in GSD.

Expected values of truncation: $E(wa_i)$ and $E(wo_i)$. Given that the probability a deal will be truncated is the product of a baseline probability, $truncp$, and the proportion of opportunists in the deal, $(1 - p_i)$, that is $(1 - p_i)(truncp)$, the expected fitness value for an altruist, given truncation, is $E(wa_i) = wa_i [1 - (1 - p_i)(truncp)]$, and that for an opportunist $E(wo_i) = wo_i [1 - (1 - p_i)(truncp)]$. For sufficiently high truncation probabilities $EV(wa_i) > EV(wo_{(i-1)})$, even though the immediate return to cooperation is less, $(wa_i < wo_i$ and $wa_i < wo_{(i-1)})$. With a baseline $truncp = 0.50$, $EV(wa_i) < EV(wo_{(i-1)})$ always.

New population values: N′, P′, ΔP after each iteration. The total "offspring" population after selection is,

$$N' = (N/n)\sum (ht_i)(n_i')$$

And the new proportion of altruistic cooperators in the global population becomes,

$$P' = \frac{\sum (ht_i)(N/n)(n_i')(p_i')}{N'} = \frac{\sum (ht_i)(n_i')(p_i')}{\sum (ht_i)(n_i')}$$

($ht_i$ is the proportion of the global population in groups of type $i$, $i = \{0, 1, 2, \ldots 7, 8, a\}$ with type $a$ groups comprising only altruists who associate only with each other, after assortative group formation, hypergeometric distribution, conditional defection and conditional cooperation, reversion, and truncation.)

Assortative group formation. Assortative group formation occurs after selection but before the next iteration of the model and depends upon the proportion of cooperators in a deal, $pi$, and a baseline probability $agfp = 0.50$, hence $(p_i)(agfp)$. Assortative group formation produces persisting all-altruist groups, $har$.
The sequence of all these changes is: $h_i \rightarrow hc_i \rightarrow ht_i \rightarrow har$.

The Price equation: within-group versus between-group selection. Sober and Wilson's formulation of Price's equation is,

$$\Delta P = ave_n{}'(\Delta p) + \mathrm{cov}_n\,(s_p\,{}'p_i)/ave_n s$$

($ave_{n'}$ is the average group size after selection, and $ave_n$ is average group size before selection). $s_i = n_i'/n$. Equivalently (including truncation selection),

$$\Delta Ps = \left[\frac{\sum ht_i s_i (p_i{}' - p_i)}{\sum ht_i s_i}\right] + \left[\frac{\sum ht_i s_i p_i - \mathrm{Phc}\sum ht_i s_i}{\sum ht_i s_i}\right]$$

where Phc is the proportion of the global population who are altruists after conditional defection and conditional cooperation, that is the proportion of altruists in the population subject to selection, and $\sum_{i=0}^{a}(ht_i)(s_i) = \text{N'/N}$. The left-hand term of the Price equation captures within-group selection, $WG$, while the right, that containing the covariance expression, measures between-group selection, $BG$. Thus, the net change in the proportion of altruists in the global population as a result of selection, $\Delta$Ps, is the sum of these two terms (in our model, $ave_{n'}(\Delta p)$ is always negative).[130]

The Price equation captures *only* the change in the proportion of altruists resulting from selection (including truncation selection), that is, from relative reproductive success:

$\Delta$Ps = P'− Phc. Total $\Delta$P in each generation, then, is the net change in the proportion of altruists in the global population resulting from the combination of conditional defection and conditional cooperation *and* selection (including truncation) within and among groups:

$$\Delta P = \Delta Ps + \Delta\ PCDL = (P'- Phc) + (Phc - P) = P'- P$$

## Box 3.7. The Present Value of an Exponentially Declining Production Stream

The discounted Present Value of an amount (funds) to be received at a future date, with continuous discounting, is: $PV = f(x)e^{-rt} = f(x)\frac{1}{e^{rt}}$

$f(x)$ = amount, $r$ = imputed discount (interest) rate, and $t$ = time.

The discounted present value (PVs) of a stream of payments $f(x)$ in the interval from $t = m$ to $t = n$ is:

$$PVs = \int_{t=m}^{n} f(x)e^{-rt}dt = \frac{f(x)}{-r}\left(\frac{1}{e^{rn}} - \frac{1}{e^{rm}}\right) = f(x)\frac{1}{r}\left(\frac{1}{e^{rm}} - \frac{1}{e^{rn}}\right)$$

For a stream of payments from oil production, $f(x)$ is variable and will decline as flow from producing wells declines. As noted, hyperbolic decline reduces to exponential decline, which can be theoretically derived for a single phase, incompressible fluid produced from a closed reservoir.[131] For our idealized illustration, assume exponential decline, and let $V$ be the volume of oil in place in a subsurface reservoir, $\gamma$ the proportion of oil in place recoverable under a given production regime, flush or prorated, and $a$ the initial instantaneous flow-rate of a well as a proportion of recoverable reserves, $\gamma$ V. The realized instantaneous rate of production, given $\gamma$, and expressed as a proportion of $V$, is:

$$R_p = (a)(\gamma)e^{-at} = (a)(\gamma)\left(\frac{1}{e^{at}}\right)$$

In this formulation, $at$ replaces $dt$ and is the constant percentage decline rate normally used by petroleum engineers, which could be estimated early in a well's life.[132]

Cumulative (total) production at time $T$, again as a proportion of $V$, from time $t = m$ to $t = n$ is:

$$\text{PT} = \int_{t=m}^{n}(a)(\gamma)e^{-at}dt = (a)(\gamma)\int_{t=m}^{n}e^{-at}dt = \frac{(a)(\gamma)}{-a}\left(\frac{1}{e^{an}} - \frac{1}{e^{am}}\right) = \gamma\left(\frac{1}{e^{am}} - \frac{1}{e^{an}}\right)$$

Total production over all intervals for either production regime is $\leq \gamma$ V.
The discounted return on exponentially declining production at a point in time is:

$$\text{PV}_{\text{prodn}} = (a)(\gamma)\left(e^{-at}\right)\left(e^{-rt}\right) = (a)(\gamma)e^{-t(a+r)} = (a)(\gamma)e^{-(a+r)t} = (a)(\gamma)\left(\frac{1}{e^{(a+r)t}}\right)$$

Cumulative payout, or the Present Value of an exponentially declining production stream $f(x)$ in a time interval $t = m$ to $t = n$ is:

$$\text{PV}_{\text{ps}} = \int_{t=m}^{n}(a)(\gamma)e^{-at}dt \bullet \int_{t=m}^{n}e^{-rt}dt = (a)(\gamma)\int_{t=m}^{n}e^{-at}dt \bullet \int_{t=m}^{n}e^{-rt}dt$$

$$= (a)(\gamma)\int_{t=m}^{n}e^{-at-rt}dt = (a)(\gamma)\int_{t=m}^{n}e^{-t(a+r)}dt = (a)(\gamma)\int_{t=m}^{n}e^{-(a+r)t}dt$$

$$= \left(\frac{(a)(\gamma)}{-(a+r)}\right)\left(\frac{1}{e^{(a+r)n}} - \frac{1}{e^{(a+r)m}}\right) = (\gamma)\left(\frac{a}{a+r}\right)\left(\frac{1}{e^{(a+r)m}} - \frac{1}{e^{(a+r)n}}\right)$$

Actual production rate would be $R_p \bullet V$ and total production $\text{PT} \bullet V$, while realized payouts would depend upon both volume in place and current price: assuming a constant price, P, instantaneously $\text{PV}_{\text{prodn}} \bullet V \bullet P$ and cumulatively $\text{PV}_{\text{ps}} \bullet V \bullet P$.

In practice, $\gamma$ and $V$ were unknown ex ante, and $r$ rarely known with certainty, while, with experience, the absolute instantaneous initial flow rate could be estimated fairly accurately. By 1906–1907, the oil fraternity had enough experience with gushers in a sequence of oil booms to acquire an intuitive understanding of the shape and character of the relevant decay curve. But $\gamma_{res}$, the substantial increase in recoverable reserves with restrained production compared to $\gamma_f$ (flush production), was only suspected early on, not really apparent until well after World

Figure 3.9. Instantaneous PV of Exponentially Declining Oil Production

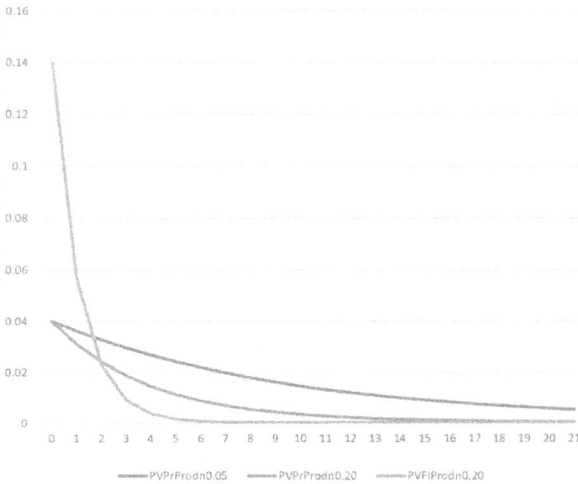

Present value of exponentially declining prorated or flush production, at time $t$:
$\mathrm{PV}_{\mathrm{prodn}} = (a)(\gamma)(e^{-at})(e^{-rt}) = (a)(\gamma)(e^{-(a+r)t})$, as a proportion of $V$, with $a_{res}$ = 0.05, $af$ = 0.70, $\gamma res$ = 0.80, $\gamma f$ = 0.20; $r$ = 0.05 or $r$ = 0.20; and t = [0,1, . . . 21].

Figure 3.10. Cumulative PV of Exponentially Declining Oil Production

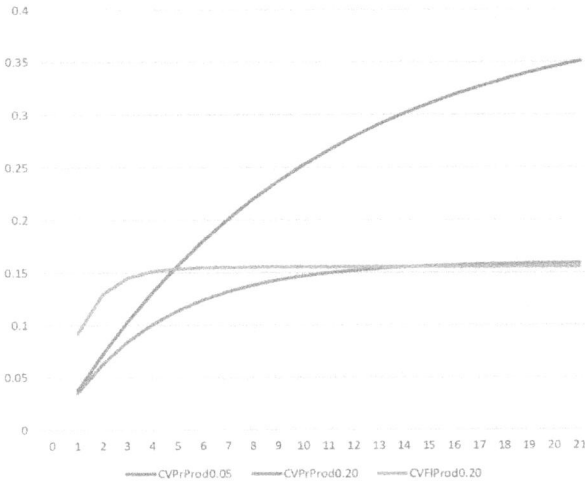

Cumulative payout of an exponentially declining prorated or flush production stream with exponential time discounting: $\mathrm{PV}_{\mathrm{ps}} = = (a)(\gamma)\int_{t=m}^{u} e^{-(a+r)t} dt$, as a proportion of $V \bullet P$, with $a_{res}$ = 0.05, $a_f$ = 0.70, $\gamma_{res}$ = 0.80, $\gamma_f$ = 0.20, $r$ = 0.05 or $r$ = 0.20; and t = [0,1, . . . 21].

War I, not securely supported by empirical evidence until the late 1920s, and remained controversial into the 1930s. Thus, in what follows, the option of restrained production was not in fact well-defined and only emerged gradually as petroleum engineering science developed and was accepted and applied by the oil fraternity in Texas.

For our contrived hypothetical comparison, $V$ will be ignored, and $\gamma_{res}$ set to four times $\gamma_f$, or 0.8 and 0.2 of oil originally in place, respectively; $a_{res}$ and $a_f$ are set, arbitrarily but plausibly, to 0.05 and 0.70 of recoverable reserves. In fact, unrestrained flush production often rendered 90 percent of the oil in place in a reservoir unrecoverable by conventional means, while well-planned prorated production could recover as much as 90 percent, so these values are conservative. Likewise, estimates for initial instantaneous flow rates are consistent with observed values. During the early years of the oil business in Texas, except for World War I, prime US interest rates hovered between 3 and 4 percent. Thus setting $r = 0.05$ for restrained production implicitly includes a risk premium of 25 to 66.7 percent, although for reserve-based loans later, responsible lenders usually also reduced loan amounts to some percentage of estimated reserves to guard against estimate errors or difficulties in development or production. In the early oil booms in Texas, however, reserve estimates were so uncertain and prices so volatile that collateralization of reserves was virtually impossible and formal credit institutions were nearly nonexistent. The imputed discount rate $r = 0.20$ for flush production therefore represents personal discount rates on the future and likely is a lower bound on perceived opportunity costs in a dynamic oil play: It includes putative income lost in forgone opportunities faster payout (flush production) could have financed. Figure 3.9. Instantaneous PV of Exponentially

## Table 3.2. Production History for Uncontrolled versus Controlled Fields[133]

| Field | Year of Discovery | Peak Annual Production as a Percent of Estimated Ultimate Recovery | 1950 Production as a Percent of Peak Annual Production | Years after Discovery for Production to Decline to 10 Percent of Peak Annual Production |
|---|---|---|---|---|
| | | Uncontrolled Fields | | |
| Seminole, OK | 1926 | 26.6 | 2.6 | 4 |
| Powell, TX | 1923 | 26.4 | 1.1 | 6 |
| Hendricks, TX | 1926 | 24.5 | 2.1 | 9 |
| | | Controlled Fields | | |
| Conroe, TX | 1931 | 3.9 | 51.2 | |
| Hastings, TX | 1934 | 4.4 | 60.2 | |
| O'Connor, TX | 1934 | 4.1 | 61.7 | |
| Magnolia, AR | 1938 | 6.7 | 63.0 | |

Declining Oil Production and Figure 3.10. Cumulative PV of Exponentially Declining Oil Production graphically depict the oil fraternity's conundrum. Table 3.2. Production History for Uncontrolled versus Controlled Fields presents historical data on select, real uncontrolled fields subject to unrestricted flush production compared to controlled or prorated fields.

# CHAPTER 4

# EVOLUTION'S EXPERIMENT: BORN OF MONOPOLY AND SIRED BY ARBITRARY POWER

In my opinion proration of oil was born of monopoly, sired by arbitrary power, and its progeny (such as these orders) is the deformed child whose playmates are graft, theft, bribery, and corruption. It is evolution's experiment.

—C. J. RILEY, JUSTICE OF THE OKLAHOMA SUPREME COURT,
DISSENTING OPINION, *H. F. WILCOX OIL & GAS CO. V. STATE*,
FEBRUARY 1933[1]

**F**rom before World War I into the mid-1920s, bureaucrats in the US Department of the Interior, the US Geological Survey, and the US Bureau of Mines (BOM), together with their academic fellow travelers, emitted increasingly shrill alarms that the United States was (or soon would be) "running out of oil." Echoing concerns expressed by their Progressive-era forebears—Gifford Pinchot and his minions in the Pennsylvania Geological Survey, for example—they touted "conservation," under their tutelage of course, as a programmatic elixir.[2] Despite these professionally self-interested jeremiads, the oil fraternity in Texas and their cousins in California, Oklahoma, and Louisiana obstinately persisted in discovering and producing oil in ever greater quantities.

Petroleum engineering science, if it could be called that, had little role in early "conservationist" programmatic proposals or in their advocacy. But both the ideological tinge of Progressive conservationist discourse and the inherent contradictions among "conservation" as nature preservation, as not-use-at-all-and-save-for-future-generations parsimony, or as "wise use" persisted in both petroleum engineering science and in oil policy disputes right through the twentieth century and into the next.[3] "Wise use" especially came in multiple varieties: "socially desirable" use, economically efficient use, microeconomic optimization, or simply "not waste." As noted, petroleum engineering did not become an established paradigmatic discipline until the mid-1920s, and even then its development was intertwined with political quarrels over "unitization," as well as the separate struggles of the oil fraternity in Texas, in cooperation with the Railroad Commission of Texas, to mitigate the supply and price chaos produced by recurrent oil booms.

## UNITIZATION AND THE EMERGENCE OF PETROLEUM ENGINEERING SCIENCE

Despite conservationist rhetoric about "running out of oil," which reached the first of several crescendos in 1924–1925, the real problem wasn't supply, but supply and therefore price, instability. Both supply and demand for oil are short-run inelastic. As flush production from newly discovered fields flooded the market, prices collapsed; as fields were depleted, supplies tightened and prices rose: shortage and high prices alternated unpredictably with glut. Among those most pained by this volatility, apparently to his surprise and dismay, was Henry L. Doherty. Doherty left school at age twelve to work for the Columbus (Ohio) Gas Company, becoming its chief engineer at twenty. By his mid-twenties he had begun to acquire control of public-utility firms, primarily natural gas.[4] Like Thomas Edison and Samuel Insull, he thrived in the golden age of utility systems building, when the technology was young and ambitious men with talent but little formal education could rise quickly.[5] Like his compadres, he was a technological enthusiast and ideologue, styling himself "Chief Engineer" rather than "President" of his eponymous holding company, thus emphasizing both his self-taught technical expertise and his technocratic proclivities.[6] He formed the Cities Service Company as an umbrella utility holding company in 1910. As an extension of its natural gas holdings, in 1912 Cities Service opportunistically bought out the oil interests, mostly in Oklahoma and Kansas, of Titusville (Pennsylvania) native Theodore N. Barnsdall. These interests were merged into a wholly-owned subsidiary, Empire Gas and Fuel Company, headquartered in Bartlesville, Oklahoma. True to his technocratic bent, Doherty promptly created the Doherty Training School, Empire Gas & Fuel, in the summer of 1916 to "train graduates from the various technical and engineering schools . . . in the oil and gas business."[7]

Initially, Doherty profited handsomely from his foray into oil, and like most oilmen in high cotton, had little use for government meddling: he told the second annual meeting of the American Petroleum Institute in 1920 that oil production should remain "a non-restricted competitive business" free of government interference, and maintained that there was plenty of oil yet to be found.[8] But as tight supplies turned to glut on the West Coast as the result of new discoveries at Signal Hill and Huntington Beach, Cities Service experienced a 50 percent decline in profits from 1920 to 1923, with one analyst estimating that earnings fell in 1921 to just one-third of those in 1920.[9]

Doherty changed his tune. He embraced what would become the dissolved-gas-drive hypothesis, which had been intermittently conjectured since the late nineteenth century but had not been pursued or developed.[10] Doherty maintained that natural gas dissolved in oil provided the means to move oil through the subsurface formation and to lift it to the surface, and that such dissolved gas reduced oil viscosity and made it easier to produce. Flush production from too many wells under the rule of capture prematurely exhausted this mechanism and resulted in rapid field depletion and underground waste in the form of oil unrecoverable without pumping. Supply and price instability naturally followed. In October 1923, Doherty offered his programmatic solution to the executive committee of the American Petroleum Institute (API), the industry trade association formed by the major integrated oil companies in 1919: Doherty wanted the major oil firms to work cooperatively to secure individual state's sanction and support for "unit operation" or "unitization of oil pools."

This notion was not new, although the strident tone of Doherty's advocacy was.[11] Doherty proposed that even before drilling commenced in a field, landowners should form a state-authorized oil district (modeled on water districts in California), which would "scientifically plan" drilling (thus minimizing the number of wells) and then allocate royalty income "on the basis of probable amount of oil under . . . [each tract] . . . before the reservoir was disturbed."[12] What Doherty proposed was both of doubtful legality and went beyond what engineering practice could do at the time. API's reception was cool but cordial: the idea was shuffled off to a study committee. Frustrated, Doherty first appealed to the Director of the Bureau of Mines (October 1923), then, still getting no positive result, to President Coolidge (August 1924), now suggesting direct federal government regulation. This agitation led to creation of the Federal Oil Conservation Board (FOCB), which would serve, usefully, as a government-industry forum for oil and gas conservation issues.

Doherty's comfort with stringent government regulation likely derived from his public utility background. Doherty was no oilman; he was a systems builder of a sort characterized by Thomas P. Hughes as "manager-entrepreneurs." He shared Samuel Insull's business philosophy (or perhaps ideology) of "accommodation, inclusion, amortization," and was not dispositionally averse to government regulation.[13] The highly leveraged holding company structure (Cities Service) Doherty built, which comprised both electric and natural gas utilities, depended, like Insull's Consolidated Edison, upon economies of scale, high throughput volume and velocity, load factor and diversity, and upon stable input supply and prices, as well as upon stable end-user prices.[14] Similarly, once he got into oil, Doherty, like Insull, promoted new consumer niches to enhance market size, load factor and diversity, stability—and profits: At the same time he was advocating new laws to conserve oil, he was urging exploitation of the emerging market for home heating oil "because householders could afford to pay more" than other customers.[15] Little wonder he found the supply and price instability in the oil business disagreeable.

This background explains much of what has been historically problematic about the reception of Doherty's ideas. Statist-oriented conservationists, bureaucrats, would-be bureaucrats, and their academic affiliates gleefully welcomed unitization: now the conservationist programme, which had had no policy, had one, and government activists had a major if not central role in it.[16] Likewise, the American Institute of Mining Engineers (AIME), sensing professional opportunity, found much to like. Reception within the oil fraternity, in contrast, varied from guarded to frosty. All were leery of overt government regulation, especially of exploration and drilling. Positive reaction to the idea of unitized production, whether voluntary or under some sort of state auspices, depended largely, but not invariably, upon the degree to which a firm's organizational capabilities, routines, history, and culture embraced integrated operations—exploration, production, transportation, refining, distribution, and marketing. The profitability of these firms depended upon managerial coordination of high-velocity throughput, not wholly unlike large public-utility systems.[17] Thus, Standard Oil of New Jersey, Humble Oil and Refining, and Socony Vacuum were eager to experiment with unitization, as were some other majors and a few larger independents. Amos Beaty, at the Texas Company, feared that Doherty's scheme would turn the oil industry into a public utility.[18] Jersey Standard's old nemesis, Sun Oil, demurred, as did Gulf Oil—its Texas operations were managed by two men who had "made Spindletop" and were constitutionally opposed to government meddling, F. C. Proctor, its chief counsel, and Underwood Nazro, president of Gulf Pipeline. Independents, with interests only in exploration and production, and whose experiences with "rational management," "orderly marketing," "efficiency," and "conservation" mostly consisted of the depredations of the Standard Oil Trust, wanted none of it.

## PETROLEUM ENGINEERING SCIENCE

As might be expected in these circumstances, the result was controversy, much heat, not much light, and a lot of dissembling and wasted energy. For all that, Doherty's immodest proposal did have sanguine and lasting consequences. Doherty had legitimated his policy panacea with a specific scientific conjecture: He claimed that reduced reservoir pressure during oil production caused natural gas dissolved in oil to come out of solution, which thereby increased oil viscosity. Greater viscosity reduced the ease with which oil could be moved by gas energy through the subsurface reservoir; too high a production rate therefore would unduly increase oil viscosity, diminish reservoir energy and pressure, and radically reduce ultimate recovery. Flush production inherently produced unconscionable waste.

Doherty's espousal of a scientific ideology was more than window-dressing. He had established the industry's first subsurface branch in the geology department of Empire Gas and Fuel in 1917, to better understand the "constitution and structure" of oil-bearing geological formations.[19] He had put up half the $50,000 the city of Bartlesville, Oklahoma, offered to induce the Bureau of Mines to locate its Petroleum Experiment Station there, and engineers from the Experiment Station, Doherty's Empire Gas and Fuel, Phillips Petroleum, also headquartered in Bartlesville, and Pure Oil, located in nearby Tulsa, came to form a tightly knit, first-person scientific and engineering community.[20] The "informal know-how trading," or collective development, in Bartlesville and Tulsa, and within the larger oil fraternity, typifies how engineering knowledge evolves.[21]

Nevertheless, petroleum engineering remained largely a "fact gathering" endeavor. Observational and anecdotal accounts of field experience, albeit often backed by quantitative data, dominated the literature. Meanwhile, geologists espoused a plethora of hypotheses about the origin and nature of petroleum deposits, with little consensus. Conjectures (speculations really) about the importance of natural gas in oil production had bubbled up intermittently from early on—the first published usage of the carbonation, or soda water, metaphor, which became de rigueur in succeeding discussions, appeared in 1865. But neither the effect of natural gas in solution nor the exact mechanism of its action were quantitatively characterized. Most concern continued to focus on the surface waste of gas after production (by venting or flaring), and on subsurface losses of oil due to poor casing and cementing and intrusion of ground water.[22] There was virtually no theoretical development regarding the behavior of petroleum reservoirs entire.

Despite the political brouhaha he kicked off with his advocacy of compulsory unitization and federal regulation, Doherty's ideological commitment to "scientific" technology bore real and tangible fruit: petroleum engineering science's first true paradigm. C. E. Beecher and Ivan P. Parkhurst, "Production Engineers" at Doherty's Empire Gas and Fuel, transformed the discipline and the profession with their seminal article, "Effect of Dissolved Gas upon the Viscosity and Surface Tension of Crude Oil," which was based upon research apparently begun in 1924 but not published until 1926.[23] Beecher and Parkhurst frankly acknowledged that when Doherty first advanced his unitization proposals with their dissolved gas rationale, he had no evidential warrant: "Prior to 1924 no information to support this contention [the effect of gas dissolved in oil] could be found in the literature or by a discussion of the subject with technical men in Government and private laboratories. So far as could be ascertained no experimental work had been carried on."[24]

Beecher and Parkhurst conducted laboratory experiments with two grades of Oklahoma crude oil, API gravities 30.2° and 35.4°, and one lighter Bradford (Pennsylvania) crude, API gravity 44.3° ("gravity" is an indicator of the hydrocarbon constituents in crude oil, with a higher gravity containing a greater proportion of lighter, more valuable fractions).[25] Beecher and Parkhurst tested the solubility of two different gas mixtures, one 52 percent methane ($CH_4$) and 39 percent heavier ethane ($C_2H_6$), the other 82.5 percent methane. Ceteris paribus, the lighter gas was less soluble than the heavier mixture, while a larger amount of either gas would dissolve in a given quantity of higher gravity oil. For a given gas, following Henry's law, the volume of gas dissolved in a given quantity of oil was proportional to pressure. The initial absolute viscosity of the lower gravity oil was "approximately 3 1/2 times greater" than that for the higher gravity oil, but even though more gas would dissolve in the higher gravity oil, the viscosity-reducing effect of dissolved gas was greater in the lower gravity (heavier) oil.

For the specific mixtures of oils and gases they investigated, Beecher and Parkhurst experimentally demonstrated that at a pressure of 500 pounds per square inch and 70 degrees Fahrenheit, the viscosity of a fully saturated oil sample would decrease about 50 percent, while the surface tension of the crude oil would be reduced by about 20 percent. Beecher and Parkhurst speculated that at much higher pressures, the amount of dissolved gas would be even greater, and resultant fluid viscosities even lower (though not proportionately), and noted that, "Pressures to 1800 lb. may exist at depths around 4100 ft. and at present a large amount of oil is being produced from much greater depths. Under such pressure sufficient gas might dissolve to make the [oil] viscosity almost equal to that of kerosene."[26]

The total volume of a fluid or gas contained in a subsurface formation is a function of the formation's porosity, or pore space (think of a sponge). The ease with which oil can move through the formation (and up and out production casing, given a pressure differential) is a function of the rock's (for example, sandstone's) porosity and permeability, that is pore size, how well the pores are connected, the size of "passages" between pores, and the "texture of the mineral-grain surfaces,"[27] as well as the oil's physical characteristics: viscosity and surface tension. Reduced viscosity increases the ease with which oil flows. Reduced surface tension lessens the oil's tendency to "cling" to the walls of the pore spaces and remain in place because of capillary action, rendering it unrecoverable, especially in "tight" formations (those with relatively small and poorly connected pore spaces).

Beecher and Parkhurst inferred from their experimental results and analyses that reduced viscosity and surface tension due to dissolved gas, as well as gas pressure, accounted for rapid, high-volume oil production early in a well's life (sometimes beginning with a gusher), that the reduction of gas in solution because of the declining reservoir pressure that resulted from unrestrained flush production accounted for observed rapid rates of production decline in oil fields, and that the resultant increased oil surface tension necessarily would result in oil left in the formation due to capillary action—underground waste. They concluded that keeping gas in solution in a producing oil reservoir for as long as possible could substantially increase long-run oil recovery and its efficiency. This argument differed from earlier concepts of the role of the expulsive force of expanding natural gas in oil production. Exhaustion of gas in either case would end the free flow of oil to the well bore. But under Beecher and Parkhurst's dissolved gas drive, not only did the expansion of dissolved gas provide the energy to move the oil through the formation, but also gas still in solution reduced the viscosity of the oil, and thereby limited water coning or gas channeling: Since both water and gas were less viscous than oil, they moved through a porous formation more easily, and would tend to bypass and strand oil, leaving it unrecoverable. Limiting the rate at which production reduced reservoir pressure would keep gas in solution longer, maintain the lower viscosity and surface tension of oil longer, and thus increase total oil recovery.

Beecher and Parkhurst's experimental results were replicated in their essentials by D. B. Dow and L. P. Calkin at the Bureau of Mines Field Office in Laramie, Wyoming (established in 1924). In their laboratory experiments, done jointly with the University of Wyoming and published also in 1926, Dow and Calkin used nearly pure methane (98.30 percent) dissolved in "two refined oils and three characteristic crudes," different from those used by Beecher and Parkhurst.[28] Dow and Calkin used pressures only up to 250 pounds per square inch and did not extrapolate further, but did reproduce Beecher and Parkhurst's findings on solubility and viscosity. Indeed, they quoted Beecher and Parkhurst's results. They did not directly investigate surface tension. And although Dow and Calkin did note that, "When the gas in the sand is exhausted, oil production ceases even though there may be enormous quantities of oil remaining unrecovered," their citation was to Theodore E. Swigart and C. R. Bopp's 1924 BOM Technical Publication, "Experiments in the Use of Backpressure on Oil Wells," which focused on maximizing utilization of the expansive force of natural gas associated with oil.[29] Dow and Calkin did not explicitly link the reduced viscosity caused by dissolved gas to greater ultimate recovery, nor did they mention unitization or unit operations at all.

The histories of Beecher and Parkhurst's and Dow and Calkin's experiments, and of Doherty's interest in dissolved gas for that matter, were intertwined. Beecher had been "one of the first three 'oil-recovery engineers' hired by the Bureau [of Mines] (in 1919), to undertake laboratory experiments on the relations between reservoir gas and oil."[30] He held BS and Civil Engineering degrees from Stanford University (1913, 1914), and had joined the Bureau of Mines after three years working in construction and two as a pilot in the United States Army Air Service. Beecher worked for the Bureau of Mines, Bartlesville Petroleum Experiment Station (created in 1917) for five years before joining Empire Gas and Fuel in 1924.[31] Dow, a chemical engineer at Bartlesville, had begun research on gas solubility in 1920, which was continued, "with the assistance of C. E. Reistle Jr., assistant petroleum chemist of the bureau." Reistle had earned a BS in Chemical Engineering from the University of Oklahoma in 1922 and immediately joined the Bureau at Bartlesville; later, in 1933, he became chairman of the East Texas Engineering Association during its most critical years, then joined Humble Oil in 1936.[32] Dow and Reistle published their preliminary gas solubility results in 1924.[33] Dow was transferred, as Engineer in Charge, to the new Laramie Field Office of the Bureau of Mines, about 150 miles south of Teapot Dome, when it opened in July of 1924. There, "work was continued" by Calkin, Assistant Petroleum Chemist. Beecher and Parkhurst noted that, "Since the experiments referred to in this paper were completed, Dow and Calkin have conducted a series of experiments to determine the solution of a practically pure methane gas and air in various crude and refined oil. The results of their work sustain those given in this paper."[34] Likewise, Dow and Calkin acknowledged that "C. E. Beecher and I. P. Parkhurst provided the writer with some of the data mentioned hereafter, and also checked the methods and apparatus used."[35] Dow and Calkin also incorporated some of Beecher and Parkhurst's data on solubility, provided by Beecher in a "personal communication."[36]

This dense web of personal and professional relationships supports several inferences. First, the Bureau of Mines Petroleum Experiment Station at Bartlesville served as a "graduate school" for petroleum engineers in the same sense that Spindletop had been the "university" for the oil fraternity writ large. Second, both Beecher's and Dow's research suffered from the "Normalcy" of budgetary stringency and the political brouhaha over Teapot Dome, as well as the bureaucratic propensity for routine to push out non-routine.[37] In addition to lack of research resources, the bureaucratic chaos engendered by Teapot Dome likely motivated the diaspora of engineering talent—including C. E. Beecher and T. E. Swigart—out of the Bureau of Mines to private employers. Doherty, with the headquarters of his Empire Gas and Fuel in Bartlesville, would have been immersed in this small, intense nascent community of petroleum engineering scientists, and drew from it his ideas on dissolved gas and viscosity, although his grander vision of unit operations, unitization, and regulation derived from his manager-entrepreneur *mentalité*.

Beecher and Parkhurst, because their experiments were completed if not published earlier, extended to higher pressure, and included surface tension, probably deserve the priority they were accorded.[38] In their inferences from their work, they asserted a great deal more than Dow and Calkin: they overtly claimed that the effects of dissolved gas on oil viscosity warranted pressure maintenance and restrained rates of oil production, and recommended unit operation of oil fields. They published in the *Transactions of the American Institute of Mining Engineers*, where their results would be more likely to be widely disseminated among technically adept oilmen. And they explicitly linked their research to Henry L. Doherty.[39] Anything associated with his proposals would likely attract attention, especially work purporting to offer a sound scientific rationale for them. Those who found unitization congenial (if certainly not federal control), including Carl Reistle, likely also influenced ex post adjudication of scientific priority. Given how much controversy, if not downright antagonism, Doherty's notions about compulsory unitization and government control had aroused in the oil fraternity, it's not surprising that petroleum engineers and scientists persisted in according priority to Beecher and Parkhurst—incontrovertibly two of their own—as would be expected of an emerging community of professional practitioners trying to produce itself, define its boundaries, and assert its social identity and authority.

Despite their originality, quality, and positive reception, Beecher and Parkhurst's experiments had limitations. Lester C. Uren, Professor of Petroleum Engineering at the University of California, Berkeley, the following October (1927), cautioned that while he supported "proper recognition" of gas-oil ratios "as a guide to producing oil wells": "It seems necessary, however, in view of the almost elemental faith with which the principle of the gas factor [gas-oil ratio] has been received in many quarters, to point out that it is a most complex factor, and one which cannot be regarded as an absolute index of production."[40] Uren explained that gas-oil ratio was only a proximate indicator of efficiency in the use of reservoir energy: a well with a higher gas-oil ratio but lower available reservoir pressure might be more energy efficient than a well with a lower gas-oil ratio but higher pressure.[41]

Strictly speaking, Beecher and Parkhurst's results applied only to gas dissolved in oil in a reservoir, and to the specific oil types and gas mixtures they investigated. Any free gas in a gas cap above the oil-saturated sand in a common reservoir, which also served to pressurize the reservoir, would differ "in composition from the dissolved gas," while "the dissolved gas may have a wide range in composition, meaning that different parts of it will leave solution at different pressure levels."[42] Moreover, little empirical work was done on the fundamental characterization of the phase behavior of mixtures of different hydrocarbon gases for "some 10 years" and the experimental program of Bruce H. Sage and William N. Lacey at the California Institute of Technology (supported by the API), while data on the surface tension of saturated oil remained "rather meager" even in 1949.[43] Secure theoretical understanding of the movement of homogeneous fluids in a hydrocarbon reservoir, and of the phase behavior of heterogeneous fluids, did not appear until Morris Muskat's and others' work in the 1930s, and did not become established, "packed down" science routinely applied by petroleum engineering practitioners until after World War II.[44]

What distinguished Beecher and Parkhurst's work at the time, and warrants its paradigmatic status, is that it not only offered a theoretical explanation for how oil behaved in subsurface reservoirs but also provided a robust, replicable, quantitative laboratory demonstration of those effects. Their experimental approach defined what would become a coherent, fruitful, and vibrant research programme, the hallmark of a successful scientific paradigm or tradition. Moreover, their insights engendered widespread empirical and practical investigations of production practices, extensive field experiments under operating conditions, and considerable enthusiasm for practical experiments with unit operations.

Beecher and Parkhurst's investigations coincided with the emergence of a differentiated petroleum engineering profession. In 1927, J. B. Umpleby observed that before the establishment of the American Institute of Mining and Metallurgical Engineers' Petroleum Division, "no forum existed for the open discussion" of the technical problem of optimizing oil recovery. Only one handbook and one textbook on "petroleum production engineering" had appeared, and, prior to 1923, virtually all engineers who worked for oil companies were engaged in "construction and location work," not analysis and management of oil production itself. And, "as late as 1926," only six schools or universities "offered special courses in petroleum production engineering."[45] Beecher and Parkhurst and the research they initiated provided the disciplinary core for the new profession.

After Doherty's insistent agitation had led to creation of the Federal Oil Conservation Board (FOCB) by presidential executive order in 1924, its first major consequence was to spook the American Petroleum Institute into initiating a "conservation" survey of oil producers. The results of that survey, released in May 1926, essentially said, "Waste? What waste? There ain't no waste here." The API survey was widely dismissed at the time, and by historians since, as a political document designed to stave off federal meddling in the oil business.[46]

Of more lasting importance, although little noted outside the oil fraternity, was a parallel joint effort, inspired by Beecher and Parkhurst's findings, between the Petroleum Division of the Bureau of Mines and technical subcommittees of the API to comprehensively understand production practices, especially the role and management of dissolved gas. API established a Gas Conservation Committee in 1927, which in turn appointed a technical subcommittee, which then appointed regional technical subcommittees who reported preliminary findings at a meeting of some 250 "oil company executives and operators, officials of the Institute [API], attorneys, petroleum engineers, Government officials, and scientists," in Ponca City, Oklahoma, in October of 1927.[47] The API also quickly published a popular elementary primer on oil production.[48] Everette Lee DeGolyer concluded in 1941, "With these actions the industry may be said to have become fully advised and to have accepted our present concept of the importance of gas in solution."[49]

The data collected by the API Gas Conservation Committee's regional subcommittees "were turned over to the newly created Division of Development and Production Engineering" of the API during 1928; an agreement was worked out between the API and the Bureau of Mines under which the BOM would organize and augment the data and issue the result as a "cooperative report" of the BOM and the API. Harold C. Miller, BOM Senior Petroleum Engineer, expanded and synthesized the assembled data, and the API published the report in 1929, under the unwieldy title, *Function of Natural Gas in the Production of Oil: A report of the U.S. Bureau of Mines, Department of Commerce, in cooperation with the Division of*

*Development and Production Engineering of the American Petroleum Institute, based on data
gathered and reported by the Kansas and Oklahoma, Pacific Coast, Rocky mountain, and
Texas and Louisiana regional committees of the Gas Conservation Committee of the American
Petroleum Institute.*[50] Everette Lee DeGolyer observed in 1961 that with Miller's report, "oilmen had come practically to the present understanding of the nature of the oil-gas solution
in the reservoir."[51] Miller added little in the way of fundamental insight or laboratory data
to what Beecher and Parkhurst and Dow and Calkin had done. What his report did do was
synthesize and interpret a great deal of empirical data and production experience that oil operators and nascent petroleum engineers had accumulated in terms of Beecher and Parkhurst's
findings and analysis.[52]

The reified Beecher and Parkhurst became the nucleus around which disparate if practically
(and personally) connected domains of inquiry could condense into a differentiated and coherent discipline and profession. Beecher and Parkhurst's insights and their derivatives provided a
systematic basis and rationale for engineering practice: for the design and operation of rational
systems of oil production, which was rather its purpose.[53] Public affirmation of Beecher and
Parkhurst's paradigm mutually authenticated petroleum engineers, their discipline, and their
autonomy and social authority: now they could not only do and describe, but also prescribe.[54]
Petroleum engineering scientists had unequivocally produced themselves and their discipline
as the arbiters of oil "conservation" and of rational oil and gas production.

As would be expected from the nature of a paradigm, the Beecher and Parkhurst paradigm
broadly conceived spawned a vibrant research tradition. As J. B. Umpleby recorded in his
survey of the field in 1932, papers in this new tradition quickly came to predominate in the
annual volumes of the Petroleum Division of AIME (Table 4.1).[55]

**Table 4.1. Papers in the Annual Volumes of the Petroleum Division of AIME**

| Year | Total Engineering Papers | Gas and Related Papers |
|------|--------------------------|------------------------|
| 1925 | 22 | 1 |
| 1926 | 25 | 5 |
| 1927 | 30 | 5 |
| 1928 | 20 | 5 |
| 1929 | 43 | 23 |
| 1930 | 35 | 22 |

As had petroleum geologists before them, petroleum engineers developed a nuanced moral
economy that governed their cooperation and competition: petroleum engineers were norma-
tively committed to sharing with their peers any information that would advance the field so
long as it did not surrender immediate proprietary advantage. Miller in his report, for example,
acknowledged "The many operators and oil companies who contributed data, some of which
are of a confidential nature."[56] Likewise, production of petroleum engineering knowledge
depended upon emergent small-world social networks. Engineers who had worked at one
time or another for the Bureau of Mines petroleum division in the 1910s and 1920s became
a closely knit elite among the first generation of petroleum engineers. Oil firms' "technical

men," the petroleum engineers, appear to have worked with their peers in the Bureau of Mines in considerably greater harmony than prevailed among government bureaucrats, politicians, and oil company executives.

Although Miller acknowledged the possible existence of water drive, he nevertheless concluded that, "There is no doubt in the minds of oil men and engineers that the predominant expulsive force that drives oil to wells is the energy stored in the compressed gas absorbed in or associated with the oil."[57] As is so often the case, this paradigm (or progressive research programme, if you wish) became conventional wisdom became orthodoxy, which served to mask, if not outright suppress, heterodox hypotheses or theories, especially water drive.[58] As J. B. Umpleby concluded in 1932, "The function of gas in the development and production of oil is the most fundamental concept in the industry."[59] Quickly developed what Donald MacKenzie calls the "trough of certainty" in which a large number of knowledgeable oilmen happily wallowed, secure in their belief in dissolved-gas drive and the paramount importance of gas-oil ratios.[60] Those few close to the research front harbored sophisticated reservations, while the much more numerous hoi polloi—the smaller independents with as yet little direct contact with petroleum engineers, and those outside the oil fraternity—knew little, cared less, and sensed in all such talk of "conserving" reservoir pressure or moderating rates of production the stench of price fixing and monopolistic predation.

In view of the pride of place accorded Miller's report among petroleum engineering practitioners, equally significant is what it did *not* do. Despite being animated by Beecher and Parkhurst's research, nowhere in Miller's text (nor in its index) do the words "unitization" or "unit operations" appear. Rather, allusion is made to "coöperative development" or "a coöperative problem," or "an ideal coöperative development plan."[61] Given how politically charged "unitization" had become, and the wide diversity of opinion about it within the oil fraternity and even among the majors, one suspects a delicate modus vivendi had been worked out between the API and the BOM. Although there was considerable talk of "cooperative development," and shutting-in pure gas wells or abandoning and plugging wells with high gas-oil ratios to conserve reservoir pressure in an oil pool, nothing in the volume addressed the truly nasty problem of allocating production, or revenue from production, among multiple interests in a field. Like "unitization" had been, "cooperation" was invoked as if it were unproblematic. More constructively, the volume made no pretense of being the final word: "Little is known of the changes which take place after drilling into reservoir formations because the complexity of the components and phases make it difficult to translate surface data back to underground conditions."[62]

When Beecher and Parkhurst's paper first appeared, its scientific approach and its implications for unit operations were greeted with considerable enthusiasm among those already frantically looking for a way to stabilize the supply and price of crude oil. It's likely that much of the "scientific" approbation it received was not wholly untainted by material interests. Beecher and Parkhurst provided a scientific rationale for Doherty's conservationist unitization proposals—their avowed intent. Among the most positive comments published with the original paper in 1926 were those of John R. Suman and Theodore E. Swigart. Suman was an engineering graduate of the University of California (1912) and had worked for the Rio Bravo Oil Company (the Southern Pacific Railroad subsidiary which managed the oil interests deriving from its considerable land holdings) in California and Texas, briefly for Roxana (Shell), and had just joined Humble Oil and Refining as its director of production. As noted, Swigart earned his engineering degree from Stanford in 1918, and had worked for the Bureau of Mines for five years before joining the Shell Company of California as a production engineer.[63] Beecher and Parkhurst's findings catalyzed major oil companies to initiate voluntary, cooperative unitization projects in some of the oil fields where they operated.

But to their professional credit, and evincing their growing professional autonomy, petroleum engineers resisted pressure from their employers, from the API, and from AIME to endorse extreme policy prescriptions. In a session organized by Earl Oliver, chairman of the Petroleum Division of AIME, at their meeting in Houston in October 1931, a panel of petroleum engineering scientists and practitioners were asked to comment on the proposition that landowners be restricted by law to recovering only the oil actually in place under their surface estates and on whether petroleum engineers had the ability "to make the necessary determinations." The engineers, true to what they could actually do, and to their scientific integrity and commitments, at best equivocated.[64] Likewise, engineers, as well as lawyers, equivocated again the next year when asked whether it was practicable to legally create individual property interests in reservoir energy. For example, Ben E. Lindsly had earned an engineering degree from MIT in 1905 and had worked briefly in California for the Honolulu Oil Company in their newly formed geology department, although he was performing tasks that would later be recognized as petroleum engineering.[65] By 1931, he was among the Bureau of Mines's most eminent petroleum engineers. He wryly noted that creation of a property right in reservoir energy "might not entirely satisfy or be equitable to the tract owner who has an immense amount of energy but little oil and only a meager market for his gas."[66] Both the oil-in-place proposal and the recognition of a private ownership right in reservoir energy were recognized as backdoors to compulsory unitization, and in both instances petroleum engineers for the most part discreetly demurred.

## PRORATIONING AND UNITIZATION: VOLUNTARY COOPERATION

Meanwhile, the Railroad Commission of Texas had been muddling toward field unitization in the form of prorationing under its statutory authority to prevent physical waste and to allocate access to pipeline transportation (see Table 4.2 for relevant laws). Consistent with its established tradition of regulation through cooperation, and with the cooperative ethos of the oil fraternity in Texas, the Railroad Commission typically intruded only upon petition of operators in a field or in response to a critical lack of transportation infrastructure.[67] Since cooperation was voluntary, little or no coercion or enforcement (other than restricted access to pipelines) was used. Under these circumstances, the Commission was willing to play dicey with the antitrust laws, reasonably confident of not being challenged or prosecuted. In these endeavors, the Commission necessarily had to consider the entire oil pool as a single unit, but it initially took little interest in subsurface waste and none in pressure maintenance or, overtly, in the price of oil: cooperative measures were legitimated as preventing surface waste, consistent with the Railroad Commission's constitutional and statutory mandates. Other than spacing regulations, exceptions to which were easily obtained and abundant, no restrictions were placed upon drilling or exploration.

The Railroad Commission's first foray into field prorationing was in the Burkburnett field in North Texas between Wichita Falls and the Red River, which had been discovered in 1918.[68] By the summer of 1919, northern Wichita County was awash in oil with no way to get it out. Upon petition of operators in the field, the Railroad Commission first tried to shut in the field (for five days) pending completion of additional pipelines, only to have that order overruled by the Attorney General of Texas, C. M. Cureton, in response to the folk belief that shutting in wells completely would kill them. Instead, Cureton's substitute order required existing pipelines to take oil proportionately from each well: prorationing. Actual implementation of the order depended upon voluntary organization by the operators: a field committee who in turn appointed "umpires" or arbitrators who allocated production among wells and determined the order in which wells would be connected to pipelines. Most, but not all, operators cooperated, and as the rate of new well completions stabilized, pipelines were completed, and the price of oil recovered, the crisis eased, and proration faded away by the late fall of 1919.[69]

The Yates field marked the first overt unitization of an oil pool in Texas, first by voluntary agreement among operators, then with Railroad Commission sanction. The Yates pool had been discovered in October 1926, following several other discoveries in the Permian Basin in West Texas, beginning with Big Lake (Santa Rita No. 1, on part of the University of Texas land grant) in 1923. Yates became the largest oil field in the United States before East Texas, 20,000 acres. One spectacular well, Yates No. 30-A (1929), initially alone could produce 200,000 barrels of oil per day.[70] But the problem, even worse than usual, was that Yates was in the middle of nowhere, with virtually no transportation infrastructure. Yet Yates differed from most other oil fields in one critical respect: most of it was covered by large-tract leases under large ranches (that of Ira Yates in particular) held by a relatively small number of major oil companies or larger independents, among them Gulf, Pure Oil, and Humble Oil & Refining.

William S. Farish, one of the Humble's founders who had become its president in 1922, had been a vociferous critic of Doherty and his unitization proposals when they were first offered. But by 1927, likely under the influence of John Suman and his petroleum engineers, Farish had become an equally impassioned advocate of unit operations. Hardly a prisoner of "consistency," that "hobgoblin of little minds,"[71] it was not Farish's first unabashed volte-face: recall his unkind words about integrated refiners in 1915, four years before he negotiated, with Ross Sterling, sale of half of Humble Oil to Jersey Standard.[72] Farish's newfound faith in unitization didn't sit well with all his peers. F. C. Proctor, Gulf's chief counsel, and a former law partner of Humble's chief counsel, E. E. Townes, described Humble's broader proration proposal as a "nostrum worthy of a blather skate politician."[73] Given the dubious antitrust niceties of unit operations, Proctor all but dared Farish to try it.[74]

Farish had done his homework. Or rather, teams of petroleum engineers under Suman and of lawyers, including Hines and Rex Baker, under E. E. Townes at Humble, together with outside counsel Robert E. Hardwicke, as well as counsel for some of the other majors, had done theirs. At an August 1927 meeting of Yates operators, Farish bemoaned the surface waste overproduction and storage caused and warned of the dangers of water intrusion bypassing the low-gravity oil in the reservoir, leaving it unrecoverable.[75] Humble had the only pipeline in that part of West Texas, and at a later meeting in Fort Worth, Farish offered to extend it to the Yates field and to guarantee purchase of 30,000 barrels of oil a day if the operators in the field would unitize their operations and restrict production. Wary, James A. Elkins advised his own client, Pure Oil, to stay out, as did Gulf, although both did *de facto* cooperate by limiting production. On September 8, the operators' committee agreed to a unitization plan, which allocated total agreed production to each lease based on its wells' potential, how much the wells on a lease could produce if flowed wide-open.[76] Operating costs and revenue were shared in proportion to production quotas. The plan was condoned by the Railroad Commission in October as being consistent with its authority to prevent physical waste.[77]

The Yates agreement's original ninety-day term was renewed and revised twice, with adjustments to the allocation formula. As Proctor pointed out, the original per-lease allocation or "allowable" favored densely drilled leases and encouraged drilling further, unnecessary wells, which promptly occurred.[78] On first renewal, in January 1928, the formula switched to an allowable based on the size of an operator's proven acreage within the established confines of the field. This iteration favored large leaseholders and promoted more drilling at the margins of the field to "prove up" new acreage and thus obtain a greater allowable. Operators agreed to a third iteration in June 1928, which divided the field into 100-acre units, with each unit allocated a proportionate share of pipeline capacity; within each unit, wells received an allowable based on their individual potentials. Upon petition of the operators, the Railroad Commission assumed supervision of the field on July 1, 1928, and issued a proration order setting allowables for each 100-acre unit, one-quarter based on acreage (100 acres divided by proven area of the field) and three-quarters on well potential.[79]

The Yates pool was widely touted by majors and the Railroad Commission at the time and by most historians since as a paradigmatic success. Even though smaller independents were eventually forced out of the field by restricted production and delated payouts, evidence does support claims that restrained production conserved reservoir energy and enhanced ultimate recovery.[80] Of more immediate and persuasive interest, proration in the Yates pool was claimed to have reduced average production costs to only four cents a barrel.[81] But despite becoming a poster child for unit operations, Yates proration never satisfactorily solved the thorny problem of production allocation within an oil field, so neatly elided by Doherty and his disciples. The problem would continue to vex the Railroad Commission for the next thirty years.[82]

The Railroad Commission's concurrent foray into prorationing in the Hendrick field, also in the Permian Basin about 90 miles northwest of Yates, didn't fare as well. Discovered also in 1926, the big difference in the Hendrick was the prevalence of many small-tract leases—five or ten acres—and an equally large number of small-scale, independent oil operators. By late 1927, in addition to the usual problem of production in excess of available pipeline capacity, wells began to produce inordinate quantities of salt water—so much that operators feared the whole field would soon be ruined. Major company engineers argued, without proof, that decline in reservoir pressure due to flush production had increased oil viscosity and promoted water coning, and that reduced rates of production would ameliorate the problem. As the problem worsened through early 1928, the operators appointed a committee of six to formulate a prorationing plan, with R. D. Parker, the chief supervisor of the Oil and Gas Division of the Railroad Commission and a degreed engineer (University of Texas, 1898) who also had worked for the US Geological Survey, as its chairman. The Railroad Commission issued a proration order at the end of April, which set an allowable of 150,000 barrels per day, one-fifth of the field's potential. In a manner similar to the Yates order, the Hendrick order divided the field into units of 40 acres, with half the fields allowable divided equally among units, and half allocated on the basis of the ratio each unit's potential production to total field potential. Within 40-acre units, all leases had to share the allowable, which, as in the Yates field, penalized very small leases. While Hendrick production was restrained, the prorationing orders were widely violated (which the Railroad Commission lacked authority to punish), and the reduction of water intrusion not as great as the engineers had projected.[83] Both advocates and opponents of prorationing drew from the experiment what they would.

Unitization of the Van pool, identified by a Pure Oil seismograph survey in 1927 and drilled in 1929, marked the first experiment in unit operations of a field in which the interests of disparate lessees were fully merged. E. E. Townes, Humble's chief counsel, had surmised that an agreement among several firms for exploration and, if oil were found, development, executed before drilling began, ought not violate the antitrust laws, but that ex post agreements to limit production from existing properties likely would. The larger interests in the Van pool were held by five major oil companies, Pure, Humble, the Texas Company, Sun Oil, and Shell, and one monumentally irascible farmer who owned a tract in the middle of the field, and for which he turned down a half-million-dollar lease bonus payment: he chose to drill his own wells, and everybody else be damned. Money didn't improve his disposition.[84] In the fall of 1929, the five majors "conveyed their lease holdings to one another and created a joint partnership giving each partner an undivided interest in the whole unit." Initially, credit for production would be allocated to each partner based upon the ratio of its lease-holdings to the whole field. After two years' production experience, petroleum engineers from the companies would evaluate the reservoir and establish the production potential of individual leases and wells, upon which a revised, more precise allocation formula would be based.[85] Pure Oil, as the largest interest holder, became unit operator. While the Railroad Commission was kept apprised of the agreement and its administration, it had no role in either approving the unitization agreement or in its management. The partners lauded unitization in the Van pool for its efficiency in both drilling and production, and for its preservation of reservoir pressure and its expected enhanced ultimate recovery.

## STATEWIDE PRORATIONING

The success of some prorationing and unitization experiments did little to resolve the worsening problem of overproduction, in Texas or out, which the deepening Depression aggravated. Following extensive hearings, the Railroad Commission of Texas issued its first statewide prorationing order August 14, 1930. The Commission summarized the need, the process, and the rationale in its annual report:

> Notwithstanding there was a considerable increase of oil production in Texas in 1929 over previous years, the year 1930 followed with the discovery of several new oil and gas fields in the State, and production greatly exceeded the existing market demand, and, upon petition of the oil fraternity, this Commission, after hearings, entered orders prorating the flush production of new fields and also entered a general proration order covering the State which provided for a maximum daily allowable production.
>
> The Commission feels that proration is necessary from a standpoint of conservation. It does not think operators should be allowed to produce oil and store it in excess of the market demand. Storing of oil results in waste by evaporation and deterioration, in addition to the waste and damage that is caused by the flowing of large wells without restraint, thereby losing gas pressure from the pay horizon which always inevitably results in water trouble and permanent injury to the pay.[86]

Easier said than done. Top down, the Railroad Commission had to estimate the "reasonable market demand" for Texas crude oil in the next period, usually 90 days. It had to fairly allocate production to meet that target demand among the many and varied oil fields in the state in such a way as to avoid "surface waste" (excess storage) and local overproduction. Field allowables had to be divvied up among individual wells or leases within fields in a manner that was both equitable and protected the correlative rights of all producers in the field.

"Correlative rights," as enunciated by the United States Supreme Court in *Ohio Oil Company v. Indiana* (1900), arise from a surface owner's absolute right to take possession of resources beneath his land and those rights' juxtaposition to the fugacious character of oil and gas and the common-pool nature of oil and gas reservoirs.[87] In such circumstances, the state (Indiana) could enact regulations to assure that no person used his property in a way that unreasonably appropriated or damaged (wasted) that of another, as in venting, or "popping," natural gas. The Texas Supreme Court offered a more succinct and practical definition in 1923: "If the owners of adjacent lands have the right to appropriate, without liability, the gas and oil underlying their neighbor's land, then their neighbor has the correlative right to appropriate, through like methods of drainage, the gas and oil under the tracts adjacent to his own." The Court also noted that in Texas the mineral realty is "often worth far more than anything else on or beneath the surface within the proprietor's boundaries."[88]

The Railroad Commission's solutions were voluntaristic and cooperative, true to the traditions of the agency since its inception.[89] The Commission appointed a Central Proration

Committee (or Central Advisory Committee), or rather accepted the nominees of major oil companies and larger independents. This committee provided "nominations" of what purchasers wanted to buy.[90] This estimated demand was divided by the sum of the production potentials of every well in the state. Each oil field was then assigned its pro rata share. The Commission then issued a proration order for the field specifying its allowed production and how it was to be allocated among wells and leases. Actual administration was delegated: In the Panhandle field, for example, as summarized by the court in the first *Danciger* case, the Commission appointed an "'umpire' . . . to have supervision of the enforcement. . . . The umpire and his assistants were paid by the producers, not by the state. Much of the information furnished to the commission and to the umpire . . . was furnished by the operators themselves. . . .[T]he commission relied to a large extent on the co-operation of the producers themselves for its information and the enforcement of its orders."[91] Even though the court in *Danciger* expressed some qualms about this privatization of quasi-judicial regulatory functions, the court also acknowledged, "No complaint is made that the agents designated in the order failed to discharge their duties, were corrupt or partial in doing so, nor that the orders themselves were not properly and impartially enforced."[92]

Appeal was built into the system: Commissioners themselves were easily accessible to the oil fraternity, and the Commission held regular, wide-open statewide and field hearings, at which anyone could be heard. In virtually all of its proceedings, formal and quasi-judicial or informal, the Commission maintained its tradition of access and openness: ". . . each interested person is given an opportunity to have his say: that is, testify under oath, to present witnesses and documents, to interrogate witnesses or speakers, and to make a statement or a rambling speech."[93] As an attorney put it in a talk in 1953, "Discussing the rules for practice and procedure before the Oil & Gas Division of the Railroad Commission is like delivering an address on snakes in Ireland. There aren't any."[94]

Beyond the Railroad Commission, the District Courts of Travis County (Austin) were designated by statute to hear all appeals of Railroad Commission orders. This arrangement's official rationale, as adopted later by the US Supreme Court, maintained that it would avoid "the confusion of multiple review of the same general issues," or, in the words of the Texas Court of Civil Appeals, quoted with approval by the US Supreme Court, "If an order of the commission . . . can be collaterally attacked in the various courts and counties of the state . . . interminable confusion would result."[95] The suspicion remains, however, that because State District Judges in Texas are elected from the Districts in which they reside (all state judgeships at any level in Texas are elective), perhaps not all could be counted on to be qualified or unbiased.

By the autumn of 1930, the oil fraternity in Texas and the Railroad Commission had willy-nilly accumulated a portfolio of trial-and-error experiments in production control and "scientific" management of individual petroleum reservoirs, ranging from voluntary unit operation and production restraint sanctioned and eventually supervised by the Railroad Commission (Yates) to Railroad Commission field prorationing at the behest and with the cooperation of most, but not all, operators in the field (Hendrick), to complete unitization independent of the Railroad Commission (Van). Efficacy varied; interpretations of the experiences varied even more. The Commission had issued its first statewide prorationing order and had been promptly hauled into court for exceeding its authority. The knowledge petroleum engineers deployed and publicly produced grew gradually but steadily with each new field, and each new dispute, and each new experiment, and each new trial. Despite considerable and often acrimonious controversy in professional society meetings and, especially, in Railroad Commission and Texas legislative hearings, the Railroad Commission's tradition of regulation through cooperation, not coercion, and its and the oil fraternity's mutual devotion to opportunity for Texans, persisted.[96]

Specific attitudes toward prorationing varied. Reactions were not systematically related to material interests. Few would benefit more from production restraint than F. C. Proctor, Underwood Nazro, J. Edgar Pew, or Hugh Roy Cullen, yet all were vehemently opposed to prorationing, although Pew did modify his stance later. Conversely, while neither H. L. Hunt, E. W. Marland, Michel T. Halbouty, nor Glenn H. McCarthy will ever enter the liberal pantheon, all were passionate advocates of conservation and state prorationing. Yet material circumstances did matter, not so much in attitudes toward prorationing but in the experiences of it. Prorationing or any form of production restraint prolonged payouts. Larger independents, and especially larger integrated firms, those with something other than the promise, or hope, of future oil production, to hock, had better access to capital markets, and lower capital costs, than smaller independent promoters or operators whose interests were confined to exploration and production. Often, smaller operators felt prorationing forced them out of prolific fields. Nevertheless, in the fall of 1930, none of these problems appeared to be something the oil fraternity and the Railroad Commission of Texas together couldn't muddle through.

Not that modern industrial society, "corporate capitalism," the "associative state," or slick "rational management" didn't still rankle old Populist sensibilities. As an Oklahoma Supreme Court Justice wrote in the expressive dissent with which this chapter began,

> In my opinion proration of oil was born of monopoly, sired by arbitrary power, and its progeny (such as these orders) is the deformed child whose playmates are graft, theft, bribery, and corruption. It is evolution's experiment.[97]

## Table 4.2. Oil and Gas Laws, Texas and US, 1876–1935

| Date | Law | Notable provisions |
| --- | --- | --- |
| 1876 | Constitution of the State of Texas | "Monopoly" prohibited (Article I Bill of Rights, Section 26).<br><br>Railroads: granted eminent domain; declared common carriers; discriminatory rates prohibited, state granted rate setting powers (Article X, Sections 1 and 2).<br><br>Subsurface mineral resources (including oil and natural gas) relinquished by state to surface owner in fee simple (Article XIV, Section 7). |
| 1889 | Texas Antitrust Act | Prohibited trusts, monopolies, and conspiracies in restraint of trade. |

| 1890 | Constitutional Amendment | Railroad commission authorized (Article XVI, Section 30). |
|------|--------------------------|----------------------------------------------------------|
| 1891 | Railroad Commission Act | Established Railroad Commission. |
| 1895 | Texas Antitrust Act | Criminalized antitrust violations with fines, possible prison sentences. |
| 1899 | Texas Antitrust Act | Prohibited pools, trusts, monopolies, and conspiracies to control business and prices; criminal penalties. |
| 1899 | Resource Conservation Act (Texas) | Water to be cased off in oil or gas wells; abandoned wells plugged; gas from gas wells not allowed to escape. No enforcement provisions. |
| 1913 | Federal Revenue Act (US) | Imposed first federal income tax; "intangible costs" of drilling a well (no salvage value, therefore not capitalized) expensed in tax year incurred; depletion allowance based upon estimated recoverable oil. |
| 1915 | Oil Conservation Act (Oklahoma) | Prohibited waste, including production in excess of reasonable market demand; state price fixing authority (never used). |
| 1917 | Constitutional Amendment (Texas) | "Conservation and development of all the natural resources" of Texas "public rights and duties"; legislature to enact laws and regulations to carry same into effect (Article XVI, Section 59a). |
| 1917 | Pipeline Severance Act (Texas) | Intrastate firms allowed to integrate two or more operations (e.g., transportation and refining), but not interstate firms. |
| 1917 | Act Regulating Pipelines (Texas) | Railroad Commission of Texas tasked with enforcement of oil and gas conservation regulations. Oil pipelines that transported others' oil declared common carriers. If production in an oil field exceeded capacity of pipelines, Railroad Commission to assure "ratable taking" from all producers without discrimination. Cost of regulation paid for by a production tax on all crude oil produced in Texas (initially one-twentieth of one percent, 0.0005). |
| 1919 | Oil and Gas Conservation Act (Texas) | Railroad Commission to promulgate rules for efficient drilling practices; pipeline companies required to obtain Commission certificate of compliance with conservation rules before connecting new wells; oil producers to keep and submit drilling logs (geological strata penetrated) and production records (neither public for six months); costs paid by production tax. Railroad Commission could only take legal action to shut in wells after waste had occurred. |
| 1920 | Cox Gas Act | Railroad Commission granted power to allocate gas from a common pool among purchasers (ratable taking). |
| 1926 | Federal Depletion Allowance (US) | 27.5 percent depletion allowance deductible from production income; estimates of percentage of recoverable reserves therefore unnecessary. |
| 1929 | Oil and Gas Conservation Act (Texas) | Restored Railroad Commission authority to impose punitive sanctions for violation of its orders (inadvertently omitted in general legal code revision, 1925); waste defined as "physical waste"; consideration of "economic waste" explicitly prohibited. |
| 1930 (March) | Common Purchaser Act (Texas) | Common carriers or "purchasers affiliated with a common carrier" to purchase oil "ratably" with no discrimination; extended to some natural gas in 1931. |

| 1931 (April 16) | Marginal Well Act (Texas) | Exempted pumping wells with less than a maximum daily production, as a function of depth, from limitation under prorationing orders (for example, a well with a depth of less than 4,000 feet was allowed up to 40 barrels per day). |
|---|---|---|
| 1933 (April 27) | Marginal Well Act (Texas) | Marginal or "stripper" well exemptions reduced (e.g., to 10 to 35 barrels per day, dependent upon depth); production in excess of allowable a felony offense. |
| 1931 (August 12) | Anti-Market Demand Act (Texas) | Definitions of physical waste substantially broadened, but waste "not to be construed to mean economic waste, and the [Railroad] Commission shall not have the power to attempt, by order or otherwise, directly or indirectly to limit the production of oil to equal the existing market demand for oil." Privately funded, voluntary field umpires and supervisors to be replaced by Railroad Commission employees. |
| 1932 (November 12) | Market Demand Act (Texas) | "Waste" defined to include "economic waste," that is, production "in excess of transportation or market facilities or reasonable market demand" as determined by the Railroad Commission. |
| 1933 (June 16) | National Industrial Recovery Act (US) | Subsection 9(c) provided a Code of Fair Competition for the oil industry; enforcement delegated to the Department of the Interior (all other codes under the authority of the National Recovery Administration); determination of state allowables left to state commissions. |
| 1935 | Connally Hot Oil Act (US) | Interstate transport of oil produced in violation of state conservation laws and regulations prohibited. |
| 1935 | Interstate Oil Compact Act (US) | Authorized voluntary cooperation among state regulatory agencies; no power to prescribe state production quotas or regulations. |

# CHAPTER 5

# MERE THEORY AND SPECULATION

The experiences in the field at different rates of flow over a period of years have confirmed rather conclusively the testimony of the Commission's experts given in the *Macmillan* and *People's* cases, and which in such cases was more or less brushed aside by the courts as being mere theory and speculation.

—ROBERT E. HARDWICKE, 1938[1]

East Texas blew the oil fraternity and Railroad Commission's genteel Texas two-step, as well as petroleum engineers' happy scientific consensus, right into the deep Piney Woods.[2] With no surface geological features that would indicate the likely presence of oil and a number of forlorn dry holes, what would become the East Texas oil field was a region of Texas mostly discounted by majors and larger independents alike. But after Spindletop, every little town in Texas, and just about every wide spot in the road, had its prophets and seers who saw oil beneath their feet. Some were right. The discovery well for the East Texas field, Daisy Bradford No. 3, came in October 3, 1930, midway between the great metropolises of Overton and Henderson, neither of which had a paved street. It was the third try on the same lease for a down-on-his-luck promoter of dubious repute, Columbus Marion "Dad" Joiner, who seemed more interested in his lessor's home cooking than in actually drilling an oil well. In cahoots with a charlatan (but occasionally lucky) "geologist," "Dr., Ph.G., M.D., C.E." A. D. "Doc" Lloyd (a.k.a. Joseph Idelbert Durham, et al., to his multiplicity of apparently simultaneous wives), Joiner discovered oil. Two more discovery wells followed, Lou Della Crim No. 1, near Kilgore, thirteen miles north of Joiner's discovery, on December 28, 1930, and Lathrop No. 1, near Longview, a further thirteen miles north, on January 26, 1931.[3] Initially, nobody knew whether these wells tapped three separate pools or one huge pool. It was one: the largest oil field ever discovered in the continental United States, ultimately forty-five miles long by twelve miles wide, 140,000 acres, and estimated to have contained something over 5.5 billion barrels of high-gravity, sulfur-free oil ideal for refining into gasoline (Figure 5.1. The East Texas Oil Field).

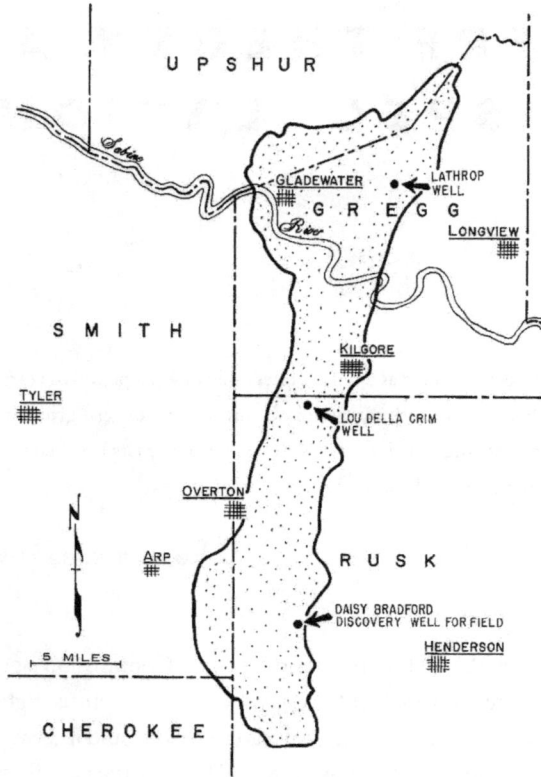

**Figure 5.1. The East Texas Oil Field**[4]

The discovery wells came in at the beginning of the second year of the Great Depression in a part of Texas already afflicted by poverty and protracted drought; for the mid-continent to the northwest, it was the first year of the human and ecological tragedy called "the Dust Bowl."[5] East Texas is relatively densely settled, and small tracts predominated. Even those could be subdivided into multiple leases. In Kilgore and other small towns, town lots and even dedicated streets were leased, as at Spindletop or Burkburnett, and some town lots held as many as six individual wells. All production came from the Woodbine, a Late Cretaceous sandstone strata named by geologists, as was customary, for the small town north of Dallas where the formation "outcropped," or reached the surface. The Woodbine underlay most of East Texas and dipped down in a northwest to southeast direction before rising slightly and "pinching out" against the Sabine Uplift, just east of where the East Texas oil field lay. In the East Texas field, the Woodbine was found at a depth of 3,600 to 3,700 feet; the oil "pay" averaged 40 feet in thickness, thinner to the west, thickest (over 100 feet) in what became the center of the field, or "fairway," before thinning out against the Sabine Uplift. The Woodbine in the East Texas oil field was therefore slightly tilted, lowest to the west, highest to the east. The deeper sections of the Woodbine (down tilt) to the west contained only salt water, which also underlay roughly half of the westernmost oil-bearing sands.

Soup to nuts, an average East Texas well costs about $26,000 to drill and complete, although wells costing as little as $9,000 were not uncommon, and a lot of shoestring promoters and operators drilled wells for way less. Free of Dad Joiner's constraints, financial and otherwise, wells could be drilled, from "spudding in" to completion, in only two or three days.[6] Long before the true magnitude of East Texas was realized, Joiner's discovery prompted a sharp rise in lease prices: Alfred MacMillan had paid a grand total of one dollar for a lease on 64 acres on April 25, 1930, but $3,000 for a lease on 93 acres on February 13, 1931, both in unproven territory.[7] Just an average well initially could produce 15,000 barrels of oil a day. Even at 10 cents a barrel, the posted price in mid-1931, a well would pay for itself in under three weeks, and it was all gravy after that. No other oil boom, not Spindletop nor Sour Lake nor Burkburnett, nor Seminole nor Oklahoma City, had ever matched the scale and intensity of East Texas.

Within a year of the completion of Daisy Bradford No. 3, the field reputedly could produce a million barrels a day, and a new well was being completed every hour. By the end of 1931, there were 3,540 wells in the field. As of January 1, 1933, there were 321,500 producing oil wells in the United States, among which were 15,000 flush wells that flowed without being pumped, at minimal cost—the valve structures atop them are called "Christmas trees" for good reason. Of those flush wells, 11,000 were in the East Texas field, and they alone had an aggregate potential production of five million barrels a day, more than twice the daily oil consumption of the country. The remaining 306,500 pumping wells nationwide could produce an average of less than five barrels of oil per day.[8] As of January 1, 1938, there were 23,950 wells in East Texas, and by 1940, there were twice as many producing oil wells in the East Texas field alone as in the state of California, and the field was already known to have more or less legally produced 1.5 billion barrels of oil.

The wide distribution of property ownership, the prevalence of small tracts, and the years of poverty in East Texas, as well as nearly nonexistent barriers to entry, all amplified by the desperation of the Great Depression, gave independents, especially smaller independents, their day, at least for a while. The majors and larger independents (such as H. L. Hunt, who had bought out Dad Joiner almost immediately) quickly began to buy up production. Even by July 10, 1931, the top nineteen operators controlled 57 percent of the acreage in the field but had produced only 36 percent of the oil; the next twenty largest operators held 12 percent of the field and had produced 15 percent. The remaining 586 operators controlled 30 percent of the acreage in the field, but 334 hadn't yet drilled their leases. The 242 others, with about 20 percent of the field's total acreage, had produced 49 percent of the oil.[9] Recall, however, that for every operator of a lease, there are multiple interests, and even more people counting on the boom: "Laissez les bons temps rouler!"

## EAST TEXAS AND THE GREATER TRAVAILS OF PRORATIONING

Prorationing in Texas was already in trouble even before East Texas came in. The Railroad Commission of Texas's first statewide order had met immediate legal challenge.[10] Joe Danciger, a Tulsa-based independent, operated forty-two oil wells in the Panhandle field with an aggregate potential of 5,200 barrels a day, which were restricted by the statewide proration order to 25 percent of that.[11] Pending trial on the merits, Danciger Oil & Refining Company had obtained a temporary injunction at the end of September 1930. In a complaint that would be replicated many times over the next twenty years, Danciger alleged that he was producing oil without waste, that he was not storing oil in excessive quantities, that he had ready market for all the oil he produced, and that the statewide order both exceeded the Railroad Commission's authority and violated the US Constitution's Fourteenth Amendment prohibition against taking property without due process of law. The Texas state district court nevertheless sustained the statewide order February 14, 1931, holding that it in fact bore a reasonable relation to the prevention of physical waste.[12] Danciger promptly appealed to the Texas Court of Civil Appeals in Austin, which did not rule on the case until March 23, 1932.[13] (The more important oil and gas cases are summarized in Table 5.1. Significant Case Law, 1866–1943.) The Court of Civil Appeals affirmed the decisions of the Railroad Commission and of the District Court, and held that even if an individual well or lease were itself operated without waste, if such operation caused "gas dissipation, or destruction of 'reservoir energy'" it would still violate the law.[14] Taking the entire reservoir as the appropriate conservation unit, the Court unequivocally held that the prorationing orders were "reasonably calculated to prevent physical waste."[15]

The immense Panhandle field, 125 miles long by an average of 15 miles wide, has a unique geological structure: It is really a set of interconnected sweet natural gas (suitable for home use), sour (sulfur-laden) gas, and oil reservoirs, in some cases overlying one another.[16] But in all instances, the oil-bearing pools were gas drive, and thus were uniformly agreed to fit the Beecher and Parkhurst paradigm. The Texas Court of Civil Appeals, ruling after the federal *MacMillan* case had been decided, noted that while much of the evidence introduced to support prorationing in East Texas might have been "largely theory and speculation," for the Panhandle field, "there was ample evidence to show . . . that proration of output or ratable taking from the field within prescribed limits was a reasonable and effective method of minimizing . . . waste."[17] Compliance, of course, was a different problem. Although not protected by the initial Danciger injunction, many operators throughout Texas blithely ignored the general prorationing orders. Even after the District Court sustained the orders in first *Danciger*, many thought that the simple fact that an appeal had been filed inoculated them from penalty and continued to produce as they saw fit.

Against this backdrop, the Texas Railroad Commission held its first special hearing on the East Texas field in Austin, February 24, 1931, attended by thirteen Pullman cars full of irate East Texas operators. Perhaps not surprisingly, given the scale and pace of East Texas development and the uncertainty surrounding it, the Railroad Commission had seemed slow to intervene—no doubt they also were loath to dampen the one bright economic beacon in all of Texas, if not the nation. The Commission issued its first proration order for East Texas April 10, 1931.[18] Likely because of the State District Court's decision in *Danciger* favorable to the Railroad Commission, Alfred E. MacMillan, a smaller East Texas independent operator and refiner, filed suit in Federal District Court in Austin.[19] MacMillan owned two wells in East

Texas capable together of producing 50,000 barrels per day; he claimed that he operated his wells "skillfully and in such a manner as to prevent waste, and to cause no injury to producing sands of adjoining properties," and that his refinery, MacMillan Petroleum Company, had binding contracts to sell his output, both in the United States and in Canada.[20] The order in effect when the case was tried limited production in East Texas to 160,000 barrels a day. Similar to the Van voluntary agreement, the East Texas field was divided into 20-acre parcels, or a fraction thereof, and production allocated based on the ratio of that parcel's potential to that of the whole field, except that no well could be restricted to less than 100 barrels a day.[21] MacMillan's total allowed production under the order was only 1,455 barrels a day.[22]

Echoing Danciger, MacMillan's complaint alleged that the Railroad Commission's orders violated the Fourteenth Amendment to the US Constitution by "depriving plaintiffs of their properties without due process of law, and denying them equal protection of the laws," that the orders had "no relation to the conservation of resources or the prevention of their waste," but were "a mere arbitrary order designed to control the output, price, and market of crude oil."[23] Since the issues were constitutional, three judges heard the case, DuVal West, District Judge, Western District of Texas, Austin, Randolph Bryant, District Judge, Eastern District of Texas, Tyler, and Joseph C. Hutcheson Jr., Fifth Circuit Judge, Houston. At the time, according to an 1842 US Supreme Court decision, *Swift v. Tyson*, federal district courts could rely on a "national common law" and were not bound by state-court interpretations of state statutes; hence federal judges enjoyed broad discretion.[24] *MacMillan* was tried de novo, and the federal court was not required to conform to the Texas District Court's decision in first *Danciger*. Trial was held June 24, and decision rendered July 28, 1931.

The Railroad Commission did not fare well. Commission witnesses made multiple and sometimes contradictory claims. The core of the Commission's argument was that if MacMillan (or other proration "violators") were to persist in producing in excess of assigned allowables, the result would be irreparable damage to the field as a whole through both sub-surface and above-ground waste. Railroad Commission prorationing would prevent such waste; any ascribed effect on the price of oil was coincidental to that legitimate purpose.[25] The cause of underground waste was claimed to be water "coning" or "channeling." Forty to 50 percent of the westernmost sections of the field was underlain by salt water, and the whole of its western edge was bordered by salt water.[26] "Excessive" rates of withdrawal from an oil well would create a volume around the well in which pressure was much lower than that of the reservoir as a whole; since salt water is less viscous than oil, it would move more easily through the formation, "coning" (from below) or "channeling" (from the edge) toward the well and bypassing pockets of oil, which would thereby become unrecoverable. Excessive rates of production would also reduce reservoir pressure, thus causing the loss of reservoir gas energy (the Beecher and Parkhurst argument).[27] Moreover, unratable taking would promote surface waste: If some operators had pipeline connections or markets and produced wide-open while others did not have outlets, those without also would have to produce at high rates to avoid drainage, and, of necessity, store their oil on the surface, with attendant losses through evaporation, flood, and, most likely, fire.

Unfortunately, the Railroad Commission could offer no substantial, factual evidence for any of these evils having occurred in East Texas. Other than evidence from the drilling logs required to be deposited with the Commission, which indicated only the depth to which wells were drilled and the producing formation, and reported (and suspect) production figures, the Railroad Commission had virtually no data on the East Texas field, neither bottom-hole pressures, gas-oil ratios, water-oil ratios, nor rates of water intrusion for MacMillan's or for any other wells. Commission witnesses' claims about water coning or rapid pressure decline therefore rested upon theory or upon analogy to other oil fields, some notably dissimilar in geological character and operating history to East Texas.[28] Railroad Commission witnesses uniformly admitted that there was no scientific basis for setting the total allowable for East Texas at 160,000 barrels a day, that it was an "arbitrary figure that they [the Railroad Commission] felt would surely prevent waste in that field," and stubbornly maintained that it was best to be conservative in setting allowables to avoid damaging the field.[29] No one could point to any scientific consensus at all: Robert Penn, testifying as an expert witness for the Railroad Commission, on redirect examination by Fred Upchurch, counsel for the Commission:

Q: I say, there is quite a diversity of opinion among these petroleum engineers and geologists about this underground waste?
A: I have never seen any one to agree on anything.[30]

Although most of the witnesses for plaintiff (MacMillan) and defendant (Railroad Commission of Texas) were well-qualified, there was no agreement on what "control" the East Texas field was under: hydrostatic (what would later be called water drive), dissolved gas (the Beecher and Parkhurst paradigm), or capillary. Most agreed that there was insufficient data to make a definitive determination.

Only Wallace Pratt, who had become Humble Oil's Chief Geologist in 1918, supervised all of Humble's oil and gas exploration, and was probably the most knowledgeable and astute of all those who testified for the Railroad Commission, maintained (as a rebuttal witness) that the 900 or so existent wells in East Texas did provide sufficient data to infer that the field was under hydraulic or "hydrostatic control." Under direct examination by Robert Hardwicke, Pratt claimed that the difference in elevation of the Woodbine sand where it "outcropped" or reached the surface at "four or five hundred feet above sea level" north of Dallas, some 100 miles away, and its depth in the East Texas field, "3,200 feet below sea level" provided a hydrostatic head, which pressurized the entire field. Pratt admitted that the friction of water moving through the Woodbine formation would dissipate its pressure. He neither claimed nor explained how pressure might be maintained in the field over so large a distance, and suggested that "expansion of gas" also had a role in bringing oil to the surface.[31] Pratt produced no hard data, either to support his hydrostatic control hypothesis, or, more critically, to warrant the specific allowables imposed upon East Texas operators, relying instead upon vaguely defined "experience" in other fields, and upon the well-worn fear of water coning.

His brief and unrepresentative testimony is the sole evidentiary basis for Robert Hardwicke's later claims about what engineers knew in 1931.

Judge Hutcheson, who questioned Pratt directly, was not impressed:

**Q:** Now, how can you determine that barrelage [total field allowable], not knowing what is going on under the ground? You cannot, can you?

**A:** We cannot determine accurately until we have determined it from experience in the field; we are beginning to learn since some of the wells have already gone to water; until we can determine we have simply got to be sure that we are not taking out too fast.

**Q:** You do like the old system of finding out if a woman was a witch, if she burned, she was a witch, and if she didn't, she wasn't; that is the way, if the field drowns, you are doing badly, and if it doesn't, you aren't?[32]

In contrast, MacMillan's case comprised a set of uncontroverted factual assertions: MacMillan's two wells, no matter what their rate of production, produced no water whatsoever. Production rates were voluntarily held to 4,000 to 5,500 barrels per day for each well, nowhere near their potentials. This production was not stored but immediately shipped via pipeline to MacMillan Refining, which had binding contracts to sell all its refined products. Gas liberated with produced oil was either used on-site or flared, in accordance with Railroad Commission regulations. No gas-oil ratios were determined, nor were they required to be. One of MacMillan's wells was completed 102 feet above the water level in the field, the other 52 feet. No salt water directly underlay either of the MacMillan leases. As determined from other wells, the closest edge water to either MacMillan well was at least 8,000 feet (one-and-a-half miles) away. MacMillan operated his wells in full compliance with best engineering practice. Given the location and absence of bottom water under MacMillan's leases, any hazard from coning or channeling was purely conjectural. In fact, the only wells in the entire East Texas field that had produced excessive salt water had been drilled through the oil-bearing sand into underlying salt water or within a few feet of it: their problem was poor drilling practice, not overproduction.[33]

Ironically, the most damning evidence for the Railroad Commission came from R. D. Parker, Chief Supervisor of its Oil and Gas Division. Parker had graduated from the University of Texas in 1898 with a degree in civil engineering and, after railroad experience, had become chief engineer of the Railroad Commission in 1908, moving to the then new gas utilities division in 1920; he was appointed Chief Supervisor of the Oil and Gas Division in 1927.[34] He had borne most of the burden of creating prorationing from 1926 on, and he was widely (and justifiably) reputed to be knowledgeable, thorough, impartial, competent, and honest. But in a letter dated May 23, 1931, addressed to "All Purchasers and Transporters of Crude Oil in the East Texas Fields," he wrote that the Railroad Commission "insistently urge prompt action in the form of fair offers to buy prorated oil in substantial amounts and as near the posted price as possible and on that stabilized basis only, and immediate offers to make the necessary [pipeline] connections to take the oil," and that the Commission asked "that transporting companies refuse to handle any oil not produced strictly in accordance with the provisions of our order."[35] Hardly surprising that prorationing to prevent physical waste might be mistaken for simple price fixing: the letter's importance is reflected in its being included verbatim as Note 1 of the *MacMillan* decision.

One Railroad Commission expert witness, E. V. Foran, did offer one of the earlier ritual self-productions of petroleum engineering and one of its first public assertions of disciplinary authority. Foran earned his degree in geology (University of Idaho, 1921), but had extensive experience with the Bureau of Mines and with the United States Geological Survey, Conservation Branch, during the 1920s; he identified himself as a "petroleum engineer":

> **Q:** What exactly is the distinction between a Petroleum Engineer and a Geologist?
> **A:** Petroleum Engineer is one whose problems and whose services are confined more to the actual production of oil, rather than the location of the reserves or the preliminary work necessary to the finding or discovery of oil. The Production Engineer is one whose problems are of a mechanical nature, more so than a Geologist.[36]

By the fall of 1932, "petroleum engineer" would be vernacular in the oil fraternity, used unproblematically and without explanation by the Railroad Commission, complainants' witnesses, and the judiciary alike: the practical imprimatur of a community's successful self-production.[37]

Whatever their expertise, Hutcheson became increasingly impatient with Railroad Commission witnesses. More than once he admonished them to quit wasting time trashing violators and to stick to facts, not speculation.[38] To Robert E. Hardwicke, counsel for the Railroad Commission: ". . . so much of this testimony, this constant pinging upon violators and that is lurid and emphatic but has a tendency to be argumentative rather than factual, and what we would like to see as we go along is some of the facts as well as the color."[39]

Since the Texas Conservation Act of 1929 authorized only the prevention of physical waste, and explicitly excluded "economic waste," the Court found the East Texas orders "unreasonable and void . . . because issued in the attempted exercise, not of delegated, but of usurped powers."[40] They therefore violated due process and resulted in unconstitutional taking. The central Findings of Fact were two: that any beneficial effect the Railroad Commission's orders for the East Texas field might have on ultimate recovery was "largely theory and speculation," and that the Railroad Commission had no scientific or evidentiary basis for setting allowables. If proration "could be made effective to prevent physical waste, it could only be properly applied in each field after careful tests and experimentation there." But,

> It was further established that the allowable for the East Texas field and for plaintiffs' wells was fixed at an arbitrary basis arrived at without test or experimentation either on plaintiffs' property or in the field generally . . . and that the order had no reasonable relation whatever to the prevention of physical waste.[41]

Hutcheson framed the opinion with his customary restraint: "Presumptively valid though such [legislative] acts are, courts, bound to give effect not to fictions, but to realities, may not, in construing them, close their eyes to what all men can see. Disregarding pretense, subterfuge, and chicane, courts must, looking through form to substance, ascertain the true purpose of a statute not from its recitals of purpose, but from the operation and effect of it as applied and enforced"—which was little more than a conspiracy "to bring and keep oil prices up."[42]

The immediate effect of the *MacMillan* ruling, especially its tone, was to convince most in the oil fraternity that proration was dead and could be safely ignored, and to stampede the state legislature, whom Governor Ross Sterling had summoned into special session on July 14, into believing that no market demand statute could pass constitutional muster. The result was the Anti-Market Demand Act, passed August 12, 1931, which prohibited consideration of economic waste in even stronger terms, stipulating that waste "shall not be construed to mean economic waste, and the Commission shall not have the power to attempt, by order or otherwise, directly or indirectly, to limit the production of oil to equal the existing market demand for oil, and the power is expressly withheld from the Commission."[43]

The Anti-Market Demand Act tried to compensate for this prohibition by outlawing every conceivable form of physical waste and by, for the first time in Texas legislation, explicitly considering the oil pool entire. The 1929 Conservation Act had prohibited forms of physical waste caused largely by poor drilling and completion practices, such as escape of natural gas or intrusion of ground water into the producing formation due to inadequate casing or cementing; its definitions of "underground waste" essentially expressed the oil fraternity's accumulated wisdom circa 1920.[44] It focused exclusively on individual wells, and neither considered the field as a whole, mentioned correlative rights or reservoir energy, nor authorized field-wide prorationing. The 1931 Anti-Market Demand Act, in contrast, in effect codified the Railroad Commission's and the Texas District court's findings in first *Danciger*, and substantially expanded the conception of physical waste to include operation of a well with an excessive gas-oil ratio; locating, drilling, or operating a well in a manner that would reduce ultimate recovery of oil or gas from the pool; inefficient or improper use of gas, gas energy, or water drive; or operation that would result in inequitable withdrawal from the common pool.[45] These definitions of waste and correlative rights radically expanded the range of issues the Railroad Commission might legally consider. Whereas the "old statute [1929] looked at waste as committed singly and individually, each well by itself. The new statute [1931] . . . recognizes waste as brought about by contribution, introduces for the first time into the statutory prohibition against waste the idea that operations are to be looked at as a whole."[46]

Now Texas legislation formally recognized and incorporated the insights of petroleum engineers, as developed in the tradition established by Beecher and Parkhurst. How much of this newfound enlightenment represented genuine conviction and how much a throw-everything-at-the-wall-and-see-what-sticks desperation isn't clear. What is clear is that in their legislative testimony, and no doubt in their lobbying, petroleum engineers, especially those from the Humble Company, had succeeded in making the most advanced petroleum engineering hypotheses and theories au courant, in the Texas legislature, as in the Railroad Commission.

Unfortunately, there is a difference between espousing and embracing concepts and theories and proving by a preponderance of evidence that, in a specific instance, they were anything more than "theory and speculation."

After the *MacMillan* opinion was filed, only a few petroleum engineers, principally at the Railroad Commission and at the Humble Company, and a handful of lawyers noticed its more hopeful subtexts. Judicial notice that a proper prorationing "plan could only be applied to each field after careful test and experimentation there," and that the commission had arrived at an arbitrary allowable "without test or experimentation," implied that engineers should do exactly that.[47] The court fully warranted the state's compelling interest in the conservation of oil; it's objection to the specific order in *MacMillan* was that it was not shown to bear any *reasonable* relationship to that purpose. Had it done so, the verdict might have been different. And as lawyers realized, first among them Jacob Wolters, the Texas Company's chief counsel, the court had not ruled on market-demand proration per se, only on the commission's order which violated the explicit statutory prohibition against it. Proration under a market-demand statute, consistent with the reasonable prevention of physical waste, could conceivably fare better.[48]

In East Texas, the number of wells, production, and chaos rose together, fast: the day the Anti-Market Demand Act was enacted, the East Texas field produced over a million barrels of oil—over one-third of US consumption—from some 1,600 completed wells, with many more drilling.[49] Befuddled, stymied by injunctions, and required to give ten days' notice before it could even hold hearings on new proration orders, much less formulate and promulgate them, the Railroad Commission intimated it would be at least the first of September before a new order for East Texas would be issued. Rumors of impending violence in the field were rife—well-founded or illusory, depending upon whom you asked. East Texas crude was swamping everybody's boat. At the urging of his fellow oil-state governors, especially William H. "Alfalfa Bill" Murray of Oklahoma, Governor Sterling declared martial law in East Texas August 16, 1931, and shut down the field.

## AN OKLAHOMA EXCURSION

Murray, confronted with rampant disregard of Oklahoma Corporation Commission prorationing orders, particularly in the Oklahoma City field, had declared martial law August 4, 1931, for an area 50 feet in radius around every oil well in twenty-seven different Oklahoma pools.[50] The law differed in Oklahoma: its oil conservation statute, passed in 1915, granted the Oklahoma Corporation Commission authority not only to prohibit "waste" in oil and gas production in the usual physical terms but also, pursuant to that purpose, to limit production to market demand and, separately, to directly fix oil prices in the State of Oklahoma, although the latter authority was never exercised and, therefore, never subjected to legal challenge.[51] The Oklahoma Supreme Court had upheld the statewide prorationing orders of the Oklahoma Corporation Commission, the state regulatory agency charged by statute with natural resource conservation, October 14, 1930, in *C. C. Julian Oil & Royalties Company v. Capshaw, et al.*, just after Daisy Bradford No. 3 was completed.[52]

The Oklahoma Supreme Court ruled that Oklahoma prorationing was well within the police powers of the state as defined by the US Supreme Court in *Ohio Oil Co. v. Indiana* (1900) and succeeding cases, and that unrestricted production in excess of market demand inevitably would result in surface waste through evaporation, leakage, and fire. The Oklahoma Supreme Court further found that if some wells produced to capacity while others were restricted or shut in, oil drainage from off-setting leases would necessarily occur, violating the established correlative rights of owners in the common pool (again relying on *Ohio Oil Co. v. Indiana* and succeeding federal cases, as well as the specific prohibitions in the Oklahoma conservation act).[53] Moreover, either excessive rates of production or disproportionate or non-ratable taking would result in underground waste: dissipation of gas energy in the common pool, water coning, and unrecoverable oil. In its majority findings, the Court relied explicitly on the expertise of petroleum engineers, and unequivocally adopted the Beecher and Parkhurst gas-drive paradigm as reified by Miller.[54]

When Murray imposed martial law, a federal challenge to Oklahoma prorationing, *Champlin Refining Co. v. Corporation Commission of Oklahoma, et al.* was before the Federal District Court for the Western District of Oklahoma, Guthrie. Champlin was a smaller but integrated operator-refiner-marketer with leases in the Greater Seminole and Oklahoma City fields. The Oklahoma City wells were on town lots of less than two acres each. They had gauged potentials of 9,600 and 6,000 barrels a day but were limited by commission order to only 267 and 167 barrels per day, respectively.[55] In view of intimations the Oklahoma Supreme Court would approve prorationing in *Julian*, Champlin, like *MacMillan*, turned to the federal judiciary.

*Champlin* hinged, as did all prorationing challenges, upon whether the conservation acts and the prorationing orders issued under them bore any "reasonable" relation to the prevention of waste. But the scientific and engineering arguments Champlin presented were very different from those offered in *MacMillan*. Witnesses for the plaintiff (Champlin) and for the defendant (the Oklahoma Corporation Commission) agreed that gas energy was critical in both Greater Seminole and Oklahoma City. The Oklahoma City field especially was animated by extremely high gas pressure, and waste or needless dissipation of gas cap or dissolved gas would reduce ultimate recovery.[56] All expert witnesses worked very much within the Beecher and Parkhurst tradition, by then broadly accepted by almost all petroleum engineers.

Given this agreement on the fundamental character of the fields, Champlin argued that the Oklahoma Corporation Commission's orders were deficient not because they were unsupported by scientific evidence but because they were not scientific enough. Champlin witnesses voiced two major objections to how wells' allowables were set: First, the Commission considered only the gauged potential production of each well. Second, Champlin claimed that the Oklahoma City field actually was underlain by at least four separate, distinct producing pools or horizons, with some wells producing from multiple horizons. In the first instance, when the Corporation Commission depended only upon well potential, it excluded from consideration lease acreage and well spacing, the thickness of the producing formation underlying the lease and its porosity and permeability, its percentage oil saturation, oil viscosity, reservoir

temperature, surface of sand grain, and a well's position on the geologic structure.[57] No attempt
was made to determine optimum gas-oil ratios for each individual well.[58] Petroleum engineers
and petroleum geologists agreed that these factors determined the recoverable oil in place
under a lease and the quantity of oil each owner was entitled to produce, and also dictated
optimum production rates to maximize total physical recovery from the lease and the pool.
About how exactly to calculate these values, especially in a relatively new and still developing
pool like Oklahoma City, there was little or no consensus.

Moreover, Champlin contended (and Oklahoma Corporation Commission witnesses did
not dispute) that the Oklahoma City field was underlain by at least four distinct, unconnected
producing pools or horizons, with some wells producing from multiple horizons. These
separate "pays" differed in geological character and gas pressure; therefore, one allowable
set on the basis of a single gauged well potential was scientifically and technically flawed.[59]
Champlin argued that these deficiencies in sum were so egregious that they negated the
Oklahoma Corporation Commission's pretense that the purpose of prorationing was to
conserve resources and protect correlative rights.

The Oklahoma Corporation Commission's counterargument was straight Beecher and
Parkhurst. Unrestrained, or excessive, or too rapid withdrawals from the common reservoir or
source would waste gas energy, dissipate gas, especially dissolved gas, increase oil viscosity, result
in gas channeling and water intrusion (coning), damage the entire reservoir, and leave massive
quantities of oil unrecoverable. Those without markets or steel storage tanks would be forced
to produce their wells wide-open anyway, which experience showed would invariably result
in storage in earthen pits or dammed creeks, with corresponding catastrophic above-ground
waste—what Oklahoma's chief umpire, Ray M. Collins, described as "a perfect saturnalia
of waste."[60] The Federal District Court, in a two-to-one opinion, affirmed the Oklahoma
Corporation Commission's orders, followed by Champlin's appeal to the US Supreme Court.

Of more interest than the published opinions themselves are the Federal District Court's
majority's Special Findings of Fact, which would be adopted in toto by the US Supreme
Court (the findings were not published with the opinions, but sent up as part of the appellate
record).[61] The District Court in turn had adopted verbatim the findings of the Oklahoma
Corporation Commission in its order of October 28, 1930, issued after lengthy hearings and
still in force, with modification, at the time of trial.[62] Critical among the District Court's
findings of fact, for Champlin and for all succeeding proration cases from whatever state,
was that "improper use of reservoir energy brings about underground waste and injuries to
the various owners," and that "unratable production results in the destruction of correlative rights."[63] Combined with the doctrine of correlative rights expressed in *Ohio Oil Co. v.
Indiana*, petroleum engineering insight entailed that all wells in a common pool should share
proportionately in demand for crude oil.

The District Court did not directly address Champlin's claims that the Oklahoma
Corporation Commission's orders and procedures were scientifically deficient. Rather
the Court concluded, "We are not concerned with the question of whether such sections
of said Act provide the best method of preventing waste and protecting such rights. The

determination of that question is a legislative prerogative." The legal standard for the legitimacy of such procedures was not that they be the best possible, but that they be "reasonable," which had been clearly enunciated in *Ohio Oil Company v. Indiana* and further reified in *Chicago, Burlington & Quincy Railway Co. v. Illinois* (1906) and succeeding cases.[64]

The US Supreme Court heard arguments in the Champlin case March 23, 1932, and unanimously affirmed (and echoed) the District Court's decision May 16. Among its findings were four that shaped prorationing and the relationship between petroleum engineering science and the oil fraternity from then on. First, waste, gas drive, reservoir pressure, and ultimate recovery were physically linked: "Uncontrolled flow of flush or semi-flush wells for any considerable period exhausts an excessive amount of pressure, wastefully uses the gas and greatly lessens ultimate recovery."[65] Second, even if plaintiff (Champlin) could produce, refine, and market oil without waste, the effect of its unrestricted production would harm the correlative rights of other operators in the field, and necessarily, by "The improvident use of natural gas pressure inevitably attending such operations," cause physical waste.[66] Third, given these facts, the Oklahoma Corporation Commission's orders served only to prevent waste, and did not fix prices: "None of the commission's orders has been made for the purpose of fixing the price of crude oil or has had that effect."[67] Fourth, prorationing orders and their procedures need only be "reasonable," not optimal.[68] Reasonable was good enough. Oklahoma had found the Goldilocks zone, just enough but not too much petroleum engineering. Beecher and Parkhurst's original demonstration of the role of dissolved gas in oil production, the critical role of reservoir pressure, and their effect on correlative rights and ultimate recovery, through the medium of developing petroleum engineering practice, had become certified and authenticated scientific and judicial fact, the test of reason.

## EAST TEXAS UNDER THE GUN

Meanwhile, Governor Ross Sterling, in Texas, was doing less well. On grounds of imminent insurrection, he proclaimed martial law throughout the four counties comprising the East Texas oil field on August 16, 1931, and ordered the field closed down from 7 a.m. August 17. Unlike Murray, who could plausibly claim to be simply enforcing legal Oklahoma Corporation Commission orders, Sterling enjoyed no such legal fig leaf. Although the *MacMillan* decision technically applied only to the plaintiffs' wells, it had effectively nullified all Railroad Commission prorationing orders. The Texas National Guard, aided by a detachment of Texas Rangers, did get the field under control, despite ongoing shenanigans.[69]

The Railroad Commission issued a new order September 2, 1931, effective September 5, which contained East Texas's first set of comprehensive field rules. Rule 6 noted that "evidence at the time was insufficient for an order fixing an oil-gas ratio for the oil wells in the field," but did suggest best use of gas energy occurred at production rates of 200 to 300 barrels per day; therefore, all wells were limited to 225 barrels per day until better evidence to fix a gas-oil ratio became available.[70] The Railroad Commission and its engineers were still thinking very much within the Beecher and Parkhurst framework. As new wells continued to be completed, compliance remained iffy at best. "Hot oil" got its name one rain-chilled night in mid-September when a National Guard sergeant, surreptitiously scouting the field, leaned up against a storage tank and found it hot—warmed by high-temperature oil as it came out of the ground. In his report, he substituted "hot oil" for the more cumbersome "oil produced in violation of military orders and in excess of allowables granted by the Texas Railroad Commission," and a new term entered the oil fraternity's lexicon.[71]

Legal challenges to Sterling's orders came fast and furious, the most significant of which was *Constantin et al. v. Smith et al.*, filed by Eugene Constantin and J. D. Wrather in the Federal Court for the Eastern District of Texas, Tyler, October 13, 1931. The case was decided February 18, 1932, by a three-judge panel headed by Joseph C. Hutcheson Jr., who wrote the opinion. After noting the histrionics about tyranny and oppression in the briefs of both plaintiffs and defendants, and expressing his reluctance to add yet another martial law opinion that was "so concerned with general principles that it says far more than it decides,"[72] Hutcheson got to the point: There was no evidence of insurrection in East Texas; both civil courts and civil government were fully functioning; therefore, Sterling had grossly and arbitrarily exceeded his authority as governor. Sterling reasoned that the resulting injunction applied only to plaintiffs' wells, and left elements of the National Guard in East Texas to continue enforcing Railroad Commission orders for the rest of the field while he appealed to the US Supreme Court, which heard arguments November 15–16, 1932, and affirmed the District Court's findings of fact and law December 12. Neither the District Court nor the Supreme Court questioned the power of the State of Texas or the Railroad Commission to conserve resources or restrict production, although the Supreme Court did, almost wryly, observe that Sterling had denied plaintiffs' lawful use of their property, "Instead of affording them the protection in the lawful exercise of their rights as determined by the courts."[73] Had it really been necessary for the governor to intervene, he should have enforced the *MacMillan* decision.

## *PEOPLE'S*: NEW LAW, SAME SONG

After the Federal District Court had overturned martial law February 18, 1932, the Railroad Commission issued a new proration order for East Texas February 25, which imposed a flat 75 barrel per day limit on each well whatever its potential production, structural location, or acreage. Total allowable for the field was limited to 325,000 barrels a day.[74] For the first time, gas-oil ratios were fixed, at 500 ft³ per barrel of oil. The Commission claimed that data on "pressures, pressure decline, oil-gas ratios, the movement of the water table" had been accumulated and compiled, "so that the engineers were enabled to testify with more assurance and more conclusively at our hearing February 12, 1932, as to what is going on underground."[75] Nevertheless, as was now customary, a small herd of federal lawsuits quickly challenged this order too. Representative is the complaint of *Bill & Dave Oil Corporation, et al. v. Lon Smith, et al.* In addition to the standard allegations of capricious and arbitrary taking in violation of the Fourteen Amendment, the complaint asserted, following the Beecher and Parkhurst gas-drive hypothesis, that the extreme production limitations imposed by the Order actually wasted reservoir gas energy, and that higher rates of production would improve the wells' gas-oil ratios and thus enhance ultimate recovery. The complaint also reported that the Railroad

Commission had never tested plaintiffs' wells and maintained that the individual well, not the pool as a whole, should be the unit of conservation.[76]

Complainants' argument had sound scientific warrant. As noted, Lester C. Uren, Professor of Petroleum Engineering at the University of California, had expressed reservations about the unrestrained enthusiasm oilmen had shown for minimum gas-oil ratios as the sole desideratum for efficient oil production. He pointed out that other key variables, such as reservoir pressure, determined efficient use of the energy contained in the dissolved gases in petroleum, which was the real objective, and for which gas-oil ratios were at best an iffy surrogate measure. Uren had founded the first four-year petroleum engineering program in the country at the University of California in 1915 and was a preeminent authority.[77] At Berkeley, he built a large, sand-filled pressure tank into which oil-gas mixtures could be introduced, then withdrawn, or "produced," at different rates with different back-pressures applied. The greater the backpressure, or production pressure, the lower the gas-oil ratio. Uren's experiments demonstrated that,

> Up to a certain point this conservation of gas [lower gas-oil ratio] results in increased ultimate recovery, but beyond this point the energy lost in gas not fully expanded, as discharged from the well, exceeds that conserved by reduction of the gas factor [gas-oil ratio], and thus the ultimate recovery is diminished. Too low a gas factor, if obtained by increasing the back-pressure [restraining production] above the critical point, may therefore be indicative of reduced production efficiency rather than the reverse.[78]

These results entailed that if the East Texas field were subsumed under the broad Beecher and Parkhurst paradigm—if it were a gas-drive field—then too severe a production restriction would waste reservoir energy and reduce ultimate oil recovery: extremely low allowables would violate the conservation rationale for prorationing. Uren also showed that, given gas drive, the production rate for most efficient energy utilization would change with field depletion, that gas-oil ratios should increase as reservoir pressure fell, and that optimum rates of production therefore could only be determined by repeated tests on each individual well.[79] Plaintiffs' expert witnesses in *People's* argued that to maximize utilization of reservoir (gas) energy and thus prevent underground physical waste and maximize ultimate recovery from the field, each individual well should be produced at its real-time, most efficient gas-oil ratio.[80]

In its answer to plaintiffs' complaints, the Railroad Commission now unequivocally embraced the water-drive hypothesis. The commission recited its findings from its February 25, 1932, order: that the field was primarily water driven, and that data showed,

> [T]here is a notable deficiency of gas, which fact demands not only the most effective plan of orderly production of oil to conserve gas energy but a proper use of the water drive by withdrawals of oil in such slow rates and under such conditions as will permit the water to come in and gradually and efficiently displace the oil . . . that the maximum ultimate recovery . . . can be attained only by the maintenance of an orderly, gradual oil production rate.[81]

On March 4, 1932, Judge Randolph J. Bryant (Eastern District of Texas, Tyler Division) issued a temporary restraining order against the Railroad Commission pending a full hearing (with the required three judges) on a temporary injunction. Multiple plaintiffs, complaining that the Railroad Commission had never tested their wells, then applied to use their own experts, at their own expense, to ascertain "actual conditions of the oil and gas ratio, bottom-hole pressure, etc. of the wells."[82] Since such tests would entail production in excess of allowables, the Railroad Commission opposed the applications. The Court granted them anyway, and plaintiffs found, sure enough, that the higher the rate of production from their wells the lower, and therefore the more efficient, the gas-oil ratio. The wells allegedly showed higher shut-in bottom-hole pressures after the tests than before.[83]

Since most of the challenges were based upon the same grounds, a number of cases were combined and heard together by Joseph C. Hutcheson Jr., Bryant, and William I. Grubb (Northern District of Alabama), as *People's Petroleum, Inc., et al. v. Sterling et al.* June 2, 1932.[84] The Court denied the plaintiffs' application for a temporary injunction July 19, 1932. Two factors predominated in the Court's reasoning. First, as noted above, the Anti-Market Demand Act radically transformed the legal basis for determining underground waste. While under the previous law, individual wells had been the unit of analysis, now the whole pool or reservoir was the appropriate unit: "here the statute not only authorizes, it directs, the commission to consider the needs of a pool as a whole."[85] Second, this time around, the Railroad Commission did present evidence specifically about East Texas, which showed sharp pressure declines under high rates of production, then relatively rapid pressure recovery following martial law enforcement of proration: In contrast to *MacMillan*, the court wrote,

> . . . we have here a mass of testimony as to tests and experiments made in the field, and as to conclusions reached which, if believed, amply support the commission's view that the allowable of 325,000 barrels for the field is a wise provision against waste in the interest of the state and of all the owners. In fact, the testimony of E. O. Buck, a witness for the commission, disputed by no one, was to the effect that, in his opinion, as the result of these regulations, every person in the field will ultimately derive more from his wells than they would derive, if there were no restrictions.[86]

For these reasons, the Court declined to temporarily enjoin the Railroad Commission order pending full trial.

When he testified, E. O. Buck was a Railroad Commission petroleum engineer assigned to East Texas. His father had been in charge of the electrical shops of the Beaumont streetcar system, and the younger Buck had worked as an apprentice electrician before enrolling in Texas A&M in the fall of 1921, with the intention of becoming an electrical engineer. He had to drop out during 1922–1923 for financial reasons and worked for the Gulf Pipeline Company before returning to A&M in the fall of 1923, now resolved to become a petroleum engineer. A&M did not yet have a dedicated petroleum engineering program, so Buck graduated with a degree in industrial education with a major in geology and a minor in engineering, and promptly went to work for Gulf Oil as a geologist, mostly in Texas.[87]

Buck stayed with Gulf until 1931, the nadir of the Depression. As the youngest and lowest man on the totem pole of Gulf geologists, Buck resigned rather than be cut back. Like so many, he drifted to East Texas, arriving the day martial law began. But by then he was woven into the fabric of the small-world social networks that characterized both the oil fraternity and the community of "technical men" within it. Through his connections, as well as fortuitously showing up to watch a burning blow-out and figuring out how to extinguish it, Buck got hired as a Railroad Commission petroleum engineer. Given his prior experience and acquaintances, and his role with the Railroad Commission, he quickly "made acquaintances with practically every engineer that was working [for] these various oil companies in there, and we began to exchange ideas and thoughts." This nascent invisible college, reminiscent of that which had coalesced in Bartlesville in the 1920s, slowly converged on the water-drive hypothesis, even though it lacked both definitive evidence and any theoretical explanation. Buck's testimony thus ritually produced the opinion of a community of petroleum engineers who were deeply familiar with and knowledgeable about East Texas.[88]

But, having dodged the bullet, or a temporary injunction anyway, in July, the Railroad Commission promptly hoisted itself on its own petard. Despite Buck's affirmative and uncontroverted testimony that the East Texas field should be limited to 325,000 barrels per day to prevent subsurface waste, on August 31, 1932, the Railroad Commission increased the allowable to 375,000 barrels a day.[89] The *People's* cases were tried on their merits less than three weeks later, September 17. Hutcheson wasn't the sort to take kindly to being played. He began his *People's* opinion with the pointed observation that after the July Court date,

> . . . at a hearing held for the purpose of fixing new allowables . . . the commission, over the protest and against the opinion of its advisers, and without any supporting evidence being offered, increased the East Texas allowable from 325,000 to 375,000 barrels per day. . . . The allowable for East Texas it arbitrarily apportioned as before equally per well among the more than 8,000 producing wells in that field, in entire disregard of the differences as to each well in productive capacity, situation on the structure, thickness, and character as to richness and yield of the underlying sands and proximity to water.[90]

The Railroad Commission's arbitrary action had given the lie to its earlier protestations that its East Texas orders were only reasonable conservation measures.

The Railroad Commission was enjoined October 24, 1932, on grounds that its flat per-well allowables did not prevent waste and were inherently inequitable and confiscatory. Plaintiffs had presented the familiar Beecher–Parkhurst–Uren scientific argument, backed by their own test data, that different rates of production yielded different gas-oil ratios in each well and that wells differed in their lowest (most efficient) gas-oil ratios. Therefore, no flat rate allowable for all wells in the field could maximize utilization of reservoir energy and prevent subsurface waste.[91] Ironically, even Robert Hardwicke, as outside counsel for the Railroad Commission, freely admitted, "In all frankness . . . I sincerely desire to find out some method of distribution in East Texas which is more equitable than a [flat per] well basis."[92] Nevertheless, the Commission maintained that the huge, and rapidly growing, number of wells in the field made the individual tests advocated by plaintiffs neither practicable nor reasonable. More familiarly, the Court found that the real purpose of the order disputed in *People's* was, quoting *MacMillan*, the forbidden one, "to control the delicate adjustment of market supply and demand, in order to bring and keep oil prices up."[93]

The real nub of the issue, however, remained the contradictory scientific hypotheses advanced by the opposing sides. The Commission's experts asserted "the field to be water driven, and that it should be produced so as to maintain a continuous follow up of the water." Plaintiffs' witnesses contended the field to be gas-driven, and that the Railroad Commission orders, which ignored gas-oil ratios, would cause great waste.[94] Unhappily, these "theoretical" arguments were indeterminate, which Hutcheson found just a little annoying:

> In fact, so radical are their differences, and so contrary their opinions, so voluble, so volatile are most of the witnesses in advancing them, and so equal are they all in cocksureness, that form of knowing which easily mistakes certitude for certainty, that, if we assume, as we suppose on this record we should, them all to have equal theoretical knowledge and an equal absence of intention to deceive, the theories might best be held to counterbalance.[95]

What Hutcheson did find persuasive was plaintiffs' claim that the extreme restrictions of the current prorationing order would result in drainage of the oil under their leases, and therefore in unconstitutional taking: ". . . if the present condition is maintained, plaintiffs will lose oil to which they are entitled to the wells on the east . . . and the sands lying to the east will produce the oil which has been driven from plaintiffs' lands to them."[96] It was another problem that would come back to haunt the Railroad Commission thirty years later.

Hutcheson's harshest words (among many) targeted the lack of empirical evidence sufficient to choose between the competing scientific hypotheses, a fault he placed squarely at the Railroad Commission's door:

> The evidence is conclusive that the commission has never made experiments, tests, or inquiries to ascertain the greatest amount of oil which each producer may take from his wells without injury to the field, and that it has never encouraged, nor indeed permitted, such inquiry or test. . . . On the contrary, it has prevented such tests.[97]

Once again, the lack of scientific consensus prevented finding the Commission's proration orders "reasonable."

The *People's* decision had two immediate salutary consequences, not obvious at the time. First, *People's*, together with the US Supreme Court's affirmation of Oklahoma's market demand statute in *Champlin*, convinced Governor Sterling that only a market demand law might preserve oil conservation and stabilization in Texas. He called another special session of the legislature at the beginning of November. After folderol extreme even for that august body, the Texas Legislature passed a Market Demand Act November 12, 1932: "Waste" was now defined explicitly to include "production of oil in excess of transportation or market facilities or reasonable market demand, and the commission may determine when excess production exists or is imminent and ascertain the reasonable market demand."

Second, the Railroad Commission issued a new proration order under the Market Demand Act November 29, 1932, which limited statewide production to 845,625 barrels a day and that of East Texas to 325,000. East Texas production would be allocated among its now 9,000 wells on a two-thirds per well, one-third acreage and bottom-hole pressure basis.[98] The inclusion of bottom-hole pressure marked the first time the Commission considered reservoir energy or pressure directly in setting allowables. The Commission reduced the statewide allowable to 789,745 barrels a day, and East Texas to 310,000 barrels, December 12.

The new orders required solid data on bottom-hole pressures throughout the East Texas field, which spurred the Railroad Commission to heed Hutcheson's harsh indictment about tests in *People's*. Sterling had appointed former Amarillo mayor Colonel Ernest O. Thompson to the Railroad Commission June 4, 1932, to replace former Governor Pat Neff. Thompson's title was genuine: he had graduated from the Virginia Military Institute and commanded a machine gun battalion during the First World War. He won election to a full six-year term on the Commission in November. After the US Supreme Court announced its decision in *Smith v. Constantin* December 12, Thompson, as the youngest and most energetic and ambitious of the commissioners, went to East Texas to assume control of the field from General Wolters, who at the time estimated 100,000 barrels a day of hot oil were still being run.

E. O. Buck urged Thompson to shut down the field for two weeks and let him borrow bottom-hole pressure gauges (from major operators) to do a systematic survey. Readings would be taken in a sample of wells running wide-open, then after they had been shut in. If reservoir pressure rose during the shutdown, it would demonstrate that the field was water-driven, and that excessive rates of production were damaging it.

If pressure did not rebound and only equilibrated, it would imply that the field was under gas drive. Tests were begun on 28 wells December 17, 1932, in the midst of East Texas's first heavy snowfall in thirty years. Results were sobering: Initially pressures on the extreme western edge of the field, where it abutted the conjectured source of salt water in the Woodbine, were 1,400 pounds per square inch. Pressures fell linearly across the field until they reached only 700 pounds per square inch on its eastern edge—where the Daisy Bradford No. 3 had shown an inferred pressure of 1,600 pounds per square inch only two years earlier. Seventy-two hours later, after the wells had been shut in, the wells on the western side of the field still showed 1,400 pounds per square inch, but those to the east had risen to 1,300 pounds.[99]

The Railroad Commission issued a new proration order January 1, 1933, which limited East Texas to 290,000 barrels a day and reverted to a flat per-well allocation (albeit with a purported minor adjustment for higher bottom-hole pressure),[100] and was promptly hit with twenty-three federal suits in the Eastern District of Texas. These suits were consolidated as *Rowan & Nichols Oil Company, et al. v. C. V. Terrell*, and heard, together with renewed complaints from the *People's* plaintiffs, January 27, 1933. The Railroad Commission changed its story again, apparently misled by lawyers who, having read *Champlin*, misdiagnosed the problem as market demand, not scientific consensus. Buck did present his test results and an explanation of the importance of bottom-hole pressure, prompting Hutcheson to harrumph, "The oil-gas ratio was the big thing in the first case, but now it's bottom-hole pressure."[101] But the Railroad Commission justified its order largely by Texas's new Market Demand statute. Once bitten, twice shy, Hutcheson wasn't buying it. After noting that "the evidence on this hearing" regarding physical waste was "not different from the former [*People's Petroleum Producers, Inc., v. Smith et al.*]," he again trashed the Railroad Commission:

> The case for defendants on the justification of the allowable was put on the proposition that the order was based not on the prevention of waste from channeling, trapping, loss of reservoir pressure, etc., the claimed basis of the old orders, but on waste from production in excess of market demand. We feel that it is not necessary upon the interlocutory application to pass upon this issue since the orders are invalid, as the others were, because [they do not allocate] the allowable to the owners in the field as they did in the *Champlin* case . . . equitably, preserving to each owner substantially that which he owns . . . it was the uncontradicted testimony of Mr. Buck, witness for the defendants, that the present order did not fairly distribute the allowable and in substance that the proper way to allocate the field was to make a difference between the best part and the worst part of it by basing the allocation of such allowable on the producing abilities of the wells, thus effecting a fair distribution.[102]

The new order flatly contradicted the scientific evidence offered by its own expert witness.[103]

In response to this setback, and to meet other criticism that Buck's December sample was too small and his survey "not scientific enough," Thompson ordered tests to be redone.[104] The Railroad Commission shut down the field April 11, 1933, so that its engineers could gauge the potential of 275 key wells. Average potential came out to be 517 barrels a day, which indicated that the by then 10,000 wells in the field could produce something over five million barrels of oil a day. Out of fear that the courts would reject any lower allowable, and over the objections of its engineers that it was way too high, the Railroad Commission issued a new proration order for East Texas April 22, with a total allowable of 15 percent of potential, or 775,800 barrels a day. Rather than the rejected flat per well basis, now allowables were based upon individual wells' potentials, although with no adjustment for lease acreage.[105] This order too immediately went sailing through the revolving door of the federal courthouse.

Now complainants alleged that the field allowable was too high and would dissipate reservoir pressure, that it favored densely drilled leases and promoted unnecessary and wasteful drilling, and that it would result in drainage from more responsibly developed leases (with wider spacing among wells). By the end of May, reservoir pressure in key wells on the west side of the field had declined precipitously, from 1,400 to 1,200 pounds per square inch, and saltwater intrusion had accelerated. Bryant consolidated the challenges into *Hunt Production Company v. Lon A. Smith, et al*. At a hearing before the requisite three judge panel, May 26, the Court "balanced conveniences" and declined to issue a temporary injunction. Hutcheson did emphatically tell plaintiffs "to get these cases out of this court to a higher court and get a definite decision—find out who is right and who is wrong—let's have an end to this constant running in and out of the federal courts."[106] The cases were never tried on their merits, rendered moot by passage of the National Industrial Recovery Act (NIRA), June 16, 1933.

## HIGH PLAINS DRIFTER: SECOND *DANCIGER*

Just how broadly the courts were willing to construe "reasonable" if an oil field fell within purview of the Beecher and Parkhurst paradigm, as well as an indication of just how crucial a role scientific uncertainty in petroleum engineering played in the East Texas decisions, are evinced by Hutcheson's very different response in *Danciger Oil and Refining Company of Texas v. Smith*, decided June 25, 1933. Joe Danciger, rebuffed by the Texas courts, had made a federal case out of it. Danciger's complaint hinged on three substantive claims: First, the optimum recovery of oil in place from the disparate reservoirs in the greater Panhandle field could be obtained only by operating each well at its most efficient gas-oil ratio. Second, exemption from prorationing of marginal wells, those with potential production below a specified, and low, limit, meant that those wells could produce their full potential, while prorated wells could not. As a result, marginal wells would eventually drain oil from adjacent prorated leases and violate those operators' correlative rights. Third, echoing Champlin's complaint, Danciger claimed that by using only individual well potential, the Railroad Commission neglected lease acreage, pay depth and oil saturation, formation permeability and porosity, and a well's structural advantage or disadvantage, all of which were known scientifically to determine recoverable oil in place.[107]

In his opinion, Hutcheson found both Danciger's scientific claims and his empirical evidence persuasive:

> Impressed as we are with the strength of plaintiffs' contention, and inclined as we might be to hold that upon the very clear, vigorous, and convincing affidavits which they have furnished the preponderance of the evidence is with them on the issue that the best, the most reasonable way to prevent waste in that field, in view of all the conditions there, would be to produce each well at its proper gas-oil ratio, abandoning all attempts at proration or ratable taking, it must be kept in mind that the case is not before us for trial de novo of that issue on the preponderance of the evidence. The issue before us is one of claimed confiscation through arbitrary and unreasonable orders.[108]

In that light, Hutcheson accepted the Railroad Commission's defense that "not every inequality in the administrative handling of a difficult situation like this . . . strikes administrative action down."[109] Given that allowables were set based upon well potential production, not per well, and that the Panhandle allowables were a substantial proportion of those potentials, the Court found that whether "wise or unwise, they [the orders] represent an exercise, neither unreasonable nor arbitrary, of the judgement and discretion of the commission in the discharge of [its] difficult administrative duties."[110] The decisive difference between *Danciger* and the East Texas cases was that prorationing in *Danciger* was consistent, if barely, with the widely accepted communal paradigm deriving from Beecher and Parkhurst: the legal determinant of "reasonable," as sanctified by the US Supreme Court in *Champlin* the year before.

## MODELING THE MORAL ECONOMY OF THE OIL FRATERNITY IN EAST TEXAS

Even as East Texas wallowed in crisis, with prorationing orders issued, legally challenged, enjoined, and widely violated in any event, Railroad Commission orders for other fields in Texas were uniformly obeyed, despite allowables for those fields being repeatedly reduced to accommodate the tsunami of production, legal and illegal, flowing out of East Texas.[111] The GSD model offered in chapter 3 provides some insight into what happened to the oil fraternity and its moral economy in these extraordinary circumstances. To portray the scale of East Texas and the explosiveness of its growth, treat East Texas as a very rapid sequence of short booms: shorten generations or iterations between invasions or colonization events—the influx of new entrants—from five to three. Assume that East Texas was initially exploited by a population approximately half of whom were altruistic cooperators and half opportunists, both drawn from the preexisting oil fraternity; new entrants comprised one-third cooperators and two-thirds opportunists, as in the original model. Because drilling wells was so cheap, and the likelihood of success so high, opportunists got lucky more often: set the probability that an all-opportunist deal would be truncated to 0.40, which means that 60 percent of even all-opportunist deals would pay off.

Results from the adapted model are shown in Figure 5.2. The increased frequency of invasion or colonization events combined with a lower baseline truncation probability sharply reduces the efficacy of the model's replicator dynamics in winnowing out opportunists. Over 28 iterations, the average proportion of opportunists remains at least 30 percent, while the proportion of assortative all-altruist deals stays below 50 percent. The output of the modified model, applied to East Texas, with the original model in chapter 3 used for the rest of Texas, is consistent with observed levels of cooperation: near complete compliance with field and statewide prorationing orders in the rest of Texas (where effective winnowing had occurred) and their frequent defiance or evasion in East Texas.

Since the whole object of the game was to not get caught, hard data on violations of Railroad Commission prorationing orders is hard to come by, and what hard data there is likely just sets a lower bound. For example, a Railroad Commission inspector, testifying at a Commission hearing in Tyler January 12, 1933, reported that of the 700 to 750 operators in the East Texas field, 98 (14 percent) committed most of the violations.[112] Likewise, as of November 30, 1933, of 4,913,960 barrels of oil in steel storage in East Texas, approximately 930,317, or a little less than 20 percent, were "according to the records, illegally produced."[113]

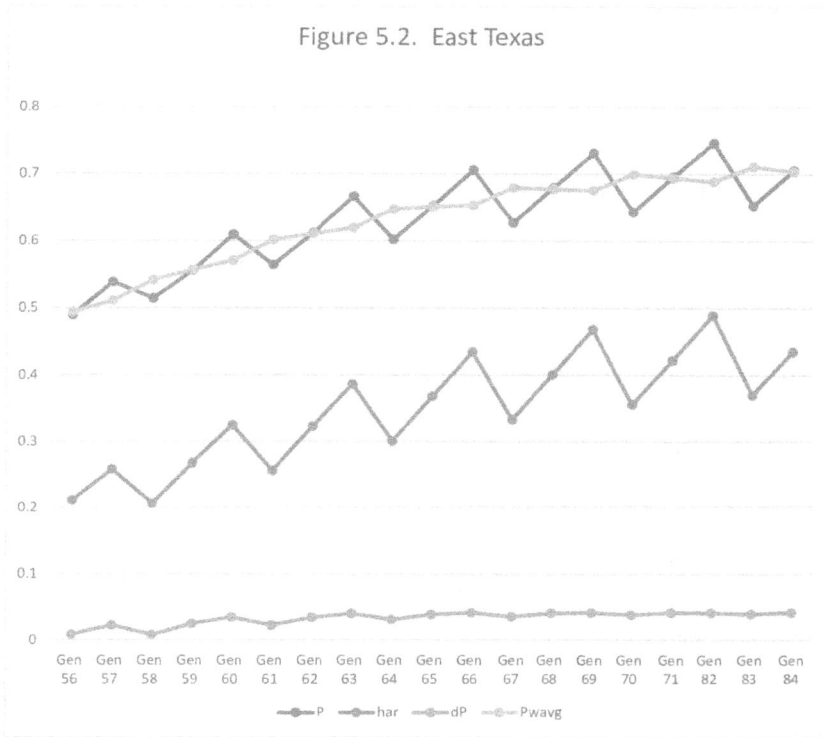

Figure 5.2. East Texas

**Figure 5.2. Evolution of Within-Deal Altruistic Cooperation with More Frequent Invasions (Colonization Events), Every Third Iteration**

P = proportion of conditional altruistic cooperators in the global population. har = proportion of global population who are altruistic cooperators and committed to all-altruist deals. dP = change in the proportion of cooperators in each iteration. Pwavg = 5 iteration moving average of the proportion of cooperators. Replicator dynamics, variables defined as in Figure 3.1.

Unrecorded hot oil production was almost certainly higher, with anecdotal estimates ranging up to a billion barrels between 1930 and 1935.[114]

Olien and Olien intimate that compliance with prorationing orders outside East Texas derived from operators' inability to find markets for excess production.[115] But given behavior in those fields earlier in their history (flush production come what may), and the desperate straits all operators experienced in the early years of the Depression, that explanation is unpersuasive. More likely is moral economy. As noted, moral economy evolves among within-deal altruistic cooperators and does not logically entail higher-level, field or statewide, cooperation. Yet, it is more likely that such a well-winnowed population would cooperate on a larger scale, while a hybrid structure combining lower-level altruistic cooperation and higher-level administration is more stable.[116] The experience in East Texas supports this surmise: it takes time for selection to eliminate opportunists and secure moral economy, and it takes time to reestablish cooperation.

E. O. Buck, the Railroad Commission's star petroleum engineer, summarized it nicely in 1981: Of the Central Proration Committee, the loose, voluntaristic association that had coordinated voluntary production restraint in prorated Texas fields before 1930 and advised the Railroad Commission, he said simply, "When [the] East Texas oil field came in, the Central Proration Committee, which was nothing more than a gentleman's agreement, ran out of gentlemen."[117]

## TWO STEPS FORWARD, ONE STEP BACK

The prorationing battles of the early 1930s left some lasting scars. W. P. Z. German, who wrote the American Bar Association "History of Conservation in Oklahoma" and had served as lead counsel in the successful defense of the Oklahoma Corporation Commission and its prorationing orders before the US Supreme Court in the epochal *Champlin* case, writing in 1938 recalled, "No one not closely associated with the activities of those hectic days can understand just how full of trial and tribulation they were for the combatants on both sides of the controversy, those who fought for ratable production and those who fought against it."[118] Commissioner Ernest O. Thompson later complained that the Railroad Commission "had been thrown out of court 212 times before it got a valid order," and that on one occasion he personally had been enjoined seven times while walking from the Capitol building to the Driskill Hotel in Austin, where a statewide prorationing hearing was to be held.[119]

But some scars healed. According to Ross Sterling's autobiography,

> Then, ten years later—in about 1946, I think it was—Judge Hutcheson and his brother Palmer came to see me, and the judge said, "I want to tell you that you were right and I was wrong about the martial law situation."
>
> I told the judge that I was "mighty glad" to hear him say that. "It confirms the opinion I've always had of your character."[120]

### Table 5.1. Significant Case Law, 1866–1943

| Date | Case | Citation | Findings |
|------|------|----------|----------|
| 1866 | *Funk v. Haldeman* | 53 Pa. St. Reports 229 (Pennsylvania Supreme Court) | Minerals *in situ* part of real estate. |
| 1875 | *Brown v. Vandergrift, et al.* | 80 Penn. St. Reports 142 (Pennsylvania Supreme Court) | Given fugacious nature of oil, lessee's failure to drill and produce oil results in lease forfeiture. |
| 1897 | *Kelly v. Ohio Oil Company* | 57 Ohio St. 317, 49 N.E. 399 (Ohio Supreme Court) | Since impossible to tell source of oil or gas produced by a well, whatever gets into the well belongs to the well owner. |
| 1899 | *Westmoreland Natural Gas Company v. DeWitt, et al.* | 130 Pa. St. 235, 18 Atl. 724 (Pennsylvania Supreme Court) | Oil *ferae naturae* (of a feral nature) and therefore, by analogy, subject to same rule of capture applied to wild game: even though landowner possessed oil so long as it was in place, once it escaped, no longer his. |

| 1900 | *Ohio Oil Company v. Indiana* | 177 U.S. 190 (US Supreme Court) | Affirmed Indiana prohibition as waste of "venting" (or "popping") natural gas incident to oil production as local regulation of real property, not unconstitutional taking without due process; framed dilemma posed by fee ownership of oil and gas *in situ* and their fugacious nature; recognized the correlative rights of interests in a common pool; articulated standard that regulation or use of state police power must bear a "reasonable" relation to public purpose (conservation). |
| --- | --- | --- | --- |
| 1904 | *Houston &Texas Central Railroad v. East* | 98 Tex. 146, 81 S.W. 279 (Texas Supreme Court) | Applied English common law rule of capture for water (*Acton v. Blundell*, 1843) to alleged water drainage in Texas. |
| 1907 | *Barnard v. Monongahela Gas Company* | 216 Pa. St. 362, 65 Atl. 801 (Pennsylvania Supreme Court) | Since no way to determine amount or location of oil originally in place, landowner's only protection against alleged drainage was to drill and produce it himself. |
| 1910 | *Bender v. Brooks* | 103 Tex. 329 (Texas Supreme Court) | Affirmed surface owner's absolute ownership of minerals in place under his land. |
| 1911 | *Herman v. Thomas* | 143 S.W. 195 (Texas Court of Civil Appeals, Austin) | Oil belongs to whoever first appropriates (produces or "captures") it; alleged drainage of adjacent tracts incurs no liability. |
| 1915 | *Texas Co. v. Daugherty* | 107 Tex. 226, 176 S.W. 717 (Texas Supreme Court) | Recognized alienable mineral estate *in situ* separable from surface estate. |
| 1930 (October 14) | *Julian Oil and Royalties Company v. Capshaw, et al.* | 145 Okla. 237 (Oklahoma Supreme Court) | Upheld Oklahoma market demand statute; effect on price incidental to prevention of physical waste. |
| 1930 (December 3) | *People v. Associated Oil Co.* | 211 Calif. 93 294 Pac. 717 (California Supreme Court) | Upheld California conservation statute preventing waste of natural gas or of gas (reservoir) energy; no proration in statute. |
| 1931 (November 23) | *Bandini Petroleum v. Superior Court of California* | 284 U.S. 8, 52 Sup. Ct. 103 (US Supreme Court) | Affirmed *People v. Associated Oil* decision, including scientific rationale for limiting waste of reservoir energy. |

| 1930 (September 30) | *Danciger Oil & Refining Company v. Railroad Commission* | (Texas District Court, Travis County) | Challenged first statewide Texas prorationing order as applied to Panhandle field; injunction against Railroad Commission granted by District Court pending trial. |
|---|---|---|---|
| 1931 (February) | *Danciger Oil & Refining Company v. Railroad Commission* | (Texas District Court, Travis County) | After trial on merits, constitutionality of proration and its "reasonable relation" to prevention of physical waste upheld. |
| 1932 (March 23) | *Danciger Oil & Refining Company v. Railroad Commission* | 49 S.W. 2d 837 (Texas Court of Civil Appeals, Austin) | Decision sustained on appeal. |
| 1931 (July 28) | *Alfred MacMillan, et al. v. Railroad Commission of Texas* | 51 F. (2d) 400 (Federal District Court, Western District of Texas, Austin) | Proration order for East Texas field enjoined; found to bear no reasonable relationship to prevention of waste. |
| 1932 (February 18) | *E. Constantin, et al. v. Lon Smith, et al.* | 57 F. (2d) 227 (Federal District Court, Eastern District of Texas, Tyler) | Martial law in East Texas field overturned; but authority of Railroad Commission to limit production to prevent waste, if order equitable, acknowledged and encouraged. |
| 1932 (December 12) | *Sterling v. Constantin* | 287 U.S. 378 77 L. Ed. 375 (US Supreme Court) | Affirmed *Constantin v. Smith* |
| 1932 (July 19) | *People's Petroleum Producers v. Sterling* | 60 F. 2d 1041 (Federal District Court, Eastern District of Texas, Tyler) | Temporary injunction against Railroad Commission prorationing order for East Texas field denied. |
| 1932 (October 24) | *People's Petroleum Producers v. Smith* | 1 F. Supp. 361 (Federal District Court, Eastern District of Texas, Tyler) | On hearing on merits, prorationing order for East Texas again enjoined as unreasonable and confiscatory. |
| 1931 (August 5) | *Champlin Refining Company v. Corporation Commission of Oklahoma* | 51 F. 2d 823 (Federal District Court, Western District of Oklahoma, Guthrie) | Oklahoma market demand prorationing orders upheld; conservation of gas pressure and reservoir energy legitimate public purposes; any effect on oil price incidental. |
| 1932 (May 16) | *Champlin Refining Company v. Corporation Commission of Oklahoma* | 286 U.S. 210 (US Supreme Court) | Lower court decision affirmed, including *verbatim* adoption of its scientific rationale. |
| 1932 (June 8) | *Texoma Natural Gas Co. v. Railroad Commission of Texas* | 59 F. 2d 750 (Federal District Court, Western District of Texas) | Common purchaser requirements for gas in Panhandle field enjoined. |

| 1933 (February 15) | *H. F. Wilcox Oil & Gas Co. v. State et al.* | 1933 OK 110, 19 P. 2d 347, 162 Okla. 89 (Oklahoma Supreme Court) | Affirmed Oklahoma Corporation Commission market demand prorationing orders; notable for vociferous dissent. |
|---|---|---|---|
| 1933 (June 25) | *Danciger Oil and Refining Company of Texas v. Smith* | 4 F. Supp. 236 (Federal District Court, Northern District of Texas, Amarillo) | Affirmed Railroad Commission statewide order as applied to the Panhandle field |
| 1934 (February 12) | *Amazon Petroleum Corporation, et al. v. Railroad Commission of Texas, et al.* | 5 F. Supp. 633 (Eastern District of Texas, Tyler) | Railroad Commission market demand prorationing order for East Texas sustained; reasonable relationship to prevention of waste. |
| 1934 (February 13) | *Panama Refining Company, et al. v. Ryan, et al.* | 5 F. Supp. 639 (Eastern District of Texas, Tyler) | Subsection 9(c) of the National Industrial Recovery Act prohibiting interstate shipment of oil produced in excess of state-set allowables, Petroleum Administration Board and Federal Tender Board established by Secretary of the Interior, and Petroleum Administrator, Harold Ickes, declared an unconstitutional delegation of legislative authority. |
| 1934 (May 22) | *Ryan v. Amazon Petroleum Corporation* | 71 F. (2d) 1, 8 (Fifth Circuit Court of Appeals, New Orleans) | *Panama Refining Company, et al. v. Ryan, et al.* reversed. |
| 1935 (January 7) | *Ryan v. Amazon Petroleum Corporation* | 293 U.S. 388 (U.S. Supreme Court) | District court ruling in *Panama Refining Company, et al. v. Ryan, et al.* affirmed. |
| 1935 (June 12) | *Brown v. Humble Oil & Refining Co.* | 126 Tex. 296 83 S.W. 2d 935; on rehearing: 87 S.W. 2d 1069 99 A.L.R. 1107 (Supreme Court of Texas) | Texas Supreme Court affirms decision of Texas Court of Civil Appeals (1934) reversing District Court (1933) approval of a R37 exception for a well on a voluntarily subdivided 1.5-acre tract. Petroleum engineering evidence of drainage critical in higher court decisions. Supreme Court justices for the first time directly question the wisdom and validity of the rule of capture. |

| 1936 (March 30) | *Consolidated Gas Utilities v. Thompson* | 14 F. Supp. 318 (Federal District Court, Western District of Texas) | Although the preponderance of the technical evidence showed that the plaintiffs had suffered drainage, not the operators without pipeline connections, the Court found that the 1935 Texas Natural Gas act only proscribed waste and did not protect correlative rights. |
|---|---|---|---|
| 1937 (February 1) | *Thompson v. Consolidated Gas Utilities* | 300 U.S. 55 (US Supreme Court) | Affirmed, although the Supreme Court expressed doubts about the lower Court's interpretation of the Texas statute's intent and affirmed the power of the state to protect correlative rights. |
| 1939 (June 12) | *Rowan & Nichols Oil Co. v. Railroad Commission of Texas* | 28 F. Supp. 131 (Federal District Court, Western District of Texas) | Rowan & Nichols challenged a prorationing order for the East Texas field which, as implemented, held their wells, among the best in the field, to barely more than the worst. The District Court found the order to be confiscatory. |
| 1939 (November 3) | *Railroad Commission of Texas et al. v. Rowan & Nichols Oil Co.* | 107 F. 2d 70 (United States Court of Appeals, Fifth Circuit) | Affirmed. |
| 1940 (June 3) | *Railroad Commission of Texas v. Rowan & Nichols Oil Co.* | 310 U.S. 573, 60 S. Ct. 1021, 84 L. Ed. 1368, 1940 U.S. Lexis 1079 (US Supreme Court) | Reversed. In view of the dynamic character of petroleum engineering knowledge, the Court declined to tamper with the expert decisions of the Railroad Commission. |
| 1941 (January 6) | *Railroad Commission of Texas v. Rowan & Nichols Oil Co.* | 311 U.S. 570, 61 S. Ct. 343, 85 L. Ed. 358, 1941 U.S. Lexis 1109 (US Supreme Court) | Rowan & Nichols challenged the Commission's follow-on order, which ostensibly considered other scientific factors, but made little practical difference. Federal District Court enjoined, but the Supreme Court, citing the "fruitful empiricism" of the Commission's "continuing administrative process," again reversed. |
| 1943 (May 24) | *Buford et al. v. Sun Oil Co.* | 319 U.S. 315, 63 S. Ct. 1098, 87 L. Ed. 1424, 1943 U.S. Lexis 1103 (US Supreme Court) | Recognized that the Railroad Commission and designated state courts were "working partners," who had developed the "specialized knowledge" necessary to fairly administer conservation statutes and regulations. |

# CHAPTER 6

# REASONABLE MINDS

. . . we are bound to say that all this vast amount of evidence, submitted in favor of the commission's findings, is too ponderable to be brushed aside as no evidence at all. We find ourselves wholly unable to say that the conclusion the commission reached is not one which reasonable minds could entertain.

—JOSEPH C. HUTCHESON JR., *AMAZON PETROLEUM CORPORATION V. RAILROAD COMMISSION OF TEXAS ET AL.* (1934)[1]

## REASONABLE MIND

Joseph C. Hutcheson Jr. was the sort of jurist who would give attorneys ulcers and strict constructionists apoplexy. One journalist said that messing with him was "like monkeying with a naked bolt of lightning."[2] Occasionally testy or not, he had a nuanced sense of fairness and justice, which he was more than happy to use the power of the court to encourage.[3] DuVal West, who served with Hutcheson on the court that heard the second (federal) *Danciger* case, praised both the quality of his jurisprudence and his "striking a balance" by "weighing threatened injuries [and] benefits received." West's letter, written as the court was formulating its decision, testifies also to Hutcheson's reputed "dominant" personality: "Believe me when I say that your discussion of the facts, in light of the difficulty of application of the law, is treated in a way that convinces me that you are a real judge."[4]

Real judge or no, Hutcheson thought deeply about what he was doing. He published an extraordinarily candid article in the *Cornell Law Review* in the spring of 1929 (while still the sole judge for the Federal Southern District of Texas), "The Judgement Intuitive: The Function of the 'Hunch' in Judicial Decision."[5] Hutcheson rejected the notion of jurisprudence as logomancy, and with it the Weberian notion of formal rationality (that all rational decisions should be determinate in laws and rules), and instead went looking for how decisions were actually made. He found that the process is not deductive, no matter how it is later written up. Rather, "I, after canvassing all the available material at my command, and duly cogitating upon it, give my imagination play, and brooding over the cause, wait for the feeling, the hunch—that intuitive flash of understanding which makes the jump-spark connection between question and decision."[6] Then, the judge, "having travailed and reached his judgement . . . struggles to bring up and pass in review before his eager mind all of the categories and concepts which he may find useful directly or by analogy, so as to select from them that which in his opinion will support his desired result."[7] Hutcheson thus distinguishes between what philosophers of science sometimes call the context of discovery and the context of justification, and sees his process as the better alternative, "since the narrow and prejudiced view of the obligations of a judge in the decision of causes prevent his resort to the judgment aleatory by the use of his 'little small dice.'"[8]

Far from advocating irrationality, Hutcheson drew explicit parallels between legal decision, scientific discovery, and mathematical invention, even citing the French polymath Henri Poincaré.[9] Compare Poincaré on mathematical creation to Hutcheson on judicial decision:

> Often when one works at a hard question, nothing good is accomplished at the first attack . . . then all of a sudden the decisive idea presents itself to the mind.
>
> Most striking at first is this appearance of sudden illumination, a manifest sign of long, unconscious prior work.[10]

Such unconscious work is "only fruitful, if it is on the one hand preceded and on the other hand followed by a period of conscious work."[11] Poincaré hypothesized that in this unconscious effort, the subliminal mind assembles random combinations of an arbitrarily large array of elements prepared by prior conscious work, and then selects "among these combinations so as to eliminate the useless ones or rather to avoid the trouble of making them" so that "only the interesting ones would break into the domain of consciousness." This crucial unconscious selection for conscious appraisal is accomplished by the mathematician's "special esthetic sensibility which plays the role of the delicate sieve."[12] It is then "necessary to put in shape the results of this inspiration, to deduce from them the immediate consequences, to arrange them, to word the demonstrations."[13]

Hutcheson's parable of the probability cubes—his "little small dice"—as a parallel to finding and following his "hunch" suggests he understood the underlying random element in his search process.[14] Despite the provocative title of his essay, it is remarkable that a sitting federal judge carrying the heaviest case load of any single-judge District in the United States should have taken the time to probe his own reasonable mind so deeply.[15]

## AMAZON, ET AL.

The penultimate federal prorationing case in Texas, *Amazon Petroleum Corporation v. Railroad Commission of Texas* (1934), provided ample scope for Hutcheson's intuition, as well as his other faculties. Of the six major Texas and Oklahoma prorationing cases prior to *Amazon*, two state cases, *Julian* in Oklahoma and first *Danciger* in Texas, and two federal cases, *Champlin* from Oklahoma and second *Danciger* from Texas, had affirmed prorationing orders. Two East Texas federal cases, *MacMillan* and *People's* (as well as a scrawny battalion of others) had enjoined prorationing orders. All four of the affirmative cases concerned gas-drive fields, about which there was uncontroverted scientific consensus: broadly, the Beecher and Parkhurst paradigm. In contrast, in the East Texas cases, there was no scientific or technical consensus.

California, which never had any form of state-mandated oil prorationing, market demand or otherwise, did in its conservation statutes prohibit both waste of natural gas alone and waste of natural gas or gas energy in oil production, as measured by gas-oil ratios—again Beecher and Parkhurst. The California Supreme Court, in *People v. Associated Oil Co.* (December 3, 1930) made the determination of such waste explicitly dependent upon petroleum engineering analyses. After reciting the factors affecting oil and gas production (extent of oil sand, reservoir pressure, porosity, and so on), the court held that natural gas "allowed to come to the surface without its lifting power having been utilized to produce the greatest quantity of oil in proportion, may be said to amount to an unreasonable waste of natural gas," citing the petroleum engineering science publications of Theodore E. Swigart, C. E. Beecher, and Henry C. Miller.[16] The US Supreme Court, in *Bandini Petroleum Company v. Superior Court of California* (November 23, 1931—prior to *Champlin*), affirmed the California state court's ruling, quoting the state court's scientific reasoning verbatim.[17]

Most legal commentaries, at the time and since, credit the simple fact that Oklahoma had passed a market-demand statute (in 1915), which explicitly prohibited "economic waste," or production in excess of market demand, while Texas had explicitly excluded economic waste in its definitions of waste in its oil and gas conservation legislation, as being the causal difference between *MacMillan* and *People's* versus *Julian* and *Champlin*. California, of course, doesn't fit the mold at all. The actual, decisive differences between these cases are technical. No American court has ever sanctioned state limitation of oil or gas production to "reasonable market demand" on its own bottom, or as a means of market stabilization or price fixing. Market-demand prorationing, or production restraint of any sort, is permissible only as a means to some other end. By the 1930s, use of the state's police power to regulate private property for the public good had a long and well-established legal precedent.[18] In all cases, the legal criterion for the legitimate use of the police power is that it bear a "reasonable" relation to a proper public purpose. For oil and gas resources, this standard meant the prevention of surface or subsurface waste, the protection of correlative rights, or the conservation of natural resources.[19] The core criterion is scientific and technical: How does production in excess of reasonable market demand cause waste or violate correlative rights, and how does prorationing prevent same? This issue in turn hinges upon the scientific character of hydrocarbon reservoirs, precisely the domain of petroleum engineering. After *Champlin*, the legal standard for production restraint had become a simple modus ponens: if de facto Beecher and Parkhurst, then de jure reasonable. If not, then "unreasonable" and void. Problem was, East Texas didn't fit.

The complaint in *Amazon Petroleum Corporation v. Railroad Commission of Texas* was filed October 26, 1933, in the Federal District Court for the Eastern District of Texas in Tyler. By agreement of all parties, *Amazon* was combined for trial with a small herd of similar cases contesting Railroad Commission prorationing orders in East Texas and also with another East Texas case, *Panama Refining Company et al., v. Ryan et al.*, which challenged the legality of the Petroleum Code of the National Industrial Recovery Act. Plaintiffs advanced two sets of claims. First were the traditional objections to Railroad Commission prorationing orders: that they were "unconstitutional, illegal, and confiscatory" under the US Constitution's

Fourteenth Amendment prohibition of taking property without due process of law, and served not the legitimate purpose of preventing waste, but instead, illegally, simply fixed prices.[20] The second set of claims maintained that the Railroad Commission had abrogated its statutory responsibility to prevent waste of Texas resources by supinely accepting "consumer demand" estimates of the US Bureau of Mines and production allocations under the Petroleum Code of the National Industrial Recovery Act, and that as a result the actions of federal agents in Texas in enforcing production restrictions were illegal.

The two sets of claims asserted "two wholly separate and distinct causes of action," the first requiring a three-judge federal panel, the second decision by a single federal district court judge. By mutual agreement, the causes were tried together, but severed for submission.[21] The due process complaint against the Railroad Commission of Texas was heard and decided as *Amazon Petroleum Corporation v. Railroad Commission of Texas et al.* by the required three-judge panel comprising Joseph C. Hutcheson Jr., Justice, Fifth Circuit Court of Appeals; Thomas Martin Kennerly, District Judge for the Southern District of Texas, Houston; and Randolph J. Bryant, District Judge for the Eastern District of Texas, Tyler. The NIRA complaint was decided by Bryant alone (*Panama Refining Co. et al. v. Ryan et al*). In *Panama*, Bryant held that the President of the United States, under Section (9c) of the National Industrial Recovery Act, had subsumed legislative powers that Congress could not constitutionally delegate; his decision was ultimately affirmed by the US Supreme Court January 7, 1935, in an eight-to-one decision, prior to *Schechter v. United States*, which extinguished the NIRA.[22] *Amazon* and *Ryan* were argued together in Houston, December 14, 1933.[23] The court's *Amazon* decision, written by Hutcheson (with Kennerly dissenting), was rendered February 12, 1934, and finally bestowed legal legitimacy on Railroad Commission prorationing in all of Texas, including East Texas. To understand how *Amazon* differed from the earlier East Texas cases requires knowing what petroleum engineers knew in the summer of 1931, when *MacMillan* was tried, in the summer and fall of 1932, when *People's* was heard, and what they had learnt by December 1933, when *Amazon* was tried. It was a bunch.

## WHAT PETROLEUM ENGINEERS KNEW AND WHEN THEY KNEW IT[24]

To reprise: When *MacMillan* was tried June 24, 1931, it had been less than nine months since Daisy Bradford No. 3 had come in. Few hard facts were known about the East Texas field. Drilling logs indicated the depth of the Woodbine formation, the depth from which oil production came, the thickness of the pay, and, taken together, the Woodbine's tilt and pinch out to the east. Core samples yielded data on formation porosity and permeability. Open-flow tests indicated well potentials, although tests were usually done by operators themselves and were of questionable accuracy. Previous dry holes to the west of the field confirmed both the depth of the Woodbine and the presence of salt water, while a few wells in the western parts of the field had drilled right through the oil sand into underlying salt water. Much about the field remained unknown: Consensus was just coalescing that it was one huge field, not three separate pools, but the field was growing rapidly and its contours were still poorly defined.

Although most petroleum engineers accepted the veracity of the Beecher and Parkhurst para-digm and assumed a gas drive, gas-oil ratios had not been systematically recorded, and usually were not measured at all.[25] Bottom-hole pressures were estimated from surface pressures at the casinghead plus allowances for depth or hydrostatic head: precious few actual bottom-hole pressure measurements had been made.

The first reliable bottom-hole pressure gauge had been designed and built by Kenneth C. Sclater, Chief Petroleum Engineer of the Marland Production Company, Ponca City, Oklahoma, in 1926, immediately following publication of Beecher and Parkhurst's definitive paper.[26] Marland actually initiated development of four devices, intended to work together: a downhole pressure gauge, a "thief," or downhole fluid sampler, which also incorporated a maximum temperature thermometer, a maximum pressure bomb or gauge, and a recording pressure bomb to track changes in pressure in the production casing. The first three instru-ments were designed to be run as a unit. Only the downhole hole pressure gauge, which was tested in the Panhandle and Yates fields, and the thermometer gave consistent readings. Sclater and B. R. Stephenson presented preliminary results of their work at the American Institute of Mining Engineers meeting in Tulsa in October 1928. They showed that downhole or reservoir pressures differed substantially from those inferred from surface casinghead measurements, while limited results with the "thief" indicated that dissolved gas and gas-oil ratios at reservoir pressures and temperatures differed even more from those at the casinghead.[27] Likewise, lim-ited results from the recording pressure gauge showed sharp discontinuities in dissolved gas, gas-oil ratios, and flow properties as reservoir fluids rose up the production casing, pressure fell, and gas came out of solution.[28]

Despite Sclater and Stephenson's promising results, development of downhole instru-ments lagged. Even in October 1930, not long before *MacMillan* came to trial, C. V. Millikan and Carroll V. Sidwell (Amerada Petroleum Corp.) complained that as important as bottom-hole pressure was, "there is less information about these pressures than about any other part of the general problem of producing oil."[29] Only a few corporate-funded bottom-hole pressure gauges existed in various stages of development and experimental use. D. G. Hawthorn, a Production Engineer for the Barnsdall Oil Company in Tulsa, reported in October 1932 that "more than a dozen companies" were "operating subsurface pressure instruments in the Mid-Continent field," while E. O. Buck estimated there were fewer than twenty downhole gauges in the entire country in late 1932.[30] In the East Texas field, no bottom-hole measurements were made until "250 or 300 wells" were already producing. The first determinations were made during May and June 1931, just as *MacMillan* was coming to trial, but were "confined to a somewhat restricted area." Data from later tests "in the outlying portions of the field," however, did suggest that initial reservoir pressure had been "not less" than 1,650 pounds per square inch.[31] But no solid information on reser-voir fluids or gases at reservoir temperatures and pressures existed. Thus, most of the basic empirical facts about East Texas were lacking.[32] When Hutcheson characterized Railroad Commission witnesses' portrayal of the East Texas field as "largely theory and speculation," he was being charitable.

The situation improved by the time of the *People's* preliminary injunction hearing in June 1932 and trial in September 1932. By August 17, 1931, when the field was shut in completely under martial law, extreme pressure differences, from 1,600 pounds per square inch in some areas to less than 800 pounds in others, were manifest.[33] The Railroad Commission issued new orders sharply restricting output September 5, 1931, with even more stringent limitations continuing into the first three months of 1932.

The big difference between *MacMillan* and *Peoples'* was the unintended natural experiment proffered by Governor Sterling's martial law decree.[34] Bottom-hole determinations showed "a marked rise in reservoir pressure in areas which had shown low or subnormal pressures just prior to the shut-down."[35] In papers presented to the American Association of Petroleum Geologists meeting in Oklahoma City, March 25, 1932, both D. R. Snow, vice-president in charge of the geological and land departments of the Barnsdall Oil Company, and C. V. Millikan argued that the limited number of bottom-hole pressure tests conducted during late 1931 and early 1932 in East Texas by private firms showed that the observed pressure behavior of the tested wells, distributed across the field, was inconsistent with a dissolved gas drive (Beecher and Parkhurst), but consistent with a water drive. Snow claimed that, "The bottom-hole pressure readings . . . indicate that the slow rate of withdrawal, which has been permitted during the last two or three months in East Texas, is allowing sufficient time for water to advance along a uniform front and to maintain reservoir pressure at very nearly a constant figure."[36] Likewise, Millikan claimed that the gas-oil ratio in East Texas "is about 350 cubic feet per barrel and all in solution at a pressure of 800 pounds or less," and that therefore gas drive could not account for observed rates of production. He conjectured that the East Texas oil field was connected along its western edge to the vast Woodbine Aquifer underlying most of East Texas, and reasoned that, "The more common belief that water is moving into the field from the west has been substantiated by bottom-hole pressures many months before such movement could be proved by wells on the west side actually going to water."[37]

Three weeks later, E. V. Foran, a consulting petroleum engineer from Fort Worth, in a paper presented April 14, 1932, at the spring meeting of the Division of Production of the American Petroleum Institute, reviewed the history of bottom-hole measurements in East Texas. Again, the decline and then increase in average pressure in the field was physically impossible in a gas-drive field: once dissolved gas came out of solution as oil was produced and pressure fell, average pressure in the field could only go down, come what may. Foran, like Snow and Millikan, concluded that East Texas was a pure water-drive field, and that severe production restraint was the only way to extend the life of the field, prevent underground waste, and maximize ultimate recovery. From these observations and analyses, many petroleum engineers agreed that East Texas was a water-drive field, or under hydrostatic control, in the older terminology. It was this logic that underlay E. O. Buck's testimony in the *People's* preliminary hearing in June 1932, which had all but convinced Judge Hutcheson.

These radical inferences required proselytizing within the community of petroleum engineers and in the broader oil fraternity, typical of how a new scientific theory or finding is socially produced and performed, or, more crassly, if accurately, "sold," to the relevant community of practitioners, and ultimately to users. The Humble Company's H. D. Wilde Jr. presented a "clear, detailed discussion of the use and interpretation of pressure contour maps" and their use in inferring reservoir characteristics in a paper before the API Mid-year Meeting in Tulsa, June 3, 1932, three weeks before the preliminary *People's* hearing.[38] D. G. Hawthorn presented detailed, comprehensive pressure contour plats (maps) of the East Texas field for February, April, June, and August 1932, in a paper at the meeting of the AIME's Petroleum Division in Ponca City, October 1932. These plats, combined with production records, showed the exact correlation between rates of withdrawal and pressure changes in the reservoir. The data for Hawthorn's plats were "compiled through the cooperation" of twelve of the major operators in the field—presumably the ones who had deployed downhole pressure gauges.[39] As Robert Boyle, the protagonist of seventeenth-century experimental science, understood, nothing like a public demonstration before an audience of gentlemen peers to convince.[40]

But all was not sugar and spice, and big lacunae remained. Not only were bottom-hole measurements still relatively few, but also even surface tests of well potentials were limited. As Hutcheson complained, the Railroad Commission had opposed operators' open-flow tests, apparently because the Commission saw them, like Japanese "scientific whaling" later, as subterfuges for commercial exploitation. Knowledge of actual reservoir contents at reservoir conditions was still restricted to inferences made from samples taken at the well head. Most critically, petroleum engineers had not yet concocted a coherent explanation for how water drive in East Texas, as an alternative to gas drive, might work. The problem was the tremendous volume of water necessary to maintain pressure that would have to flow from the Woodbine Aquifer into the East Texas field to replace the huge volumes of oil being withdrawn. If fresh water were assumed to be entering the aquifer at or near its outcrop, frictional energy losses in moving through the porous medium would be too great to permit the flow necessary to sustain observed pressures.[41] Combined with Uren's 1927 demonstration that too great a restraint on production rates in a gas-drive field would actually waste reservoir energy and result in unrecoverable oil, as well as the still sparse data on East Texas, these considerations posed a ponderable challenge to the water-drive hypothesis, and to draconian prorationing, as successfully articulated by plaintiffs in *People's*.

But petroleum engineers knew what they needed to know, or at least a lot of it, partly due to Hutcheson's unsubtle nudge in the final *People's* decision and to his harsh condemnation in *Rowan & Nichols* (January 1933). First, engineers needed much more comprehensive, quantitative data on precisely what the pressure changes in the East Texas field were in response to different rates of production and as a result of varying drilling density. They needed to know what exactly the fluids in the subsurface reservoir comprised at reservoir temperatures and pressures, and how those constituents reacted to pressure changes within the reservoir and within the production casing of wells as the fluids rose to the surface. And once these data were in hand, and the requisite empirical constants defined, petroleum engineering scientists needed to develop a robust quantitative, dynamic, and predictive theory-cum-explanation for how the field behaved. Only then could they claim to have an empirically adequate scientific understanding and offer persuasive and reliable engineering prescriptions.[42] To meet these demands necessitated invention and development of new experimental apparatus and practices, intensive, cooperative research and experimentation, creation of new institutions and cooperative relationships, and novel theoretical innovations.

It took a while, but the engineers delivered. As noted, the Railroad Commission had mandated the first systematic, but limited, bottom-hole pressure tests in East Texas in December 1932.[43] Potential production of a larger sample of 275 wells (out of 10,000) was measured before and after a Commission-ordered shutdown in April 1933. In response to criticisms that these shut-in tests did not accurately measure the real potential production of a well while other wells in the field were also flowing, Railroad Commission engineers ran new tests under normal field operating conditions in June 1933. Engineers took repeat downhole pressure measurements late in 1933, just prior to the *Amazon* trial: tests showed actual increases in reservoir pressure whenever production was restrained or the field shut in.[44] This evidence convinced many petroleum engineers that East Texas could not be a gas-drive field of the sort envisaged in the Beecher and Parkhurst paradigm. The conflict between the empirically observed pressure-production relationship in East Texas and the nevertheless significant gas-oil ratios exhibited by some of its wells focused attention on two interrelated phenomenological issues, both of which had theoretical implications: the actual contents of and conditions within the subsurface reservoir, and the behavior of reservoir fluids as they rose to the surface within the production casing.

Ben E. Lindsly, Senior Petroleum Engineer, and Carl E. Reistle Jr., Petroleum Engineer, and E. P. Hayes, Associate Gas Engineer, all at the Bureau of Mines Petroleum Experiment Station, Bartlesville, conducted the fundamental field research which finally allowed complete characterization of the fluids in the East Texas reservoir and their behavior in the production process. Lindsly, and Reistle and Hayes, reported their results in papers read before the American Petroleum Institute Division of Production meeting in Tulsa, May 19, 1933. The papers were published as *United States Bureau of Mines Reports of Investigations* 3212 and 3211 respectively, May 1933.[45] Lindsly used a just-developed BOM downhole fluid sampler to obtain direct samples of East Texas oil at reservoir temperature and pressure—the device's design had not even been completed when *MacMillan* came to trial.[46] This work was part of a larger program on "gas-solubility research"—clearly within the Beecher and Parkhurst tradition—begun in January 1930, before East Texas came in. This program aimed "to gain a more secure knowledge of the physical and chemical characteristics of the oil-gas mixtures that exist in natural reservoirs."[47] As is characteristic of radical paradigm shifts, on close inspection, an older research programme (dissolved gas drive) seems to blend nearly seamlessly in real time into a newer programme (water drive), while from a more distant perspective the replacement of old by new seems complete, abrupt, and revolutionary.

Carl Reistle had designed and developed the BOM bottom-hole sampler, which also could continuously measure temperature and pressure "at any point in a flowing or shut-in well."[48] The new BOM fluid sampler underwent its first field tests in March 1932, in Oklahoma's Seminole field. In East Texas, the Bureau of Mines engineers then took bottom-hole samples and ran extensive pressure and production tests on six wells thought to be representative of the field. The wells belonged to four different operators: Shell Petroleum, Barnsdall Oil, Sinclair Prairie Oil and Gas, and Sun Oil; these operators also supplied labor and material to conduct the tests.[49] The tests were run in the summer and fall of 1932.[50]

In his presentation, after a detailed discussion of the design and construction of the bottom-hole sampler and the procedures to take samples and to secure, process, and evaluate them in the laboratory, Lindsly reviewed his results. They were decisive. First, East Texas crude oil at reservoir pressure, 1,400 lbs./in.$^2$, and temperture,146°F, when Lindsly ran his tests, was unequivocally undersaturated, and remained so until pressure was reduced to 755 lbs./in.$^2$: Dissolved gas would not begin to come out of solution in the reservoir until pressure fell below 755 lbs./in.$^2$. Therefore, the expansion of dissolved gas could not provide the energy necessary to move oil through the formation to the well bores or to maintain reservoir pressure. The Beecher and Parkhurst gas-drive paradigm did not apply to the primary production mechanism in East Texas. Second, between subsurface reservoir pressure and temperature and the well-head at atmospheric pressure and 90°F, East Texas crude oil shrank: a barrel of oil at the surface, after dissolved gas had been liberated and temperature fell, would have occupied a volume at least 20.6 percent greater when in the reservoir. Production records based upon oil coming out of the gas-oil separators and flowing into the stock tanks on the surface understated the total volume of reservoir fluid being withdrawn by at least that amount. Given the depth of East Texas production, the volume of dissolved gas and its saturation, and reservoir pressure, combined with Reistle and Hayes' results (see below), Lindsly calculated the energy available in the East Texas field to lift the oil and found it in excess of what was needed by nearly 20 percent.

As the view that East Texas oil was undersaturated at reservoir conditions gained traction among petroleum engineers in late 1931 and early 1932, some proposed, as an alternative to water drive, that the oil in the field itself was compressible and provided the energy to move fluids through the formation.[51] This hypothesis entailed, as did the gas-drive hypothesis, that severe production restrictions or prorationing would do no good and might even cause harm by reducing ultimate recovery. Lindsly refuted this possibility. Lindsly calibrated his experimental apparatus with kerosene and got results that matched Percy Bridgman's much earlier determination.[52] Using Bridgman's results for the compressibility or expansibility of kerosene, Lindsly showed that when pressure was reduced from 1,400 lbs./in.$^2$ to 700 lbs./in.$^2$, which corresponded "roughly with the drop from the existing [reservoir] pressure to the pressure at which the East Texas oil begins to liberate gas," pure kerosene volume would increase 0.3 percent. For East Texas crude at 146°F (reservoir temperature), at any pressure above 755 lbs./in.$^2$, expansion (or compressibility) was 0.6 percent, which compared favorably with other laboratory measurements which showed that the compressibility of "all hydrocarbon liquids is of the order of 0.5 percent."[53]

Lindsly then offered a classic reductio ad absurdum: "Using a total shrinkage factor of 20 per cent" for East Texas oil between the reservoir and the surface, the 262 million barrels of total production recorded at the surface as of April 1, 1933, would equate to 327 million barrels withdrawn from reservoir. Replacing this volume by the expansion of reservoir oil alone, given 0.6 percent expansibility, entailed that the East Texas field originally must have contained 163 *billion* barrels of oil, which, Lindsly dryly observed, "completely disagrees with reliable estimates of the original reservoir content of the East Texas field, based upon such factors as areal extent, sand thickness, and porosity."[54]

Lindsly concluded that because neither gas drive nor oil expansion could account for production and observed pressure changes in East Texas, a (nearly pure) water drive was the most probable explanation, especially since estimates of "the volume of water encroachment is in close accord with the amount of reservoir oil that has been removed." Moreover, those estimates also "indicate that the displacement of oil by water in the East Texas field has been efficient and only small quantities of oil have been trapped and left behind in the water flooded area."[55] Lindsly's results convinced the community of petroleum engineering scientists, most practicing petroleum engineers, and ultimately the oil fraternity in Texas.[56] As Gulf Oil's Ralph D. Wyckoff acknowledged in October 1934, although he had expressed strong objections to the water-drive hypothesis in 1932, "We have since had available the gas-saturation measurements for the East Texas field obtained by Mr. Lindsly. . . . We are forced by this later evidence to obtain a 100 per cent water drive."[57]

Reistle and Hayes in their API paper and in their BOM *Report of Investigations* announced results that addressed any lingering doubts about water drive in East Texas. The two engineers analytically separated flow phenomena within the reservoir, radial flow, from those within the production casing, vertical flow in pipes. Reistle and Hayes began by noting three earlier attempts to understand vertical flow in oil well production casing, all of which initially were inspired by the use of artificial gas lifts (the introduction of additional gas at the bottom of a producing well to lift oil). Jan Versluys, a research engineer at Bataafsche Petroleum Maatschappij (Batavian Oil Company, the Dutch East Indies operating subsidiary of Royal Dutch Shell), Thomas V. Moore (Production Research Engineer) and H. D. Wilde Jr. (Director of Production Research), Humble Oil and Refining Co., Houston, and Moore and Ralph J. Schilthuis (also Humble Oil) had presented theoretical papers during 1929–1932 on the problem (see below), but all three suffered from a paucity of data.

Reistle and Hayes had the advantage of Lindsly's concurrent and definitive demonstration that East Texas oil was undersaturated at reservoir conditions and that the natural gas dissolved in it did not begin to come out of solution until pressure fell below 755 lbs./in.[2] Together with the new data on pressure decline contours in the East Texas field, Lindsly's results meant that Reistle and Hayes could be sure the fluid entering an East Texas well was a single-phase, homogeneous fluid—analytically a much simpler problem than multiphase flow—and that whatever gas appeared in the production casing or at the surface had come out of solution: There was no free gas in the reservoir.

Reistle and Hayes first addressed the relationship between (shut-in) reservoir pressure, pressure and temperature at the face of the producing sand (downhole), and rates of production. For these tests, they produced wells "at different rates of production for a sufficient length of time so that equilibrium conditions were established," while "a continuous pressure and temperature record was obtained at the face of the sand with the Bureau of Mines recording gage [*sic*], and the gas and oil production was carefully measured at the surface." Reistle and

Hayes confirmed that for a single-phase fluid such as that in East Texas, "the rate of production was a linear function of the pressure at the face of the sand," which agreed with both Lester Uren's experimental findings on flow into an oil well and the application of Poiseuille's law for flow in capillary tubes.[58] Moreover, "Assuming that the drainage radius of these wells will not change as the reservoir pressure declines, the rate of production of the wells will remain constant for the same difference in pressure between the reservoir pressure and the pressure at the face of the sand, regardless of the decline in reservoir pressure, so long as the reservoir pressure is above that pressure at which gas is liberated from solution in the oil."[59] From rates of production and pressure differentials, they also could calculate the relative permeability of the sand surrounding each well (which had implications for setting allowables). And by using pressure and temperature readings taken with a recording pressure and temperature gauge "at intervals of 100 feet" in the production casing, they could calculate the density of the oil in the reservoir, in the flow column, and at the surface, which in turn allowed them to calculate oil shrinkage as gas was liberated. Their calculations corroborated Lindsly's separate determination.[60]

Reistle and Hayes next analyzed the behavior of East Texas oil within the production casing or flow column. Pressure and temperature "at different points in the flow string were measured" with the BOM recording gauge. From these data, Reistle and Hayes concluded that there were "two definite types of flow, with possibly a third or transition condition." A "foam flow" prevailed when the gas-to-oil volume ratio in the flow string was less than one and the "relative linear velocity of the gas [bubbles] and oil is small," which became a "stable foam" with equal flow velocities, and near 100 percent lifting efficiency as gas-to-oil ratios approached one. If the gas-to-oil ratio were substantially greater than one, the gas velocity became much greater than that of the oil, and, therefore, the use of gas energy for lifting oil much less efficient.[61]

Not only did operators in the East Texas field provide the wells, labor, and material to conduct Lindsly's and Reistle and Hayes's research, but collectively they ended up funding the entire enterprise. In the severe budgetary stringency of 1931–1933, the Bureau of Mines ran out of chips and decided it could not support continued work in East Texas. Some forty-one different operators, majors and independents alike, contributed to a communal fund to build a laboratory near Kilgore "for the exclusive use of the Bureau of Mines." This laboratory became the nucleus of the East Texas Engineering Association, which briefly employed Carl Reistle as its chairman, and continues to do research on the field. Among the forty-one contributors were some of the most vehement and persistent opponents of prorationing.[62] When Bureau of Mines appropriations were even more "drastically reduced on July 1, 1932," and all work in East Texas was to be "suspended indefinitely," a group of operators raised the money, including salary support, to continue.[63] These efforts extended prior oil fraternity voluntary cooperation to support field umpires and their staffs in pro-rated oil and gas fields, and traced back to even earlier field safety and vigilance committees. This collective endeavor marked the functional merger of the moral economies of the oil fraternity and of petroleum engineering science.

Lindsly and Reistle themselves still had much to do. While the elegance of their logic and the quality of their data might be more than enough to convince their fellow petroleum engineering scientists, persuasion of the larger oil fraternity required that their results be socially reproduced in multiple arenas. Also, their practical relevance and implications had to be made explicit and persuasive. Lindsly and Reistle followed up their May 1933 presentations and publication of their BOM *Report of Investigations* by giving papers at the October 1933 Meeting of the Petroleum Division, American Institute of Mining and Metallurgical Engineers, in Dallas. Lindsly focused on the role of gas in reservoir dynamics, drawn from the results of the research he had completed just prior to the East Texas study. There he extended Beecher and Parkhurst's and their successors' experiments to naturally occurring oil-gas mixtures.[64] In addition, corroborating the results of Beecher and Parkhurst, Dow and Calkin, and of the later experiments at the California Institute of Technology by W. N. Lacey and his colleagues, Lindsly showed that when dissolved gas came out of solution it sharply decreased the API gravity of the oil, as well as its viscosity and surface tension.[65] In his discussion of Lindsly's paper (the only one published), Carl Reistle seconded the conclusions: "I believe that our present-day knowledge of reservoir conditions and the movement of oil and gas within the reservoir sand clearly indicates the value of retaining a maximum quantity of gas in solution in order to maintain a low viscosity and surface tension for the oil."[66] It is as if Lindsly and Reistle had realized they had all but overthrown the Beecher and Parkhurst gas-drive paradigm and wanted to head off throwing the baby out with the bathwater: keeping gas in solution was still critical. The Railroad Commission certainly got the message. In Special Rules adopted for East Texas on October 17, 1933, the commission required that each operator "shall equip his well or wells with necessary equipment for metering gas and taking bottom hole pressure as, if, and when required by the Railroad Commission" (Rule 8), and that gas measurements be reported with daily oil production at least once a month (Rule 9).[67]

Nye and Reistle in their companion paper offered detailed data on the East Texas field as of September 27, 1933, and specifically applied their and Lindsly's insights to it. Presciently, they stressed that to understand the behavior of the field it was "necessary also to have knowledge of the entire reservoir system, which may extend for several hundred miles." After presenting production and weighted average reservoir pressure, October 25, 1932, to September 27, 1933, the two researchers concluded that the "decline in reservoir pressure is a function of the rate of production. The sharp increase in reservoir pressure during the period Dec. 18–31, 1932, and April 6 to April 23, 1933, are the result of closing the field completely."[68] Nye and Reistle hammered home the empirical fact that East Texas was a water-drive field, and physically could not be gas drive:

[The] increase in the average pressure of the part of the reservoir system containing oil, as the rate of oil production was decreased but not stopped, is perhaps the most definite proof that the withdrawal of oil from the reservoir is accompanied by replacement with water . . .

Furthermore, the expansion of fluids, whether oil or gas, or a combination of these substances could hardly result in an increase in the average pressure of a reservoir system. It is this movement of water into the reservoir that supplies the pressure energy that has controlled the reservoir pressure within the oil-bearing part of the reservoir system.[69]

Given these facts, Nye and Reistle argued that stringent limitation of rates of production was necessary in East Texas and in all water-drive oil fields.[70]

By November 1933, water drive in East Texas, the importance of limited rates of withdrawal, and the efficacy of Railroad Commission prorationing were settled facts of the matter for most petroleum engineers and for much, but not all, of the oil fraternity in Texas. Theoretical explanation, too, had kept pace.

Development of theories of reservoir behavior in general, and of East Texas in particular, was interleaved in real time with accumulating empirical evidence and field experience, as well as intensive programmes of laboratory experimentation. Theoretical invention, innovation, and refutation or corroboration in petroleum engineering science emerged in the same dialectical pas de deux with experimental innovation and empirical insight as in any other natural science. Theory is separated out here for narrative convenience and because theoretical advances tended to be published—performed, if you will—in separate papers, for different audiences, and in different journals, from those in which experimental results and field data were presented. Many papers did conjoin new experimental results or empirical data with new theoretical developments, and sorting a given paper into an empirical or theoretical category is arbitrary. This account of theoretical developments is anything but a complete history of the evolution of petroleum engineering science. Criteria for inclusion here are that the particular contribution be retrospectively significant, that it be relevant to understanding the East Texas oil field, especially during the period around the *Amazon* trial, and that, taken together, these selections illustrate the historical process by which understanding in petroleum engineering science, and its effect upon petroleum engineering practice, oil production technology, and the oil fraternity in Texas emerged.

From 1926 until 1931–1932 and the spectacular advent of the East Texas oil field, virtually all theoretical work in petroleum engineering science was done within the broad Beecher and Parkhurst gas-drive paradigm, concentrated at one or the other end of a macro-micro continuum, from whole reservoir dynamics to the hydrodynamics of fluid flow in porous media. Stewart Coleman (Manager of Development Department), H. D. Wilde Jr. (Director of Production Research), and Thomas W. Moore (Production Research Engineer), Humble Oil and Refining Company, Houston, offered an analysis of whole reservoir dynamics soon after Beecher and Parkhurst's original paper. Coleman, et al. presented their paper at the Tulsa Meeting of the Petroleum Division, American Institute of Mining and Metallurgical Engineers, in October 1929.[71] They used material balance calculations to portray changes in aggregate reservoir contents and subsurface pressure, which occurred as oil and gas were withdrawn from an oil field, and derived "a mathematical relation between the amount of gas wasted from the reservoir and the average rock pressure [the older term for reservoir pressure] after a given fraction of the [gas-free] oil has been removed."[72] Their expressions assumed reliable, time-series data on bottom-hole (reservoir) pressure, reservoir fluid constituents at reservoir temperature and pressure, and total oil and gas production. As Ralph Schilthuis observed in 1936, none of these data existed in 1929, and little additional work could be done.[73] Nevertheless, Coleman, Wilde, and Moore did point to what petroleum engineers needed to find out and provided a coherent mathematical framework for analysis.

At the other end of the spectrum, the other major tributary of theoretical research concentrated on the dynamics of fluid flow in porous media and the search for fruitful physical analogies and mathematical techniques. Although a variety of researchers made substantial contributions, the most prominent players in this game were the groups centered around Morris Muskat and Ralph D. Wyckoff at Gulf Oil Research and Development in Pittsburgh, and research engineers, including William Hurst and Ralph Schilthuis, at Humble Oil and Refining in Houston. That these cooperative yet competitive efforts prospered despite the parent firms frequently landing on opposite sides of prorationing disputes in Texas testifies to the efficacy of the ideology of science and of petroleum engineering science's particular moral economy.

The foundation for understanding fluid flow in porous, subsurface oil formations was Darcy's law for water flow in sand-filled pipes, published in 1856.[74] Henry Darcy, educated at the École Polytechnique and the École des Ponts et Chaussées, served as chief engineer for the French *département* of which Dijon is the capital. The city had a spring-fed, pressurized (gravity) municipal water system which Darcy was asked to analyze. Darcy's law applies to laminar or non-turbulent water flow in a pipe of constant cross-section containing a porous medium (sand used as a filter in Darcy's experiments). Fluid viscosity is assumed to be one, and the fluid is assumed to be incompressible. For a given pipe diameter, flow was found to be proportional to the pressure gradient and to depend upon the permeability of the medium, the viscosity of the fluid, and length through which flow occurred.[75] Darcy's law became the starting point for virtually all subsequent work on water movement in subsurface aquifers, and for the science of hydrology.[76] Moreover, Darcy's law took the same mathematical form as Fourier's law for heat conduction and Ohm's law for electrical conduction, so there were not only experimental analogies at hand but also useful mathematical techniques.[77]

Not surprisingly, petroleum engineering scientists started with Darcy's law. For example, Jan Versluys, who had done his thesis on "Capillarity in Soils" in 1916, produced two definitive if idealized theoretical papers on hydrocarbons.[78] First, at the Tulsa meeting of the American Institute of Mining Engineers, Petroleum Division, in October 1929, he offered a paper which examined "gas-lift," or, as Versluys construed the usage more broadly, "a vertical tube in which the energy of gas under pressure, and of dissolved gas, is utilized for raising a liquid."[79] While Versluys provided a thorough mathematical treatment of work and energy relations, which Reistle and Hayes later exploited, as they, and Versluys himself, pointed out, "there are factors, particularly friction, slippage, and solubility of the gas in the oil that must be known or determined experimentally before the equations . . . can be applied to actual problems."[80] Second, Versluys presented a paper to the Akademie van Weenschappen, Amsterdam at the end of

June 1930, titled, "The equation of flow of oil and gas to a well after dynamic equilibrium has been established."[81] Versluys's equations applied only to steady-state flow, and, critically, assumed that the viscosity of mixed oil and natural gas could be defined separately. This latter assumption proved mistaken: the flow of hydrocarbon mixtures in porous media entailed complex problems of multiphase behavior. While suggestive, lack of basic data meant Versluys's theoretical work, like that of Coleman, Wilde, and Hurst, had little practical influence.

More fruitful were hierarchical decompositions of the problem that pursued discrete, empirically testable models. At the beginning of May 1931, Morris Muskat and Holbrook G. Botset submitted to the newly established journal *Physics* the first of what would become an extraordinary series of papers, "Flow of Gas through Porous Materials."[82] Muskat was born in 1906 in Latvia and immigrated with his family to the United States in 1911. He earned his PhD in physics from the California Institute of Technology in 1929 and immediately joined the Gulf Oil Research & Development Company in Pittsburgh as a Research Engineer. Botset was born in Indiana in 1900 and had worked as a chemist at the Mellon Institute in Pittsburgh before moving to Gulf Oil.

Muskat and Botset used an experimental apparatus consisting of a steel tube with five manometers attached along its length to measure pressure changes. Inlet and outlet thermometers indicated that flow in all experimental conditions was isothermal: no significant temperature changes occurred in the flowing gas. They used air for their experimental gas and examined flow through a variety of porous media: fine, uniform glass beads of various sizes, unconsolidated and consolidated homogeneous as well as heterogeneous sands, and sandstone samples representative of a natural gas–producing subsurface formation. Unlike previous work, they examined the effects of higher pressures (up to 4,000 pounds per square inch).[83]

Muskat and Botset reasoned that the complexity of the tortuous flow paths in porous media precluded application of classical hydrodynamic theory (the Navier-Stokes equations), so they began with Darcy's "empirically established" law for one-dimensional flow of incompressible fluids in pipes, in which flow velocity was directly proportional to pressure.[84] In their linear flow-experiments, they found "that the gradient of the squares of the pressures is proportional to a power of the mass velocity,"[85] with the power of the mass velocity varying between 1 and 2, corresponding to purely viscous (laminar) flow or turbulent flow. Since turbulent flow requires more energy, the greater the turbulence, the lower the mass velocity or mass flow. In most natural gas reservoirs, flow velocities are modest enough that flow remains viscous, with exponents near 1.

In a modified apparatus, Muskat and Botset also examined two-dimensional radial flow through a core sample cut from sandstone. Muskat and Botset found that, for a single well tapping a subsurface gas reservoir, the "mass outflow [from the formation] in unit time per unit thickness" would be proportional to the square of the pressure difference between that at the outer diameter of the area the well is assumed to drain and the interior well face.[86] From this relationship, pressure decline in the formation and production decline could be calculated. Their calculations matched the field measurements Muskat and Botset could glean as well as the experiments on a sample of 400 gas wells reported by H. R. Pierce and E. Rawlins in a Bureau of Mines *Report of Investigations* in 1929.[87] Ten months after submitting the original gas flow paper, at the beginning of March 1932, Muskat followed up this analysis of two-dimensional radial gas flow with a purely theoretical paper, also in *Physics*, using an electrical analogy for three-dimensional radial flow of an incompressible fluid.[88] Muskat realized that equations for electrical resistance along an electrode partially penetrating a conductive disk (which had never concerned electrical engineers) could be used to calculate "flux or production passing through the system."[89]

The backstory of how these seminal papers appeared in *Physics* says a lot about discipline formation, boundary maintenance, competition and cooperation, and contested institution building in physics and in petroleum engineering science. From a letter Harry C. Fowler, then director of the petroleum division of the US Bureau of Mines in Washington, wrote to Joseph Chalmers, at the BOM Petroleum Experiment Station in Bartlesville, December 18, 1931:

> I do not know whether I told you or not, but at the time I visited Dr. Foote [Paul D. Foote, Director of Research at Gulf Research and Development] in Pittsburgh last month, he told me that the Muskat and Botset paper had been promised and was to appear in the first issue of a new journal which is being launched by the American Society of Physicists. I understand that Dr. Foote is President of this organization. Personally, I believe it is a mistake for another group of scientists to come forth with a competing technical paper at this time. How much better it would have been to have the Muskat-Botset paper in the Transactions of the A. I. M. E. where it is readily obtainable and could have become a definite part of the petroleum engineering literature dealing with flow through porous media. However, it is not my business to suggest what Dr. Foote should or should not do; merely a personal observation in passing, but I thought you might be interested in knowing where to look for the printed paper of Muskat and Botset which has caused so much difficulty in reconciling with the findings of the Oil Recovery Laboratory and the Natural Gas Section work.[90]

There seem to have been differing visions about whether industrial or engineering research, however idealized, should be included within the nascent "pure" physics community, or whether it belonged in engineering institutions. A similar sort of tension reappeared later, in the late 1940s, in the form of a row about whether what would become the semi-autonomous Petroleum Branch of AIME should remain affiliated with AIME or should move to the American Association of Petroleum Geologists (AAPG). Ironically, Everette Lee DeGolyer, who was trained as a geologist and was one of the founders of AAPG, was the decisive voice in assuring continued association with AIME, and for defining petroleum engineering science as engineering rather than see it subsumed under geology.[91]

The Gulf research group, Ralph D. Wyckoff, Holbrook G. Botset, Morris Muskat, and D. W. Reed, next submitted a paper to the *Bulletin of the American Association of Petroleum Geologists*, "Measurement of Permeability of Porous Media," in August 1932, although it did not appear in print until the beginning of 1934.[92] Wyckoff had received a BS degree in electrical engineering from Michigan State in 1920, then worked for the US Bureau of Standards before joining Gulf Research and Development. He is best known for pioneering the use, in the mid-1930s, of torsion balance and magnetometer geophysical surveys to locate subsurface anomalies that might indicate oil bearing structures. After rehearsing the difference between viscous and turbulent flow, Wyckoff, et al. rigorously defined porosity and permeability, which theretofore had suffered a variety of usages: Porosity is "the ratio of the volume of the pore space to the total bulk volume of the material," usually expressed as a percentage. Permeability "of a porous medium is the volume of a fluid of unit viscosity passing through a unit cross section of the material under a unit pressure gradient in unit time."[93] Then, true to the tradition of eponymy in natural science, they argued that, "Since the D'Arcy's [*sic*] law is the basic law in porous flow," the name "darcy" should be adopted for the standard of permeability.[94] After summarizing Muskat and Botset's and Muskat's earlier theoretical derivations for gas (compressible) and fluid (incompressible) steady-state flow, Wyckoff, et al. presented detailed procedures for permeability determinations in the laboratory and suggestions for how field data might be gathered and interpreted for comparison. They also noted "that, since the flow laws for gas-liquid mixtures are not known, it is impossible, at this time, to present rigorous formulae applying to such mixtures."[95] Wyckoff, et al.'s efforts helped stabilize the common nomenclature, procedure, and practice necessary for effective collective work in the nascent community of petroleum engineers.

In an API *Production Bulletin* apparently written early in 1932, although it did not appear until the beginning of 1934, H. D. Wilde Jr. and T. V. Moore (Humble Oil) began by complaining that development (i.e., where to drill within a field) and production decisions were still guided almost exclusively by "rules of thumb and analogies to other fields," but that "experience alone can reach only the conclusion that every well and every field is governed by laws known only to itself, and [the engineer] will be faced by as many problems as there are wells."[96] In the first part of their paper, Wilde and Moore parsed the broader problem of the hydrodynamics of reservoir drainage into more manageable sub-problems. Among the possibilities, flow of simple fluids through porous media was the most easily understood. Wilde and Moore pointedly observed that, "The development in East Texas has demonstrated the possibility of important fields in which the oil is under-saturated with gas, and in such fields the laws of flow of simple fluids are more important than the laws of flow of mixtures."[97]

With regard to the flow of oil and gas mixtures in porous media, however, they commented that, "Little work has been published which would be of value," and that only enough was known to show that the laws of simple fluids could not be applied. But Wilde and Moore noted that for unsteady-state flow of simple (single-phase) fluids, there was "a sufficiently close analogy to heat to permit the use of the heat equations," although little work applying those equations to oil or gas had yet been done. To understand large-scale reservoir phenomena, such as drainage from different regions of the reservoir or pressure drop throughout the reservoir, Wilde and Moore stressed the need for much better data, especially bottom-hole pressures and bottom-hole fluid samples, neither of which were then available.[98] In the quantitative second part of their paper, Wilde and Moore independently derived, from Poiseuille's law, Muskat and Botset's results for the flow of gas in porous media, as well as results for liquid flow. Use of capillary flow theory required an assumption of an "effective pore diameter," a purely "hypothetical quantity."[99] Steady-state flow in turn entailed that "all fluid flowing to the well be supplied from some extraneous source to the reservoir . . . and is not obtained by the drainage or depletion of the reservoir itself."[100] This condition was not met in a closed gas-drive (Beecher and Parkhurst) field, but was valid for East Texas water drive, if production rates were restrained.

Four papers, published or submitted between mid-1933 and the end of September 1934, extended and developed these results to include unsteady-state flow and, explicitly, to explain the character and to retrodict and predict the behavior of the East Texas oil field. First, in an American Petroleum Institute Production Bulletin, also published in *Oil Weekly*, Thomas V. Moore, Ralph J. Schilthuis, and William Hurst (Humble) exploited data from the new continuously recording bottom-hole pressure gauge to develop "a theoretical relation of bottom-hole pressure data, rate of production, and permeability." Theirs was the first "public presentation of an unsteady-state radial flow equation for expandable reservoir fluids."[101] Due to the complexity of the mathematics involved, they offered graphical solutions for the most important relationships, which was the standard way of packaging theoretical results for practicing petroleum engineers until after World War II.[102]

William Hurst separately published a formal mathematical derivation of the unsteady-state equation in *Physics*, submitted October 30, 1933.[103] Starting from the seminal material balance paper published by Coleman, Wilde, and Moore in 1929, Hurst used the equation of continuity to develop "a fluid balance over an infinitesimal element of sand" analogous to "the unsteady flow of heat through a homogeneous solid," which he solved using Bessel functions. He portrayed unsteady radial flow for two ideal cases: a closed reservoir in which the pressure at the well remained constant but the pressure in the reservoir declined (approximating a classic gas-drive field), and a reservoir in which the rate of production remained constant and flow converged to a steady state (approximating a water-drive field into which additional fluid flowed, such as East Texas).

A month and a half later, December 18, 1933, Morris Muskat submitted to *Physics* an elegant, crystalline paper providing a unified derivation of more comprehensive equations of flow for steady or unsteady-state conditions for single-phase compressible liquids or gases.[104] He derived the appropriate equations for both linear and radial flow in porous media. After noting the similarity to classic heat flow problems but deriving solutions for conditions not found in the heat flow literature, Muskat developed "non-steady state" equations for compressible liquids specifically applicable to the extraordinary conditions discovered in the East Texas field.[105] He presented detailed calculations with parameter values extracted from East Texas data after deriving general solutions for "the problem of an oil field produced by a water drive, the water extending over a large area outside the oil pool."[106] The retrodictions calculated from his theoretical analysis matched the behavior of East Texas pressure through September 1933, using data that Ralph Wyckoff had obtained.[107] Muskat did observe that the match between theory and data required a compressibility of the Woodbine Aquifer's contents "some 10 times that of ordinary water," which "would imply that there is an equivalent of 4 1/2 percent free gas dispersed in the water horizon, which is in itself of some geological significance, although we cannot enter further into it here."[108] Muskat also extended his theoretical treatment to interference effects among wells tapping a common reservoir.[109] Although he had not yet reached the goal, Muskat made it clear that his whole research programme from 1931 on was directed toward a comprehensive, rigorous theory of multiphase flow in porous media.[110] Like Babe Ruth's famous gesture, it seems unlikely Muskat would have pointed to exactly where he intended to go without a lot of confidence he could get there.

Ralph Schilthuis and William Hurst, in a paper presented at the Tulsa Meeting of the Petroleum Division of the American Institute of Mining Engineers in October 1934, synthesized all of these results, empirical and theoretical, into a single explanation of the East Texas oil field.[111] After noting the advent of systematically collected subsurface pressure data and Lindsly's analysis of bottom-hole fluid samples, Schilthuis and Hurst summarized the established facts about the field and presented a detailed depth-contour plat for the entire Woodbine formation. From these facts plus the minimal expansibility of East Texas reservoir oil that Lindsly had demonstrated, and that pressure in the field had never declined "as rapidly as would be consistent with the expansibility of the remaining undersaturated oil and the quantities of oil removed," Schilthuis and Hurst inferred that only "about 5 per cent of the vacated space" could have been filled by expansion of oil left in the reservoir" while 95 percent of the space was filled by water intrusion from the larger Woodbine formation.[112] They knew from others' analyses that surface water entering the Woodbine at its outcrop would experience "such great friction losses" and pressure drop that it "could not be supplied at the required rate." Therefore, they reasoned that only expansion of salt water already in the Woodbine Aquifer could account for the observed rate of water intrusion.[113]

Schilthuis and Hurst then offered a rigorous quantitative account based upon Hurst's published material balance equations. They derived a single integral equation which expressed "the variations in average reservoir pressure with oil production over a certain period of time."[114] To match their ideal analysis with observed data, Schilthuis and Hurst found that it was necessary to adjust only one constant in the equation, "namely, the value for the expandability of the water in the Woodbine sand." An adjusted value of "a little over 10 times the actual value for pure water" (as determined by Percy Bridgman in 1912) yielded a production-pressure relationship that matched the historical data on the field nearly perfectly.[115] The precise match of their quantitative theory to observed pressure behavior in the field drowned out any qualms about the water expansibility assumption.

In his "Discussion" of Schilthuis and Hurst's paper, Gulf's Ralph Wyckoff reported that calculations which followed "the theory presented by Dr. Muskat's paper on compressible fluid flow and represent results obtained by a slightly different method than that used by the authors" yielded a near-perfect match between Muskat's idealized, quantitative radial flow theory and observed field performance.[116] "Thus, being an independent check, it may aid in stressing the importance of the compressible fluid theory in consideration of water drives," and, "Quite independently, we obtain about the same physical constants as have been presented in this paper, and the apparent effective compressibility of the water in the Woodbine horizon adjacent to the East Texas field is about 12 times the normal value for water."[117] E. A. Stephenson, Professor of Petroleum Engineering at the Missouri School of Mines and Metallurgy, Rolla, Missouri, voiced the community's authentication:

> In this instance the problem was approached by individuals who originally had totally different viewpoints regarding the phenomena involved. However, in both cases their independent mathematical analyses led to predictions concerning conditions at the East Texas field which subsequently were substantiated by the facts.[118]

Hurst's and Muskat's separate derivations, and Schilthuis and Hurst's and Wyckoff's independent corroborations conferred on single-phase radial-flow theory and on water drive in East Texas, what Donald T. Campbell calls "convergent validity," which substantially enhances credibility.[119] East Texas was water drive, and by the end of 1934 virtually all petroleum engineering scientists, and most petroleum engineering practitioners, agreed.

This closure was social, not logical. It is always possible to concoct alternative, ad hoc explanations for any experimental or observational result. When G. L. Nye and Carl Reistle presented their comprehensive analysis of the East Texas field in October 1933, Ralph Wyckoff argued, "If it is assumed that the East Texas field is operating on a combined water and gas drive it is possible to calculate a pressure-decline curve that fits quite accurately the observed curve over the early history of the field. However, on this basis it is impossible to explain adequately the rapid increases in pressure that have been observed during 'shut-in' periods unless complicated explanations involving gas supersaturation effects are introduced." Wyckoff himself, perhaps animated by a physicist's aesthetic sensibility, or a fondness for Occam's razor (the simpler theory is the more likely to be true), backed off: "On the other hand, using the compressible fluid theory introduced by Wilde and Moore and applying it to the water horizon on the assumption of a complete water drive, remarkably close agreement is obtainable between calculated and observed effective pressures," although he did express some queasiness about the order of magnitude greater water compressibility that theory required.[120]

A completely different alternative was advanced by D. C. Barton, Geologist and Geophysicist, Humble Oil & Refining Company, and Joseph B. Umpleby, Geologist and Petroleum Engineer, Norman, Oklahoma: dead-weight compression of reservoir volume as fluids were withdrawn.[121] In March of the next year, J. S. Hundall, in an unpublished paper read before the American Association of Petroleum Geologists annual meeting in Dallas, claimed to have experimentally demonstrated such effects, but he was rebutted in May by J. O. Lewis, who pointed out that Hundall's tests on unconsolidated sands "have no quantitative value" for consolidated sands such as those in East Texas.[122] Reistle had in fact done his own (unreported) compressibility or compaction tests on cores of Woodbine sandstone at pressures of up to 4,000 lbs./in.$^2$, and found what he considered trivial effects: "reduction in pore space . . . between 0.050 and 0.056 percent per 100 pounds per square inch increase in pressure."[123]

This competing conjecture was not completely suppressed until C. B. Carpenter and G. B. Spencer reported the results of their own meticulous experiments in 1940, from which they concluded, "The tests of the sandstone cores indicate a small degree of compressibility, but it is not considered to be large enough to affect estimates of reservoir content and ultimate recovery, fluid-energy relations, or performance of petroleum reservoirs."[124]

In February 1932, Joseph Umpleby, although acknowledging that saltwater intrusion might play some role sometimes in oil production, still could confidently write, "In general, however, expanding gas is the dominant force of expulsion."[125] But in October 1934, Schilthuis and Hurst surmised: "The effect of the expansive power of water underlying the oil is probably an important factor in a great many fields, yet this factor has been overlooked by most geologists and engineers in their investigations of the mechanism of oil production."[126] Seven years later, in 1941, in a paper titled "Production Under Effective Water Drive as a Standard for Conservation Practice," the eminent geophysicist and geologist Everette Lee DeGolyer proclaimed, "I believe that some degree of water drive is of all but universal occurrence."[127] He went on to assert that "water drive . . . exists to a degree great enough to be made effective in more than 80 percent of the fields in the United States," and that the one best way to keep dissolved gas in solution and minimize gas-oil ratios, and thus to maintain reduced oil viscosity and maximize long-run recovery, was to restrain rates of production sufficiently to fully exploit that water drive.

These developments are recounted in some detail to emphasize the communal, cooperative, and competitive research that produced them, the simultaneously incremental and revolutionary nature of innovative research programmes, and the contested evolution and social character of even the most widely accepted scientific beliefs. If not wholly overturned, the Beecher and Parkhurst gas-drive paradigm was subsumed under the broader water-drive paradigm and its concern with total reservoir energy relations. By 1941, as classical Newtonian mechanics is to general relativity, so gas drive is to water drive.

## AMAZON: THE TRIAL

A lot of oil operators in East Texas, if they noticed at all, looked at these splendid scientific achievements the way an old hog looks at a new gate, and figured they weren't going to be hemmed up for long. Injunction suits against the Railroad Commission's prorationing orders for East Texas went on unabated. The complaints of Amazon Petroleum, et al., largely the same plaintiffs who had prevailed in *People's*, started with the standard assertion of unconstitutional taking without due process of law, claimed that proration to "reasonable market demand" under Texas's new Market Demand Act was simply the old price fixing conspiracy in new accoutrements, and further asserted that the Railroad Commission's supine submission to the dictates of Harold Ickes, as administrator of the Petroleum Code of the National Industrial Recovery Act (NIRA), and to the demand determinations of the US Bureau of Mines, represented a gross surrender of state prerogatives. As noted, although the case against the Railroad Commission and that against the NIRA were heard together, the NIRA issues were decided separately by Randolph Bryant in *Panama Refining Company, et al. v. Ryan*.[128]

The crux of the *Amazon* complainants' case against the Railroad Commission, as before, was gas drive versus water drive. Hutcheson in his *Amazon* opinion ignored the Market Demand Act that had stirred so much controversy in the Texas legislature. Instead, the Court disposed of the case "on other grounds," namely whether the order "stands . . . on considerations of physical waste," that is, on the scientific character of the East Texas oil field.[129]

On this essential issue, the *Amazon* complainants determined, in Darrel Royal's apt phrase, to "Dance with them that brung ya." They reiterated the same scientific evidence and arguments they had presented in *People's*, "with some little additions."[130] Indeed, their attorneys entered into the *Amazon* record, verbatim, all of plaintiffs' expert witnesses' testimony from *People's*.[131] According to plaintiffs, East Texas was "unquestionably" a gas-drive field, and the water-drive "theory" was "ridiculous."[132] The core of their technical argument, on which the whole case hinged, was summarized in a technical appendix to Complainants' Trial Brief by W. C. Spooner, a consultant who had served as chief geologist for Arkansas Natural Gas Company and the Arkansas Fuel Oil Company, Shreveport, from 1920 to 1925:

> The water drive theory . . . demands: That as the oil is removed from the reservoir it is replaced by water; *and that the entire body of water contained in the Woodbine sand from the field to its outcropping, 100 to 150 miles distant, moves as a body to replace the oil removed from the sand*, [emphasis in original] otherwise the water, obviously could not exert a hydrostatic pressure on the oil equal to a column of water from the surface outcropping to the bottom of the hole.[133]

Citing the standard studies of flow in artesian basins, Spooner argued that frictional forces would preclude such a massive movement of water in the Woodbine.[134] No plaintiffs' witness directly addressed the critical issue of the increase in average reservoir pressure when the field was shut in or produced at sharply restricted rates, but, artfully dodging, argued that over the long run, field pressure decline paced total oil production.[135]

Stanley Gill served as an expert witness in *Amazon*, as he had in *People's*. He held an engineering degree from the University of Washington (1920) and a professional degree in chemical engineering from Columbia University (1924), after which he worked for the Mellon Institute of Industrial Research in Pittsburgh, "assigned to the Gulf Production Company," before transferring directly to Gulf. He left Gulf at the end of 1931, after which he "engaged in independent consulting practice" in Houston, "concerned almost exclusively with problems of petroleum production engineering."[136] Gill largely reprised his *People's* testimony, but now also directly assailed "Lindsley's" (Lindsly's) results, which by the time of the trial enjoyed wide circulation : "the curves on gas solubility which are constructed by Mr. Lindsley [*sic*] and upon which he based his conclusions that free gas could not exist . . . are not in accord with physical fact, and do not truly represent the facts."[137] Gill emphasized that "expulsion of oil is accomplished primarily through the agency of expansion of gas," and that greatest ultimate recovery would result from "operation of all wells" at their individual "minimum" gas-oil ratio.

Plaintiffs seem to have let their devotion to gas drive carry them overboard. The Texas Anti-Market Demand Act of 1931 and the volte-face Market Demand Act of 1932 both clearly stipulated the pool or reservoir as the appropriate conservation unit, as did the US Supreme Court in *Champlin* and the District Court in *People's*, while correlative rights had been long established. Plaintiffs maintained, correctly, that full utilization of gas drive required that minimum (actually, most efficient) gas-oil ratios be determined by tests on each individual well, with frequent (at least monthly) retests of each and every well as field pressure declined. If production allowables were adjusted accordingly, even if very high for some wells, pool conservation would be served. But plaintiffs' attorneys reasoned that individual well regulation meant correlative rights be damned: "It is no answer to this proposition to say that the operator with a market will recover oil belonging to his neighbor without a market. This he has a right to do."[138]

In contrast, the testimony of the Railroad Commission's expert scientific witnesses recited in detail the empirical facts about East Texas petroleum engineers and scientists had discovered and published since *MacMillan*. They claimed that Lindsly's bottom-hole samples proved conclusively that reservoir fluids in the field were undersaturated at reservoir conditions and that energy in the reservoir could not come from expansion of gas, dissolved or free. Furthermore, the measured increases in average reservoir bottom-hole pressures when the field was shut in or operated at restricted rates entailed that it was physically impossible for East Texas to be a gas-drive field and implied that sharp restriction on the rate of production of every well in the field was the only way to preserve reservoir pressure and maximize ultimate recovery. East Texas was a water-drive field, and the energy necessary to move the oil through the formation and lift it to the surface came from the intrusion of salt water from the immense Woodbine basin. To sustain subsurface pressure, keep dissolved gas in solution and inhibit increases in subsurface oil viscosity, and therefore to prevent underground waste, required taking oil out no faster than the salt water could flow in.[139]

This time the Railroad Commission's case was thorough almost to the point of being sadistic. For example, T. V. Moore and C. E. Reistle Jr., in a deposition given December 12, 1933, offered another clever and persuasive reductio ad absurdum, which went uncontroverted at trial.[140] Moore and Reistle claimed that through November 18, 1933, had only 75 percent of the pore volume in the field vacated by produced reservoir fluids (oil and dissolved gas) not been replaced by Woodbine salt water, rather than the observed 96 percent, the reservoir pressure necessarily would have fallen to 530 lbs./in$^2$, less than half that actually observed.[141]

Nevertheless, once burned, twice shy, Railroad Commission counsel apparently made a strategic decision not to introduce anything that smacked of "mere theory and speculation." No witness mentioned the theoretical innovations of Morris Muskat or William Hurst. Railroad Commission lawyers and their expert witnesses stuck to the "facts," which Lindsly and Reistle had so clearly demonstrated, just as Hutcheson had so often admonished them to do.

Unfortunately for the court, consensus among petroleum engineering scientists did not entail agreement among all expert witnesses: "We are impressed, but not aided, by the fact that scientific men of standing, integrity, and ability can so radically disagree. However much we may be disappointed in finding ourselves unable to determine to what extent interest in the result makes for these differences of opinion."[142] But Hutcheson did note that, given the immense volume of "experiments and tests . . . under field and under laboratory conditions" since 1932, "Everybody admits freely now what was only grudgingly admitted then, that restriction is essential in the production of that field."[143] The issue was how much restriction, which depended upon whether the field was gas drive or water drive. In the previous three years, Hutcheson had seen quite enough of "the Two-and-Seventy jarring Sects confute"[144] to be wary of committing the US Supreme Court's (and the Oklahoma Supreme Court's) Aquinian error.[145] He declined to bestow legal imprimatur on a specific paradigm or theory in petroleum engineering science, opting instead to render the only determination he was required to make—whether the Railroad Commission's order was reasonable or not.

Although Hutcheson avoided committing to a specific scientific hypothesis, he was not one on whom nuances of language were lost and was himself a master of saying without saying. In their Complainants Trial Brief, Saye, Smead, and Saye had harped on the unreasonableness of Railroad Commission orders for East Texas: "I do not believe that there is a man of ordinary intelligence in the whole United States that could reasonably conclude that there is any necessity, postulated on preventing waste, for curtailing the production of East Texas . . . to such a ludicrous amount."[146] Or later, ". . . no reasonable mind, giving due consideration to the evidence, and weighing it in the light of experience, matters of common knowledge, and with a decent respect for the opinion of mankind, could possibly reach the conclusion that the purpose or the effect of the orders of the Commission could be the prevention of waste or conservation of oil or gas."[147] Hutcheson's *Amazon* opinion:

> We have examined with the greatest care this mass of evidence, voluminous, ponderable, pretending to scientific accuracy, as to physical conditions in East Texas, and the relation of the restrictions to the prevention of waste there . . . we do find ourselves so situated that we are bound to say that all of this vast amount of evidence, submitted in favor of the commission's findings, is too ponderable to be brushed aside as no evidence at all. We find ourselves wholly unable to say that the conclusion the commission reached is not one which reasonable minds could entertain.[148]

## PRATFALLS OF WHIGGERY

However reasonable, Hutcheson took his share of knocks. In some quarters, he attracted considerable ire for his earlier East Texas decisions. Robert E. Hardwicke, on whose magisterial chapters on Texas for the American Bar Association's definitive 1938 and 1948 legal histories of conservation of oil and gas we all rely, had served as outside counsel to the Railroad Commission in a number of critical prorationing cases.[149] Judge Hutcheson had not always treated him gently. Hardwicke was an advocate of compulsory unitization under state aegis, and his 1938 history is a classically Whig interpretation which dismissed any qualms about or opposition to either prorationing or compulsory unitization as, at best, antediluvian (of oil, in this case) obstructionism:

> This outline will reflect the stubborn antipathy of many people to new ideas, and their high degree of intolerance toward any principle or course of conduct which involves a change from customary practices. It will give some indication of the power of organized minorities, particularly when they can use as weapons the ancient distrust of science and experts. In this story may be found proof of the evils which may arise in our modern economic world from unrestricted individualism, and proof of the necessity for cooperative effort for the common good.[150]

Hardwicke justified this roundhouse condemnation with a further bit of Whiggery, a snippet of which began our previous chapter:

> Events were taking place in complete harmony with what the engineers predicted would happen under such conditions, and which they had hoped to avoid by maintaining a relatively low allowable for the [East Texas] field. *Here, indeed, on a vast scale and at large expense,* [emphasis in original] was experimentation which unquestionably tested the soundness of the opinions advanced by the engineers as early as the summer of 1931, with respect to the effect of high rates of flow in the field, and with respect to concentrated production. Similar opinions had been repeated, time after time, by engineers and operators, before the Commission and before the courts, as bases for limiting the production of the East Texas Field to about 400,000 barrels daily. The wreck of the price structure and the physical damage from excessive production were high fees to pay for an experiment which added little to the scientific data on the field.
>
> No doubt the decision in the *Amazon* case [1934] gave comfort to the Commission and to the technical men who, from the very beginning and with only meager data available in 1931 (at the time of the *Macmillan* Case), were able to calculate and determine with a high degree of accuracy the efficient top allowable for the field. The experiences of the field at different rates of flow over a period of years have confirmed rather conclusively the testimony of the Commission's experts given in the *Macmillan* and *People's* Cases, and which in such cases was more or less brushed aside by the courts as being mere theory and speculation.[151]

Virtually all lawyers or historians, including myself, who have worked on oil and gas during the period have taken Hardwicke's claims pretty much at face value.[152] They fit so nicely with Whig narratives of progress. But if evidence matters, Hardwicke's second claim, regarding what engineers knew and could do in 1931 or 1932, which is the premise for the first, is false. Petroleum engineers did not know in June 1931 what they knew in December 1933, and the reasons they articulated for their opinions in 1931 or 1932 were not the ones they articulated in 1933. It took a great deal of cooperative theoretical development and empirical investigation to get from what was in fact "mere theory and speculation" in 1931 to well-founded scientific belief based upon evidence in 1933. What was "reasonable" in late 1933 was unsupported by either reason or evidence in summer 1931. To claim otherwise is to ignore how engineering science and practice, and the technologies based upon them, evolve.

# CHAPTER 7

# USEFUL RELATIONS

One of the principal distinguishing features of a bona fide engineering doctrine is that the behavior of a system within the scope of the doctrine can be more or less completely described by mathematical equations relating several measurable properties of the system . . . useful relations are discovered empirically and applied successfully for years before their theoretical significance and justification are exposed and/or understood.

—CARL E. REISTLE JR.[1]

## USEFUL RELATIONS: PETROLEUM ENGINEERING SCIENCE

*Amazon* and the world petroleum engineering remade marked a new beginning, not a happy ending. Progress, problems, and surprises continued in petroleum engineering science. By the mid-1930s, the quest to develop theoretical understanding of the fundamental dynamics of subsurface hydrocarbon reservoirs yielded sound and usable theoretical insight into the flow of homogeneous fluids—gases alone or gas-free liquids—in porous media. These results emerged from a collective and cooperative, if also fiercely competitive, effort initiated by Wilde, et al., in 1929, and pursued with such vigor by Morris Muskat, Ralph Wyckoff, Holbrook Botset, Ralph Schilthuis, William Hurst, Thomas Moore, Ben Lindsly, and Carl Reistle, and a growing community of other researchers and practitioners. As Morris Muskat wrote in 1937, "The subject of the flow of homogeneous fluids . . . may be considered at the present time to be fairly well completed. That is, analytical, graphical, numerical, or experimental methods are now available for obtaining detailed descriptions of such homogeneous fluid systems of any reasonable geometry." In addition, steady-state theory could be extended "to that of the transient state . . . such as has been carried through in detail in describing the history of the pressure decline of the East Texas field." The success of homogeneous flow theory even fed back into Henry Darcy's mother disciplines, hydrology and civil engineering: ". . . it is now possible to discuss quantitatively such questions as the drainage of water-logged lands by tile drains, the seepage of water out of canals or ditches, the uplift pressures on dams with and without sheet piling and the seepage flux through dams of arbitrary geometry."[2]

As profound and useful as homogeneous flow theory was, as Muskat himself stressed, it made no pretense "of solving what is really the fundamental physical problem in most actual cases of oil production, namely the nature of the flow of gas-oil *mixtures* [emphasis in original] through porous media."[3] Critical problems in oil production, "pressure and production decline, ultimate recovery, well spacing, and gas repressuring," remained beyond its purview.[4] Oil rarely came so neatly packaged as in East Texas. More often, oil production entailed dealing with more complex and analytically intractable multiphase flow, the comingled flow through porous media of oil, natural gas, and salt water.[5]

## Heterogeneous Flow

The next step, understanding heterogeneous flow, was daunting. This was the grail to which Muskat had so clearly pointed in 1931–1932, and toward which theoretically inclined petroleum engineering scientists worked throughout the 1930s and 1940s and into the early 1950s. They followed the same strategy that had worked for homogeneous flow: parse and conquer. In their experiments and analyses, they replicated their approach to homogeneous flow, first considering unconsolidated media (packed loose sand) and steady-state flow, then unconsolidated media and unsteady-state flow, each in the sequence linear flow, radial flow, and spherical flow, followed by consolidated porous media (such as sandstone) with steady-state then unsteady-state flow.

It was no easy procession. Physical complexity amplified mathematical truculence. As Muskat observed in 1937, had it been possible theoretically to treat a gas-liquid mixture "as the composite result of a combined flow of a homogeneous gas and a homogeneous liquid . . . it would have spared the authors three years of experimental and analytical effort spent in what at times seemed to be hopelessly futile attempts to obtain the results to be reported here." In contradistinction to the flow of a single homogeneous fluid, the flow of a heterogeneous gas-liquid mixture required taking "account of the constant evolution of the gas from the liquid phase in terms of homogeneous fluids," and finding a way "to express the permeability of either phase in the presence of the other."[6]

Just as so many other innovative scientists have done when stymied, Muskat and his colleagues changed the question. Archimedes, challenged to nondestructively determine the material composition of a purported golden crown, used liquid displacement to measure density instead. Copernicus, annoyed by the proliferation of Ptolemaic cycles and epicycles, moved the Earth. Isaac Newton, inspired by his unpublished alchemical and astrological researches, ushered action at a distance back into natural philosophy as universal gravitational attraction.[7] Charles Darwin replaced the divine harmony of design with the material struggle for existence. Albert Einstein, fascinated by simultaneity, changed indestructible matter and energy, invariant space, and immutable time into interconvertible mass and energy in an elastic space-time continuum.[8] As Stuart Kauffman recommends, on a rugged fitness landscape, when stuck on a local fitness peak, search far away.[9]

Muskat and company, confronted by the intractable analytical complexity of heterogeneous flow in porous media, made rock-solid permeability variable. Prior to 1936, the permeability of porous media (unconsolidated packed sand or consolidated porous rock such as sandstone) had been an invariant, fixed property of the medium, measured empirically and mathematically portrayed as a constant.[10] In a quest for scientific generality and broad applicability, core or other experimental samples of porous media were thoroughly cleaned and dried. They were then subjected to flow tests using a homogeneous gas or fluid of known viscosity and density—air, carbon dioxide, water, kerosene—under specified pressure differentials to determine a single permeability value.[11] Extension of Darcy's law, which originally defined permeability for a fluid, pure water, of unitary viscosity ($\mu = 1$), made calculation of the permeability of a sample to homogeneous fluids with different viscosities straightforward. Petroleum engineers assumed that samples so prepared and permeabilities so measured veridically represented the same porous media in a subsurface reservoir. Permeability was fixed, and all variance in flow loaded conceptually into fluids' differing viscosities and ambient pressure differentials.

Muskat and his gang upended these conventions. In two papers submitted together to *Physics* in late July 1936 and published together in September, Ralph D. Wyckoff and Holbrook G. Botset reported the results of their experimental programme on the flow of gas-liquid mixtures, while Morris Muskat and Milan W. Meres produced a quantitative theory of the flow of heterogeneous fluids through porous media.[12] All four worked at Gulf Research and Development in Pittsburgh, and together they presented a paper which was a "fusion and abbreviation" of their earlier separate papers at the New York meeting of the Petroleum Division, American Institute of Mining and Metallurgical Engineers, in February 1937. This paper was duly published in the Institute's *Transactions: Petroleum Development and Technology* in December 1937.[13]

The key conceptual leap in this research programme was the recognition that in a flowing two-phase gas-liquid system, "the permeability of the medium, which was a constant characterizing the medium in the case of homogeneous fluids, in the case of mixtures, becomes a variable and is dependent upon the relative volumes of gas and liquid present in the pores of the medium."[14] "[L]ocal liquid saturations will change from point to point in the system as the pressure varies," which implied that "the porous medium must be considered as changing from point to point, and can therefore no longer be defined by a single invariable permeability." So generalized, the new conception of permeability entailed assigning "to the medium a *local* [emphasis in original] structure defined by the local saturation, the latter in turn uniquely determining the local permeabilities for both the gas and liquid phases of the mixture." Once "the fundamental empirical element" and "the relation between the permeabilities of the gas and liquid phases and the volume composition of the mixture, or the liquid saturation" were determined experimentally, "all the necessary elements for formulating mathematically the hydrodynamics of heterogeneous fluid systems analogous to that of homogeneous fluid systems" would be in hand.[15]

To establish the essential empirical relationship, Wyckoff and Botset conducted a series of steady-state (constant flow) experiments using unconsolidated (loose) sand of three grain sizes, plus one consolidated sand, each packed into a sealed tube 10 feet long by two inches in diameter connected to a pressure tank. Their experimental fluid was distilled water saturated with carbon dioxide ($CO_2$) under pressure. A small quantity of salt (NaCl) added to the water made it conducting. Each foot along the experimental tube, an insulated brass ring contained a pressure-measuring piezometer; the ring also served as an electrode in an AC bridge, which measured the conductivity (or resistance) in each section. Since the gas was non-conducting, changes in electrical resistance indicated changes in the relative proportion of liquid and gas in the flow. Once a slow, steady rate of flow was established, pressure in the tank was

> reduced to a value just slightly above the saturation pressure of the liquid and the inflow and outflow needle valves on the sand column adjusted so that pressure within the flow tube was slightly below saturation pressure. A small amount of gas was thus released continuously within the sand body and flowed through the system along with the liquid.[16]

The rate of liquid and gas flow could be measured at the efflux end of the tube. Within the tube resistivity measurements indicated the proportion of the pore space in a section occupied by liquid, while simultaneous pressure readings, together with flow rate, yielded the liquid permeability of the sand within that section. The known solubility of $CO_2$ in water at a given pressure permitted inference of volume and rate of dissolved gas flow.[17] The volume of free gas flowing through a section could be calculated from its amount at the outflow face of the tube, its solubility, and mean pressure in each experimental section.[18]

The next challenge was finding a way to express these data so that they were commensurable and could be readily incorporated into a hydrodynamic theory. The Muskat groups' innovative solution was to calculate a medium's separate effective (and variable) permeability to gas and to liquid (*kg* and *kl*) as a function of the percent of liquid saturation in the medium (percent gas saturation would have worked as well, just reversed). Both *kg* and *kl* were expressed as a percent of *ko*, an initial or specific permeability, a property of the medium (unconsolidated sand) defined for a fluid of unitary viscosity, that is, pure water. In Figure 7.1. Relation between Gas and Liquid Permeabilities and Liquid Saturations, "the sum of the ordinates [y-axis value] of the *kg* and *kl* curves at any liquid saturation [x-axis] represents the *total* effective permeability of the sand under steady state conditions."[19] For example, at a 70 percent liquid saturation, *kg* = 10 percent and *kl* = 30 percent, for a total effective permeability of 40 percent, substantially less than that for homogeneous flow of either gas or liquid alone. In this figure, the solid lines represent stabilized steady-state flow, the dotted lines unsteady-state flow regimes. At very low gas-liquid ratios, above *ke*, gas bubbles collect in the smaller pore spaces and impede flow until the pressure builds up and flushes them, producing sharp variations, or transient flow. When the liquid saturation falls below about 40 percent, very high gas-liquid ratios prevail: at 30 percent saturation, the gas-liquid ratio is 23,000 ft³ per barrel. Below 10 percent liquid saturation, liquid is completely bypassed and only gas flows.

These results were remarkably robust. The curves were of the same form and exhibited nearly the same values, for all of the different sand grain sizes, homogeneities, porosities, and specific permeabilities tested (specific permeabilities varied by 1,400 percent).[21] Results were "not materially changed by variations in the viscosity of the liquid" (tested by adding sugar

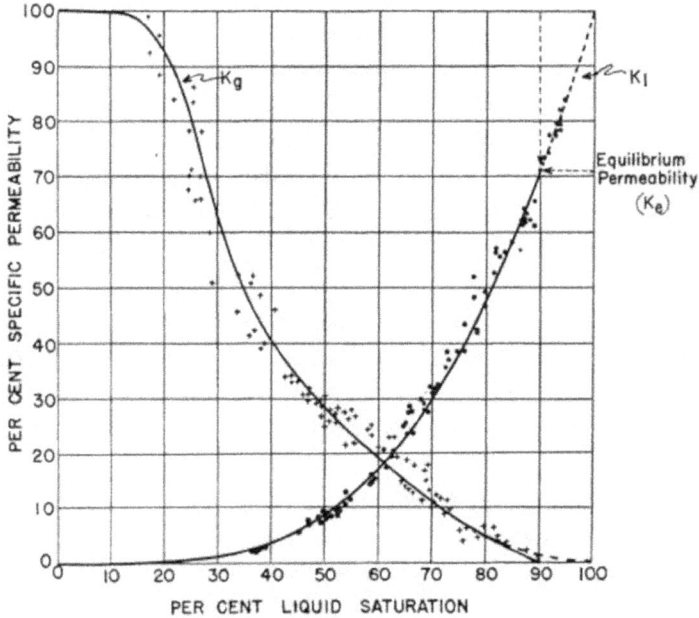

**Figure 7.1. Relation between Gas and Liquid Permeabilities and Liquid Saturations**
$kg$ = relative permeability to gas. $kl$ = relative permeability to liquid.
$ke$ = equilibrium relative permeability for liquid, at which flow becomes unsteady, all expressed as a percent of specific permeability, $k0$.[20]

to the water or by substituting kerosene for water), nor did the surface tension of the liquid have a significant effect (tested by using ethyl alcohol in place of water). Experiments with immiscible, incompressible fluids—oil and water—produced similar curves.[22] Wyckoff and Botset therefore concluded that liquid saturation was the key variable in heterogeneous flow, nearly independent of other factors.

In their companion paper, Muskat and Meres used these results to fabricate a comprehensive theory of heterogeneous flow in porous media.[23] Application of their general flow equations depended upon empirical determination of $kg$ and $kl$, precisely what Wyckoff and Botset had provided. Other constants were assumed to be already known: solubility of the gas in the liquid ($s$), gas viscosity ($\mu g$), and liquid viscosity ($\mu l$). Pressure ($p$) was to be measured, from which could be calculated pressure gradient vectors ($\nabla$). From these elements, Muskat and Meres developed an analytical theory applicable to all instances of heterogeneous flow in porous media. Unfortunately, "the nonlinear and complex character" of their equations made it "impossible to treat or solve them generally." They instead derived explicit analytical solutions for steady-state linear, radial, and spherical flow. For linear flow, their solution yielded the volume of gas flow (at atmospheric pressure) per unit area of the porous medium ($Qg$) and the volume of liquid flow per unit area of the porous medium ($Ql$), and could be used to calculate the complete history of pressure decline in a reservoir.

Extension of the theory to transient, or unsteady-state, flows resulted in a set of non-linear partial differential equations, the analytic solution of which presented a "formidable task." Muskat and Meres therefore offered numerical approximations consisting of difference equations for a "continuous succession of steady states." They worked through a numerical solution for linear flow, with only the "general physical requirements that both the pressure and the saturation must monotonically decrease with the time, and that they . . . increase" with distance from the outlet (the well bore). Strictly speaking, the results of Wyckoff and Botset and of Muskat and Meres applied only to unconsolidated sands and to two-component systems. But as all four researchers pointed out, heterogeneous flow in consolidated sands was unlikely in principle to behave very differently, nor would three-component systems, such as one containing gas, oil, and water.

Even in its restricted form, Muskat and Meres's theory had profound implications for petroleum engineering practice, especially for long-running disputes about optimal well-spacing. The conventional and intuitively appealing wisdom in the oil fraternity, that more and more densely drilled wells in a pool entailed greater ultimate recovery, had been officially sanctioned in 1924 in Willard W. Cutler Jr., "Estimation of Underground Oil Reserves by Oil-Well Production Curves," *U.S. Bureau of Mines Bulletin 228*. Cutler used production and well-spacing data from relatively shallow fields in Pennsylvania, Oklahoma, and California to infer the relationship between spacing and recovery. Ceteris paribus, he found, "a decided increase of production per well in favor of the wide spacing. This increase of production per well, is, however, accompanied by a lessened recovery per acre." Cutler stated his findings as a "tentative rule": "*The ultimate productions for wells of equal size in the same pool, where there is interference (shown by a difference in the production decline curves for different spacing), seem approximately to vary directly as the square roots of the areas* drained by the wells [emphasis in original]." Phrased differently, "doubling the distance between wells doubles the ultimate recovery per well and halves the ultimate recovery per acre." Cutler reasoned that this relationship seemed "to rest on the fundamental mechanical law that the energy required to move a fluid (either liquid or gas) through a pipe . . . is proportional to the distance." Thus, "recovery will be approximately proportional to spacing whenever the primary expulsive force is gas and the reservoir rock is continuous throughout the area under consideration."[24]

While most early complaints about dense well spacing had focused on surface water intrusion, subsurface water coning, or fire hazard, as perceived oil shortage turned to glut, the paramount concern of petroleum engineers, oil company executives, and some oil operators shifted to economic waste: too high a cost for too many "unnecessary" wells for too little return. John R. Suman voiced one of the first scientifically animated critiques in 1934, claiming, without detailed analysis, that "wider spacing might give more oil recovery because of reduced reservoir-energy dissipation."[25] But because Suman and the Humble Company had been such insistent advocates of prorationing and unitization, as well as wider well-spacing, Suman's claim was discounted by many operators as merely self-interested.

Muskat and Meres provided "perhaps the first quantitative theoretical evidence" regarding "the much-debated question" of an oil well's "so-called 'radius of drainage,' that is, the extent of the zone of appreciable liquid depletion." Their conclusions were startling. There was *no* physically limited radius of drainage: "the ultimate [liquid] saturation is for all practical purposes quite uniform over the whole flow system. Or conversely, the ultimate depletion at the closed external boundary [of the system] is practically as great as that of the rest of the system except only in the immediate vicinity of the outflow surface [well], this depletion being determined only by the ratio of the original saturation pressure to the pressure representing the state of physical depletion." Therefore, assuming a continuous and homogeneous closed reservoir, "ultimate recovery will be essentially independent of the well spacing," although the time it would take to reach depletion was "directly proportional to the square of this spacing." Optimal well spacing came down to drilling and completion costs and an imputed discount rate: At low assumed interest rates, fewer wells and prolonged production paid; at higher assumed interest rates, more wells (if not too expensive) and faster production were economically superior.[26] Below the rate limitations dictated by production mechanism (gas or water drive), ultimate recovery at depletion would be the same.

Muskat and Meres's theory had other significant operating implications. First, on the "much debated question" of gas-oil ratios, "it is clear from physical considerations, the [liquid] saturation at an outflow surface which is maintained at a pressure lower than that of the original saturation pressure must necessarily decrease monotonically with time . . . it then follows . . . that the gas-liquid ratio must . . . necessarily increase monotonically with time." There was thus no optimal gas-oil ratio invariant through the entire production history of a reservoir.[27] Similarly, application of vacuum to oil wells or so-called gas lift systems would be useless. Only gas introduced into wells at the very top of the producing structure might enhance oil recovery. Finally, and again contra the received orthodoxy, as well as existing prorationing doctrine and the body of case law, the rate of production from a closed (gas-drive) system had minimal effect on ultimate recovery from the field: "The manner of production will affect mainly that part of the sand that is near the well, and to this extent the ultimate recovery will depend upon whether or not the sand is allowed to produce by wide open flow or by gradual pressure release. At distant points, however, the theory indicates that the initial gas-saturation pressure will be the predominating factor."[28]

Muskat and Meres, and Wyckoff and Botset, were notably sanguine about the generality of their theory. While it was highly idealized, Muskat and Meres argued that it would apply to multi-well fields where neighboring wells provided "artificially created closure." Three-component (gas-oil-water) systems were unlikely to differ in principle from two-component systems. Muskat and Meres maintained that any difference caused by considering consolidated rather than unconsolidated sand would be "physically only a matter of detail," to be empirically resolved as in Wyckoff and Botset's experiments.[29] Overall, they claimed, "every problem of flow of a heterogeneous fluid system can in principle be treated by the basic differential equations."[30]

## Articulation

Major conceptual innovations and their experimental or empirical reification, in engineering science as in natural science, typically are followed by replication, sometimes explicitly, more often by use in further research or in practice.[31] Successful research programmes, paradigms, or communal practices entail theoretical and empirical extension, elaboration, development, correction or qualification, and modification. Muskat and company's recognition that considering permeability a variable was the theoretical key to understanding multiphase or heterogeneous flow through porous media—what goes on in most oil fields—and their presentation of a full analytical theory for that flow, had those effects. Since application of Muskat's multiphase flow theory depended upon empirical determination of the relative permeability of a porous medium to each component of flow in the presence of the other components—conventionally, as a function of liquid saturation—the most immediate problems were corroboration of the experimental techniques themselves and verification of the basic theory for different flow components (recall that Wyckoff and Botset, upon whose experiments Muskat and Meres relied, mostly used water and carbon dioxide). Still to be demonstrated was the applicability of phase-flow analysis to oil and natural gas, or to two immiscible liquids (oil and water).

Initial investigations explored only unconsolidated media. But oil and natural gas are most often found in consolidated porous media (sandstone, for example), to a greater or lesser extent "cemented," or solidified. Next step was finding out for sure whether the theory, with its clever innovation of variable permeability, really applied to consolidated media. Likewise, many, if not most, oil fields are three component systems: oil, natural gas, and salt water. Extension of experimental techniques and theoretical application to those systems were essential. Finally, in a time before abundant and cheap computer power and prepackaged application programs, a lot of effort went into developing more tractable, easier to use, and more economical methods of analysis and, especially, calculation.

The heterogeneous or multiphase flow paradigm (the "phase-flow formula" in oil fraternity speak) disseminated quickly through the community of petroleum engineering scientists. Its evolution and development became the province of a well-defined research front, that is, a community, or "invisible college," of researchers. Experimental corroborations of heterogeneous flow theory began to appear in the literature within a year after publication of the original four articles in *Physics*. The first, L. S. Reid (Phillips Petroleum) and R. L. Huntington (University of Oklahoma), experimentally replicated and theoretically confirmed Muskat, et al.'s results, but also, contra Muskat, resurrected Lester Uren's 1927 demonstration that an optimal rate of production in gas-drive fields did exist, and that "ultimate recovery may be reduced by producing the reservoir at a very low rate, as well as at a very high rate."[32] Extension and articulation of the theory to other fluid mixtures and more realistic conditions quickly followed. M. C. Leverett (Humble Oil) in 1938 showed that the permeability of unconsolidated sands to oil-water mixtures could be characterized as a function of water saturation, just as permeability to gas-liquid mixtures is portrayed as a function of liquid saturation.[33] Holbrook Botset (Gulf Oil) demonstrated (1939) that the same analyses applied to the flow of gas-liquid mixtures through consolidated sands (Figure 7.2. Permeability-Saturation Curves

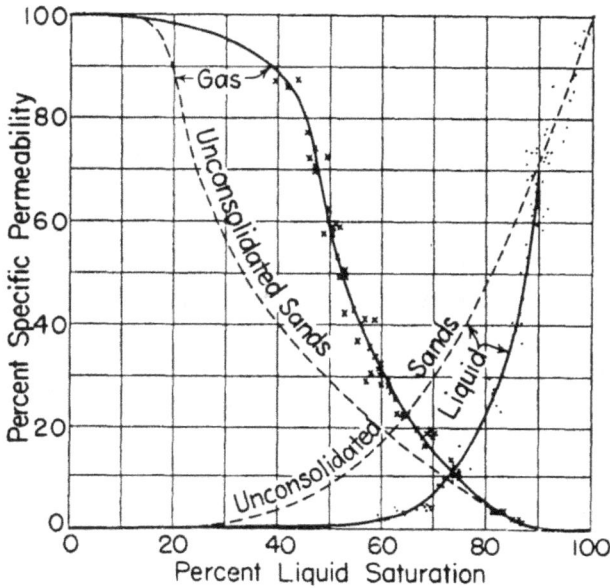

**Figure 7.2. Permeability-Saturation Curves for Consolidated and Unconsolidated Sands**
*kg* = relative permeability to gas. *kl* = relative permeability to liquid.[36]

for Consolidated and Unconsolidated Sands).[34] In commenting on Botset's paper, Stuart E. Buckley (Humble Oil) argued, in apparent pushback against "practical" engineers, geologists, and executives who denigrated the importance and winced at the costs of theoretical development and laboratory investigation:

> Regardless of the geometrical complexity of the conditions met in actual practice and the resulting difficulty in drawing exact quantitative conclusions, the underlying physical principles may be discovered through laboratory investigation, and once delineated and understood may be relied upon for the interpretation of the behavior of individual wells and complete reservoirs.[35]

M. C. Leverett and W. B. Lewis (both Humble Oil) extended multiphase theory, which thus far had explicitly treated only two-phase flow, to the steady-state flow of a three-phase gas-oil-water mixture through unconsolidated sands (1940).[37] They found that prior theoretical analyses, beginning with Muskat and Meres, applied also to three phases, and that essentially the same laboratory techniques established the necessary parameter values. Leverett and Lewis's experiments demonstrated that relative permeability to water is determined only by water saturation, and that relative permeability to gas is slightly less for three than for two-phase systems. They found that "relative permeability to oil varies in a more complex manner, being in some regions less and in others more than for the same oil saturations in two-phase flow," although it still remained independent of oil viscosity; and that the "presence of appreciable amounts of all three phases in a flowing stream . . . is limited to a relatively small region" of phase space.[38] Perhaps surprisingly, Leverett and Lewis discovered that the presence of modest volumes of connate water actually increased a system's permeability to oil.[39]

Heterogeneous flow theory ultimately had even more profound effects upon petroleum engineering science and practice than had understanding of homogeneous flow but demanded greater and more protracted efforts for its articulation and assimilation. In a fundamental paper at the AIME New York Meeting in February 1941, Stuart E. Buckley and M. C. Leverett recast understanding of the "Mechanism of Fluid Displacement in Sands" in light of heterogeneous flow theory.[41] Buckley and Leverett combined material balance analysis with heterogeneous flow theory to portray how injected, expanding, or inflowing fluids moved through a producing subsurface reservoir and affected its constituents. Buckley and Leverett derived a set of equations which depict the way a "front" of water (or gas) moves through the reservoir, hopefully sweeping oil toward producing wells. Their analysis explained the relative inefficiency of gas drive (gas displacement very quickly lost its efficacy as gas saturation rose) and supported their strong preference for water flooding in secondary recovery projects.[42] These considerations also explained why primary production in closed gas-drive fields typically recovered only a fraction of the oil in place, 20 to 40 percent, while effective water drive could produce several times as much, 50 to 95 percent.

Applications of heterogeneous flow theory could refute established beliefs, some enshrined in legal doctrine. The "productivity index" or "factor" for oil wells had first been clearly described in 1930 and was defined as the rate of production per pound drop in bottom-hole pressure. It had been cited in multiple prorationing cases and opinions all through the 1930s as a measure of production efficiency but at least tacitly treated as invariant over the life of the well. H. H. Evinger and Morris Muskat showed (1941, 1942) that for either two-phase or three-phase heterogeneous flow, no single productivity index existed if fluid composition (saturation), and hence effective permeability, continuously changed.[43] To obviate the need for complex mathematical integrations (and the skill and time they demanded) and facilitate practical application, Evinger and Muskat presented their analyses in graphical format.

Not all the news was good. Use of heterogeneous flow theory provoked its share of controversies, some noticeably acrimonious. For example, in a paper read at the Austin meeting of the Petroleum Division of AIME in 1942 and published the next year, R. E. Old Jr. (Core Laboratories, Dallas) claimed that it was possible to calculate both water encroachment (water drive) and initial oil in place without geological knowledge of reservoir characteristics (size, permeability, and so on). His proposed method elicited, from Morris Muskat and Rex Woods (Gulf Oil, Tulsa), one of the harshest rebukes in the engineering literature.[44] Most arguments were more civil, but no less impassioned.

Worse, as Evinger and Muskat acknowledged in their 1942 paper, results from heterogeneous flow theory, at least out of the box, did not match empirical measurements. Two-phase (oil and gas) theory did not account for the discrepancy between homogeneous (single phase) flow theory and "the empirical curve . . . relating the productivity index with the average sand permeability" reported by other researchers. Use of three-phase (oil-gas-water) theory didn't resolve the discrepancy either, and empirical results more nearly matched the predictions of the simpler homogeneous flow theory, despite wells' actual multiphase flow. Evinger and Muskat therefore conceded, "under field conditions, the heterogeneous character of the flow may affect the productivity factor far less, rather than more" than their calculations indicated.[45] Heterogeneous flow theory could not begin to fulfill its promise until the widespread adoption of more advanced electrical logging techniques and core analyses after World War II provided better and more complete data.

Articulation also entails revision and correction, especially when extrapolation leads to false inferences. Based upon heterogeneous flow theory, petroleum engineers had assumed that because gas flows through porous strata so much more easily than any liquid (oil or water) with which it might be mixed, water (whether natural water drive or injected) would be highly efficient in maintaining reservoir pressure and displacing gas. Not so, as T. M. Geffen, D. R. Parrish, G. W. Haynes, and R. A. Morse (Stanolind Oil and Gas, Tulsa) demonstrated in 1951 using both laboratory experiments and samples obtained with a (still experimental) Carter Oil Company pressure-core barrel borrowed from their crosstown commercial rivals. Counterintuitively, pressure maintenance, so essential in undersaturated water-drive oil reservoirs such as East Texas, was precisely the wrong thing to do in natural "dry" gas reservoirs, which would yield 80 to 90 percent of their gas by straight pressure depletion.[46]

As persuasive and definitive as the experiments were, and as elegant and general as Muskat and Meres' theoretical analysis was, the new paradigm in petroleum engineering science was just that, *engineering* science. Unshielded by the prophylactic of a pure science ideal, engineering sciences have to earn their keep, which means be useful, or at least be perceived to be useful, by their corporate, institutional, or bureaucratic sponsors. As H. H. Evinger and Morris Muskat, somewhat defensively, wrote in 1942, "Sufficient laboratory research has been carried out in the last several years on the fundamental features of the flow of heterogeneous fluids through porous media to call for consideration of what can be done with this type of information." Quantitative applications of these results were few, in part because of "the lack of sufficient laboratory data relating to practical fluids and sand systems such as would occur in actual oil reservoirs." Therein lay the rub: "unless it can be shown that if such data were obtained quantitative applications could be made, serious doubt may be cast upon the desirability of carrying on further extensive experimental work on fluid flow."[47]

While the hallmarks of great scientific theories are commonly claimed to be their quantitative precision, breadth, explanatory power, and elegance or aesthetic appeal, their great practical virtue is their ability to probe, or vicariously explore, a world of seemingly arbitrarily large complexity. They isolate and purportedly control that messy complexity: They are existential experience's barbed wire, keeping chaos out, while letting tame complexity in. So too are theories in engineering science.[48]

Morris Muskat's second book, *Physical Principles of Oil Production*, was published in 1949. It was and remains the definitive statement of fundamental theory for oil and gas production. Muskat reiterated the particular epistemological plight of petroleum engineering science:

The physics of oil production is unique among all applications of physics. It deals with objects—the oil reservoirs—that are variable continua, subject only to infinitesimal sampling. Their physical histories are irreversible transients that can be observed only at isolated points "on the fly." They are controlled by a trinity of forces—hydrodynamic stress gradient, gravity, and interfacial—superposed on a trinity of phases—oil, gas, and water. Except for relatively microscopic laboratory investigation, any major and significant field experimentation creates an irreversible change in the state of the system and destroys the possibility of repeat or comparative experimentation on the same reservoir. And finally there are no identical specimens in the ensemble of the objects of study.[49]

Nevertheless, Muskat summarized the great deal that had been learnt in the preceding twenty-five years: knowledge of the "fundamental physical properties" of hydrocarbon fluids and of oil-bearing rocks and understanding homogeneous and heterogeneous fluid flow through porous media. In addition, Muskat presented specific chapters devoted to gas-drive and water-drive reservoirs, secondary recovery, condensate reservoirs, well spacing, and estimating recoverable reserves. It was an exemplary tour de force, so much so that it seems fair to say that Morris Muskat is to petroleum engineering science as Isaac Newton is to classical physics. By the mid-1950s, petroleum engineering's theoretical foundation was all but complete.[50]

But petroleum engineering science, like other engineering sciences, is supposed to yield reliable, useful, practical knowledge. That means not only mathematically and theoretically robust, but also cost-effective and within the comfort zone of practicing engineers and their clients. As good as theories of homogeneous and heterogeneous flow through porous media were, analyses based upon them remained costly and demanding. As a result, a lot of effort went into the hunt for alternatives to the time-consuming calculations they required. Petroleum engineers explored three orthogonal avenues: exploitation of simpler to calculate (but still sophisticated) mathematics, the results of which were normally reduced to graphical or tabular form for practical use; physical simulacra in the form of (electrical) reservoir models; and dedicated, punched-card digital computer programs.[51]

Henry J. Welge at the Carter Oil Co. (the Standard Oil subsidiary in Oklahoma), Tulsa, for example, devised graphical techniques that sharply reduced the amount of field data and laboratory core analyses required for reservoir studies, and short-circuited the "tedious and time-consuming" integrations and calculations that had been required, dropping time necessary for a reservoir analysis from "an estimated man-week if the conventional integration technique is used" to a single man-day. John von Neumann, as a consultant, mathematically verified Welge's derivation.[52] These efforts themselves also produced theoretical innovations: for example, J. S. Aronofsky and R. Jenkins (Magnolia Petroleum, Dallas, the wholly-owned Texas subsidiary of Socony Vacuum) introduced (1953) a hypothetical "effective drainage radius" for a producing well to simplify computations.[53]

Seventy-five years after Buckley and Leverett published their original "Fluid Displacement" paper, and fifty-five years after Welge had offered his "simplified method for computing oil recovery,"

MMbbls Subsurface Consultants made available a free, downloadable computer application implementing Buckley and Leverett's equations.[54] The protracted journey theories of fluid flow through porous media made from leading-edge applied physics through contested articulation and extension in petroleum engineering science, to readily usable engineering doctrine, to, now, freely available, black-boxed application, nicely illustrates how—and with how great a creative effort—science and engineering science may be packed down for ready technological exploitation.[55]

## ONTOLOGICAL SURPRISES

Secure and complete theoretical understanding of multiphase flow did not exhaust the ontological possibilities. The underworld still held surprises: retrograde condensation, connate water, and—ground zero for the revolution—East Texas, again.

### Retrograde Condensation

A single-component gas at constant temperature ordinarily expands as its pressure decreases, more or less in accord with (Robert) Boyle's law (1662): $PV$ = constant.[56] Gas mixtures with multiple components (such as natural gas containing methane and butane or propane) may behave differently and exhibit "retrograde condensation." Gas containing such "condensate" is a single-phase fluid (gas) at original reservoir conditions, but as reservoir pressure falls (at constant temperature), liquid forms "and the amount of the liquid phase increases with pressure drop. The system reaches a point where, as pressure continues to decline, the liquid revaporizes."[57]

A number of European scientists had observed this effect in the second half of the nineteenth century, but by the time oil operators independently bumbled into the phenomenon in the early 1930s, "much of the early work was buried in the literature and not included" in textbooks.[58] Even so astute a scientist as Ben Lindsly was initially flummoxed by the problem, choosing to exclude its discussion from his definitive BOM Technical Paper, "Solubility and Liberation of Gas from Natural Oil-Gas Solutions" (1933).[59] As he wrote to Harry Fowler, Director of the Petroleum Division of the US Bureau of Mines, at the end of December 1931, "Regarding retrograde condensation, my present reaction is to omit reference to this phenomenon in the present paper. The reason being that I am too hazy on the subject myself, and although it might explain some of my funny curves, I believe it would be best to hold this up until we get 'more on the ball.'"[60]

The "physical phenomena" were clarified by B. H. Sage, Jan G. Schaafsma, and W. N. Lacey in one of the first API research Project 37 (California Institute of Technology) reports in 1933.[61] But understanding the issue in the context of natural reservoirs remained problematic, and within the closely-knit community of petroleum engineering scientists, appropriate experimental procedures were contested. For example, Lacey's early results came from remixing gas in "dead" oil (free of dissolved gas content), since his initial interest was repressurization, while Lindsly was concerned with how gas came out of solution in a producing oil field. The two research groups also used different experimental gases. Apparently in one discussion, Lacey suggested shaking Lindsly's experimental pressure "bomb" like a seltzer bottle to liberate gases.

In his BOM monthly letter to Fowler, Lindsly, after observing that he, Fowler, and Harold Miller probably "appreciate Dr. Lacey's problem a whole lot better than Dr. Lacey appreciates our problem," retorted, "Who in hell is going to shake the oil reservoir to make it come to equilibrium? (Of course, I will have to back up on this statement a little when the California earthquakes are taken into consideration)."[62]

What oil operators knew for sure, at the latest by the end of 1932, was that some deep, high-pressure gas reservoirs produced a "wet gas" that in surface separators might release up to 150 barrels of very high API gravity "distillate" or "condensate" per million cubic feet of gas. This distillate, natural gasoline, was the most valuable substance to come out of a gas well, so good that in many cases you could run it from stock tank to your car's gas tank and drive off, no refining necessary. Watershed (or distillate-shed) was the 1932 analysis of the 8,200-foot-deep gas reservoir in the Big Lake Field in West Texas. Subsequent recognition that many deeper gas reservoirs also produced distillate refocused the attention of petroleum engineers and petroleum engineering scientists on its properties and on optimal production and recovery strategies. By the late 1930s, the phase relations of naturally occurring hydrocarbons, especially the heavier fractions such as butane, pentane, and heptane, were well-understood empirically.[63]

But engineers realized that as reservoir pressure fell with gas production, retrograde condensation could occur within the subsurface reservoir itself. In that case, as entailed by heterogeneous flow theory, gas would flow easily around the condensed liquid hydrocarbons, leaving them trapped and unrecoverable. Thus, distillate production became a conservation issue subject to regulation as well as a technical and economic issue.[64] The obvious solution was repressurization: reinject gas stripped of its distillate at the surface back into the reservoir to maintain pressure. So long as there was no ready market for the "dry" or stripped natural gas, using it for repressurization was unproblematic, but such a program did present two other problems: Reinjection compressors, valves, and collection piping were expensive, and effective reinjection required careful analysis, planning, selection, and siting of both injection and production wells. Both these factors made unit operation of the entire field compelling.

In the 1980s, removal of federal price controls on natural gas sold in interstate commerce finally scotched the absurd result of severe gas shortages in interstate markets, and substantially higher prices, and surpluses, in intrastate markets. But now whether to recycle dry gas to conserve distillate, the price of which had become as volatile as that of crude oil with the rise of OPEC, became a dicey economic proposition. Even by 1949, with the opening of large interstate gas markets, the economic calculus had become complicated, as Morris Muskat illustrated with a detailed example in his *Physical Principles of Oil Production*.[65] With much higher gas prices, immediate production of large quantities of natural gas could very well be worth more than the sale of distillate combined with the discounted present value of delayed production of reinjected dry gas.

An economic choice to produce gas by straight depletion (without recycling) brought its own set of problems. Since pressure drop is largest near the well bore and in the production casing itself, retrograde condensation is greatest in those regions, often so great that, again in accord with heterogeneous flow theory, it can retard or even block the flow of natural gas. Not helpful if the object is the rapid production of gas, and distillate be damned. Petroleum engineers devised a variety of stratagems to combat this problem: hydraulic fracturing, drilling horizontal or inclined wells, injection of a variety of solvents. Simplest was the imaginatively but appropriately named "huff and puff"—alternately injecting dry gas into a well to revaporize condensate, then producing gas from the well until it blocked up again.[66]

## Connate Water

Geologists, then petroleum engineers, assumed that over eons of geological time, gravity would cause gas, oil, and water within a closed subsurface porous formation to segregate into relatively well-defined strata: thus, the standard depiction of distinct layers—gas cap, oil, and salt water—which still is presented in virtually all introductory discussions and texts. When geologists first began examining oil well core samples from rotary rigs, they continued the tradition established in their paleontological examinations and stratigraphic correlation efforts for cable-tool wells. Their goal was to characterize particular specimens in a way applicable to all similar rocks, wherever found. Given that they were mostly petroleum geologists, or at least employed by oil operators, in addition to the geological identity of the particular strata from which a sample came and the depth at which it was found, primary interest was in how much space the rock contained that could hold fluids (porosity) and how easily fluids might move through the medium (permeability). To assure generality and comparability, geologists evolved standard procedures for examining and measuring these salient properties of their samples. As noted, cores were thoroughly cleaned, which stripped them of any fluid or gas content; this procedure was thought to be not only good scientific methodology but also necessary, since most rotary rig cores were contaminated by drilling mud.[67]

The first published intimation that this procedure might be flawed, and the inferences based upon it prone to error, appeared in 1926. Charles R. Fettke, an associate professor of geology and mineralogy at Carnegie Institute of Technology in Pittsburgh and a consultant to the Pennsylvania Geological Survey, reported the results of his analysis of core samples from the Bradford oil field; both core sampling and analysis at the time were uncommon and still in their infancy. He found that all his samples contained "as much water as oil (and in most cases more)," even though "the producing wells in the area where the core was taken give practically no water." Water saturation varied from 26 to 48 percent, while oil filled 16 to 40 percent of pore space. Fettke himself attributed the apparent excessive amount of water to loss of oil and water contamination in the coring process, and his findings were largely ignored.[68]

Likewise, N. T. Lindtrop and V. M. Nicolaef, in a paper in the *American Association of Petroleum Geologists Bulletin* in 1929, discussed their analysis of "the displacement of oil by water, in sands of different size of grain" in the Grozny oil field, Russia. They noted "the relatively large amount of water retained in the reservoir sand after ordinary production of the wells ceases," and inferred that, "It may be that more oil is produced by present methods than is generally estimated, and that estimations of unrecovered oil are too large if water is retained in the pore space of the sand in such important quantities."[69] From quite a different direction, in 1932 Jan Versluys concluded on theoretical grounds that "'connate' water must exist in oil sands because the surface forces between water and sand are greater than those between oil and water and thus preclude complete displacement of the original interstitial water by accumulating oil."[70] These omens too fell on deaf ears.[71]

Ralph J. Schilthuis (Humble Oil) presented the definitive paper on the topic at the Oklahoma City Meeting of the Petroleum Division of AIME in October 1937. He began by iterating the accumulating anomalies: the repeated discovery of water and salt in core samples from wells producing only oil; electrical logs which exhibited "unexpectedly low resistivity" in strata supposedly containing mostly oil and gas, implying the presence of something more con-ductive, such as salt water; his own "calculations of the oil and gas content of reservoirs based upon production and reservoir-pressure decline data" that indicated "appreciably lower values than are arrived at under the assumption that the pore space is occupied completely by oil and gas," which had theretofore been the customary way of estimating oil in place. Schilthuis adopted the reasoning of Versluys and others, that ancient connate water likely was held in place in the smaller pore spaces by capillary forces, even in supposed mostly oil-bearing strata.[72]

What Schilthuis brought new to the table was experimental technique and apparatus. In place of water-based drilling mud, he persuaded Humble to use a drilling mud based on the oil from the wells he was investigating. While this approach would render the resulting core sample useless for determining oil content, the object was to measure connate water content without external water contamination. Schilthuis maintained that, except in extremely permeable for-mations, connate water held in place by capillary forces would remain in the core. To determine connate water trapped in the core, a sample was crushed, then subjected to vaporization and combustion. Water condensed out of the effluent directly measured original water content, while sodium chloride residue in the apparatus, compared with its concentration in native salt water from the same well, provided a check on the measured quantity of water in the sample. Conventional analysis techniques applied to an adjacent sample of the core yielded its porosity, permeability, and bulk density.[73] Schilthuis presented detailed analyses of cores from the oil and gas bearing zones of two East Texas wells, and one each from the Anahuac and Tomball oil fields. The East Texas cores showed an average connate water saturation of only about 10 percent, but the Anahuac measurements ranged from 16 to 98 percent, with an average of 50 percent, while the two intact cores from the Tomball field indicated 49 and 59 percent.[74]

Just as Ben Lindsly's gas saturation results had been decisive in establishing the dominance of water drive in East Texas, Schilthuis's connate water experiments convinced the community of petroleum engineers. The implications of his findings were serious. First, the common practice of estimating potential oil recoveries or reserves "purely upon acre-foot yields experienced in fields considered to be similar" was "likely to be a dangerous procedure" without more information on the specific field in question: "Two fields may differ only slightly in operating characteristics in early life but may have distinctly different unit oil saturations and thus may yield entirely different unit ultimate recoveries."[75] No more fast and loose use of "experience." Second, the presence of essentially immobile connate water in producing oil strata altered permeability, and therefore application of homogeneous or heterogeneous flow theory: "it is known that the

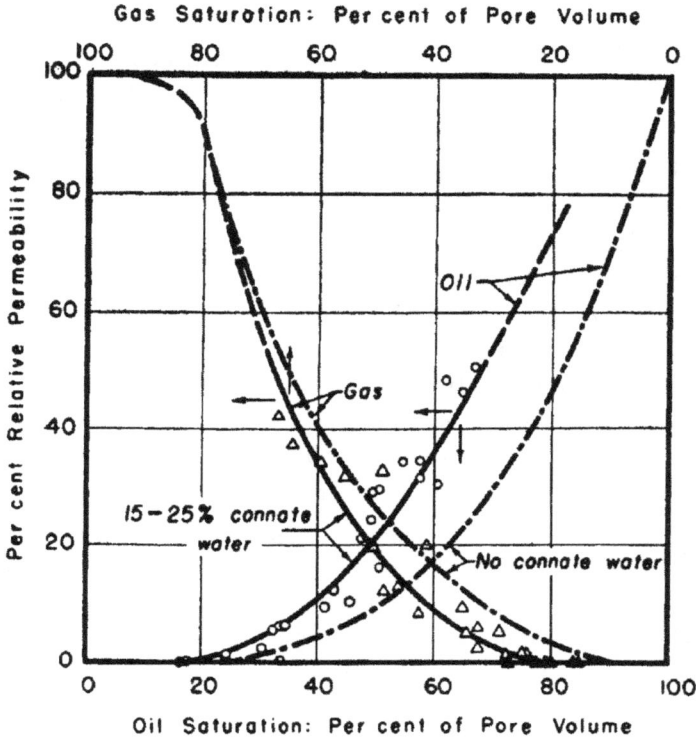

Figure 7.3. Permeability-Saturation Curves for Oil and Gas at 15 to 25 Percent Connate Water[40]

permeability of sand to oil is considerably reduced when relatively large amounts of water are contained within the pore spaces. Water saturations up to 10 or 15 per cent have little effect, but saturations exceeding these limits markedly deter flow."[76] This conclusion was revised two and a half years later by Leverett and Lewis's investigation of three-phase heterogeneous flow.[77] Nevertheless, connate water saturation became a critical boundary condition, determining baseline "effective permeability" for all reservoir flow calculations.

Although a great deal of work remained to be done to precisely characterize connate waters and their behavior, the underworld had changed again.[78] A lot. As Everette Lee DeGolyer observed in 1940,

As a result of this discovery [the prevalence of connate water], we may note that all previous estimates of the percentage of oil recovered from sand and the percentage of oil left in the sand and not recoverable are wrong. Also that many of the laboratory experiments upon the passage of oil through dry sand are valueless as representing a condition apparently not met with under natural conditions.[79]

More vividly, "Assuming 25 per cent interstitial water and 25 per cent increased volume of oil in place arising out of gas in solution, averages which are regarded as low rather than high, estimates of oil originally in place would have been reduced by 43.75 per cent and estimates of ratio of oil recovered to that originally in place would almost have doubled."[80] A half-century of Progressive doom and gloom was wrong. You cannot have wasted what was never there.[81]

## East Texas Redux

In the way that the world does when confronted by elegant scientific theories well corroborated by substantial evidence and widely embraced by knowledgeable communities of practitioners, East Texas misbehaved. Subsurface pressure in the East Texas field "declined at an accelerated rate from the beginning of October 1937."[82] This decline initially was attributed to excess production of salt water with oil,[83] but by "the latter part of 1939 and the beginning of 1940," "large discrepancies between calculated and observed values were found."[84] As adherents of established theories or paradigms typically do, petroleum engineers and petroleum engineering scientists initially dismissed these problems as ad hoc anomalies of no consequence.[85] But such communal hubris did not make the bad behavior go away.

As noted, application of the theoretical homogeneous flow equations derived independently by Muskat and by Schilthuis and Hurst in 1934 assumed that the East Texas field was connected along its western edge to an infinite aquifer, and that this condition was empirically satisfied by the compressibility of the immense volume of salt water contained in the Woodbine basin. The value necessarily imputed to the compressibility of aquifer salt water was an order of magnitude greater than that experimentally established for pure water.[86] East Texas paid none of these theoretical embellishments no nevermind: By 1943, "the calculated pressures [had] deviated progressively from the observed pressures to such a degree that . . . it [had] not been possible to predict accurately the future pressure of the field using the equation constants that were employed in the past."[87]

There were several possibilities for what had gone wrong: the extraordinary assumption for the water compressibility value, which had always seemed a little too convenient, might be wrong; the geometric configuration of the aquifer might be oddly irregular, or its permeability might vary; or production from multiple Woodbine fields might affect the subsurface pressures in all of them. As mightily as they had struggled to establish the water-drive paradigm for East Texas, petroleum engineers faced up to the issue and did what scientists or engineers do: gather more empirical data to better define and clarify the problem and its extent, develop new analytical approaches, and contrive new, or modified or extended theory to accommodate the new facts.

The subsurface Woodbine basin underlies forty-eight counties and 31,500 square miles in East Texas, an area slightly larger than Scotland. Most production was concentrated in two areas, the Mexia-Talco (or Powell) fault line to the west, where production had begun at Mexia in 1920, and against the Sabine uplift to the east, where the East Texas pinch out occurred, although other significant fields (Van, Hawkins) were scattered through the intervening basin.[88] The first solid clue as to what was amiss cropped up not in the East Texas field itself but in the Hawkins field, about 25 miles as the crow flies to the northwest. Discovered in 1940, Hawkins produced from a local anticline in the same Woodbine formation as East Texas, but also was divided internally by a sharp fault line. Initial downhole pressure tests, which were by then routine in newly discovered fields, indicated that Hawkins wells on the west side of the fault had a significantly higher pressure than those on the east side.[89] Engineers considered several possible explanations: that the wells on the eastern side tapped a separate, closed reservoir isolated from the larger Woodbine Aquifer; that some pressure had dissipated from the eastern side through the fault; or that west Hawkins was isolated by an impermeable barrier. West Hawkins "approached what would have been considered a normal original pressure for its depth," and was insulated by a 40-foot-thick "heavy asphalt deposit . . . at the base of the producible oil column." Therefore, the engineers inferred that "initial pressure in East Hawkins was some 280 lb. below normal [that in west Hawkins] because of East Texas Field withdrawals made during the ten year period prior to the Hawkins discovery."[90]

The realization that production from different Woodbine oil reservoirs might interfere with one another at considerable distances, or affect pressures throughout the aquifer, inspired petroleum engineers at the Humble Oil Company to initiate a detailed study of subsurface pressure and pressure changes in the whole Woodbine basin. From 1942 on, Humble, "in its wildcat program within the area . . . measured the pressure in clean porous water-bearing sand of the Woodbine formation in 57 wells located in 18 counties." These tests meant incurring the costs of dedicated drill-stem tests, as well as taking downhole fluid samples, as each wildcat passed through the Woodbine.[91] These data were combined with historical production and pressure records from the 42 Woodbine oil fields, 28 of which were still producing in 1951.

John S. Bell and J. M. Shepherd, Humble Company petroleum-research engineers, concluded unequivocally, in results published in 1951, that this information "demonstrate[s] that East Texas Field withdrawals have been the major cause of pressure losses in the basin, although fault-line field withdrawals have contributed significantly to such losses."[92] They constructed a detailed pressure contour map of the entire Woodbine basin east of the Mexia-Talco fault zone that illustrated precisely the pressure differences across the underlying aquifer caused by oil production in the multiple oil fields. They suggested that the pressure changes could be depicted accurately "considering radial flow principles" in which pressure varied "as the log of distance from hypothetical focal points of withdrawals from each of the two [major producing] areas." These calculated values matched observed values except for local anomalies where faulting created a barrier or restriction to drainage and produced local deviations.[93] Both William Hurst and Morris Muskat voiced approval of Bell and Shepherd's paper and complimented the Humble Company for developing and sharing the data.[94] Muskat expressed the hope that petroleum engineers would go on to "make quantitative analysis of observations using electrical analyzer or similar techniques" to "determine quantitative correlations of the data with the theoretical predictions," and, specifically, the "macroscopic expansion capacity of the Woodbine aquifer."[95]

They did. R. C. Rumble, H. H. Spain, and H. E. Stamm III, all also at Humble Oil and Refining, Houston, built an electric reservoir analyzer, or analog simulator, using electric capacitors and resistors, for the entire subsurface Woodbine basin, with special attention to the East Texas field.[96] Their design was similar to that W. A. Bruce at Carter Oil Company had built in 1942, although considerably larger and more elaborate. Oddly, they did not cite or acknowledge Bruce's earlier analyzer work at their sister firm in Oklahoma, even though it had appeared in the open literature, and Carl Reistle, by then Manager of Production Operations at Humble, had presided at Bruce's original presentation.[97] Rumble et al.'s apparatus successfully retrodicted, that is, accurately reconstructed, both the production and pressure history of East Texas and of the other fields producing from the Woodbine, and captured observed pressure changes throughout the basin. The analyzer, like the Greek Antikythera planetary simulacrum (circa 80 BC) or the Rittenhouse orrery (1771), could be run backward or forward in time, and thus generate retrodiction (historical reconstruction) or prediction with equal facility.[98]

Rumble, Spain, and Stamm (1951) finally resolved the puzzle of Woodbine saltwater compressibility. They assumed that initial porosity for the entire formation was 25 percent, and, on this assumption, their analyzer results indicated an apparent water compressibility of $5.3 \times 10^{-8}$ vol/vol/lb./sq ft, "somewhat higher" than the "$1.85 \times 10^{-8}$ vol/vol/lb./sq ft that was measured on samples of Woodbine water taken during drill stem tests."[99] Earlier, in order to refute the formation subsidence or compaction hypothesis that some geologists had advanced to explain subsurface pressure behavior in East Texas, the Bureau of Mines had undertaken a study of the compressibility of consolidated sandstones. As noted, the results had been published in 1940: C. B. Carpenter and G. B. Spencer reported that "the measured effect of pressure loss on Woodbine sand was to reduce the size of the pore volume by an amount equal to $2.85 \times 10^{-8}$ vol/vol/lb./sq ft."[100] Rumble, Spain, and Stamm therefore concluded,

> On the assumption that this effect would be additive with the compressibility of the Woodbine water, the effective compressibility of the Woodbine aquifer would be $4.7 \times 10^{-8}$ vol/vol/lb./sq ft. The agreement between this value and that experimentally determined with the analyzer is rather close and is well within the over-all accuracy of the analyzer study.[101]

On this analysis, the effect of formation compaction was some 50 percent greater than that of Woodbine saltwater compressibility. Ironic that a hypothesis emphatically rejected in 1934 and conclusively and decisively refuted in 1940 should be subtly resurrected in 1951 to save the phenomenon. Only now, like the expansibility of the entire Woodbine's brine content, the formation compaction hypothesis applied to the entire basin (not just the East Texas field itself), thus eliminating the likelihood of observable local subsidence. Rumble, Spain, and Stamm's analysis, together with Bell and Shepherd's and Carpenter and Spencer's data, validated the theretofore assumed compressibility of the Woodbine Aquifer, now both expansion of its salt water and the compaction of its pore spaces. East Texas, and the other Woodbine fields (save one) remained unquestionably water drive.

Rumble, Spain, and Stamm's simulation also validated use of the by-then standard radial flow equations as derived by Morris Muskat and William Hurst. Those equations, modified to account for interference among fields in the Woodbine basin, not only matched analyzer results but also continued to give accurate and reliable predictions through 1961.[102] The problem of controlling and prorating the East Texas field was no longer that of managing individual wells or leases, nor even the giant oil pool itself, but of moderating and coordinating withdrawals, and, after 1942, reinjection of salt water produced with oil, across the entire Woodbine.[103] It had taken a generation, and the trial by oil of a new engineering science, but East Texas might be thought to be finally understood—for now.

## THE PROCESS OF PETROLEUM ENGINEERING SCIENCE

What we have recounted in considerable detail has been chosen to warrant several inferences: Invention and innovation in engineering science is always at some level unforesighted. What keeps this unforesighted variation in bounds, and makes it fruitful, is selection. Selection is messy: claims and achievements are contentious and contested, sometimes with noticeable heat. Scientists, like litigants, argue, bicker, and change their stories.[104] Routine petroleum engineering isn't. Work in engineering science is a collaborative, collective endeavor, virtually never the sole province of one person. Specific areas of research are typically pursued by well-defined invisible colleges or research fronts but are also embedded in a larger expert community of practitioners, which serves as audience, resource, and source of authentication.

Research is characterized by a delicate balance of competition and cooperation: the moral economy of engineering science. Moral economy entails free publication and cooperative research rather than appropriation, although in petroleum engineering, advancing the scientific state of the art changes the basis for regulation and also may serve material interests. This shared state of the art also reduces, but does not eliminate, knowledge and power disparities between large, integrated firms and smaller independents. Consulting petroleum engineers provided au courant, cost-effective analyses, and in their practice extended sophisticated petroleum engineering science across the whole social spectrum of the oil fraternity, all of which, incidentally, made the Railroad Commission's job a bunch easier and externalized a lot of its costs.

Engineering science is contingent, path dependent, and corrigible, as all scientific inquiry is. The endeavor is open-ended. There is no final theory, at least not yet. The world is full of surprises, which, when more or less tamed, are called "anomalies." Best case, they give rise to new insight, but only after the whole process is reprised, sometimes not very congenially. It is how petroleum engineering science as a social process works.

## USEFUL RELATIONS: ALL IN THE FAMILY[105]

The useful relations for engineering practice that petroleum engineering science produced became the province of the first two or three generations of professionally trained petroleum engineers, who were literally the useful relations of the oil fraternity in Texas. They were the fraternity's sons and nephews (and later daughters and nieces) and cousins and in-laws and grandchildren. They were usually trained in oil-state universities and initially hired by oil-state firms. They went and learnt and came home. Their dual descent, biological and cultural, meant that underneath their technical expertise, they largely shared the heritage and values—the moral economy—of the oil fraternity in Texas.[106] They understood, and spoke, the language. They were not alien "experts" or "outsiders" sent to tell the aborigines how to live, most often in ignorance of indigenous practices and accumulated wisdom; they didn't come home to implement the fiat of some distant, and usually ill-informed and misguided, bureaucracy, which frequently yields disastrous "interventions" and rabid resistance.[107] Disagreements, even quarrels, were all in the family. Petroleum engineers already belonged, which gives the dissemination of petroleum engineering within the oil fraternity a wholly different character from that often encountered.

## Knowledge Production

So far, focus has been on specifics of petroleum engineering science and the process by which it changed practice, law, policy, and property in Texas and the sometimes recalcitrant and begrudging adaptation of the oil fraternity to their world remade. As for other engineering disciplines, the episteme of petroleum engineering emerged from a nexus of research institutions (public and private), trade group and voluntaristic organizations, elite universities (public and private), and corporate research and development efforts, sometimes in dedicated research-and-development (R&D) departments, but more often, especially before World War II, in functional production departments. Professional engineering societies, through their publications and conferences, provided primary means of dissemination and diffusion, and furnished the public arenas in which petroleum engineers could ritually produce their knowledge, their discipline, their profession, themselves, and their social authority. And given the idiosyncratic nature of all petroleum reservoirs and of individual wells, petroleum engineering practitioners in relatively narrow investigations frequently made profoundly useful contributions.

The American Petroleum Institute (API) began supporting basic petroleum engineering research in 1926, the year Beecher and Parkhurst's paper appeared, with a million-dollar grant from none other than John Davison Rockefeller. Although supervised by API advisory committees, grantees did the actual research, usually with the assistance of graduate students. For example, the longest running Project, number 37, "The Fundamentals of Hydrocarbon Behavior," at the California Institute of Technology, under the direction of Bruce H. Sage and William N. Lacey, had "as its objective the study of the behavior of pure hydrocarbons and their mixtures at elevated pressures."[108] The whole API research programme was truly fundamental, looking at the chemical composition and behavior of hydrocarbon compounds, and it was, by its nature, multidisciplinary. This research was conducted in elite educational or research institutions. But, true to the scientific ideology underpinning it, API research results were widely and freely disseminated. Two other facets of the API basic research programme also stand out: things that it was not. It was not conducted by petroleum engineering practitioners. And although some oil-state universities did host significant projects, most of the money did not go to the schools that trained the majority of practicing petroleum engineers: less than 5 percent of API research funds flowed in any form to Texas or Oklahoma.

Connecting the most basic research of the sort funded by the API in its fundamental research programme and practitioners' field investigations was the US Bureau of Mines, and especially its Petroleum Experiment Station in Bartlesville, which not only produced theoretical knowledge on its own account but also did a great deal of the empirical research that grounded the whole enterprise. Likewise, ad hoc voluntary, cooperative organizations and individual field committees developed critical reservoir data and, in the case of the East Texas Engineering Association, supported fundamental, paradigm-shifting research.

In contrast, petroleum engineering practitioners dominated the technical literature: Contributions by employees of operating oil companies to the three most important professional publications for the dissemination of petroleum engineering information (the *Transactions of the American Institute of Mining Engineers*, those of the American Petroleum Institute, and the American Association of Petroleum Geologists journal) rose from 40 percent between 1914 and 1927 to almost two-thirds for 1938–1958. Even in the latter period, authors identified as working in formal R&D departments published only 14 percent of all papers.[109] Engineers were more interested in doing, and learning how to do, than in knowing for its own sake.

## Petroleum Engineering Practitioners

The character of petroleum engineering education changed radically after 1926 and the emergence of a paradigmatic, self-conscious profession. Nevertheless, university-educated men were a rarity in the oil patch until late in the 1930s, and the idea of petroleum engineering itself was a hard sell: Oil operators' impression that a purported "petroleum engineer" was nothing more than "a geologist who had seen an oil well" receded only slowly.[110]

The programs that would dominate practitioners' education all appeared about the same time as or soon after Beecher and Parkhurst's classic paper. In contrast to the production of petroleum engineering knowledge, the production of petroleum engineers was almost exclusively in the halls of oil-state universities. In those institutions, a small number of doctoral faculty led a much larger cadre of practitioners and master's level students in the instruction of undergraduates. Most doctoral-level faculty themselves had also been at one time or another practitioners and frequently came originally from oil states.

Three sources of data on the training of petroleum engineers are membership records of the Society of Petroleum Engineers, comprehensive surveys of petroleum engineering education conducted by the Education Committee of the Society of Petroleum Engineers in 1958 and 1965, and registration records and university bulletins from three of the foremost state institutions teaching petroleum engineers—the University of Texas (UT), the Agricultural and Mechanical College of Texas (A&M), and the University of Oklahoma (OU).

The Society of Petroleum Engineers (SPE) originated as a section, then division, of the American Institute of Mining Engineers and became a separate, but still affiliated, society with its own headquarters, membership standards, budget and journal in 1938; SPE became a separately chartered society in 1957. In 1957, of 10,312 members, 4,002 listed their primary field as petroleum engineering, and, of those, 1,723 specified reservoir engineering.[111] Of a random sample of members who joined between 1930 and 1960, 83 percent held Bachelor's (BS) degrees; fewer than 2 percent held degrees beyond the master's level. In this sample, a third graduated from the University of Oklahoma (OU), the University of Texas (UT), or Texas A&M (A&M). Nearly three-quarters graduated from universities in major oil-producing states, plus the Colorado School of Mines. The major professional society for the dissemination of petroleum engineering knowledge was dominated by practicing engineers, less than half of whom held degrees specifically in the field, but nearly three-quarters of whom had earned degrees in major oil-producing states.[112]

Oil-state preeminence in petroleum engineering education was even greater. Of all holders of petroleum engineering degrees, at any level, through 1964 (N=16,826), the top three universities (OU, UT, A&M) graduated over one-third. Over two-thirds graduated from just seven schools. Nearly two-thirds earned their degrees from schools in Texas, Oklahoma, and Louisiana. Nearly 90 percent graduated from schools in major oil-producing states.[113] As an engineering discipline, petroleum engineering was the province of a small number of schools located mostly in the major oil-producing states of the midcontinent.

In these institutions, petroleum engineering was conceived, sold, and run as a practical, practitioner-oriented program. What little available direct evidence there is suggests that oil-state universities saw in petroleum engineering an opportunity to expand their turf, build their institutional constituency among a strong (economically, politically, and socially) fraternity, and possibly develop a substantial research position. The surviving notes of the president of the University of Texas for the meeting that led to the creation of the petroleum engineering program in 1928 are revealing: "Ag [agriculture] enrollment down [that is, archrival A&M was hurting]; Eng [engineering] up; states fighting duplication [of programs]; opportunity for us."[114] In other words, UT could steal a march on its historic nemesis, and co-beneficiary of the State's Permanent University Fund, A&M, by latching on to a new and rapidly developing field. Nevertheless, both the craft traditions from which petroleum engineering evolved and the paternalistic affection the oil fraternity had for petroleum engineering and engineers is nicely expressed in the inscription on a large, scale model cable-tool drilling rig the Texas Company presented to the new petroleum engineering program at UT:

> Scale model of standard cable tool rig used to drill the Texas Company—J. W. Parks No. 1 Well, Section 3359, T. E. & L. Co. Survey. This well was completed as the first producing oil well in Stephens County, Texas on November 2, 1916, with an initial production of 75 barrels of oil per day at a total depth of 3220 feet.
>
> This model was constructed by employees of the Texas Company at Parks, Texas and has been presented to the University of Texas for exhibition and for instructional use in petroleum engineering classes.

Because petroleum engineering is a synthetic engineering science, what its practitioners learned is a combination of basic science and mathematics, traditional engineering (mechanical, civil, chemical and, after World War II, electrical), and domain-specific engineering theory and technique.[115] Curricula for petroleum engineers were virtually standardized by the eve of World War II.[116] What variance there was largely reflected slight institutional differences in interests (California schools, for example, with no in-state deposits of natural gas, offered little coursework on that resource). More commonly, inter-institutional differences reflected purely intra-institutional turf issues among departments. UT, for example, instituted a petroleum geology program within the geology department that differed by one course from the simultaneously introduced degree program in the new Department of Petroleum Production Engineering.

The general structure of petroleum engineering education, the distribution and character of its basic course requirements, changed very little between the mid-1930s and the mid-1960s. But courses specifically in petroleum engineering, always a minority of total credit hours, did change drastically over time to reflect the changing content and the increasing abstraction of petroleum engineering theory and practice: as W. R. Woolrich noted in 1964, "The so-called descriptive technologic courses have been replaced by those employing the engineering or quantitative approach."[117]

As petroleum engineering matured as an engineering science, its emphasis on practical experience and craft empiricism correspondingly diminished. In 1924, the year of its founding, the University of Oklahoma's program emphatically recommended "that the student interested in this work secure experience in actual operations."[118] Although all programs continued to urge their students to get practical experience (which most seem to have done), by World War II the recommendation had softened considerably, and for the first time alternative coursework was offered which could, at A&M, "be taken in place of field work as required of all graduates in petroleum engineering."[119] By the mid-1950s, the requirement for practical experience had withered away to a course which consisted "primarily of visits to oil and gas fields," in other words, to a glorified "get acquainted" series of field trips. That A&M students did not take such activities quite as seriously as their elders would have preferred is perhaps expressed by the next sentence in the course description, "The requirements of this course will not permit a student to take any other course at the same time."[120]

What then did petroleum engineering practitioners know? From the beginning in the 1920s, mathematics that would have been the envy of Newton or Leibniz, physical theories beyond anything Lagrange or Laplace touched upon, organic chemistry that would have amazed Lavoisier or Dalton, engineering methodology that would have made Leonardo da Vinci proud: a tribute to historical time and the efficacy of formal education. In their laboratory experience, they had learned both the explicit and tacit elements of fundamental scientific methodology, as well as specific engineering variants thereof, such as permutations of control volume theory and parameter variation. Finally, they had come to possess a positivist, utilitarian scientific ideology which guided their engineering practice, justified it to themselves, animated their self-productions, and described and legitimated it to the oil fraternity at large.

Petroleum engineers carried their engineering science, and their scientific ideology and values, unproblematically back into the oil fraternity for the simple reason that they were its own come home from college to do things better, and, not incidentally, more profitably. In time they did change the values and outlook of the oil fraternity, so that by the late 1950s, to be an oil man meant to believe (or at least to profess to believe) in conservation and to abide faithfully by best engineering practice as defined by petroleum engineers and sanctioned by state-regulatory agencies: oilmen had become "reasonable." But that change came from within, from sons who spoke in familiar accents and had been "away" only to the state university, and maybe the Army.

Petroleum engineers matched the oil fraternity's social profile almost exactly. Petroleum engineers were sons; very few women appear to have earned petroleum engineering degrees prior to the 1970s. Before 1964 Texas A&M was an all-male institution and, additionally, required membership in the Corps of Cadets (Army ROTC). Petroleum engineers were white: universities in Texas and Louisiana before 1960 were racially segregated by law, as were those in Oklahoma de facto. Petroleum engineers were overwhelmingly from rural or small-town origins. Their parents were most likely small businessmen or self-employed professionals, although many had blue-collar backgrounds. A substantial minority were sons of people in the oil business. Their religious preferences were overwhelmingly Protestant, with the more fundamentalist (but still mainline) Baptist and Methodist denominations predominating. Petroleum engineers were part and parcel of the oil fraternity.

Aside from explicit racial exclusion (based on segregated schools) and what may be presumed to have been substantial implicit sexual bias, there is no evidence other groups were systematically recruited to or excluded from petroleum engineering programs. Petroleum engineering enrollments (as well as all engineering enrollments) were volatile. Petroleum engineering graduates at OU, UT, and A&M never exceeded 30 percent of all engineering graduates, and petroleum engineering enrollments responded strongly to perceived job opportunities: World War II aside, graduation figures reflect, slightly lagged, the periods of most rapid expansion of domestic oil exploration and discovery.[121] Petroleum engineers were largely self-selected, and their choice reflects both their cultural background and responsiveness to perceived economic opportunities. On every sociometric dimension for which there is reliable data, petroleum engineers were the sons of the oil fraternity.[122]

As one would expect of people who chose to become petroleum engineers and to attend the universities they did and who came from the places they did, most (for whom data exist) went right back into their native society and culture. As Table 7.1 shows, of those who belonged to the Society of Petroleum Engineers, most worked for major oil companies. A substantial number worked for service companies, especially the well-logging and core-analysis firms that grew rapidly after World War II.[123] These engineers were an especially credible and salient source of petroleum engineering knowledge for the oil fraternity at large: after 1947, almost all wells were electrically logged and had their cores systematically analyzed in corporate or, more often, in specialized service company laboratories. And it was those analyses that told people, and their bankers, whether they were rich or poor. Except for sex or large bass, few things focus attention quite so well.

Table 7.1. Employment by Sector, Members of the Society of Petroleum
Engineers (N=180)

| Sector | All Members | | Petroleum Engineers | |
|---|---|---|---|---|
| | Number | Percent | Number | Percent |
| Major | 77 | 43% | 73 | 54% |
| Independent | 35 | 19% | 19 | 24% |
| Consultant | 12 | 7% | 4 | 5% |
| Service | 44 | 24% | 11 | 14% |
| Other | 12 | 7% | 2 | 3% |
| | | | | |
| Total | 180 | | 79 | |

Note: All nine electrical engineers and two of the three physicists in the sample worked for well-logging firms.

There remains an asymmetry in the data. At any given time, the total number of holders of petroleum engineering degrees and the number of members of the Society of Petroleum Engineers is about equal. But slightly fewer than half of the members of that Society actually hold petroleum engineering degrees. Where do the other half of the petroleum engineers, the ones who do not join their professional society, go? Anecdotal evidence suggests that those petroleum engineers stayed in the oil fraternity—often as independent oil operators or producers, or as employees of smaller producers, or as owners or employees of related businesses. Given the social backgrounds of petroleum engineers, it seems unlikely most of them strayed very far. Membership in an organization like the Society of Petroleum Engineers is likely correlated with someone else's willingness to pay dues and fees and travel expenses, which major oil companies or larger independents are probably more disposed to do. For this reason, major company employees are likely overrepresented in the Society of Petroleum Engineers, and the number of petroleum engineers feeding back into the oil fraternity is probably larger than can be definitively demonstrated with available data. Moreover, that invisible half is likely to have been even better integrated into the broader fraternity. By virtue of their being both certified petroleum engineers and not associated with any "outside" institution, such as a major integrated producer-refiner, and by virtue of their nativity, these individuals served not only to disseminate the substantive knowledge of petroleum engineering but also to legitimate it to the fraternity, even as they produced themselves and their social authority.

In July of 1952, O. G. Lawson, who had spent his life in a variety of capacities typical of the early Texas oil fields, was interviewed in Cisco, Texas. He summed it all up this way:

> We used to produce the oil and put it in tanks on top of the ground to evaporate and burn just because there could uh—be no corporation [sic] among oilmen and we had no laws to help them. Was more every body for himself . . . each one had to get all the oil out that he could first . . .

We weren't getting out over a fourth or a third of the oil, and we recognized it as far back as my youth that we weren't getting all of the oil out, but we knew no other way to do it. . . . Now in later years, I'd say within the last twenty years [we] have learned how to produce oil and now I would say that we probably would get out two-thirds of the oil; with one method of recovery that number would cover as much as three-fourths. I think the time will come when we get 90 percent or more, as we learn more and more . . . and more and more about secondary recovery.[124]

Everyman's conversion, his perception of the efficacy of petroleum engineering, and his new-found faith in petroleum engineering science, more than science indicators, citation indices, discipline formation, professionalization, science budgets, R&D expenditures, or competing ideologies of pure versus applied science, portray petroleum engineering science *in* the oil fraternity in Texas. Useful relations indeed.

# CHAPTER 8

# FRUITFUL EMPIRICISM

We rejected these arguments as an attempt to substitute a judicial judgment for the expert process invented by the state in a field so peculiarly dependent on specialized judgment. We said in effect that the basis of present knowledge touching proration was so uncertain and developing, that sounder foundations are only to be achieved through the fruitful empiricism of a continuous administrative process.

—FELIX FRANKFURTER, MAJORITY OPINION, *RAILROAD COMMISSION OF TEXAS V. ROWAN & NICHOLS OIL COMPANY*, 1941[1]

The maturation of petroleum engineering science fortuitously coincided with complementary currents in American jurisprudence. Across the board, both federal and Texas courts became less inclined to inquire closely into the technical bases of decisions made by quasi-judicial regulatory agencies. This reticence enhanced both the authority and the autonomy of the Railroad Commission of Texas. Legitimated by its own success, backed by the moral economy of the oil fraternity and its ideology, supported, through benign neglect if nothing else, by the legislature and people of Texas, and led by the tireless and charismatic Ernest O. Thompson, the Commission became more assertive and venturesome, although still (mostly) true to its tradition of cooperation rather than coercion. Simultaneously, the assurance provided by increasingly efficacious petroleum engineering practice, combined with the price and supply stability assured by Railroad Commission prorationing, revolutionized oil-and-gas finance and opened new vistas of opportunity for the oil fraternity in Texas.

Yet knotty and noisome problems remained. The empiricism of the Railroad Commission of Texas's continuing administrative process was interminably vexed by "mine and thine." Even the moral economy of the oil fraternity in Texas, with its core tenets never challenged no matter how much transformed, got a little frayed around the edges. Unlike field and statewide production restraints with a reasonable and substantial scientific rationale, which arguably could be a positive-sum game (payoffs to all could be increased by cooperation and conservation), mine-and-thine problems are inherently zero-sum (what one guy gets, the other loses), which presented the most intractable conundrums faced by petroleum engineering practitioners, the Railroad Commission of Texas, and the oil fraternity in Texas.

The oil fraternity in Texas: Claude K. McCan, rancher, Victoria; James S. Abercrombie, J. S. Abercrombie Company, Houston; Ernest O. Thompson, Railroad Commission of Texas, Austin; H. G. Nelms, Independent oil operator, Houston; Herman Brown, Brown and Root, Houston; circa 1959. (Photo courtesy of Mrs. E. C. Japhet)

## THE LAW AND THE PROFITS

The great Progressive crusade to fumigate and rationalize government at all levels found expression in creation of a host of quasi-judicial regulatory agencies. Special-purpose boards, districts, and authorities had been around since colonial times and had blossomed in the nineteenth century: port authorities, levee districts, water and sanitation districts, reclamation authorities. What Progressivism brought new was an Enlightenment faith that application of Scientific Principles could make a rapidly changing and industrializing society rational, efficient, and fair. From the 1880s on, quasi-judicial agencies, federal and state, sprouted like crab grass: Interstate Commerce Commission (ICC), Texas Railroad Commission, Food and Drug Administration (FDA), Federal Trade Commission (FTC), Federal Reserve, Federal Power Commission (FPC), plus the alphabet soup of New Deal initiatives.

This parade of progress ran headlong into judicial review. The principle was about as settled as law gets: federal courts could subject both legislation, whether from the US Congress or state legislatures, and acts of governments and their agencies, federal or state, to constitutional review (*Marbury v. Madison*, 1803), while Texas courts could review both Texas legislation and the actions of state agencies. But states, as well as the federal government, had the power to regulate private businesses that affected the "common good" (*Munn v. Illinois*, 1877).[2] Yet the broad discretion granted to quasi-judicial agencies, combined with the often highly technical character of their domains, the evidence they considered, and the orders they issued, made trouble: To what extent should courts delve into the bases of those agencies' decisions? Should agencies' decisions be "black boxed," with the courts uncritically accepting their findings of fact? (Recall that even so acute an intellect as Joseph C. Hutcheson Jr. had trouble following the nuances of developing petroleum engineering science.) Or should courts conduct a trial de novo, and decide the facts for themselves? When fundamental issues of due process of law or unconstitutional taking were at stake, courts continued to probe more deeply. But the courts also adopted a satisficing standard for legitimacy: a regulatory agency did not have to show that its action was optimal or the best possible or even that a different result would not have been better, only that its decision bore a "reasonable" relation to some proper public purpose. In litigation, however, it is always in the agency's interest, ever mindful of its authority and autonomy, and in the interests of successful supplicants, to keep the black box nailed shut, while plaintiffs and those hungry for billable hours try to pry it open. The problem is analytically insoluble and remains one of the most noisome in jurisprudence.

During the first half of the twentieth century, the tide was running against judicial mucking about. The key precept became "substantial evidence." The notion first appeared in federal statute law, although not in those words, in the Interstate Commerce Act of 1887. By 1938, "nineteen administrative acts [contained] the substantial evidence rule in express form," although E. Blythe Stason, in his 1941 analysis of substantial evidence in federal administrative law, griped, "'Substantial evidence' is one of those numerous legal terms of doubtful content which may mean almost anything."[3] The closest thing to a definitive statement of the federal standard had come from Chief Justice Charles Evans Hughes in *Consolidated Edison Co. v. National Labor Relations Board* in 1938: "Substantial evidence is more than a mere scintilla. It means such relevant evidence as a reasonable mind might accept as adequate to support a conclusion."[4] This standard pretty much matched, in word and deed, what Hutcheson had done in *Amazon* in 1934.

These currents in federal jurisprudence swept the federal judiciary out of the business of overseeing Texas Railroad Commission prorationing. In its two *Rowan & Nichols* decisions, 1940 and 1941, the US Supreme Court ceded to the Railroad Commission of Texas and designated Texas courts nearly unlimited authority to regulate oil production. The cases originated in a Rowan & Nichols Oil Company challenge to a Railroad Commission prorationing order for East Texas (again), in 1938. Rowan & Nichols operated five oil wells located on some 24 acres in the central, or "Fairway," section of the field, where the productive sands were thickest and their permeability and porosity most favorable. Both well density and potentials varied radically across the field: Some wells were drilled one well to every five acres, other wells on a fraction of an acre; some had potentials of six or seven barrels a day, some over 25,000.[5]

Although the Railroad Commission determined each of Rowan and Nichols's five wells to be capable of producing over 20,000 barrels a day, the challenged order limited their output to 22 barrels a day each, a mere "pittance of about 2 barrels a day" above what the worst (stripper) wells in the field were permitted.[6] The Federal District Court found that the order was confiscatory and violated the Fourteenth Amendment, and the Fifth Circuit Court of Appeals affirmed.[7] The Railroad Commission appealed to the US Supreme Court.[8] Rowan & Nichols fared about as well as the British and French Armies at the time. The 6–3 majority opinion, delivered June 3, 1940, by Felix Frankfurter, explicitly recognized the dynamic character of petroleum engineering knowledge that Hutcheson had characterized in the *Amazon* decision in 1934. The court acknowledged that prorationing presented "as thorny a problem as has challenged the ingenuity and wisdom of legislatures," and that its administration was "full of perplexities." As a result, the Court sustained the Commission's order, holding that, "It is not for the federal courts to supplant the Commission's judgment even in the face of convincing proof that a different result would have been better."[9]

A new prorationing order for East Texas had already been issued September 11, 1939. This order too was duly challenged and enjoined. Hence *Railroad Commission of Texas v. Rowan & Nichols*, 1941.[10] Although the new order ostensibly "took into consideration two other factors—bottomhole pressure and the quality of the surrounding sand of the wells—as well as hourly potential," for Rowan and Nichols, these changes meant their total allowable production for all five of their wells rose from 154 barrels a day to 260. As the Supreme Court wryly observed, "This comparison of the practical operation of the two orders exposes the emptiness of the claim that a constitutional line can be drawn between them."[11] The Court reiterated its findings from the 1940 decision, concluding again "that sounder foundations are only to be achieved through the fruitful empiricism of a continuous administrative process."[12] Whenever recourse to state courts was available, the federal judiciary was not to intervene. This federal commitment to fruitful empiricism was reaffirmed two years later in *Burford et al. v. Sun Oil*.[13]

In contrast to federal law, the phrase "substantial evidence" appeared rarely in any state's statute law and was wholly absent in the various acts defining the responsibilities and procedures of the Texas Railroad Commission. But the same issues and conflicts appeared in Texas as in federal venues. By 1939, Texas courts largely acquiesced in Railroad Commission findings based upon its assessment of petroleum engineering evidence.[14] Nevertheless, like a pack of 'coon dogs who've lost the scent, Texas courts then hunted around some (for about ten years) before settling on a definitive meaning for substantial evidence, testimony both to the elusiveness and contentiousness of the concept and to the diversity and contingencies of specific cases.[15] With the Feds mostly out of the game, Texas jurisprudence eventually did assure the Railroad Commission of Texas considerable autonomy in its regulatory processes and orders. Appeal was still possible, but the bar to get into state court was higher and a prospective plaintiff's burden of proof heavier.

The Commission had carved out considerable legal latitude in which to vouchsafe the moral economy of the oil fraternity in Texas, although it still moved with its traditional circumspection and solicitude when it could, and only exercised its substantial authority when it felt it must, which was rarely. Some problems were solved expeditiously, sometimes agreeably (saltwater disposal in East Texas), others not so agreeably (flaring of casinghead gas from oil wells). The efficacy of petroleum engineering and the price stability assured by Railroad Commission prorationing finally made large-scale collateralization of proven oil and gas reserves possible. Yet other issues—mostly of the mine and thine variety—persisted and festered: marginal or stripper well exemptions from prorationing, and especially exceptions to Rule 37 (R37) spacing regulations.

## Salt Water and Flare Gas

Like love and children, oil is a mixed blessing. In the case of oil, it is invariably mixed with salt water and natural gas. Both were big nuisances. Salt water, inevitably produced in some quantity with oil, has to go somewhere once it's brought to the surface: it gets into ponds, creeks, and streams, and pollutes. Naturally odorless, clear, and tasteless, gas collects. If not contained or burned off, it sneaks up and suffocates people and livestock, or blows stuff up. Even in the early years, field safety committees tried to limit saltwater pollution, but the standard method of saltwater disposal was evaporation pits, and one good gully washer made a mess. Without gas pipeline connections, rare until the 1920s and still absent in most oil fields until after World War II, natural gas had no commercial value, and when produced with oil typically was flared (burned off). The Second World War marked a turning point in the handling of both issues.

Salt water had been a problem in Texas oil fields from Spindletop on and became especially noisome in East Texas. By 1941, East Texas was producing more saltwater than oil, something over 400,000 barrels a day. Largely at Ernest O. Thompson's behest, the Railroad Commission provided increasingly generous production allowables for salt water reinjection.[16] As a result of a mass meeting in January 1942, operators organized the East Texas Salt Water Disposal Company, funded by voluntary subscription of operators, but dedicated to reinjecting salt water into the Woodbine, at cost, from any well in the field.[17] Since maintaining production, and therefore pressure, in East Texas was considered essential to the war effort, the Petroleum Administration for War (PAW) quickly signed on, and the East Texas Salt Water Disposal Company began reinjection October 1, 1942.

The design of the collection, treatment, and reinjection systems posed major engineering challenges. With the field sprawled over 130,000 acres and three separate watersheds, and containing 25,987 producing wells, gravity-feed saltwater collection, adopted to contain costs, demanded careful surveying and construction. Unfiltered and untreated warm salt water turned out to provide an ideal growth medium for microflora, which quickly clogged reinjection pumps and wells; expensive preinjection treatment was required. Engineers initially expected to use abandoned oil wells for reinjection, but the necessity of precise siting for gravity collection, and exact location relative to the advancing subsurface saltwater front in the Woodbine, resulted in most reinjection wells being purpose-drilled.[18]

Nevertheless, the plan worked. By 1952, 98 percent of all salt water produced in the field was returned to the reservoir.[19] East Texas Salt Water Disposal became the paradigm for effective water reinjection. In 1953, the East Texas Salt Water Disposal Company, jointly with the Petroleum Extension Service of The University of Texas Division of Extension, and the Texas Education Agency Trade and Industrial Service, published a 116-page how-to guide.

Got salt water, got a manual for that. The company even produced and distributed a 16mm color movie, with soundtrack, to promote its achievements.[20]

In contrast to its characteristically cooperative approach to saltwater disposal, the Railroad Commission's decision to ban the flaring of casinghead gas produced with oil was the most abrupt and coercive in its history. Ironically, the proximate causes of this precipitate action lay not in the oil fields of Texas but with German submarines in the Gulf of Mexico and in the remote boardrooms of New York life insurance companies.

With already an excess supply of accessible natural gas from pure gas fields, casinghead gas, that necessarily produced with oil, was flared in increasingly large volumes.[21] The Railroad Commission of Texas was not unaware of the problem: William J. Murray, who was appointed to the Commission in January 1947, had been a long-time critic of gas flaring and had chaired an engineering committee that had issued a scathing report in 1945, claiming that underreported flaring was wasting a billion and a half cubic feet of casinghead gas a day.[22] But still, no market, no value.

World War II had changed the equation. The war put tremendous stresses on American energy supplies. Worse, attacks by German submarines off the Gulf and Atlantic Coasts during the first half of 1942 brought tanker traffic that ordinarily carried crude oil and gasoline to the Northeast to a virtual halt.[23] This crisis led to the construction of the Big Inch (24-inch diameter) and Little Big Inch (20-inch diameter) pipelines from Texas to eastern Pennsylvania and New York.[24] The end of the war kicked off a political brouhaha about what to do with the two pipelines. The government, which owned the lines, did not want to see them scrapped, but rather preserved or used in some form such that they could be reactivated in an emergency. Consensus among oil companies, although far from universal, was that tanker transport of oil under peacetime conditions would be cheaper. That left conversion to natural gas. After considerable if obscure political skullduggery, Texas Eastern Transmission Corporation purchased both pipelines for $143,127,000, the largest single postwar disposal of government property, and only a few million dollars less than their inflated wartime construction costs.[25]

It was a ton of money. But the Texas Eastern group, centered around James A. Elkins (Vinson, Elkins, and City National Bank, Houston) and Herman and George Brown (Brown and Root), had a pretty good idea which money tree to shake. Pipeline promoters turned to the most cash-flush institutions in the country, the big life insurance companies. They had the money, but they lived on actuarial time. That, and the regulations and canons of fiduciary responsibility under which they operated, largely determined the structure of pipeline finance. It may not have been a match made in hell, but it was one of which Mammon could have been proud. Life insurance companies were in a bind. They had seen the average yield on their investment portfolios drop from 5.05 percent in 1930 to 2.88 percent in 1947, during a period in which corporate bonded indebtedness, the preferred investment for life insurance companies, had decreased by $6 billion while life insurance assets rose by $19 billion. Nearly a third of life insurance company investments were earning less than the contractual interest rates on outstanding policies.[26]

Pipeline bonds were ideal investments. Transaction costs were minimal: one agreement might cover upward of $100 million.[27] Life insurance managers knew not with the certainty of a somnambulist but with that of an actuary precisely what their financial needs were: adequate rates of return, precisely defined payout schedules, and near-absolute security. And with long-term supply contracts for gas on one end of the pipeline and long-term delivery contracts on the other, and with a Federal Power Commission (FPC) certificate of public convenience and necessity granting them eminent domain in between, gas-pipeline companies had about as close to a license to print money as American capitalism had to offer. The match suited the pipelines even better. Their bonds were, in the parlance of the financial community, "unseasoned," and would have been difficult to place through public offerings. Moreover, then current FPC rules limited total return on pipeline investments to 6 percent. With other common stocks averaging returns of 8 to 9 percent, the pipelines had to be highly leveraged to attract any equity investment at all. And they were highly leveraged: 80 to 90 percent debt financing was the norm during the postwar period.

By 1947, the future of natural gas was bright as a blue flame. The postwar boom drove prices of competing fuels sharply higher: Texas natural gas in 1947 sold at the well head for an average of $37 per million BTU; West Virginia natural gas for $154, Texas crude oil for $325, and West Virginia bituminous coal at the mine mouth for $172. Cost to transport gas from Texas to New York via a 24-inch pipeline at a load factor of 80 percent was about $20 per million BTU; cost to transport coal from West Virginia by rail was $12 per million BTU. Even though the well-head price of natural gas in Texas had doubled between 1940 and 1947, it still could be delivered in New York for less than a third of the cost of coal, and could be sold for much less than gas manufactured from coal.[28] And just as predecessor direct current local distribution systems and their parent companies provided the distribution infrastructure for high-voltage regional alternating current networks (once reliable and efficient transformers and rotary converters were available), the presence in most eastern cities of manufactured (coal) gas utilities also promised both the physical distribution system and the organizational, marketing, and conversion capability to quickly and efficiently sell large quantities of natural gas.[29]

The fly doing the backstroke in this cheery cocktail was a brute fact about natural gas pipelines. Unlike other utilities, pipelines depended directly upon a specific wasting asset: when the gas runs out, the pipeline is worthless. As a result, prudent financial practice requires that the bonded debt secured by the pipeline be fully amortized before the gas supply disappears. That period normally was presumed to be twenty years, which was also the rule of thumb used by the FPC for granting a certificate of public convenience and necessity. Life insurance companies quickly realized that they needed independent verification of gas supplies. After 1947, which was the first year in which the ratio of gas reserves to consumption had declined, they required certification of available gas by an independent consulting petroleum engineer. "Available gas" was defined in indenture agreements as "the minimum volume of natural gas, which, by reason of the existence of proved natural gas reserves (including gas in solution or in a common reservoir with oil or distillate and to be produced with such oil or distillate in the form of casinghead gas) . . . the Company can . . . reasonably expect to produce, or to purchase."[30]

Enter the Railroad Commission of Texas. After a series of informational and show cause hearings beginning in January of 1946, the Railroad Commission dropped the hammer. On

March 17, 1947, effective April Fools' Day, the Railroad Commission ordered all 614 wells in the Seeligson oil field shut in until "measures were taken" either to use casinghead gas for an approved purpose or to reinject it into the producing formations.[31] Seeligson operators had known for over a year that the Railroad Commission would no longer tolerate the 38,588,000 cubic feet of unprocessed casinghead gas they were flaring each day.[32] Fed up with the operators' "lack of diligence and bickering," the Commission shut 'em down.[33] The operators, with Shell Oil as the lead plaintiff, obtained a temporary restraining order in District Court.[34] The Commission appealed directly to the Supreme Court of Texas.

In a startlingly assertive and broad opinion, the Texas Supreme Court held, "Whatever the dictates of reason, fairness, and good judgement under all the facts would lead one to conclude is a wasteful practice in the production, storage or transportation of oil and gas, must be held to have been denounced by the legislature as unlawful." After reviewing Texas conservation legislation since 1899, the Supreme Court found that in all instances specifically enumerated prescriptions and proscriptions were prefaced by "*including*" (emphasis in original), which "clearly manifests the intention that the adoption of rules for the . . . purposes specified in the act should not exhaust the power of the Commission nor limit its responsibility to prevent waste, but should be included in all other rules which might be found to be reasonable and necessary to that end." The Court therefore reasoned "that oil and gas may be produced within the efficient oil-gas ratio prescribed by the Commission and still be wasteful. The statute merely determines that the operation of an oil well with an inefficient oil-gas ratio is wasteful as a matter of law. It does not state conversely that production and flaring of casinghead gas within efficient ratios is not wasteful."[35]

In its Seeligson decision, the Texas Supreme Court granted to the Railroad Commission of Texas a latitude in preventing waste that made the "elastic clause" of the US Constitution look like handcuffs.[36] Succeeding decisions made clear just how loose the shackles were. At the time of the District Court proceedings on the Seeligson shut-in order, the Commission had already called show cause hearings for sixteen other fields.[37] All sixteen resulted in no-flare orders. Of the sixteen orders, only two faced legal challenges: *Sterling Oil and Refining* and *Flour Bluff Oil Corporation et al.* Both backfired: Texas appellate courts broadened rather than limited the Railroad Commission's power.[38] The *Flour Bluff* decision gave explicit expression to the historical conservationist ideological commitment to gross future value, not discounted present value, as the appropriate measure of "conservation."[39] Operators in none of the other fourteen fields facing shut-in orders filed legal challenges. Neither did those in twenty-six other fields for which "show cause" hearings were called at the same time. Dark took back the night in oil fields all across Texas, but hearths and stoves would be warm.

That the Railroad Commission of Texas should have struck like a rattlesnake somebody stepped on came as a surprise to all, not the least the oil fraternity in Texas. But why the Commission acted so precipitously is not hard to figure. Connate gas (dissolved in oil, and released as oil is produced) and associated gas (in free-gas caps in oil fields and in condensate reservoirs) constituted some 30 percent of all proven natural gas reserves. With the stroke of a pen that bright March morning in 1947, the Railroad Commission resolved natural gas's catch-22: without a pipeline, natural gas had no market and no value; without assured "available gas," a pipeline could not be built. By banning flaring natural gas in one major oil field, and by making it abundantly clear that they would move forcefully to end all flaring in the State of Texas, the Railroad Commission promised to increase the long-term supply of "available gas" by 40 percent, thereby assuring the financial integrity of pipeline indentures and encouraging rapid pipeline construction. It was about as slick a demonstration of political and legal legerdemain as had been seen since the United States Congress conjured European capital out of a perceived-to-be-empty continent by granting the Union Pacific Railroad alternate sections of federal public lands right across the Great Plains, which promoters and underwriters quickly collateralized in railroad bonds and sold to European investors.[40]

## Collateralization

In contrast to their richly textured if often embellished storytelling, which was really their métier, by 1940 oilmen's public pronouncements usually took the form of ritualized invocations of a conservationist credo (as wise use), now guided by petroleum engineering science and guaranteed by the Railroad Commission of Texas, combined with a paean to the universal superiority of free enterprise. Yet what people do is generally the more reliable guide to their true beliefs—what economists like to call "revealed preferences"—than what they say. On this criterion, collateralization signaled a major change in what oilmen believed. Forever long on hope and often lease-poor, oilmen hunt money like tomcats on the scent. Early on, some banks would lend to individuals or companies based upon the known (or believed) quality of the person, but not upon specific oil properties, purported reserves, undeveloped leases, or prospects. Not many independents or even good-sized firms qualified for asset-backed loans. Hence the oil fraternity swipe: "Banks are in the business of lending money to people who don't need it." Many and varied forms of de facto credit did evolve within the oil fraternity: farm-out deals, bottom-hole money, sale, purchase, or swap of interests in deals, sale of short-term production payments, credit extended by drilling and supply companies, friend helping friend, often in kind: informal mutuality and cooperation. Some "credit" was nothing more than the apparently appetizing sustenance of the sort Daisy Bradford provided for months to Dad Joiner and his sometimes crew.

Collateralization of proven reserves democratized bank and other institutional credit, and redefined opportunity for the oil fraternity in Texas. The calculus of collateralization is straightforward:

> Underwriting reserve based loans to oil and gas companies requires assessing three fundamental risks: whether the properties contain the projected amount of resource (reserve risk), whether the borrower can efficiently produce the reserves within the timeframe and cost estimate (operational risk) and whether the borrower would receive a price for his production sufficient to make a profit and repay the loan (price risk).[41]

Even by the late 1930s, petroleum engineering science and its reification in petroleum engineering practice were more than good enough to mostly eliminate, or at least constrain, reserve risk, especially at the 50–60 percent discount from Present Value normally, and conservatively, used by banks to establish a borrowing base. Evaluation of operational risks depended upon both petroleum engineering assessments and the knowledgeability of oil and gas loan officers, themselves usually either petroleum engineers or widely experienced in the oil business. The dissemination of petroleum engineering knowledge made evaluation of production practices and field operations more robust, transparent, and reliable. And with Railroad Commission of Texas and its prorationing policies astride the world price of oil, relative price stability was assured from the mid-1930s through the end of the 1960s, although severe production restrictions in the late 1950s caused some difficulties.

With effective prorationing, stable prices, and the reliable estimates of recoverable reserves petroleum engineering provided, banks and other financial institutions burdened by fiduciary responsibilities could rationally make loans secured by production or reserves. Like it or not, oilmen had to play the game. In the fiercely competitive environment of an oil boom, being "fustest with the mostest" mattered. Or as Napoleon is said to have remarked (but probably didn't), "God is on the side of the big battalions." In a developing oil play, the guy with the biggest pile of ready cash had an overwhelming advantage; how much of it was borrowed mattered not at all to prospective lessors. Every time an oilman borrowed money—and virtually all did at one time or another—or hired a consulting reservoir engineer, read an engineering report, talked to a loan officer, or bought or sold production or a production payment, he testified to his faith in petroleum engineering as surely as devotees of Ma Tsu testify to their faith when they burn incense or ritually recount the beneficence of the goddess.

Collateralization and scientific and technological change alter the selection dynamics among oil deals in similar ways. Biological variants which benefit from mutations of large effect that confer competitive advantages in resource acquisition produce selective sweeps that eliminate less efficacious "wild" (original) types. Likewise, deals (and individuals) who, in their evaluation of prospects and in their drilling programs, adopted micropaleontological analyses, seismic exploration, analyses of well cores from nearby wells (or from others thought to be similar), petroleum engineering analyses, electric logs, fracking, horizontal drilling, or computer simulations, as well as collateralization, enjoyed large competitive advantages over those who did not embrace such innovations. The innovators (mutants) quickly came to dominate the oil fraternity. This process is separate from but complementary to the competitive advantages generated by altruistic cooperation within deals.

First National Bank in Dallas became a pioneer in oil finance, apparently being the first to make reserve-based production payment loans (with repayments assured by a dedicated percentage of oil as it was produced). Republic and other Dallas and Oklahoma banks quickly followed.[42] Houston banks, more remote from the East Texas action, lagged, but began to catch up in the mid-1930s. The information mid-continent banks had initially used in making loans was informal and qualitative, usually opinions solicited from existing clients in Tulsa or Dallas or Houston, or from bank directors who were in the oil business. Banks, already partially integrated into the oil fraternity via their clients and directors, tapped into the close-knit fraternity's "truthful gossip" to make lending decisions.[43] Only as petroleum engineering science and practice developed and disseminated rapidly in the mid-1930s did collateral evaluation become more formal and objective. This evolution increased the pool of available credit, induced additional banks to enter the market, and expanded the number of oil fraternity members who could tap into institutional credit. It made collateralization the norm in the pursuit of opportunity.

After World War II, management of tax obligations became an increasingly critical part of oil and gas finance, indeed, of being an oilman. The object was not so much to evade taxes as to move them around, both temporally and among various parties to an oil and gas transaction, to facilitate further exploration and production. To this end, so-called ABC or "elevator" deals became the predominant form of "E&P" (exploration and production) finance—according to one estimate, comprising 65 percent of all oil loans by 1965.[44] Ascribing invention of ABC financing to a single individual (or even a few) is a little like figuring out who invented BBQ. As with many technological innovations and practices, ABC transactions emerged willy-nilly within a community of oilmen and bankers immediately after the Second World War.[45] The ABC deal evolved from earlier reserve-based lending, which in turn had evolved from traditional banking practices of lending to persons or firms of high repute. The essential feature of ABC finance, the use of an intermediary to finesse tax obligations, was used as early as 1935 by Houston's First National Bank in the sale of the Yount-Lee Oil Company to Stanolind Oil and Gas (a subsidiary of Standard of Indiana, later Amoco).[46] Harold Wineburgh, in his biography of Fred Florence, long-time president of Republic National Bank, Dallas, credits Algur H. Meadows, chairman of the General American Oil Company (and Florence's across-the-street neighbor), and his attorney, J. W. Bullion, with contriving the ABC mechanism.[47]

ABC deals involved three players, a seller, designated A, a purchaser, B, and an intermediary, C, or the "elevator attendant," who held the paper while the two principals rode to the top. A was typically an independent who wished to free up capital for further exploration. Following the maxim "A mule and a jackrabbit can't plow together," independents tended to specialize in finding oil, not in long-term development and production, which better fit integrated producer-refiners' "visible hand" preferences for predictable throughputs and costs. In an ABC deal, A sold his working interest in a lease or leases to B, usually a major or larger independent who wanted to increase his reserves at minimal risk. A's sale to B was subject to a reserved production payment, which A then sold to C. The gross value of the reserved production payment was the sum of the collateral loan amount, already agreed between C and the underwriting bank, plus an interest rate 0.05 percent to 0.0625 percent above the bank's rate of interest. A received the collateral loan amount from the production payment purchaser, C. The unencumbered portion of the property went directly to the primary purchaser, B, for which he paid A in cash. C normally put in no cash but borrowed 100 percent of the value of the production payment from a cooperating bank or other financial institution.[48]

It was a win-win-win. B got his reserves and operated the lease; income from the unencumbered portion of the property's production was designed to equal the expenses of operating and, if required, developing, the lease; therefore, B need pay no or minimal taxes on income from the lease until the production payment was satisfied. B usually could purchase the oil due C if desired. Once the production payment was released, B owned the interest outright and could take the depletion allowance on total continuing oil production. A paid capital gains taxes on the proceeds of the sale to B and on the loan value of the production payment he received from C. During the period, the capital gains tax was 25 percent, while the corporate tax rate was 52 percent, and the top personal income tax rate 91 percent (little wonder most good-sized independents incorporated). If A had substantial deductions, such as intangible drilling expenses (all those with no salvage value), or losses on dry holes during the year, or carry-forward losses from previous years, he might pay no taxes at all on the transaction. Since C's production payment was considered a return of his capital investment, he could claim cost depletion and pay no tax on the remitted collateral value; C's interest paid to the bank was also deductible. C's only taxable income thus was his arbitrage, the 0.05 percent to 0.0625 percent premium on the loan balance; if C were a tax-exempt charity, he paid no tax either. The bank, of course, got its interest. Given the efficacy of petroleum engineering evaluations and the price stability assured by the Railroad Commission of Texas, neither B, C, nor the bank incurred significant risks. ABC deals added about 20 percent to the value of oil interests sold in these transactions.[49] Bankers, lawyers, and oilmen invented several variants of the ABC deal, mostly to reduce tax obligations.[50]

This gravy train got derailed in the Tax Reform Act of 1969. Although oilmen argued that ABC financing actually increased total tax revenue in the long run, it usually did postpone coughing up the money.[51] Critics claimed these deals, as well as the depletion allowance, subsidized the drilling and operation of inherently inefficient and unnecessary oil wells in the United States. The 1969 Act accordingly not only eliminated the tax advantages of ABC deals but also reduced the percentage depletion allowance from the sacred 27 1/2 percent to 22 percent. This bit of economically rational policy wonkery anticipated the Arab Oil Embargo of October 1973 by four years, during which time American oil imports soared some 250 percent.[52]

It took some time, but the oil fraternity and their bankers and lawyers adjusted to the new rules. After the demise of ABC financing, oilmen and bankers reverted to reserve-backed term loans. But bankers' data showed just how safe and secure reserve-based loans were: defaults or losses were minimal, a tribute to efficacious petroleum engineering.[53] Sometime between 1974 and 1978, Continental Illinois Bank began using "borrowing-base, revolving-loan facilities." The problem with regular term loans was that although they typically were rolled over, completely new loan documentation, title authentication and lien searches, engineering analyses, and loan agreements had to be done de novo, which entailed renegotiation and substantial transaction costs, as well as some uncertainty for borrowers. Worse, their payments included principal. Under a borrowing-base revolver, borrowers repaid no principal, only interest. Bank petroleum engineers determined a borrowing base, usually about half of the present value of a customer's proven reserves. Every six months or a year, the bank engineers reevaluated the borrower's reserves: if they had shrunk, his borrowing base would decrease and he might have to repay part of the principal of the loan; if they had increased, which was more common, the borrowing base would increase. Other than the engineering analyses, no other transaction costs were incurred. By 1978, the borrowing-base revolver, only very rarely used for financial suicide, had been adopted by virtually all energy lenders. Collateralization remained the sine qua non of being an oilman.[54]

Cumulative engineering progress in understanding, predicting, and controlling subsurface hydrocarbon reservoirs drove collateralization. In hiring experienced petroleum engineering practitioners, banks not only acquired their professional expertise but also tapped into their personal and professional social networks and into the moral economies of petroleum engineering and of the oil fraternity in Texas.[55] For example, when Jesse Jones's ultraconservative National Bank of Commerce in Houston finally got serious about production-based lending in 1947, Jones hired E. O. Buck, one of the most accomplished and widely acquainted petroleum engineers in Texas, to head the department. In 1952, Second National Bank of Houston, soon to become Bank of the Southwest, hired Harold Vance to run its oil loan department. Vance, who had earned a BS degree in Petroleum Engineering from the University of California in 1923, had worked for the US Bureau of Mines and then for Marland and Continental Oil before becoming an independent oil operator and consulting engineer in East Texas in 1931. He became chairman of the Petroleum Engineering Department at Texas A&M in 1934. By 1952, he'd likely taught a goodly proportion of the petroleum engineers in Texas and knew just about everybody.[56]

For oil operators, to collateralize meant presenting technical engineering analyses to the banks' oil department engineers. Consulting petroleum engineers, in the early 1930s mostly occupied with legal testimony in prorationing cases and in Railroad Commission hearings, rapidly and enormously expanded their practice to include evaluation of oil reserves for presentation to the banks. As was his habit, Everette Lee DeGolyer saw the future early, organizing what became the preeminent consulting petroleum engineering firm in the country, McNaughton and DeGolyer, in 1936.[57] Circa 1957, Harold Vance estimated that 10 to 20 percent of the petroleum engineers in large companies were engaged in "valuation work," which also constituted at least 50 percent of the work done by consulting engineers.[58] In recognition of the increasingly specialized character of oil and gas property evaluation, the Society of Petroleum Evaluation Engineers (SPEE) was founded in 1962, with Harold Vance, William Hurst, and Herbert F. Poyner Jr. as original directors.

Whatever the oil fraternity professed, and by 1950 most oilmen professed enthusiastic allegiance to conservation, to prorationing, to the "fruitful empiricism" of the Railroad Commission's "continuing administrative process," and to scientific rationality, their form of life depended upon petroleum engineering practice. Their collective and individual self-productions portrayed themselves, without contradiction, as both "the greatest gamblers," the ultimate risk takers, and as rational, scientific producers. Petroleum engineers reciprocated, portraying themselves and their profession as paragons of scientific rationality, innovation, and technical and economic efficiency in oil and gas production. The world that the oil fraternity in Texas, petroleum engineers, and fruitful empiricism made was very different from that birthed at Spindletop and radically transformed at full flood in East Texas. By the eve of World War II, practicing petroleum engineers were warranted to produce that most obdurate, persuasive, and meaningful of all "social facts," money in the bank. Nevertheless, although unified in professed defense of opportunity, autonomy, privileged access, and rational production, the oil fraternity could still put on pretty good mine-and-thine family feuds, some of which flummoxed the Railroad Commission for thirty years. Petroleum engineers got caught in the crossfire.

## MINE AND THINE: TAKING ONE MAN'S PROPERTY AND GIVING IT TO ANOTHER

Mine-and-thine conflicts came down to who got what when and who decided. They derived from the inherent contradictions among conservation as not use, conservation as wise use, wise use as, only, economically efficient use, and the sacrosanct prerogatives of private property, all bollixed up with the Texas constitutional paradox of mineral fee ownership in place and the rule of capture. The worst of mine and thine afflicted four sometimes intertwined issues: natural gas production, marginal well exemptions, unit operations, and Rule 37 (R37) well-spacing exceptions.

### Gas Pains

Control of natural gas production lacked the scientific, conservationist rationale of oil prorationing. Unlike oil production, there was no petroleum engineering evidence that restrained rates of production in natural-gas reservoirs, at least in those not containing condensate, could increase ultimate recovery: Ultimate gas recovery was independent of rate of production. Railroad Commission regulation of natural gas reservoirs therefore became a very expensive, ten-year-long mine-and-thine brawl. The issue never much mattered until the advent of the giant Panhandle gas (and oil) field. Really a population of separate but often interconnected pools, the Panhandle "field" is some 125 miles long and 15 to 20 miles wide and underlies more than one and a half million acres across eight counties in the Texas Panhandle and portions of Oklahoma and Kansas. Within this area, over a million acres held sweet gas suitable for home heating with minimal treatment; the remaining half a million acres contained sour (sulfur-laden) gas not usable for home use without treatment. Some of the natural-gas deposits, sweet or sour, also held extractable condensate or distillate. About 50 oil pools dotted the northern portion of the field, and the world's largest known helium deposit also is located there.

By 1935, nine major pipeline companies controlled 80 percent of the sweet gas areas and produced gas for their own distribution systems, which served domestic and industrial customers as distant as Minneapolis–St. Paul, Denver, and Indianapolis. Non-integrated operators, with no market, controlled the remaining 20 percent of the sweet gas.[59] It was a perfect recipe for vicious, zero-sum conflict. The Railroad Commission, aided by the Texas legislature, tried a number of stratagems to bring some sort of equity, consistent with the moral economy of the oil fraternity in Texas, to the greater Panhandle field. Opposed by alien interstate pipeline companies, none of these efforts were successful and some were disastrous.

Repeated attempts to impose some form of market-demand prorationing for natural gas, lacking the conservationist or waste prevention rationale petroleum engineering provided for oil prorationing, were rejected by the federal courts.[60] Apparently with the intent of intimidating the pipelines into ratable taking, the Commission launched an almost scorched-earth policy, allowing operators to strip gas of condensate and then "pop" the residue: By 1935, a billion cubic feet a day of perfectly usable gas was being blown into the air. Didn't work. Nor did the courts approve subsequent Texas legislation explicitly authorizing the Commission to impose gas prorationing to protect correlative rights. Both the Federal District Court, in an opinion written by Joseph C. Hutcheson Jr., and the US Supreme Court, in an opinion authored by Louis D. Brandeis affirming the lower court decision, rejected the State's rationale. As Brandeis wrote, "Our law reports present no more glaring instance of taking one man's property and giving it to another."[61] The sweet gas problem in the Panhandle never was resolved by regulation. Instead, as the use of natural gas expanded nationwide, new pipeline connections finally provided the market—and the opportunity—the previously unconnected operators and the Railroad Commission had sought all along.

## A Stripper for Everyman

Had conservation as "wise use" been encapsulated in a pithy aphorism, it might have been, "Waste not, want not." Conservationism in America had been provoked in the nineteenth century by appalling waste and landscapes despoiled and often nearly uninhabitable—"Mister Peabody's coal train has hauled it away."[62] Little wonder waste found such central place in conservationist ideology. But what got left out was the costs of not waste. The problem still plagues conservation, however defined, and is likely analytically insoluble, not least because scientific and technological change continually shift recognized costs and benefits.

For the oil fraternity in Texas, the operational conservationist credo became maximizing the ultimate recovery of the oil and gas resources with which they had been blessed. Of course, like "Thou shalt not kill," this commandment quickly got hedged about with all sorts of modifiers and qualifications. Maximum ultimate recovery was construed to mean "by conventional means" or, more contentiously, maximum efficient, that is, economical, recovery or its related technical expression, production within a Maximum Efficient Rate. Again, these operationalizations were subject to continual scientific and technological, as well as market and political, redefinitions. Marginal or stripper well exemptions were designed to prevent pumping wells with small production from being prorated to the point that their production would no longer cover their costs of operation: They would be abandoned and any remaining oil in the ground left forever unrecoverable—the final and irreversible form of waste. Texas's 1931 Marginal Well statute exempted from prorationing limitations wells capable of producing less than up to 40 barrels per day, depending upon well depth; a revised act in 1933 both adjusted exempt production downward and provided less generous depth schedules.

As oil fields in Texas aged and began to be depleted, the proportion of stripper wells, even in rationally managed fields, necessarily increased and sopped up an increasing proportion of the market demand for Texas crude. To preserve opportunity in the oil fraternity and encourage full, wise use of the state's resources, the Railroad Commission ended up playing a game of regulatory whack-a-mole. As would be expected when a regulatory agency pursues multiple goals, complexity and contradictions emerged together: adjustments in one place caused another problem to pop up somewhere else. The Commission mandated a growing number and variety of categories of wells exempt or partially exempt from production limitations. In addition to marginal well exemptions, regulations provided discovery allowables for new wells, production bonuses for water-flooding, gas reinjection, or other pressure maintenance projects, saltwater disposal, or for wells located on salt domes. By 1963, exempt categories were absorbing 1,213,853 barrels a day of a statewide market demand of 2,818,812, or 43 percent; marginal wells accounted for 412,260, or about one-third, of those exempt barrels.[63]

To reduce regulatory and transaction costs, in 1947 the Railroad Commission introduced a standardized depth-acreage "Yardstick" to determine baseline field allowables in new fields until sufficient information accumulated to define specific Field Rules (the Yardstick was revised in 1965). These schedules bore no known exact relationship to efficient physical recovery or waste prevention. One of the ironies of advances in engineering science is that the more sophisticated, esoteric, and precise knowledge becomes, the more expensive and demanding its application becomes. Just as the transition from interchangeable parts in the early nineteenth century to mass production later in the century was marked by backing off precision and reducing the quality of fit, so the Yardsticks substituted "good enough" for analytically precise.[64] But the Yardsticks were generated by the oil fraternity itself, by expert advisory committees comprising petroleum engineers, geologists, and experienced oil operators, as well as Railroad Commission engineers: while satisficing, they were legally "reasonable," not capricious.[65] And supplicants dissatisfied with Yardstick values or later Field Rules could always appear before the Commission for hearings or rehearings—petroleum engineering analyses in hand—or appeal to the designated courts of the State of Texas.

## One for All and All for One

The nostrums most often advanced for mine-and-thine conflicts in oil and gas fields were unit operations and their little brother, pooling. Both solutions came in two varieties: strong—that is, compulsory—and weak, or voluntary. From the early 1930s, unitization was the favored solution advocated by some major integrated oil companies, the majors' trade association, the American Petroleum Institute, the American Institute of Mining Engineers, and various study committees of the American Bar Association, as well as leading attorneys.[66] Pooling referred to the merger of smaller tracts to assemble a leasehold large enough to meet Railroad Commission–mandated minimum spacing requirements and drill one well, with costs, and proceeds, if any, to be shared as agreed by pool participants. Opposition was greater to compulsory variants of either proposal, with unitization drawing greater ire than pooling.

Texas has never had a comprehensive, compulsory unitization law, although bills providing for such have been introduced in virtually every legislative session since the early 1930s. The legislature did enact a limited voluntary unitization statute in 1949 but did not pass a circumspect compulsory pooling law until 1965. Nevertheless, given its broad authority and discretion, the Railroad Commission was not without resources to encourage cooperation where it thought desirable. True to its traditions, it habitually did so by proffering the carrot rather than wielding the stick. Virtually all Railroad Commission–sanctioned unitization directly served legislatively mandated conservation, not broader and fuzzier goals of "rationalization." But unitization, as Henry Doherty foresaw, meant turning the determination of property, and of wealth, over to professional petroleum engineers.

In contrast to voluntary unitization under some circumstances, voluntary pooling was a nonstarter as long as Railroad Commission prorationing formulae included a significant per-well or well potential factor, and the "more wells, more oil" hypothesis offered any hope of getting an R37 (Rule 37) spacing exception. For folks whose idea of an oil well is summarized in the old blues phrase, "She walks like she got oil wells in her backyard," spacing regulations and pooling, not to mention unitization, looked like a swindle.[67] The published literature is dominated by engineering rationalism, lawyerly pontification, and the large producer view, which, whatever their conservationist or field-safety glosses, unremittingly pushed wide well spacing and pooling or unitization to reduce transaction, exploration, and production costs—in their interests. What all this rationalization talk masked was what small tract owners and small producers saw all too clearly: proration was all about who got what when—mine and thine. This view from Everyman was rarely heard, even more rarely recorded, and almost never heeded, and is therefore hard to find.[68]

One clear example of its expression appeared during a Railroad Commission field hearing on the newly discovered Old Ocean field in Brazoria and Matagorda Counties, held in Austin October 28, 1937.[69] A few small landowners in the field made the trip to Austin to argue against 40-acre spacing. They were noticeably abashed by their surroundings and self-conscious about their relative technical ignorance. As A. W. Pollard, "a layman from the backwoods" who owned a 20-acre tract in the field with his sister (on which their father lived), put it:

> From my own experiences with them [Abercrombie & Harrison, who discovered and developed the field], they have expressed and exerted a power over me; they wanted our property at what we thought was of no value, and they have gone even as far, on one occasion, as to tell us that they had no intention what ever of drilling on small tracts of land . . . and that smacks of oppression to me.[70]

Pollard's nephew, J. B. Dailey, who also testified, went right to the nitty-gritty:

> . . . there is another angle that I would like to present it from; that is from the standpoint of the interest of the landowner. On a 40-acre spacing rule it is not in his interest, and everybody knows why. You don't even have to bring it up.
> Commissioner Thompson: Everybody knows why, you say?
> Yes, sir; I assume they do.
> What is it, if every everybody [*sic*] knows about it?

Well, in effect it means that a small landowner must cut his royalty in two. Now, normally royalty is one-eighth. In the pooling propositions that we have heard from Harrison & Abercrombie they states [*sic*] one-sixteenth. We personally don't care about the pooling, but we don't like to cut our royalty in two and accept one-sixteenth.[71]

Neither Thompson nor anyone else contradicted or even addressed Dailey's numbers. Thompson didn't exactly browbeat the dissenters, but he wasn't what might be called overly solicitous either.

Everyman had a point. A landowner with 20 acres in a field should be indifferent to whether he received a 1/8 OR (overriding royalty, 1/8 of the oil produced with no costs deducted) on his 20 acres or a 1/16 OR on a pooled 40-acre spacing or proration unit: he would be entitled to $(1/8)(20) = 2.5$ or $(1/16)(40) = 2.5$ acres of oil in place in either case. But specific field rules actually determined who got what when. For example, if production were allocated 1/2 on acreage, 1/2 on well potential, or 1/2 on acreage and 1/2 on well depth, the owner with a 1/8 OR on 20 acres would get 50 percent more than if he had a 1/16 OR on 40 acres. Only if allowables were allocated purely on a per-acre basis or some other surrogate for oil in place would the payoffs be equal. Such scientific rationalization was a long time coming, and not without costs.

## (Nearly) Everyone's Special: "Making Rules and Granting Exceptions to Them"

Nowhere did the latent conflicts between conservation, scientific rationality, economic efficiency, interests, private property, and opportunity in the oil fraternity become more manifest than in the granting of exceptions to Rule 37 (R37), the Railroad Commission regulation governing spacing of oil wells in Texas fields. Within a month of the enactment of the 1919 conservation act, the Railroad Commission issued forty Statewide Rules and Regulations, among which was Rule 37 (R37), which limited how closely wells could be drilled to each other and to property lines.[72]

When R37 was first implemented, under the rule of capture the only protection a landowner or lessee had against drainage by his neighbor was to "go and do likewise"—to drill offsetting wells and get his oil out first. In these circumstances, the Railroad Commission fully anticipated granting exceptions to R37 in order to assure Everyman the opportunity to recover his property.[73] As a result, R37 exceptions were easier to get than a Mexican divorce and spawned even more litigation in Texas courts. Since most R37 exceptions were for wells in already proven fields, almost none came up dry—they were the least risky oil wells in Texas. With the advent of statewide prorationing in 1930 and the discovery of the giant East Texas field in an area where small tracts predominated, applications for R37 exceptions skyrocketed. It was the sheer volume of R37 exceptions that provoked York Y. Willbern's ungenerous, Weberian-rationalist jibe that "the work of the commission consists of making rules and granting exceptions to them."[74]

Many cases concerned voluntary segregation, the subdivision of a larger tract which would be permitted one well under R37, into smaller tracts, lessees of each of which would then apply for an R37 exception, resulting in more wells than would have been allowed on the original tract. Most were out-and-out scams, and the Commission and the courts quickly extinguished R37 exceptions for any tract voluntarily subdivided after 1919.[75] But other cases involved title disputes or sometimes convoluted heirship and partition. There is nothing like oil to flush kinfolks out of the woodwork or resurrect the dead. In one instance, in 1921, six siblings had divvied up a 240-acre tract in Rusk County (East Texas) upon the demise of their parents. Ten years later, after oil was discovered, a seventh sibling, presumed dead in 1921, reappeared to claim his one-seventh, including a 2.59-acre parcel, 33 feet wide and 3,342 feet long, upon which his lessees obtained an R37 exception to drill three wells. Sun Oil sued, and the Court of Civil Appeals (1933) decided the tract in question was voluntarily subdivided and that, therefore, the R37 exceptions were null and void.[76]

The most critical and persisting R37 quarrels were about drainage from adjacent leases to R37 exception wells: How much would occur? How much was "reasonable" or tolerable? What evidence was adequate or appropriate to determine the relevant values? How else, without recourse to coercion and compulsory pooling or unitization, might the property rights of small tract owners and lessees be secured? Scientific progress in understanding, and in acquiring information about, subsurface reservoirs provoked a thirty years' war between engineering and technical rationality versus untrammeled opportunity and historical property rights and prerogatives.

The Railroad Commission itself had difficulty in defining unambiguous criteria and in articulating a theoretical basis for its spacing decisions. A lot of the mischief derived from W. W. Cutler's 1924 Bureau of Mines Bulletin, which maintained that more wells meant more ultimate oil recovery.[77] For the next twenty years, Cutler's claim became a mainstay of litigants' expert testimony when seeking R37 exceptions. In a Rule announced August 25, 1936, the Commission explicitly adopted Cutler's hypothesis, only to frantically backtrack five months later.[78]

Courts didn't do much better. Litigants, especially major companies, repeatedly tried to get the courts to ordain petroleum engineering as arbiter of subsurface property rights, that is, to give petroleum engineering analyses the final say in determining the oil in place that a mineral owner or lessee was entitled by law to recover. In effect, without quite saying so out loud, petroleum engineering would supplant the rule of capture. R37 exception cases, which invariably concerned small tracts alleged to drain adjacent leases, became the focal battleground for these efforts. But this "rationalization" programme repeatedly ran into the still established "more wells, more oil" dogma, and into the reluctance of the Railroad Commission and of the Texas courts to stifle opportunity in the oil fraternity.

After *Amazon*, the first major challenge to R37 exceptions to reach appellate courts in Texas was *Brown v. Humble Oil & Refining Co.*[79] This case, originating in the East Texas field as usual, contested an application for a well on a 1.5-acre tract split off from a 3-acre tract, which already contained one well. Humble operated eight wells on the adjacent 99.1-acre lease, and, not surprisingly, cried foul. Humble's argument rested on the scientific testimony of three technical experts qualified by the District Court as expert witnesses, which meant that by virtue of their specialized training and knowledge, they could present opinions and draw inferences and conclusions without having witnessed any specific occurrence relating to the suit.[80] The crux of Humble's case was alleged drainage from the Humble's 99.1 acres to the 3-acre tract, with either one or, soon, two wells. Humble's witnesses estimated that each acre underlying the entire tract originally held 40,000 barrels of oil. At current allowables, it would take Humble's eight wells 15.65 years to produce all the oil underlying their lease, while Brown's two wells would produce all the oil underlying the three acres in 1.91 years—after which they would poach their neighbors' (including Humble's) oil.[81]

The Court of Civil Appeals of Texas reversed the District Court and enjoined the second R37 well. Brown appealed to the Supreme Court of Texas, which affirmed the decision of the Court of Civil Appeals June 12, 1935. During oral arguments, Justices of the Supreme Court had asked opposing counsels for a "discussion of the rule of capture, its wisdom and equities."[82] Brown et al.'s (Plaintiffs') argument, filed April 19, 1935, consisted in a rousing rendition of the ideological representation, past, present, and future, of the oil fraternity in Texas. Brown maintained that the rule of capture was "not only wise, but just and equitable, as near as may be, considering the fugitive nature of oil and gas." The rule of capture was necessary, not just "convenient," to reward risk-taking.[83]

Hines Baker, on behalf of Humble, presented a more comprehensive and subtle argument April 24, 1935. Baker asserted that under Texas law ownership of oil in place was primary and constitutionally mandated, subject only to the reasonable exercise of the State's police power, which included conservation of natural resources. Based upon his review of common law going back to English water rights, Baker argued that the rule of capture was secondary, purely a rule of convenience adopted by the courts and necessitated by ignorance, the inability to ascertain the origin or ownership of oil when it came out of the ground.

Baker's solution was novel and radical. He reasoned that since advances in petroleum engineering science and practice made it possible to determine oil in place, the Railroad Commission had the authority to adjust the allowables it prescribed to equitably protect all vested interests, whether the tract be small or large. Justice Richard Critz had asked explicitly why it would not be "perfectly proper" to allow the infinite subdivision of tracts and to grant R37 exceptions to drill wells on them, "if, by the adjustment of the allowable each owner may be entitled to recover a quantity of oil substantially equivalent in amount to the recoverable oil under his land." Baker's answer was that, "Clearly, the Railroad Commission, in order to prevent waste, has the right to restrict the rate of flow in the same way that it has the right to regulate spacing. . . . This power to control the rate of flow, in order to prevent waste, also enables the Commission to overcome the advantage obtained by one who is given an exception to the spacing rule, by restricting his allowable production to the extent necessary to offset this advantage."[84]

Baker founded his immodest proposal upon "the laws of Physics," that is, petroleum engineering science.[85] Oddly, the scientific rationale for his proposal was obsolescent the day he offered it, which perhaps illustrates the difficulty and delay in diffusing leading edge engineering science even within a remarkably capable and integrated firm such as Humble Oil. Baker invoked the water-percolation theory of pressure maintenance in the East Texas field from the trial two years earlier, even though it had been discounted even then.[86] Baker's primary scientific citation was Henry Miller's 1929 *Function of Natural Gas in the Production of Oil* even though by 1935 it had been largely supplanted by new theoretical and empirical results—ironically even more supportive of his legal argument.

The Texas Supreme Court, in an opinion written by Associate Justice John H. Sharp published June 12, 1935, not only affirmed the Court of Civil Appeals decision in favor of Humble Oil and enjoined Brown's second well, but also adopted Hines Baker's argument nearly verbatim:

It is now, however, recognized that when an oil field has been fairly tested and developed, experts can determine approximately the amount of oil and gas in place in a common pool, and can also equitably determine the amount of oil and gas recoverable by the owner of each tract of land under certain operating conditions.[87]

The Court didn't shy away from the implications:

The commission, in order to prevent waste, has the power to limit the rate of flow in the same way that it has the power to regulate spacing. . . . This right to control the rate of flow . . . also enables the commission to offset the advantage of one who is given an exception to the spacing rule by limiting his allowable production to the extent necessary to overcome this advantage.[88]

The decision in *Brown v. Humble* set the oil fraternity atwitter, if not alight. Neither elation, especially among the majors, nor outrage, especially among the smaller independents, was hard to find. The expected *Brown et al.* motion for a rehearing unleased a riptide of amici curiae briefs, most of which were hostile to vituperative. The typical complaint was premised on the same ideological profession as *Brown et al.*'s trial brief: Overturning the rule of capture would deprive plaintiffs, and the oil fraternity, "of their property and their rights" and "deny future generations the right individually to pursue the exploration and development of the oil resources of the State." It would "subordinate individualism to mass property rights and paternalism."[89]

Thoroughly spooked by the genie it had uncorked, the Texas Supreme Court beat a retreat faster than Santa Anna's army after San Jacinto, and just about as orderly. In its opinion refusing a rehearing and affirming its injunction in *Brown v. Humble*, delivered November 27, 1935, the Court recanted everything else: "The language [of the Court] was used, not for the purpose of prescribing rules or standards, but merely in a discussion of the validity and purpose of the rule and its exception."[90] The Railroad Commission wasn't much abashed. On remand, Brown's permit for his well on the disputed 1.5-acre tract was again granted, this time in order to prevent waste, presumably on the "more wells, more oil" theory. On appeal, the courts did not interfere.[91]

So long as the "more wells, more oil" dogma had legal traction and provided a waste prevention rationale for R37 exceptions, Texas courts vacillated, sometimes allowing, sometimes enjoining Railroad Commission permits. Decisions usually rested on the details of the specific cases.[92] Not until 1941 did Texas courts start to rein in invocation of the "more wells, more oil" argument, beginning with *Railroad Commission et al. v. Shell Oil, Inc. et al.* (usually referred to as "Trem Carr"), another East Texas R37 exception case.[93] After a generation, "more wells, more oil" had finally begun to lose its charm.

Nevertheless, five years later, after equivocating for the better part of a decade, the Texas courts backtracked again and finally came down on the side of opportunity for Everyman: Moral economy trumped property and scientific rationality. In *Railroad Commission v. Humble Oil & Refining Company*, March 6, 1946, usually known as "the Hawkins decision" because the case concerned the Hawkins field, a Woodbine discovery in Wood County, Texas, the Court of Civil Appeals of Texas announced,

> . . . it is held that the owner of an "involuntarily" segregated tract cannot be denied the right to drill at least one well on his tract however small it may be. From which it would seem that his allowable can not be cut down to the point where his well would no longer produce, nor below the point where it could not be drilled and operated at a reasonable profit.[94]

One man, one tract, one well, profit guaranteed.

The Hawkins dictum was repeatedly reaffirmed, most notably in *Ryan Consolidated Petroleum Corp. v. W. L. Pickens* (1955). But some light was beginning to dawn. Will R. Wilson (later Attorney General of Texas) filed a prescient dissent in *Ryan*, in which he was joined by two other Justices. Wilson went right at the rule of capture and the refusal of the courts to inquire into subsurface realities as petroleum engineers defined them:

> This impulse to stop judicial inquiry at the mouth of the well springs from a judicial conviction that the subterranean movement of liquids is too "occult and secret" to be the proper subject matter of proof. . . . There was more justification for this feeling fifty years ago than there is now. I went to the trouble of writing a dissent . . . because of my belief that the principles of geology and reservoir dynamics will eventually win the same acceptance in judicial thought that they are accorded in business and financial thinking.[95]

The handwriting was on the wall, although a lot of folks in the oil fraternity, at the Railroad Commission and in the Texas courts, chose not to see it. As Frank Vandiver wrote of Southern leaders before the War Between the States, "They hewed hard and fast to the old virtues while the verities went adrift."[96]

# CHAPTER 9

# JUSTICE, SWEET JUSTICE

Was it somethin' I did, Lord, a lifetime ago?
Am I just now repayin' a debt that I owe?
Justice, sweet justice, you travel so slow.

—WILLIE NELSON[1]

Like a two-year-old, the problem with science is that it doesn't stay put. Today's gospel truth becomes tomorrow's false idol. From current perspective, once perfectly respectable scientific beliefs range from quaint to ludicrous: geocentric cosmos, cycles and epicycles, phlogiston, humoral and miasma theories of disease, phrenology, fixed continents, Great Chain of Being, invariant space and time. Worse, like a two-year-old climbing out of her crib, scientific beliefs don't stay put in their ivory tower either. They futz about in the real world and have consequences: how to produce a petroleum reservoir, treat a disease, build a flying machine, make a bigger bomb. Consequences affect interests, which gets folks' attention and more often than not riles them up. Transitions—which means all the time—are messy and contested, as evinced by the turbulent history of petroleum engineering. The legal dictum "one man, one tract, one well, profit guaranteed" was crossways with petroleum engineering science the day it was anointed. Continuing progress in petroleum engineering science and practice, and especially in acquiring, processing, and analyzing subsurface information, only made the disjuncture greater and accelerated the creative destruction of ancestral ways.

## RIGHTEOUSLY MINE: THE TRIUMPH OF PETROLEUM ENGINEERING

The Railroad Commission's and the Texas courts' privileging of opportunity for Everyman over scientific rationality and private property became increasingly unstable as petroleum engineering science and practice gave every indication of approaching closer and closer to the grail of *knowing* what was underground. Rule 37 (R37) exceptions, in which the consequences of this rank ordering were most apparent, brought the issue to a head. Despite the Hawkins and *Ryan* decisions, scientifically and technologically adept operators kept trying.[2] The dam burst in 1961–1964 in three Supreme Court of Texas cases, two for gas and distillate, the third for oil: *Atlantic Refining Company v. Railroad Commission,* known as Normanna, *Halbouty v. Railroad Commission,* known as Port Acres, and *Railroad Commission of Texas v. Shell Oil Company,* sometimes known as the Quitman case.[3] As Hines Baker had argued in 1935, ownership of oil in place—property—became primary, and the rule of capture secondary, a "rule of convenience." Now a lessor or his lessee were entitled to capture only what was there and theirs in the first place—and petroleum engineers were warranted as a matter of law to determine what and how much that was.

In Normanna, the efficacy and authority of petroleum engineering practice, and of petroleum engineers, were explicitly on trial. The case concerned a natural gas and condensate field partially underlying the Normanna townsite in Bee County, southwest of Victoria.[4] The Railroad Commission determined that one well could drain 320 acres and defined that as the spacing rule and proration unit for the field. This finding rested largely upon the expert technical testimony of Jack K. Baumel, a consulting petroleum and natural gas engineer who had been chief engineer of the Railroad Commission's Oil and Gas Division. Electric logs (Schlumbergers) were the key empirical data upon which Baumel depended.[5]

Electrical well logging had been invented and developed by French brothers Conrad and Marcel Schlumberger, who founded their eponymous firm in 1926. Their original intent was to use electrical conductivity or resistivity surveys on the surface to probe for likely subsurface oil-bearing formations, in competition with torsion balance, seismic, and conventional geological techniques. They claimed to have successfully used their method in Romania and Russia, but the presence of underground pipelines or wells in developed areas vitiated their approach. Resistivity surveys could, however, point to the presence of other deposits near the surface: for example, the Minnesota Department of Highways used them through the 1930s to locate sand and gravel deposits.[6]

But the brothers Schlumberger quickly realized that measurements taken vertically in an already drilled well could indicate the character and contents (including oil, water, or gas) of the strata through which the well passed and ran their first electrical resistivity well log in the Pechelbronn oil field in Alsace in 1927. Their technology was introduced experimentally to the United States in 1929 in California, with the Schlumberger Well Surveying company opening in Houston in 1935. But the use of such logs was still limited in the late 1930s, and their diffusion was all but halted by World War II.[7] Hence the Court's implicit doubt in the Hawkins decision (1946) expressed in the phrase, "called Schlumbergers."[8]

Initially, Schlumberger well surveys could be used only "in a qualitative way to correlate formations penetrated by the drill . . . and to provide some indication of reservoir content." This limitation was not remedied until publication of Gustave E. "Gus" Archie's "The Electrical Resistivity Log as an Aid in Determining Some Reservoir Characteristics" at the beginning of 1942. Archie (Shell Oil) presented a set of empirically derived equations that, given adequate background information about the reservoir, allowed rigorous quantitative interpretation of electric logs.[9] To establish the resistivity of a formation when its pore space was partially filled with brine (salt water) and the remainder with less-conductive hydrocarbons (natural gas or oil), Archie exploited prior experimental techniques on multiphase flow, especially those of Wyckoff and Leverett. Schlumberger was a science-and-technology-driven enterprise and subsequently introduced instruments that offered both higher and more accurate resolution and could operate under a wider range of downhole conditions, from within cased wells, for example.[10] By the mid-1950s, electrical logging, by then offered by Schlumberger and several competing firms, was a standard tool engineers used to evaluate oil and gas wells and fields.

Because each of the Normanna field's four productive horizons contained significant volumes of distillates, prompt definition of field rules, including spacing rules and production limitations, were essential, lest reservoir pressures fall and cause retrograde condensation and irretrievable loss of both distillate and gas.[11] At the Railroad Commission field hearing to establish rules for Normanna, Baumel proposed, in addition to 320-acre spacing, that production among wells be prorated on the basis of acreage times bottom-hole pressure. Multiple small tract owners and lessees asked for the customary 1/3 per well, 2/3 acreage allocation (based upon the ratio of the acreage assigned to each well to the total proved acreage in the field).[12]

Small tract owners voiced the same discontent and sense of injustice as others similarly situated had twenty years earlier at Old Ocean. A. A. Nance, "representing two royalty owners of a 28-acre tract" likely to be within the field:

> Now, we have been contacted by both Atlantic and Tidewater, not on one occasion but several, and that's to lease the tract, and they told us without question there would have to be a pooling agreement in each and every case. . . . We've been in pooling agreements . . . and we find our land is tied up. . . . One field we're in will last for eighty years, and, God knows, I'll be gone by that time . . . and my mother, who is advanced, who owns the other share . . .
>
> Now, we think we own the land, we bought it, we held it, we paid for it. Now, if we want to drill it on an equitable basis, we will get something out of it within our lifetimes . . . but if we go solely on acreage, it will be highly discriminatory against those of us who have small tracts.[13]

When J. A. Rauhut questioned Nance, the historical faith came to the fore. Like all the bounty of nature God had created, oil and gas were there for Everyman:

> MR. RAUHUT: Now, Mr. Nance, you want the Railroad Commission to adopt a rule that will let you drain your reserves under your tract?
> MR. NANCE: No, I don't. I don't want a separate well rule for me and a separate well rule for you. I want Atlantic [one of the major operators in the field] to have a 1/3–2/3, too. I want them to get rich just like me.[14]

In the field orders it issued, the Commission allocated production among wells in the Normanna field on a 1/3 per well and 2/3 acreage formula, as was its established course of conduct in fields with a mixture of large and small tracts.[15]

The partnership of Bright & Schiff had obtained a lease on the equivalent of one town lot, 79 feet wide and 130 feet long, comprising less than 3/10 acre, for which they secured an R37 exception permit to drill one well. Atlantic fought Bright & Schiff every step of the way, even allegedly seeking "to intimidate drilling contractors," and Bright & Schiff had obtained a Bee County Court injunction against Atlantic's interference.[16] As the Normanna field developed, Atlantic Refining and Tidewater Oil Company, together with five others, had gradually acquired control of nearly the entire field and had voluntarily pooled their interests. Bright & Schiff's lease was the only one not pooled. By the time the case reached the Supreme Court of Texas, appellants had completed six wells, and planned to drill two or possibly four more within the then-known extent of the field.[17]

In their suit seeking to invalidate the Railroad Commission's 1/3–2/3 allocation formula, plaintiffs (Atlantic et al.) alleged that enormous drainage would occur and claimed that "a reasonable estimate of the value of the gas in place" under the Bright & Schiff town lot was $7,000, but that over the expected 20-year life of the field, under the Railroad Commission's formula, their single well would produce $2,500,000 worth of gas, which would come from guess where.[18] Although the District Court's Findings of Fact supported Atlantic's claims, it upheld the Railroad Commission order, citing the decisions in Hawkins and *Ryan*. Atlantic et al. appealed directly to the Supreme Court of Texas.

Despite Atlantic's earlier shenanigans, the Normanna decision ultimately depended upon whether Atlantic et al.'s petroleum engineering evidence was sufficient to rebut the prima facie validity of Railroad Commission orders, demonstrate that the Commission's allocation formula was unsupported by substantial evidence and thus overcome the evidentiary bar to court intervention, show that the order was so unreasonable as to constitute confiscation in violation of due process, and overturn the Hawkins and *Ryan* decisions—and the rule of capture in Texas. The crux of the appellants' case rested upon the expert testimony of their primary scientific witnesses, Donald D. Lewis and John W. Crutchfield.

Lewis, a "qualified geologist" with wide experience in Wilcox reservoirs such as Normanna, who worked for the consulting firm of Crutchfield and Pruett, presented structural maps (which depict the contours of the productive formation) derived from "electric logs of the various wells to determine at what depths the top of the particular sand was encountered by the wells, drill stem test data to determine whether at the point the test was taken the sand contained oil, water, or gas, completion data to show in which reservoir or sand each well is completed and potential [production] test data." The structural maps also showed two faults that delimited the field to the northwest and to the southeast. Lewis introduced isopach maps, which reflect "the net thickness of the hydrocarbon saturated sand at any point on the particular structure." Lewis employed "one of the most accurate methods of isopaching, using micrologs of the wells." Micrologs, introduced in 1948, were a refinement of electric logs or Schlumbergers, and used electrodes mounted on a pad pressed against the borehole wall to give fine measurements of resistivity, which allowed more accurate inferences about formation contents. Together, "By using the structure map of a reservoir to determine its areal extent and the isopach map of the same reservoir to determine its net sand thickness, the volume of the sand that contains hydrocarbons may be computed."[19]

John Crutchfield "testified as to the gas and condensate reserves contained in each of the three reservoirs thought to be productive." At the time. Crutchfield was a consulting petroleum engineer, a principal in the firm of Crutchfield and Pruett in Corpus Christi. He had earned BS engineering and petroleum engineering degrees from the University of Oklahoma in 1940 and 1941, after which he worked for Humble Oil from 1941 to 1946, rising to the position of Senior Resident Engineer in Southwest Texas; he formed Crutchfield and Pruett in 1946. The firm did mostly "estimation and evaluation of oil and gas reserves for banks, insurance companies, buyers and sellers." Crutchfield had studied numerous Wilcox formation fields. Among his bank clients were most of the first movers in reserve collateralization: First National of Dallas, Republic National of Dallas, First City National of Houston, Bank of the Southwest in Houston, as well as First National City of New York and First National of Chicago. Atlantic had hired him to determine the recoverable reserves in the Normanna field, and specifically "to determine the participation of a well drilled on a three-tenths acre town lot, in such reserves, under the Commission's proration formula," and to determine the gross value of that production.[20]

For his analyses, in addition to all the data used and developed by Lewis, Crutchfield had bottom-hole pressure data, both inferred from drill stem tests and measured directly by downhole pressure gauges (the two differed by less than 2 percent), and downhole fluid samples.[21] He used "the volumetric method to determine the recoverable reserves of gas and condensate" for the upper three of the Normanna field's four reservoirs (data for the fourth and deepest was not yet available).[22] Crutchfield even wheeled a blackboard into the courtroom to show the volumetric equations and illustrate exactly how he did his calculations.[23] His study generated robust estimates not only of the distillate and the dry "pipeline" gas that could be recovered from the entire field, but also the volumes that could be recovered from only beneath a 3/10-acre tract. Assuming eleven wells total in the reservoir, under the Railroad Commission 1/3–2/3 formula, Bright & Schiff's lone R37 exception well would recover 9,533 million cubic feet of gas, compared to the 25 million cubic feet underlying the tract, which meant that 9,508 million, or just over 9.5 billion, cubic feet of gas would be drained from other leases in the field.[24]

Once his volumetric analysis was in hand, Crutchfield could then calculate the value of each well's production under various Railroad Commission–mandated allocation formulae, at different prices, and over different production time frames. Crutchfield's procedures were standard practice for evaluation work. His volumetric analyses showed that although Bright & Schiff's lease originally had only $7,000 worth of hydrocarbons in place beneath it, under the Railroad Commission formula, it would produce gas and distillate worth over $2,500,000, the estimate adopted by the Supreme Court of Texas.[25]

Bright & Schiff offered two expert petroleum engineering witnesses: Joseph Ballanfonte and William H. Price.[26] Under discovery, Atlantic et al. had provided Ballanfonte and Price with all the data and other information used by their expert witnesses. But Bright & Schiff not only refused to turn over to the appellants any information whatsoever about their well, but also refused to share that information with their own expert witnesses, Ballanfonte and Price. As a result, Bright & Schiff presented no affirmative factual case at all, and Bright & Schiff's counsel's (Payne's) examination of his own witnesses consisted largely of a series of "let's assume" questions.[27] Ballanfonte and Price challenged none of the methodologies used by Atlantic's expert witnesses, nor did they dispute any of their deductions or inferences. They offered "no estimate of the field reserves and made no estimate of reserves in place under the town lot well." They contended instead that "from the evidence introduced at trial an estimate of such reserves could not be made," and, rather than denying the plaintiffs' estimate of $2,500,000 in drainage, maintained that there was insufficient data upon which "to base such estimate."[28] Ballanfonte and Price presented no data and no engineering analyses of their own, and none specifically about the Normanna field.[29] Appellees' cross-examination of appellants' witnesses, especially Crutchfield, likewise had consisted mostly of a whole school of red herrings with little or no relevance to the material issues at hand, even to the point that it began to try the patience of the District Judge.[30]

The Supreme Court of Texas was not impressed, although it did ponder the case a full year before announcing its decision.[31] After observing that appellee's witnesses had not said that appellants' (Atlantic's) witnesses were wrong in their estimate that only $7,000 worth of gas underlay Bright & Schiff's lease or in their contention that Bright & Schiff's well would drain $2,500,000 worth of gas from adjacent leases, but had only whined that data were insufficient, the Court concluded:

> It appears from the record that whether the estimate of field reserves made by appellants' witnesses was over estimated or under estimated, that there is not the slightest doubt but what there would be enormous drainage from the other tracts in the field, including those of appellants, to the Bright & Schiff tract under the 1/3–2/3 proration formula.[32]

Bright & Schiff's case was likely poorly served by testimony that when Atlantic had offered to buy out the 3/10-acre town lot, H. R. Bright had responded that he "expected to make a 'profit in excess of a million dollars out of this well.'"[33]

Bright & Schiff seem to have bet the farm on the shield provided by Hawkins, *Ryan*, and the "law" of capture. Appellees argued that the rule of capture was of equal standing and dignity in law to ownership of oil and gas in place. Drainage was part of the deal and was not sufficient to render a Railroad Commission order unreasonable or invalid. Bright & Schiff claimed that even if Atlantic's estimates were true, under the law of capture, and under the precedents established in Hawkins and sustained in *Ryan* and other cases, they were only taking what they were legally entitled to: whatever its magnitude, drainage was sanctioned by law if it were necessary to allow a small tract owner to recover the oil and gas under his property and make a profit. Bright & Schiff rehashed the historic "natural" advantages accorded small tracts before prorationing, and maintained, as had the Court of Civil Appeals in Hawkins, that the Railroad Commission could legitimately take those advantages into account in its allocation formulae.[34]

In their brief, Atlantic et al. argued, as had Hines Baker in 1935, that the "law" of capture was and had always been only a rule of convenience adopted in the face of ignorance and was and always had been only a rule of non-liability for drainage, not a license to poach. After reviewing cases from *Westmoreland & Cambria Natural Gas Company v. De Witt* (Pennsylvania Supreme Court, 1889) on, as well as more recent Texas cases from *Brown v. Humble Oil & Refining Company* (Supreme Court of Texas, 1935) through *Marrs v. Railroad Commission* (Supreme Court of Texas, 1944), Atlantic contended that capture had consistently been regarded as secondary to ownership of oil and gas in place. Since petroleum engineers were no longer ignorant of subsurface phenomena and could determine oil or gas in place beneath a surface tract with reasonable precision, this rule of convenience was no longer necessary nor appropriate. Atlantic stressed that under Section 6008 of the Natural Gas Act of 1935 and its subsequent construals, the Railroad Commission had an explicit duty to protect correlative rights and assure each mineral owner a fair opportunity to recover his property and claimed that this mandate trumped any residue of the rule of capture, Hawkins and *Ryan* notwithstanding.[35]

Having found unequivocally that as a matter of fact "enormous drainage" would occur from appellants' leases to the Bright & Schiff 3/10-acre lease, the Supreme Court of Texas's majority opinion, written by Robert W. Hamilton, after reviewing in detail both the statutory requirements of Section 6008 and the Texas precedents, wholly agreed with Atlantic et al.'s argument.[36] The Court dismissed the relevance of Hawkins and other cases to the facts of the Normanna case by reasoning that in those cases the appellants had long assented to the established field rules they were now contesting: "We are in agreement . . . that where producers have acquiesced in and have failed to complain of the Commission's proration orders for a long period, during which time other operators have expended vast sums in exploration and drilling operations, such producers should not be heard to complain."[37] The Texas Supreme Court's majority opinion concluded, "we feel compelled to hold, under the facts of this case, that the 1/3–2/3 proration formula promulgated by the Railroad Commission for the Normanna field is invalid. . . . It does not come close to compelling ratable production; neither does it afford each producer in the field an opportunity to produce his fair share of the gas from the reservoir."[38]

The majority opinion and subsequent denial of appellee's Motion for Rehearing induced a lengthy and energetic dissent from Justice Clyde E. Smith, who had written the *Ryan* decision. Smith rejected Atlantic's petroleum engineering evidence in toto as well as the trial court's finding that "drainage of a tremendous quantity of gas and condensate from other leases and tracts" to Bright & Schiff 3/10 acre would occur, "because plaintiffs [Atlantic et al.] wholly failed to prove their theory."[39] But the real nub of Smith's dissent was his unshaken conviction that "The Texas law of property in oil and gas has been defined by this court as embracing two parts of equal force and dignity: the law of ownership in place, and the law of capture."[40]

Despite the usual small herd of impassioned amicus curiae briefs, the Supreme Court of Texas denied a Motion for Rehearing May 31, 1961, and the majority Normanna opinion stood. After further hearings, on September 27, 1961, the Railroad Commission issued a new Normanna field order. It didn't mince any words about what the problem was, or what its solution entailed:

> . . . the Commission is of the opinion and finds that an acreage allocation formula would reasonably provide each party an opportunity to produce its fair share of the gas in a reservoir, but because of the additional responsibility placed upon the Commission, that is, to arrive at a means of preventing confiscation of a small tract's reserves while allowing each operator to produce its fair share of the gas in a reservoir, as knotty a problem as has ever been placed before the Commission, must be solved, that such solution in all probability is best resolved through the use of a special allowable that would encourage a small tract owner to negotiate with his neighbors for fair and just treatment, but would also provide a sufficient allowable to such small tract to encourage a reasonable attitude in such neighbors so that they would endeavor to work out this common problem.[41]

Specifically, the order mandated a 100 percent acreage allowable for wells in the Normanna field, but:

> Rule 3b. Any operator having the right to drill a gas well upon a tract containing less than 100 acres, after application, notice and hearing prior to the drilling of the well, may, if in the judgment of the Commission the facts so justify, be given a definite and fixed special allowable, provided the well is completed as a producing well, higher than the allowable that would be fixed for any well completed upon such tract by reference to said allocation formula upon proof to the Commission (1) that the drilling and completion of a well upon said tract is not economically feasible; and (2) that each and every owner of the right to drill upon acreage immediately adjacent to said tract of less than 100 acres has refused to allow the pooling of said tract with enough of said immediately adjacent acreage to create a drilling unit of at least 100 acres upon fair and reasonable terms . . . provided further, however, that any well drilled before the adoption of this rule will be treated, for purposes of this rule, as though it had not yet been drilled.[42]

Bright & Schiff's well was granted a special allowable that permitted it to continue to produce more than its fair share of gas, but not nearly as much more; it was acceptable to all parties and litigation ended. The Railroad Commission had brokered a modus vivendi between petroleum engineering rationality and the opportunity ethos of the oil fraternity in Texas. Doubtful if Solomon his own self could have done much better.

The Railroad Commission got the message. In a Memorandum prepared for Ernest O. Thompson, September 8, 1961, the Commission's Senior Legal Examiner, Fred Young, concluded:

It appears that the Court has held that the Rule of Capture is nothing more than a rule of non-liability for drainage where such drainage is caused by reasonable, prudent and legal operations. . . .

The language of the decisions cited by the Court in reaching this conclusion indicates that *fair share* means the recoverable oil in place or its equivalent [emphasis in original].[43]

The deluge petroleum engineering science and practice unleashed wasn't long in coming.

Revolutions ramify, and even its critics portrayed Normanna as "revolutionary."[44] Very much as heterogeneous flow theory had propagated through petroleum engineering science and practice by articulation and extension, so Normanna propagated through oil and gas law. The next major case to feel and feed the tremors, Port Acres (*Michel T. Halbouty v. Railroad Commission of Texas*, 1962), differed materially in its particulars.[45] First, its lead plaintiff and then appellant, Michel T. Halbouty, was an independent rather than a major. Given the reputation the Railroad Commission had for protecting the interests of independents, Halbouty's suit was, as the *Dallas Morning News* described it, "like biting the hand that feeds you." Second, and more importantly, Port Acres, like Normanna, was a gas-condensate field, but it was much more fully developed than Normanna had been. The Railroad Commission established a 160-acre spacing pattern, and allocated allowables according to its customary 1/3 per well, 2/3 surface acreage formula.[46] Halbouty et al. had begun contesting this formula before the Railroad Commission in mid-1959 (before the Normanna decision), claiming that it was confiscatory, and asked that it be replaced by a net acre-foot formula. The Commission refused and continued to authorize R37 exception permits.[47] The twenty-two R37 wells (out of forty-seven in the field) were located on tracts ranging from 2/10 of an acre to five acres.

The technical crux, and import, of the case remained, as in Normanna, alleged egregious drainage to small tracts, the determination of which depended upon petroleum engineering analyses. The character of engineering testimony in Port Acres pretty much replicated that in Normanna. Although witnesses differed in their specific estimates of producible reserves in the field, about the exact extent of the field and about whether a water drive might (appellees) or did not exist (Halbouty et al.), the Court observed that " the basic facts are not substantially in dispute."[48] The material fact of enormous drainage was not contested: Each small-tract acre would depend upon drainage "for 92.8% of its gas recovery and 90.63% of its condensate recovery."[49]

The Supreme Court of Texas wasn't anymore buying one tract, one well, unlimited profit guaranteed. Once again citing Section 6008, the Court found that the legislated obligations of the Railroad Commission were to prevent waste of gas and condensate and to "give each well its fair share of the gas to be produced from the reservoir."[50] To the contention that rigorous application of the oil and gas in place doctrine would have dire consequences for small tract owners, operators, and local economies, the Supreme Court of Texas now in effect answered, "Tough. You shoulda known better."[51] As in Normanna, after additional hearings the Railroad Commission issued a new order for Port Acres in February 1963, providing for production allocation based upon the net productive acre feet of sand attributable to each well. The small tract owners hired Price Daniel, Governor of Texas 1956–1962, to represent them, and the small tract "protection clause" was duly included in the new Port Acres order, as at Normanna. This new order too went unchallenged.[52]

*Railroad Commission of Texas v. Shell Oil Company* (1964), the Quitman case, was the penultimate decision in establishing the primacy of property rights in oil and gas—and the authority of petroleum engineers to determine them. It too was a crucial precedent, since it concerned an oil field to which the Texas Natural Gas Act did not apply. A decision curtailing production from R37 wells would require extending the logic of Section 6008 of the Natural Gas Act, and of the Normanna and Port Acres decisions, to oil. Shell Oil, Amerada Petroleum, and Sun Oil brought the original suit challenging the Railroad Commission's field order establishing a 50 percent per well, 50 percent acreage allocation formula for four deep (4,200 to 8,500 feet), separate, and unconnected oil reservoirs on the Quitman structure in Wood County.[53] Upon appeal to the Supreme Court of Texas, the case again hinged on material findings of fact, which in turn depended upon petroleum engineering evidence. In Quitman, that evidence showed that large-scale drainage from the plaintiffs' to the small-tract wells had occurred and would continue to occur so long as the 50–50 order were in force, and decided the critical technical issue before the Court.

The District Court, then the Court of Civil Appeals, then the Supreme Court of Texas adopted respondents' (Shell et al.'s) factual claims as veridical. Combined with its findings of law, the Supreme Court ruled that there was no substantial evidence to support the Railroad Commission's Quitman field order, that it failed to protect correlative rights and assure to each mineral rights owner a fair and reasonable opportunity to recover his property, and that, "The conclusions reached in Normanna are equally applicable to these facts."[54] The order was therefore invalid and its enforcement enjoined.

After the Quitman decision, it was clear that the special status accorded small tracts in prorationing since the 1930s was dead, whether for gas wells or for oil wells. But the question remained of how far back the change would reach. As noted, citing as precedents the Hawkins and Standard Oil Cases, the Texas Supreme Court had held in Normanna "that where producers have acquiesced in and have failed to complain of the commission's proration orders for a long period . . . such producers should not be heard to complain."[55] In *Railroad Commission of Texas et al. v. Aluminum Company of America* (1964), the court made it clear that the courts of the State of Texas had no intention of revisiting the field rules for all 1,700 oil fields in Texas and had no intention of requiring that the Railroad Commission do so either: "The theory of the Normanna decision . . . is that its holding is to be applied prospective and not retrospectively."[56]

Largely in response to Normanna, Port Acres, and Quitman, the Texas legislature passed the Mineral Interest Compulsory Pooling Act on August 30, 1965, retroactive to March 8, 1961, the date of the Normanna decision. This act essentially codified the modus vivendi the Railroad Commission had invented for the Normanna field: the act mandated compulsory pooling of small tracts and it protected the correlative rights of small tract owners just in case the Commission determined that larger neighbors were not dealing fairly.[57] While the act was intended to, and did, encourage pooling of small tracts, the act also explicitly authorized the Commission to put its thumb on the scale on the side of Everyman and use what John Kenneth Galbraith in another context called its "countervailing power" to equalize the bargaining positions of big and small.[58]

It had been not quite forty years since Beecher and Parkhurst had published their seminal paper on the role of dissolved gas in oil production. Despite Henry L. Doherty's vaulting ambition and his erstwhile disciples' hype and overreaching, Beecher and Parkhurst's work marked the beginning of a long and tortuous, and expensive, journey, not a terminus. But by the mid-1960s, petroleum engineering scientists could plausibly claim to possess well-winnowed, robust, and reliable theories of multiphase flow in porous media and to have discovered and accommodated unexpected physical phenomena in oil and gas reservoirs. By the mid-1960s petroleum engineers could plausibly claim to understand the behavior of entire subsurface geological formations, such as the Woodbine basin, the detailed characteristics and performance of individual oil and gas reservoirs, and to reasonably allocate ownership rights and prerogatives within individual fields and among individual leases and wells. Abetted by extraordinary innovations in acquiring and processing subsurface information—core analysis, downhole pressure measurement and fluid sampling, and electric logging and its derivatives, such as sonic logging and nuclear logging, plus, increasingly, digital data processing and simulation, petroleum engineers deployed a concatenation of techniques and theoretical insights that provided convergent validity for inferences about what was there and how it would behave. They had reduced the law of capture to a mere rule of convenience and had made themselves the unchallenged practical and legal arbiters of who owned, and who got, what, when, where, and how.

## REPOING MORAL ECONOMY

On a cloudy, warm, humid East Texas spring day, April 17, 1961, the same day as the ill-fated Bay of Pigs invasion of Cuba, a Shell Oil crew sent to work over a well on the western edge of the East Texas oil field that had been producing for twenty-five years made a discovery unprecedented in the annals of the oil fraternity in Texas or anywhere else. When the crew opened the well's outlet valve, it commenced a high-volume flow of pure drilling mud. The mud's rhythmic pulsation told the crew its source was nearby, driven by a piston mud pump. Sure enough, a rig some half mile away had directionally drilled right into the Shell well, piercing its production casing. This remarkable discovery lit a slow fuse that eventually ignited the slant-hole "scandal," the most serious to afflict the Railroad Commission and the oil fraternity in Texas in their mutual history.

Directional drilling is one of those unusual instances in the evolution of technology when what initially is an unmitigated nuisance is incrementally transmuted, by ingenuity and development, into a useful and valuable technique (Figure 10.4. Vertical, Deviated, and Horizontal Wells). Unintentionally deviated, or crooked, holes had been a problem with rotary drilling from its beginning, and often resulted in stuck drill pipe, lost holes, or the inability to set casing. It's likely that almost all wells drilled with rotaries into the 1930s were crooked to some degree, and until the 1920s deviation ordinarily was thought a bad thing.[59] The first deliberately, if imprecisely and illicitly, deviated wells were apparently drilled at Huntington Beach in Southern California soon after the field's discovery in 1919. In deference to the recreational value of the beaches, the State of California forbade building offshore platforms but did allow onshore drilling behind the beach; some onshore wells at Huntington Beach were found to be bottomed under the ocean, where even larger oil deposits were thought to lie.[60]

This discovery led engineers to the realization that many oil wells deviated from vertical and spawned a mini boom for the legal fraternity, who filed multiple suits across the county which alleged that deviated wells crossed property lines and inflicted damage on the victims of this trespass. Early downhole well-survey devices were crude. Inclination surveys, which used an instrument consisting of a small camera and a miniature plumb bob lowered down the hole, could indicate whether the well bore was vertical or not but could not reveal the direction of its inclination. The first deviation survey instruments, which could also tell direction of inclination, combined a magnetic compass, a plumb bob, and a camera, but the compass was compromised by the metal content of casing and other equipment. These devices were supplanted after 1926 by an instrument in which a Sperry gyroscope, derived from Sperry's aircraft gyrocompasses, replaced the magnetic compass. These instruments were manufactured and marketed under the Sperry-Sun name by a partnership of Sperry and Sun Oil, which had been involved in one of the deviated-well lawsuits and partly funded development of the devices. Because of their more expensive and delicate instruments and the skill required to operate them, deviation surveys were more costly and time-consuming to run than inclination surveys, and therefore rare.[61]

The desire to exploit legal offshore California oil leases without building offshore platforms induced H. John Eastman of Long Beach to develop precision directional drilling: the State of California, as lessor, was persnickety about which wells tapped which offshore reservoirs. Eastman developed both the technology (whipstocks, which deflect the drill bit and drill string away from vertical in a desired direction, knuckle joints to allow the drill stem to follow the new path, and specialized spudding bits and reamers), and, more importantly, the complex operating techniques required for directional drilling.[62] Eastman's innovations, together with those of others, set off a second oil boom at Huntington Beach in 1933.

When directional drilling came to Texas in 1934, it came as precision directional drilling, and it came to solve a dire emergency, cost no object. In 1932, independent oil operators Dan Harrison and Jim Abercrombie (Harrison & Abercrombie) had managed to take a 15-acre lease in the big middle of the 19,000-acre Conroe field, which George Strake had discovered in 1931. The field was mostly controlled by Strake, Humble Oil, and the Texas Company. Harrison & Abercrombie completed their first well in 1932. The field's wells were extraordinarily prolific, but notorious for their high and unpredictable pressures. In January 1933, a Standard of Kansas well on a lease adjacent to the Harrison & Abercrombie well blew out and caught fire, as did another Standard of Kansas well close by. It was these blowouts that provoked Harold Ickes's ignorant and intemperate telegram to Ernest O. Thompson, who was on scene. Due to the geological instability of the field, the Harrison & Abercrombie well also blew out and cratered but didn't catch fire. Seeing what was coming, Abercrombie had built a moat around his well to prevent fire encroaching on the well, but the moat also served nicely to collect the oil flowing into the well crater at a rate of some 10,000 barrels a day. After three months of trying standard solutions, twelve 600-foot relief wells reduced gas pressure in the two burning blowouts enough to allow them to be extinguished. But Harrison & Abercrombie's well continued to flow unabated.

Humble Oil, in the person of John R. Suman, was watching its oil field drain away. Conroe was a water-drive field, and as pressure declined production from other wells in the field fell to less than a hundred barrels a day, with every likelihood the entire field soon would be ruined. At Suman's instigation, Humble bought out Harrison and Abercrombie's lease for $300,000. Suman, originally from California, was familiar with the new directional wells being drilled at Huntington Beach. Even though Eastman's methods were still largely unproven, Suman contracted with him to drill a directional well to kill the Harrison & Abercrombie blowout. Eastman put his bit within 22 feet of the bottom of the Harrison & Abercrombie well, and it was finally killed January 9, 1934. Precision directional drilling had come to Texas and gradually became part of the standard repertoire of sophisticated drilling contractors. In their sales agreement with Humble, Harrison & Abercrombie had retained the 1.5 million barrels of oil they had accumulated, which, under the rule of capture and under the circumstances, were not subject to proration; they sold it off at an average price of $1.10 a barrel.

The spectacular success of directional drilling at Conroe defined what it was to the oil fraternity in Texas, to petroleum engineers, and to the Railroad Commission of Texas. As is so often the case, users defined the technology: In directional drilling, precision became the primary desideratum, and largely determined the status and reputation of practitioners. As with digital computers before the late 1960s, this communal perception led to a rigid and narrow notion of what the technology was and what it was for. But what users do with a technology is open-ended, and, as with the telephone or the internet, users found or invented novel, and sometimes illicit, uses.[63]

Quite apart from their number and flagrantly illegal status, the slant-holes implicated in the East Texas scandal were notably imprecise, which blindsided majors and the Railroad Commission alike. Primary sources on the slant-hole scandal are scarce. Most information comes, as it does for the moral economy of the English crowd in the eighteenth century or the Vietnamese peasantry in the twentieth, from statist or other "respectable" sources with an ax to grind and interests in play. While ex post well-deviation surveys done by engineers give an accurate physical portrayal of the illicit slant-hole wells that were discovered, the rest of the who, what, where, and when comes from establishment media or court records. Understandably, for most of the protagonists in East Texas, discretion is the better part of valor.[64] As for the English crowd in the eighteenth century, much inference depends upon evidence of "the dog that didn't bark" variety.[65]

The ruckus in East Texas was hardly the result of mom-and-pop operators engaged in petty theft. Ultimately, 380 deviated wells were identified in the East Texas field; another twenty-three turned up in the nearby Hawkins field, most drilled by the same people active in

East Texas. East Texas was easy pickings for directional drillers: the field had a straight, clearly defined eastern boundary 45 miles long where the Woodbine pinched out against the Sabine Uplift, with the richest and longest-lasting productive areas of the field just to the west. On the western edge of the field, drilling was iffier, but the line of saltwater intrusion was well-known and easily identified; all an operator had to do was pick up a lease with dead or watering-out wells and bottom his hole off to the east. Of the 380 deviated wells uncovered in East Texas, 88 percent were along its eastern edge, 5 percent to the west, the rest scattered.

Majors (not exactly disinterested) estimated that 60,000 barrels a day were being filched in 1961. Estimates of the dollar value of the illicitly produced oil ranged from $40 million to $1 billion (also from interested sources), with the most common estimate being about $100 million. Whatever the exact figures might have been, drilling, completing, testing, documenting, and producing the wells in question required the participation and cooperation, or at least the complicity, of a large number of engineers, lessors, drilling contractors, drillers, crews, suppliers, oil well testing and cementing firms, Railroad Commission and contract inspectors, lawyers, and bankers. Employees of major oil companies, most folks in the Railroad Commission field offices in East Texas, the staff of the Federal Petroleum Board in Kilgore (responsible for enforcing the Connally Hot Oil Act), even the vaunted East Texas Engineering Association (also in Kilgore), sat right on top of the scandal for a good fifteen years and managed not to notice.[66]

There had been intimations something was amiss in East Texas for some time. As a Railroad Commission employee later testified, rumors about deliberately deviated wells had begun to circulate in the field as early as 1946. The Commission was concerned enough about the problem by the beginning 1947 to insert in all R37 permits issued for the east side of the East Texas field a straight-hole clause, which required an inclination survey of newly completed wells. Persisting rumors about East Texas, combined with the increasing popularity of directional drilling for perfectly legitimate purposes, motivated the Commission to issue its Statewide Rule 54 in April 1949. This Rule required a special permit, obtained after a Commission hearing, to drill a deliberately deviated well and also required that a more expensive directional, not just an inclination, survey be run, witnessed by a Commission employee and filed with the Commission, after the well was completed.

In April 1950, Edwin G. Stanley, the Railroad Commission's district engineer in Kilgore and the son of the Texas Ranger Captain who had worked with E. O. Thompson to get the field under control in 1933, wrote the Commission in Austin that the rumors of deviated wells continued and that many of the inclination surveys filed with the Commission might be bogus. In May 1950, the Railroad Commission notified East Texas operators that its Kilgore office was authorized to require a directional survey of any well at any time. That summer, a Commission engineer who had just witnessed a survey, which showed a straight hole, heard a rig running the next night at the same location: a resurvey revealed a second, deviated well. The Railroad Commission progressively tightened its regulations, which either solved the problem (the Commission's belief) or substantially enlarged the circle of those involved in the grand deception.

The slant-hole operators were as innovative and resourceful as their forebears had been a generation earlier during the first years of prorationing and martial law in East Texas. Sometimes they used plastic pipes (which metal detectors couldn't find) to connect surveyed straight wells that had gone to water with producing slant-hole wells. A slant hole with good production could be connected to multiple dead wells, and flow regulated to keep the reported production of the several dummy wells under the twenty barrels a day allowed marginal wells.[67] More directly and more commonly, after the Railroad Commission began requiring timely deviation surveys for all new wells, they simply coopted the folks contracted to conduct the surveys or, maybe, Commission employees on site who read and reported them.

Shell's miraculous mud well really got the attention of the Railroad Commission as well as operators in East Texas, especially the majors, although the full scandal didn't break for another year. Petroleum engineer and Railroad Commissioner William J. Murray, likely the most technically astute person to serve on the Commission, reevaluated the problem. First, precision directional drilling (such as at Conroe in 1934) had allowed a deviation from vertical of about 25 degrees (Figure 10.4. Vertical, Deviated, and Horizontal Wells). But slant holes in East Texas didn't require precision. Anybody who could manage an azimuth of 45 or 50 degrees could hit productive Woodbine sand, and this lessened accuracy permitted greater inclination angles: one well was later found to be slanted a full 65 degrees. Thus, slant holes could be drilled from further away by less technically adept crews. What the Shell fiasco had revealed was unlikely to be a singleton.

Murray's second insight, in January 1962, was even more helpful and derived from the perverse effects of the Marginal Well Act. Most wells in the East Texas field whose potential exceeded the marginal well threshold of twenty barrels a day were still limited to about twenty-one barrels a day, except that market demand prorationing allowed them to produce the equivalent of only eight days, or 168 barrels, a month. A marginal well was exempt from market demand restriction and could produce nineteen barrels a day all month long, or 570-plus barrels. By 1962, there were 3,895 marginal wells in the East Texas field, too many to test. But Murray reasoned that slant-hole operators did not want for greed. The production from a true marginal well slowly declines year on year. All Murray needed to do was identify the purported marginal wells which reported production close to twenty barrels a day year after year to find the slant holes, which he hired a UT engineering student to do. The slant holes stuck out like proverbial sore thumbs.

When the scandal went public big time in the summer of 1962, all hell broke loose: multiple formal investigations, lawsuits, indictments, recriminations, and grandstanding enough to go around. It even got more coverage than the concurrent Billy Sol Estes scandal afflicting the federal and state departments of agriculture. Attorney General Will Wilson sent along twenty-five Texas Rangers, augmented by twenty-five highway patrolmen, to "assist" Railroad Commission employees in running inclination and directional tests. It was a reprise of 1932 on a smaller scale. The appearance of the Rangers set off rumors of martial law again among those who remembered 1931–1932, but that was never a real possibility. Humble, Continental Oil, and Pan American Petroleum, later joined by Texaco, sought multiple court orders to enjoin operators from interfering with surveys on suspect wells. Even though the Railroad

Commission had issued an Emergency Field Order June 2, 1962, banning plugging wells in the East Texas field, a lot of operators had begun putting cement plugs in their wells, or filling them with junk, to block the tests. The majors had filed multiple civil damage suits which would depend upon the evidence the surveys produced.[68]

The aggrieved suspects responded in kind. Five of them filed a $10,000,000 damage suit against Humble, Texaco, Hunt Oil, Mobil, Ohio Oil, and Pan American Petroleum, "alleging injury to their reputations, and that they have been exposed to 'public hatred, contempt and ridicule.'" Some eighty or so of the good citizens and businessmen of East Texas did as their ancestors had done in the 1930s (or in the 1880s) and organized the East Texas Producers Group, which then generously offered to "mediate" the dispute. One of their number, G. U. Yoachum, wrote the Attorney General of the United States, Robert F. Kennedy, asking that he launch an antitrust probe of the major oil firms. Kennedy declined. Sometimes the reaction was more direct, with majors claiming some of their leases had been vandalized.[69]

East Texans harbored a long tradition of believing themselves unjustly and illegitimately abused—by cotton factors, railroads, banks, "the money power," revenuers, Standard Oil, big oil companies, the Texas Railroad Commission, or the NIRA Petroleum Administrator. An undercurrent of resistance ran straight through from Robin Hood to Sam Bass to the Farmers' Alliance to Populism to moonshiners to East Texas oil operators. The wells of discontent were old and deep.[70] The group of eighty rehearsed familiar themes: The "big boys," the majors, the monopolists, were out to get them, abetted by the Railroad Commission, the Attorney General of Texas, and the big city newspapers beholden to the majors. The majors had manipulated the prorationing system and used their financial power to buy up the better leases in the East Texas field and had thereby denied to the citizens of Texas their privileged access to the God-given resources of Texas. With the connivance of the Railroad Commission, they had systematically drained the oil, west to east, from under the property of others. All the accused in East Texas had done was repossess what was rightfully theirs. This sometimes-noisy campaign gained little traction outside East Texas, where it seems to have generated wide if not publicly expressed assent, which may have been its purpose all along. Rituals of self-production, communal authentication, and reproduction are like that; it's where solidarity comes from.

The State of Texas did manage to collect just over a million dollars in civil penalties. The bereaved majors filed multiple damage suits claiming drainage, but under the statute of limitations, the courts, mostly in East Texas, restricted damages to two years. Most such suits were settled out of court on undisclosed, likely paltry, terms. The few that did go to trial had problems. In one case, Texaco sued an operator for $2,000,000 and was awarded $16,000 by a local jury. It had been well established since *Amazon* in 1934 and the research of Ben Lindsly and Carl Reistle that water from the greater Woodbine formation was intruding into the East Texas field from the west. In a perfect world, the salt water would have risen as a level horizon throughout the field, although, since the field was tilted up from west to east, the wells on the west side would have watered out first. But in practice, the water influx also pushed oil from under the westernmost tracts eastward. This theory had been adopted as a fact of the matter even by the US Supreme Court in the second *Rowan & Nichols* case in 1941 and replicated in virtually every discussion of the field thereafter.[71]

These physical facts of the matter, together with the messy history of the field in the early 1930s and the enormous volumes of unreported hot oil it no doubt produced, made it difficult to establish who had stolen what from whom when. Albert Dawsey, a petroleum engineer testifying as a defense witness for H. L. Long of Tyler, who had been sued for damages by Humble Oil, claimed, "Humble leases already have produced considerably more oil than was in place when oil was discovered in the early 1930s." He then recited the widely acknowledged facts of oil migration from west to east in the field.[72]

Victimization of the big and rich leaseholders in the prime center and eastern parts of the field was a hard sell and real damages to them nearly impossible to demonstrate. As a later Railroad Commissioner, Jim Langdon, recounted, an expert petroleum engineering witness for a plaintiff in an East Texas damage suit (not cited) offered this testimony:

"How many barrels of oil were there under plaintiff's tract before development drilling began?"
"One million barrels," the engineer replied.
"And how much oil has plaintiff produced from the tract?" the lawyer asked.
"Two million barrels," was the engineer's answer.
"How much oil remains under the tract?" came the lawyer's next question.
"One million barrels," the engineer answered.[73]

The slant-hole scandal generated 163 criminal indictments. Some of those indicted already had colorful histories. But most of those indicted were more respectable, including State District Court Judge David Moore of Gladewater and County Judge Earl Sharp of Longview, as well as one county district attorney.[74] This deluge of criminal cases resulted in exactly one conviction, and that was overturned on appeal. Since criminal offenses are tried in the county in which they occur, it's no surprise that jury nullification had its day in East Texas. Although there were widespread allegations that Railroad Commission personnel had accepted bribes, when push came to shove, "the courts were able to come up with little information about their extent."[75] The Railroad Commission did shut down deviated wells as it discovered them, which cost their operators, but did not shut in other, straight wells on the same leases. Very little overtly punitive action was taken. According to the law and the profits in Texas, venality, even to the point of larceny, turns out not to be mortal.

Respectable opinion at the time was not, and historical commentaries since have not been, kind to the slant-hole perpetrators or to the people of East Texas. The orthodox view is simply that the whole thing was nothing more than graft, theft, bribery, and corruption on a grand scale. Consider the two best sources on the slant-hole scandal: Clark and Halbouty's "The Slant-holers," ten years after (1972), and David Prindle's "Slant Wells," twenty years after (1981).[76] Both most commonly refer to the slant-hole operators and their confederates as "thieves." Clark and Halbouty reported that "not a single oil thief had spent a single day in prison." And "the miscreants walked among their fellow citizens with heads high." Slant holes meant "trespass, theft, bribery and a willful, intentional violation of the laws." "It means . . . corruption and bribery."[77] Prindle claimed that "Kilgore might almost be said to be a community that for nearly twenty years was based on well-organized larceny." He complained that "Some of the principal thieves still live comfortably in Kilgore, as respected members of the community" and "There are even today rumors in East Texas about slant wells still pumping happily almost two decades after they were supposed to have been eliminated."[78] Clark and Halbouty as well as Prindle express some combination of astonishment and outrage not just at the lack of opprobrium accorded the perpetrators but at the sympathy, even admiration, they elicited within their communities. Both marvel, with considerable disgust, at what had to be the extraordinary breadth and support "the conspiracy" must have had, and at how long it had persisted, if not undetected, at least concealed. Any legitimation the perpetrators might have offered, whether in terms of the geological and engineering characteristics of the field, the majors' abuse of their economic power, or the rights of Everyman, are dismissed as at best mere cant or more likely as gross obfuscations and self-serving fabrications.

But compare E. P. Thompson on the moral economy of the English crowd in the eighteenth century:

> It is possible to detect in almost every eighteenth-century crowd action some legitimizing notion. By the notion of legitimation I mean that the men and women in the crowd were informed by the belief that they were defending traditional rights or customs; and in general, that they were supported by the wider consensus of the community.
>
> The food riot in eighteenth-century England was a highly complex form of direct popular action, disciplined and with clear objectives.
>
> The economy of the poor was still local and regional, derivative from a subsistence-economy. Corn should be consumed in the region it was grown, especially in times of scarcity. . . . What is remarkable about these "insurrections" is, first, their discipline, and, second, the fact that they exhibit a pattern of behavior for whose origin we must look back several hundreds of years. . . .
>
> It is the restraint, rather than the disorder, which is remarkable; and there can be no doubt that the actions were approved by an overwhelming popular consensus.[79]

Moral economy offers a better explanation for the East Texas slant-hole scandal and accounts for its otherwise problematic features: its apparent widespread popular support and participation, its longevity and concealment, the absence of evidence of bribery or other coercion, the absence of social opprobrium, during or after, the lack of criminal convictions, the poor results of damage suits, and its harmony with known historical traditions.[80] Just as the English crowd in their food "riots" in the eighteenth century enforced ancestral communal norms of just price, the right of even the meekest members of the community to subsistence and the privileged access of the local community to locally produced foodstuffs, East Texas slant-hole operators, their compadres, and their communities repossessed the moral economy of the oil fraternity in Texas and its guarantees of opportunity for Everyman and privileged access for Texans to the natural resources of Texas. In resisting the tyranny of science, the depredations of market capitalism, the machinations of the "visible hand of management," and the might of the big and rich, and in trying to bring what they saw as economic justice to their communities, they had unknowingly recapitulated the moral economy of the English crowd in the eighteenth century.[81]

It would be interesting to know what E. P. Thompson would have made of a bunch of renegade East Texas oil operators who recreated a functioning moral economy deep in the Piney Woods, or what they would have made of a devout English Marxist who is their unwitting historical champion. No matter. The ghost of Sam Bass walked tall and proud the streets of Kilgore and Henderson.

# CHAPTER 10

# MY HEROES HAVE
# ALWAYS BEEN COWBOYS

My heroes have always been cowboys
And they still are, it seems
Sadly in search of, but one step in back of
Themselves and their slow-movin' dreams

—WAYLON JENNINGS[1]

The Faustian bargain that is science very nearly cost the oil fraternity its soul. Almost but not quite. Salvation was Santa Rita, the patron saint of the impossible. The triumph of petroleum engineering unarguably denied opportunity, the soul of the oil fraternity in Texas and its moral economy, to many who, by choice, proclivity, talent, or circumstance, remained ignorant of its principles and practices. But on the southeast corner of the campus of the University of Texas at Austin stands Santa Rita's shrine, a now decrepit and rarely visited cable-tool draw works and wooden pump jack that purports to be Santa Rita No. 1, the discovery well for the great Permian Basin oil field in West Texas. A good part of the Permian Basin field underlies land reserved to the State of Texas when the Union joined Texas in 1845, land later dedicated to the support of state universities (originally the University of Texas and the Agricultural and Mechanical College of Texas). The money Santa Rita spawned and the educations the University of Texas, Texas A&M, and their sister land-grant institutions in the midcontinent provided saved the oil fraternity in Texas.

## TRIBULATIONS

For all its scientific and technological acumen and flamboyant entrepreneurship, the oil fraternity in Texas still experienced unhappy trials and tribulations: what it called its "long depression" from the mid-1950s through the 1960s, then violent price and supply gyrations after 1973, which were more reminiscent of the decades from 1901 through the early 1930s than the years of Railroad Commission ascendency.[2] According to Michel T. Halbouty's estimate, the number of independent oil operators in the United States fell from 30,000 to 12,000, or

Figure 1: The cost of a barrel of oil in nominal terms

Source : Deutsche Bank, Global Financial Data

Figure 2: The cost of a barrel of oil in real USD terms

Source: Deutsche Bank, Global Financial Data

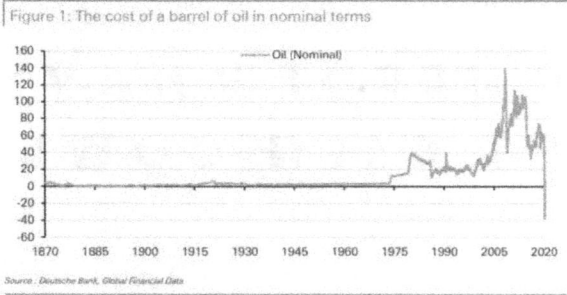

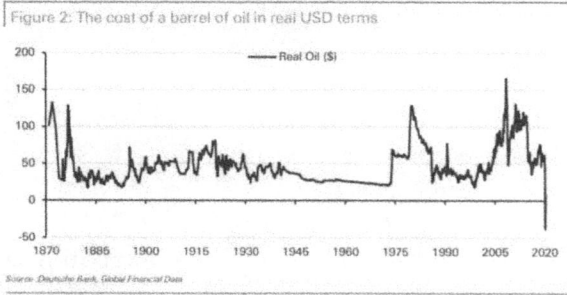

**Figure 10.1. Oil Prices, 1870–2020, in Nominal and Real US Dollars**

**U.S. petroleum consumption, production, imports, exports, and net imports, 1950-2019**

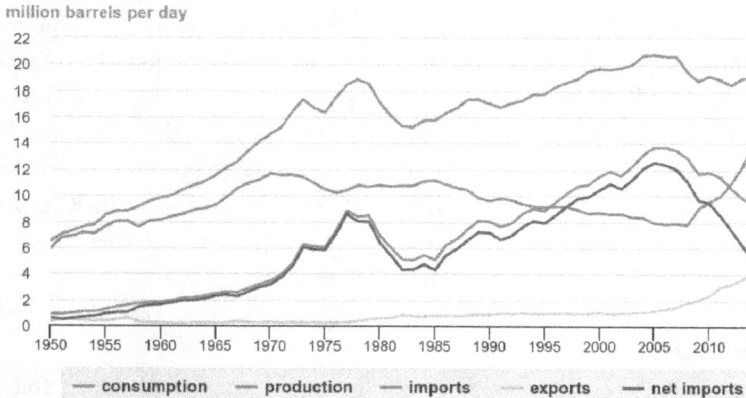

million barrels per day

consumption — production — imports — exports — net imports

Source: U.S. Energy Information Administration, *Monthly Energy Review*, Table 3.1, March 2020, preliminary data for 2019

**Figure 10.2 US Oil Consumption and Source**

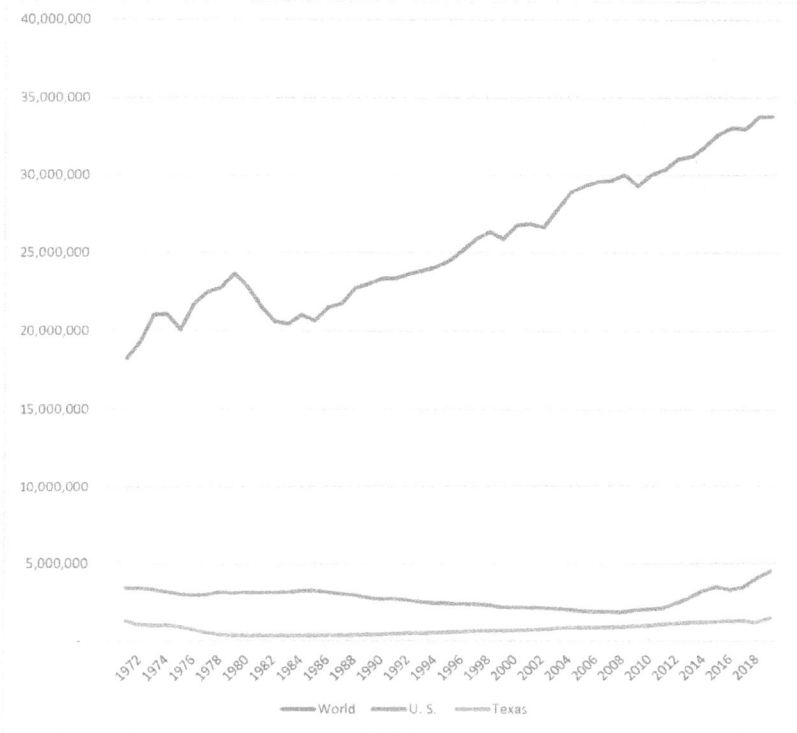

**Figure 10.3. Annual World, US, and Texas Crude Oil Production, 1971–2019, in 1,000s of Barrels**

by 60 percent, in the seven years between 1957 and 1964.[3] A very different oil fraternity only fully recovered in the second decade of the twenty-first century. The most prolific American wells moved incrementally but steadily into deeper and deeper water, while cheap imported crude could easily undercut the prices necessary to keep many marginal wells, and some pretty good ones, in production in Texas. (Figures 1.1. and 1.2. summarize Texas oil production.) Prices, now set on a world market rather than stabilized by Railroad Commission of Texas prorationing, ranged from a high of about $140 a barrel to a low of -$40 a barrel. The latter negative value is a single-day spot price which occurred when commodity speculators realized, as consumption and demand tanked early in the COVID-19 pandemic, that they held futures contracts for more oil than there was capacity to store. You couldn't give the stuff away: you had to pay somebody to take it. Even East Texas at its worst never got that bad. (Figure 10.1. shows crude oil prices 1870–2020 in nominal and real dollars.[4]) Even though world oil production was dominated by sources outside the United States, the fracking boom enabled the United States, in early 2020 just before the pandemic, to become a net oil exporter—to export more oil than it imported—for the first time since World War II. (Figure 10.2. depicts US oil consumption, production, imports, exports, and net imports, 1950–2019, in millions of barrels per day[5]; Figure 10.3. shows annual world, US, and Texas oil production, 1971–2019.[6])

The shift of world preeminence in oil production stimulated the internationalization of petroleum engineering science and practice, which the Society of Petroleum Engineers advocated and promoted. True to its ideological representation of itself, and that representation's Enlightenment roots, the Society's president in 1971, L. B. Curtis, proclaimed, "Petroleum engineering technology is faceless and knows no national boundaries." Much of the proselytizing abroad came from the diaspora of American petroleum engineers, whose companies sent them wherever there might be oil to be got. But much of the growth of petroleum engineering education and practice outside the United States also owed a debt to Santa Rita's blessing: as John C. Calhoun Jr., Distinguished Professor of Petroleum Engineering Emeritus at Texas A&M, put it, internationalization depended upon a "preceding lengthy period of time during which students from other countries came to the US to study petroleum engineering." The seeds were well and truly planted: SPE membership in 1957 stood at 12,658, including students; 11,643 were from within the United States, or 92 percent. Fifty years later, in 2007, the Society had 73,235 members, including 16,739 student members; 41,012 were from outside the United States, and 32,223 from the US, or 44 percent.[7]

In 1976, the Society of Petroleum Engineers deleted the term "foreign" from all publications and references. Responding to Saudi Arabia's and Venezuela's nationalization of concessions to foreign oil companies in 1976 and to the Iranian revolution in February 1979, the SPE's *Journal of Petroleum Technology* editorial in May 1979 reaffirmed the apolitical character of petroleum engineering science and practice: "SPE comprises engineers and managers who hold dissimilar political theories about petroleum's role in national and international politics. The tenet that links them in a cooperative goal is SPE's only objective: to share, update, and foster application of improved technology for producing energy." Although internationalization was not without opposition within SPE, the sons and daughters of the Enlightenment won.[8]

As a result, all petroleum engineers share much the same interdisciplinary pool of knowledge and practice. They are equally competent: "US and UK engineers now work side by side with engineers from Asia, Africa, and Latin America. This change took place gradually but accelerated during the 1980s." As C. O. Stokley reported, "Years ago, the only engineers considered credible for employment by international operators were American or Canadian-trained engineers. Around 1982, that resistance began to change. Now the oil field has evolved into a much more accepting arena for engineers trained in other countries, many of whom are topnotch. They are knowledgeable, well schooled, enthusiastic, and willing to work anywhere in the world." Petroleum engineers share much the same alienated representation of themselves and their profession, and much the same moral economy: As Sadad Al-Husseini, then (2006) recently retired Saudi Aramco's Executive Vice President for Exploration and Production, argued,

> The move of investments and production capacity away from North America starting in the mid-1970s made it necessary for US-based professionals to redirect their focus abroad. As their jobs moved, they took with them their social traditions and networks, including their network of academic centers, their professional values,

and their access to professional societies. This was a real boost for professionalism throughout the world.

SPE selected its first non-US president in 1993, Jacques Bosio, an executive at Elf Aquitaine in Paris, who quipped, "[T]he least you can say about a Frenchman is that he is a true foreigner!"[9]

## TECHNOLOGICAL OPPORTUNITIES

The oil fraternity in Texas faced a much more volatile economic and political environment after 1970. But, as a MITRE manager (in a very different business) once explained to his team, "We don't have any problems here, only new opportunities to excel." Through all the turbulence the world after 1970 brought, the oil fraternity in Texas, now armed with the educations and competencies Santa Rita gifted, persisted. As it had since Spindletop, the fraternity ultimately prospered again through an uncanny concatenation of technological innovation, obstinate entrepreneurship, and moral economy.

Some innovations are more than a radical change within a system. Rather, these innovations are themselves systemic, synergistic, and holistic. They combine subsystems or elements, each of which may have a long and convoluted history, into a new system that exhibits novel, sometimes revolutionary capabilities. For example, HMS *Dreadnought* (1907) combined existing large-caliber, breech-loading naval guns and armor-piercing shells, incremental developments in steel armor, revolutionary steam-turbine engines, and well-established naval design and construction techniques to produce the first modern, all big-gun battleship.[10] The innovations that enabled the renaissance of the oil fraternity in Texas were similarly systemic and synergistic, culminating in horizontal drilling (the ultimate slant hole) and hydraulic fracturing (fracking). Vic Rao, Senior Vice President and Chief Technology Officer, Halliburton, claimed in 2007 that horizontal wells were the "single biggest productivity-enhancing technique in the business of developing and lifting hydrocarbons." He maintained that in general "disruptive technologies require convergence of three factors: the right combination of enabling technologies, compelling economics that highlighted a niche play at first, and industry risk takers and/or a new organizational dynamic."[11]

## Horizontal Drilling

The critical enabling technologies for horizontal drilling are 3D seismic and related reservoir computer simulations (which offer much improved subsurface target identification, acquisition, and evaluation), and steerable drilling systems. These drilling systems comprise mud motors and hard monolithic polycrystalline diamond bits, which permit continuous drilling without so often pulling the drill string to change bits; MWD (Measurement While Drilling), which gives continuous, real-time information on drill bit position and orientation as well as borehole trajectory; and quantitative LWD (Logging While Drilling), which provides real-time formation data as drilling progresses. All these developments rest upon extraordinary, technologically convergent developments in computation, microcircuits, miniaturization, ruggedization, and computer programming.[12]

Vertical Well    Deviated Well    Horizontal Well

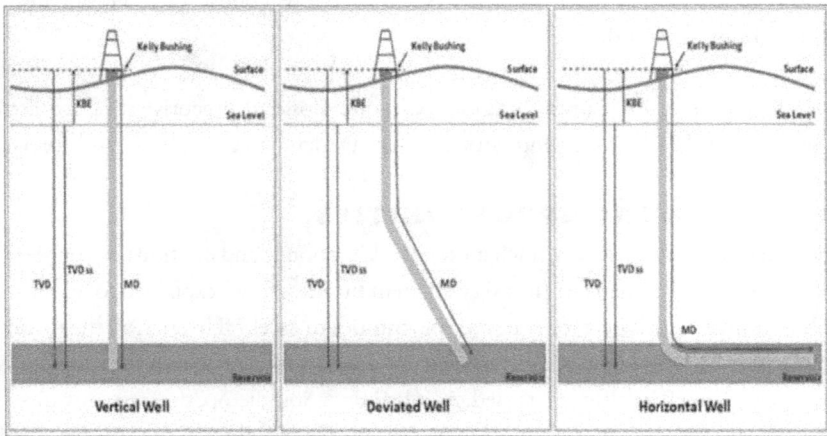

**Figure 10.4. Vertical, Deviated, and Horizontal Wells**

Rao noted that from the 1980s these "disruptive" technologies did not for the most part come from academic research institutions or oil companies themselves but rather were invented, developed, and often deployed by oil-industry service firms. R&D expenditures by major US exploration and production companies declined sharply (with the oil price collapse) after 1985, and by 2000 were running at less than half their 1985 level. In contrast, R&D spending by "traditional oil field service companies" (Baker Hughes, Halliburton, Schlumberger, Smith, Weatherford) accelerated in the mid-1990s.[13] Nathan Rosenberg, in his seminal article on the US machine tool industry between 1840 and 1920, introduced the concepts of vertical disintegration and technological convergence. Machine tool production began as an adjunct to textile manufacture, usually in textile mill shops dedicated to building and repairing textile machinery. But as the competencies of these operations grew, they differentiated out of the mills (vertically disintegrated) to become specialized, free-standing machine tool manufacturers, some further specializing in specific types of tools. As for computers or microchips now, these machine tools were technologically convergent, applicable to a wide variety of metal-working industries producing an even wider variety of final products.[14] The same dynamic appears in oil field services: specialized firms compete to provide specialized capabilities across the spectrum of oil and gas exploration and production firms, from small to the very largest.[15]

The paths of such innovations are twisting and contingent, stretch back in time, and often involve unforesighted synergies. As seismic sensors became more precise, the "log jam" in using seismic information moved to data processing and evaluation, particularly visualization.[16] More capable computers and algorithms enabled 3D rather than 2D seismic analysis, the first of which was reported in 1967. But 3D seismic didn't become de rigueur until 1982, when Veritas introduced standardized programs for seismic processing, followed by new-generation visualization tools in 1998. 3D seismic radically changed the calculus of oil exploration: "Wildcat" success rates rose from 1 in 7 to 1 in 3.[17]

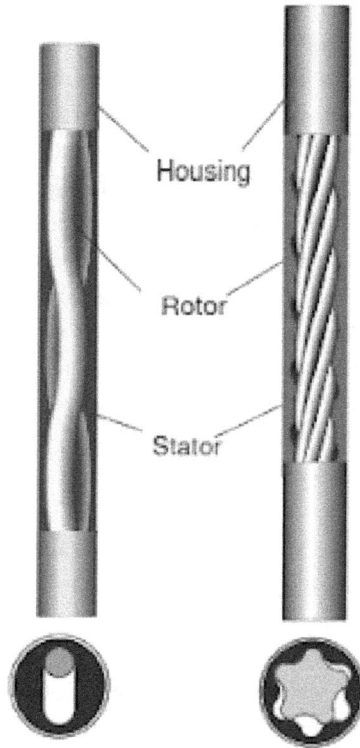

**Figure 10.5. Mudmotor Power Section, Longitudinal View**

The earliest oil wells simply dug down to, or slightly into, a productive formation, and usually were not cased at all. Production casing and cementing (sealing around the bottom of the casing, near the top of the productive strata) initially served to prevent surface water intrusion into the producing formation or the escape of gas and oil. With the advent of shaped charge perforating after World War II, wells could be drilled into the oil-bearing layer of a porous stratum (below a gas cap, above underlying salt water), and casing set vertically through much of its productive depth, then perforated. 3D seismic, with its detailed picture of reservoir shape and volume, offered the tantalizing possibility of drilling horizontally through a productive stratum, thus enormously expanding the perforated area through which oil or gas could enter the well (Figure 10.4. Vertical versus horizontal drilling).[18] The trick was to drill horizontally from a vertical well and keep the horizontal segment within the productive strata.

The horizontal drilling system, bottom to top, comprises a bit and mud motor just below a bent sub, which is used for directional control, modules for MWD (Measurement While Drilling) and LWD (Logging While Drilling), a mud telemetry processor and valves, all stacked into a Bottom Hole Assembly (BHA) at the terminus of the drill string.[19] Neither horizontal drilling nor mud-powered motors were new ideas, but their practical development and deployment did not occur until the 1980s. Mud motors are positive-displacement, eccentric-shaft types based upon a reverse application of the Moineau pump, invented by the French engineer René Moineau in 1931. Positive displacement motors, pumps, and compressors use rotating, interlocking, "twisted" lobed rotors, or rotating rotors and fixed stators, to change the flow volume or velocity and to increase or decrease the pressure of a working fluid, or to add or extract energy from it. Mud motors, Moineau pumps, Lysholm compressors, Roots blowers, and Wankel engines are all rotary, positive-displacement types. Downhole mud motors extract energy from the circulating drilling mud pumped down the drill string to cool the drill bit and remove cuttings. The mud motor obviates the need to rotate the entire drill string. Contemporary mud motors use an elastomeric (hard rubber-like) lobed stator, inside of which a steel lobed rotor spins (Figure 10.5. Mudmotor Power Section).[20] The idea of horizontal drilling was not instantaneously accepted. John Thorogood: "As a new graduate in 1973, I recall a petroleum engineer . . . self-consciously [remarking] how it would be much better if we could drill down to the formation and then simply turn the well horizontally to maximize exposure of the reservoir. We all laughed at him, because it was obviously quite impossible to do."[21]

MWD uses three orthogonal fluxgate magnetometers and three orthogonal accelerometers to detect orientation and motion of the BHA.[22] Electrical power is supplied to the measuring units by batteries, or, more commonly, by a small generator turned by a small, mudflow-powered turbine.[23] The BHA also includes more complex and capable LWD instrumentation. Depending upon the well and geological formation being drilled, different logging instruments can be installed.[24] None of the information acquired by MWD or LWD is of much use unless it is transmitted to the surface for processing and evaluation. Mud-pulse telemetry provides this link. The drilling mud or fluid being pumped down the inside of the drill string is a medium. By using a valve in the drill collar just above the drill bit to slightly modulate the rate of flow and pressure of the mud, a pulse is sent up the moving column of mud, which can be detected by a pressure transducer mounted on the mud standpipe at the surface. The signal travels "at the speed of sound in mud, or between 4,000 and 5,000 ft./sec.," although the medium's bandwidth is limited, a few to about 140 bits per second. First demonstrated in 1962, mud-pulse telemetry did not provide sufficient value-added to warrant widespread use without the systemic synergies created by horizontal drilling.[25] All of this instrumentation has to fit within a BHA less than 10 inches in diameter and operate in a hostile, high-temperature, shock and vibration afflicted environment.

The effect of horizontal drilling was profound: it drove lifting costs per barrel of oil down by more than half even as the cost of individual wells increased substantially.[26] By 2015, extended-reach horizontal wells included the world's longest borehole, 44,294 feet, with a horizontal displacement of 39,469 feet, or 7.5 miles (in the Chayvo field, offshore Russia's Sakhalin Island in the Pacific; it was drilled by a consortium operated by Exxon Neftegas). Horizontal drilling changed how petroleum engineers go about being petroleum engineers. As Vic Rao recounted, "A Shell executive once confided to me that, in the early days of that period [1984–1994], one needed permission to plan horizontal wells, but by the late 1990s, one needed permission not to plan one."[27] By 2018, horizontal drilling dominated exploration and production in the United States.[28]

## Fracking

The other stream of innovation that fed into the shale revolution and the salvation of the oil fraternity in Texas is hydrafracturing, or fracking, which also has a long and convoluted history. "Fracking" (sometimes spelled "fracing," or "fraccing"), as the name implies, fractures a producing formation to facilitate the flow of oil or gas into the well, although the term's exact meaning has changed over time. Operators in the Pennsylvania oil fields in the nineteenth century "shot," or "fracked," their wells with nitroglycerin to loosen the reservoir rock and enhance production. Producers in West Texas in the 1920s used copious amounts of nitro for the same purpose. Fracking is one of several well-stimulation technologies, including acidizing, which uses acid to clean out or etch channels in the producing formation.[29]

In lieu of dangerous and temperamental nitroglycerin, a number of operators and service firms experimented with high-pressure fracturing, or hydraulic fracturing, also called fracking. Once casing is set and perforated, high-pressure piston pumps, similar to those used for pumping mud or for well cementing, inject a fracking fluid into the well. This fracking fluid is viscous: in the first experiments, napalm, later, water containing a gel or viscosity-increasing agent, such as guar. In addition to fracturing the formation rock, the fluid carries a proppant, most often sand, which keeps the fissures open once the fracking fluid is withdrawn. Hydraulic pressure creates or enlarges fissures in a tight producing formation; the fracking fluid flows in, carrying the proppant with it. Addition of a "gel breaker" to reduce the viscosity of the fluid allows reservoir pressure to force the fracking fluid back out, hopefully leaving the proppant behind to hold the fissures open and increase the flow of oil or gas. The fracking fluid must be tailored to the specific chemistry of the producing formation and its contents.

Modern hydraulic fracking got its start in 1947, when Floyd Farris, an engineer with Stanolind Oil and Gas (Standard of Indiana), analyzed the relationship between treatment pressures during acidizing, water injection, or cementing, and well performance. Farris's study persuaded Stanolind to experiment with fracturing a gas well in the Dolomite (primarily limestone) reservoir in the Hugoton field in southwestern Kansas; results were promising enough that 22 other wells were then experimentally fracked.[30] Stanolind obtained a patent on its Hydrafrack process in 1949. As so often occurs in vertical disintegration, since Stanolind was not an oil well-service company, it granted an exclusive license for Hydrafrack to Halliburton, who had the specialized competencies, markets, and infrastructure to exploit and develop it.[31] By the end of 1955, operators had performed over 100,000 hydraulic fracturing operations in conventional vertical wells.[32]

In 1956, M. King Hubbert and D. G. Willis, both at Shell Development Company in Houston, published the fundamental paper theoretically analyzing hydrafracking. Hubbert was an outstanding geologist and geophysicist and is better known outside petroleum engineering science for his stream, beginning in 1949, of ill-fated "peak oil" prognostications. In their 1956 paper, Hubbert and Willis investigated the mechanics of hydraulic fracturing and its relationship to the geological stresses in subsurface formations.[33] They offered an idealized, quantitative theory relating the magnitude of those stresses and their spatial orientation to the orientation and propagation of induced fractures, as well as the minimum pressures required under different conditions. As had Beecher and Parkhurst's 1926 paper, Hubbert and Willis's paper initiated a research programme, not concluded it. Like all such empirical inquiries in engineering science, the investigations they launched are open-ended, necessarily so given the endless variety nature affords.

By the early 1980s, most of the key innovations that would create the shale revolution were, separately, in place, at least in nascent form: 3D seismic, horizontal drilling and its enabling technologies, hydrafracking. These elements still needed to be synergistically combined, which would demand both creative development and imaginative, obstinate entrepreneurship.

## AUSTIN CHALK

What set off the fracking boom then horizontal drilling was the second phase of the third Austin Chalk play in Texas, which erupted in 1985. The Austin Chalk is a soft, white limestone formation that stretches in an arc from Mexico across Central and into East Texas and on into Louisiana and southern Mississippi.[34] Some oil had been produced from natural fractures in the Austin Chalk since the 1920s, but despite intermittent plays in the 1930s and late 1950s, the wells, at best, came in strong but then petered out, leading one operator to describe the formation as "the most perverse, contrary, incorrigible oil field known to man." Nevertheless, as the price of oil rose following the 1973 OPEC oil boycott, Texas independents discovered that conventional vertical wells (the only kind common at the time) drilled into fractured zones of the Austin Chalk, then augmented by fracking, could be profitable.

As has happened so often in the oil fraternity in Texas, stubborn and improbable entrepreneurship birthed the first (or second, third, or fourth, depending upon how you count) Austin Chalk boom in the late 1970s. Its protagonists were as unlikely as Patillo Higgins (Spindletop), Rupert Ricker, Frank Pickrell, and Haymon Krupp (Santa Rita No. 1), or Dad Joiner (Daisy Bradford No. 3). The motley crew in the Austin Chalk included a small independent, C. W. "Chuck" Alcorn, who specialized in buying up marginal production, often for little more than the salvage value of the wells' equipment, a couple of Dallas real estate promoters, a Dallas attorney who was third-generation oil, assorted other independents, large and small, and one experienced and, in retrospect, brilliant consulting geologist, Ray Holifield.

The Austin Chalk play generated the customary oil fraternity drama. But despite considerable competitive rancor among some operators, the fundamental moral economy of the oil fraternity (and of the frontier) held: "If one of them had a blowout, the others would send men to help get it under control. If one operator was short of mud or pipe or equipment—and shortages were the rule in the early days—the others would swap and share."[35] Starting in 1985 and accelerating after 1989, Texas independents, as well as now some majors, picked up developing horizontal technology and began drilling horizontally through multiple, seismically identified sweet spots in the Chalk, thus minimizing drilling costs and making it cost-effective to exploit smaller and smaller oil and gas deposits. Again production rose. After another decline, a new revival came after 2013 as operators began not only to drill horizontal wells but also to use more advanced, multistage fracking techniques to make new fractures in relatively porous intervals in the Chalk.[36]

## SHALE AND SLICKWATER

Natural gas or oil from shale was the oil fraternity's Fata Morgana, luring more than one otherwise sensible operator onto (or into) the rocks of financial ruin. No doubt gas and oil were really there, somewhere. Getting it out was the problem. Shales are sedimentary rock composed mostly of clays with traces of other minerals, and typically are laid down in primordial lakes, lagoons, river deltas, or other slow-moving water. The particles that form shales are finer than sand grains, which precipitate out first; in undisturbed sedimentary sequences shales therefore usually overlie sandstones. Shales constitute 75 percent of the sedimentary rock in the earth's crust, but contain 95 percent of all the organic matter, which still amounts to less than one percent by mass in the average shale. These shales are thought to be the source rock for the hydrocarbons found in conventional porous formations. But the shale itself notoriously lacks both porosity and, especially, permeability. Except on geological time scales, what happens in shale stays in shale.[37]

One of the oil operators who fell under shale's spell was George Phydias Mitchell. Born to Greek immigrant parents in Galveston in 1919, Mitchell earned a petroleum engineering degree (with emphasis in geology) and graduated as valedictorian of his class from Texas A&M in 1940. Like so many in the oil fraternity, he came up the hard way, starting out, after service in World War II, as a wildcatter and independent oil operator. He and his brother, Johnny, initially made deals and trades in what became Houston's informal oil exchange, the Esperson Building drugstore.[38]

Mitchell made his first big score in the Boonsville Bend field in Wise County, just northwest of Fort Worth. Several majors had drilled in the area and found no oil (a little was found later) and some gas, but nothing they considered commercially promising. The most productive part of the field is at a depth of about 5,000 feet, but reservoirs are thin and discontinuous, and in tight, relatively impermeable conglomerate sandstone. Reportedly on a tip from a Chicago bookie, Mitchell acquired acreage in the field. The majors had never fracked their wells, and Mitchell reasoned that then-emerging hydrafracking technology could produce decent wells. He drilled his first Boonsville well in 1952, fracked it, got a good gas well, and continued drilling. Realizing that he had tapped into a large stratigraphic trap, with backing from Houston investors, he quickly leased 300,000 acres for three dollars an acre.[39]

Now he needed money to drill. Seeking to move aggressively into oil lending, Bank of the Southwest had just hired Harold Vance to run its oil and gas department. As noted, Vance held a petroleum engineering and geology degree from the University of California (1923) and had headed Texas A&M's Department of Petroleum Engineering from 1934 to 1953. He had taught Mitchell at A&M. Confident in his own, and in Mitchell's, judgment, Vance was willing to take a risk unusual for even a Texas bank: he lent money on the basis of electric well logs indicating the reserves behind Mitchell's wells, even though they were not currently producing. Boonsville became one of the largest natural gas fields in the United States, and the foundation for Mitchell Energy's growth. Mitchell negotiated a long-term, fixed-price contract with Natural Gas Pipeline Company of America, which supplied the Chicago area.[40]

But by the late 1970s, Mitchell Energy was running out of gas, and Mitchell faced the prospect of failing to meet his contractual obligations. Mitchell developed the same obsession with the Barnett Shale that Patillo Higgins had with the Big Hill south of Beaumont. Mitchell convinced himself, if few others, that fracking could free up the gas in the Barnett Shale, which underlay his Boonsville reservoir at 7,500 feet and was the source rock for its gas and oil. Mitchell drilled the first of many Barnett Shale wells in 1981. To reduce drilling costs, the first wells redrilled and deepened existing wells in the Boonsville Bend field, which were then dual-completed. These wells continued to produce revenue, although the foray into the Barnett was not commercially successful. An experimental well, drilled from the surface, would have been prohibitively expensive.[41] In Boonsville, Mitchell already had a cut-rate lab.

Orthodoxy held that minimal water should be used to frack shale, since shale's high clay content would cause it to absorb water, swell up, and close any fractures fracking might produce. Mitchell and his petroleum engineers tried multiple variations of fracking fluids and volumes through the next ten years, at a total program cost of an estimated $250 million. Most of the wells barely covered their costs, and none were especially prolific. By 1996, Mitchell Energy was close to bankruptcy.[42]

As if Santa Rita smiled, serendipity brought salvation—and an energy revolution. Despite acrimonious dissention within the firm, at Mitchell's insistence, Mitchell Energy persevered in its quest. A young Mitchell engineer, Nick Steinsberger, who had graduated from the University of Texas in 1987 with a petroleum engineering degree, was completions manager in the Boonsville Bend field and was supervising the fracking operations. Steinsberger began reducing the concentration of gelling polymer chemicals in the fracking fluid to reduce costs, but the fluid tended to lose consistency. (Fracking jobs by then cost $750,000 to $850,000, with about 40 percent going for the exotic gel solutions.) In the summer of 1996, Steinsberger noticed that a low-concentration fluid wasn't cross-linking properly, and appeared more like what he later would call a "slickwater solution." They went ahead and used it anyway and got surprisingly good results. At an industry-sponsored outing to a Texas Rangers baseball game, Steinsberger ran into a friend, Mike Mayerhofer, who worked for Union Pacific Resources (spun out of Union Pacific Railroad), who told him that his outfit was drilling and fracking the deep (8,000 to 12,000 foot), very tight Cotton Valley sandstone formation in East Texas, using a mostly water fracking fluid with minimal polymers to reduce friction. Results in the sandstone were promising.[43]

Steinsberger recalled his experience in the Barnett nearly a year earlier and resolved to deliberately try injecting a watery fracking fluid with fewer chemicals and a lower proportion of proppant at higher pressure and a higher flow rate. The "slickwater" solution flowed more easily and faster, with lower friction, and therefore reached higher pressures in the reservoir and pushed deeper into the formation for the same surface pump pressure. Initially, the new technique showed no advantage over conventional gel fracks. Pretty much as a last gambit before the whole Barnett project got shut down, Steinsberger fracked six test wells with slick-water carrying varying proportions of proppant, all introduced gradually to avoid proppant "drop out" or separation. One, S. H. Griffin No. 4, was phenomenally successful, and by 1998 was still flowing over a million cubic feet of gas per day, even after ninety days, better than any other Mitchell Energy well had ever done. Not only did slickwater produce prolific wells, but also, because it required lower quantities of expensive chemicals, reduced the cost of fracking by $75,000 to $100,000 per well.[44]

Meanwhile, Mitchell Energy had hired a geologist, Kent Bowker, who had intensively studied the Barnett at Chevron before it shut down its own shale project. Bowker had found that Mitchell's, as well as everyone else's, estimates of how much gas the Barnett Shale might hold were based upon earlier estimates from the US Department of Energy and the Gas Research Institute for geologically different Appalachian shales. Bowker's geological analysis of the Barnett, plus his "desorption" test of a core sample from a newly drilled Barnett well, indicated that the Barnett Shale contained at least four times as much gas as Mitchell Energy's highest estimates. In ordinary permeable formations, reservoir fluids exit core samples as pressure is reduced, and therefore cannot be measured directly from the core—the problem sealing downhole core barrels had been intended to solve. But shales are so impermeable that if their cores are sealed and tested promptly upon recovery, a reasonably accurate measure of their fluid contents (gas, oil, or water) is possible.[45]

It took a while to convince skeptics within Mitchell Energy, but the combination of Bowker's data and Steinberger's successful slickwater-fracked wells persuaded George Mitchell to go all-in by mid-1998. Although Mitchell Energy nearly failed in late 1998, the application of slickwater fracking to more and more of its shale wells raised their production by 2001 to 365 million cubic feet of gas a day, up 250 percent in two years. As word of Mitchell's shale gas triumph got out, more and more operators piled into the Barnett. What had been a Mitchell Energy show (in 1995, Mitchell completed seventy wells to other operators' three) by 2001 became a cast of thousands spectacular worthy of Cecil B. DeMille. With less favorable leases, the newcomers brought with them the horizontal drilling technology that had proved so successful in the Austin Chalk. During 2003, of 780 new wells in the Barnett, 130 were horizontally drilled. By 2005, horizontal wells outnumbered vertical wells, and by 2008, operators completed 2,901 horizontal wells, compared to only 183 vertical wells.[46] The new technological paradigm for exploiting shale and other tight formation gas and, soon, oil, was established: 3D seismic, horizontal drilling, and slickwater fracking. Moreover, with the development of improved LWD and directional control, multiphase perforating in horizontal wells became the norm (this technique perforates at as many as thirty specific intervals along the horizontal reach of the well to tap LWD-identified "sweet spots" in a reservoir).[47]

Persistent, determined, expensive, and sometimes desperate empirical experimentation, conducted mostly by petroleum engineers and geologists—practitioners, not academic researchers or R&D establishments—backed by entrepreneurial, risk-trophic independents, created horizontal drilling and fracking, separately and then together. These protagonists' painstaking and protracted trial-and-error experimentation stood much closer to Thomas

Edison's determined and wide-ranging search for a high-resistance filament material for his incandescent light bulb than to Archimedes's "Eureka" moment or to Isaac Newton's being conked on the head by a falling apple.[48]

Once the new paradigm was well-defined and widely shared, which didn't take long given the moral economies and dense social networks in the oil fraternity and in petroleum engineering, shale gas and oil booms replicated across the county. Most of the action occurred in areas with previous production from conventional sandstone or limestone formations, which were already familiar to the oil fraternity. Among the more spectacular plays were Fayetteville, gas, northern Arkansas, 2004, which was especially critical, since it demonstrated that the Barnett was not some kind of freak, and that slickwater fracking and horizontal drilling would work in other shales, and Marcellus, gas, in Pennsylvania, West Virginia, and Ohio, 2005. Two thousand eight was the annus mirabilis, with not only continuing natural gas plays but also the first forays into extracting oil from shale formations by fracking, the Bakken in North Dakota and Montana, then the Eagle Ford in Texas, followed in 2010 by oldie but goodie, new oil and gas from the Permian Basin.

Each of these successor booms exhibited much the same pattern. A single or a few especially venturesome, agile, and risk-trophic individuals or smaller independent oil or gas operators would hear of Mitchell's success in the Barnett Shale or of others' success in later shale booms, intuit or find out what their technical approach was, and reason that an area they had operated in or were familiar with ought to contain shale gas or oil deposits amenable to similar technological exploitation. These entrepreneurs would begin to assemble as large a lease block as they could and drill an exploratory well or wells. More often than not the increased leasing activity, obvious from deed record filings if not directly observed by scouts, would set off a leasing boom, driving lease costs sharply higher. What had been learnt in the Austin Chalk or the Barnett Shale could not be simply "applied." Given the extent of shale deposits across the country, and the variety of the geological conditions in which they existed—depth, thickness, overlying and underlying strata, specific chemical composition—each succeeding play presented unique challenges that required technological adaptations and innovations, especially in fracking fluids and procedures.

## FINANCING THE STONE

As disruptive technologies habitually do, the advent of slickwater fracking and horizontal drilling, and the big-time oil and gas plays they animated, both exploited and induced innovations in oil and gas financing. After 1970, the oil fraternity in Texas was buffeted by recurrent oil and gas price fluctuations, some extreme, and recurrent financial crises, some partly of its own making, others inflicted. Changes in tax laws and banking regulations fostered innovative new forms of E&P (exploration and production) financing, several of which were sooner or later abused or subverted. All E&P financing remained founded upon increasingly accurate and comprehensive petroleum engineering analyses and upon geological and geophysical assessments of prospective oil and gas reservoirs. But the very financial innovations that expanded the pool of capital available to the oil fraternity also attenuated the information received, much less understood, by those shelling out the cash: multiple levels of intermediation inadvertently created powerful filters, which worked to everyone's disadvantage when the expected unexpected happened.

The first of the recent oil price collapses—unknown since the early 1930s—hit in 1985 when the Saudis finally tired of playing Texas Railroad Commission and restraining their own output to support prices for their economic, sectarian, ideological, and political foes. Overextended Texas banks failed. Between 1987 and 1990, 7 of 10 of the largest Texas banks collapsed, while two more were taken over by out-of-state institutions. Included in the casualty list were banks that had been the mainstays of the oil fraternity: First National and Republic National in Dallas, First City National, Bank of the Southwest, and National Bank of Commerce in Houston. The concurrent savings and loan (S&L) scandal only amplified the crisis. The effects for the oil fraternity were unhelpful: When Republic Bank of Midland was taken over by NCNB (North Carolina National Bank), Midland oilmen translated its new name as "No Cash for No Body."

In the tradition of hair of the dog that bit you, the oil fraternity turned to further financial innovations to bail itself out of its slough of despond. The biggest change to E&P financing, the greatest since the introduction of borrowing-base revolvers, came with the emergence of a robust futures market in oil and gas. Oil exchanges had existed before—until Standard Oil announced in 1895 that it would pay its posted prices, not any exchange set price. The new futures market appeared when a minor, slightly tawdry exchange, the New York Mercantile Exchange (Nymex), started trading heating-oil contracts in 1978. Nymex introduced oil options in 1987 and natural-gas futures in 1990. Although supplanted by the digital Intercontinental Exchange after 2000, a widely accepted, used, and quoted oil futures market all but eliminated at least short-term risks from price fluctuations in oil and gas loans. By judicious use of futures contracts, price stability for the life of a loan could be assured, which meant interest rates could be lower and loan amounts higher.

The 1990s also brought new opportunities for independents as the majors began divesting themselves of "non-core" assets and of the talented petroleum engineers and geologists who could exploit them and as new wrinkles in non-bank financing appeared. Specialty firms that assembled loan packages and aggregated capital from large institutional investors and private equity firms and investors developed "mezzanine" lending—investments that were junior to senior secured loans but ahead of ordinary equity holders. Although this financing form withered away after Enron's abuses and ultimate collapse in 2001, it served as a gateway to a panoply of new oil and gas financing arrangements.

The full price of purely market-mediated credit for the oil fraternity didn't become apparent until the 2008–2009 real-estate provoked financial crisis. Much second-tier oil and gas debt had ended up in the hands of remote third-party investors and speculators not only well beyond the pale of moral economy but also bereft of enough knowledge of the oil business to make even their self-interest effective. Like folks getting a divorce who can no longer communicate in any language other than dollars and cents, problem oil and gas investments meant endless maneuvering and bickering among various classes of debt holders and their lawyers—not helpful in a business where speed and agility matter. The days of the simple handshake deal were gone.

## MINE AND THINE REVISITED: THE RETURN OF THE RULE OF CAPTURE?

With the advent of fracking, where a well was bottomed, so clearly defined in law and measured with precision by directional surveys, was open to question. Where exactly do all of those little fractures and fissures go? Do they cross lease lines and drain adjacent tracts? Did fracking constitute trespass, and, if so, were damages presumed, so that compensation could be claimed without showing actual damages, or did damages to real property, the subsurface estate, have to be proven?

The Texas Railroad Commission has yet to issue orders directly regulating fracking—apparently out of solicitude for a critical and rapidly evolving technology. The Commission did, however, promulgate Statewide Rule 29 (R29), Hydraulic Fracturing Chemical Disclosure Requirements, January 2, 2012, which mandated that fracking contractors inform operators of all chemical substances, their concentrations, and their suppliers that were intentionally used in fracking fluids.[49] The Commission also had quickly moved to harmonize horizontal drilling with its regulation of well spacing. The Commission issued Statewide Rule 86 (R86), Horizontal Drainhole Wells, June 1, 1990,[50] which required complete and detailed reporting of well depth, orientation, and subsurface trace, as well as the number and location of "take points" (perforations) and nonperforation zones (NPZs). Later amendments provided for the adjustment of proration unit sizes and for unique field rules for fracked reservoirs.[51]

Given the Railroad Commission's reticence to trifle with hydrafracking, it was left to the Texas courts to try to define its legal limits. Although the Texas Supreme Court had explicitly extended trespass to include fracking in a pair of companion cases, *Gregg v. Delhi-Taylor Oil Corporation* and *Delhi-Taylor Oil Corporation v. Holmes* in 1961, a number of muddled and partially contradictory decisions emerged as the new fracking technology took hold across Texas.[52] For now, the definitive case in Texas is *Coastal Oil & Gas Corporation v. Garza Energy Trust* (2008).[53] From 1978, Coastal, with a well-deserved reputation as one of the sleaziest outfits in Texas, held leases on three tracts in the Vicksburg T formation in Hidalgo County from various members of the Salinas and Garza families, who themselves got into protracted title and boundary disputes that lasted more than twenty years (1978–1999) and spawned multiple lawsuits.[54]

Coastal purchased one of the tracts in fee simple in 1995, so it owed no overriding royalty on production from that tract. In 1996, Coastal drilled a well on its fee tract 467 feet from adjoining leases, shutting down another producing well on its own land to meet Railroad Commission spacing regulations. Garza Energy Trust sued, claiming that Coastal's fracking so close to the property line trespassed on its property and inflicted damage from drainage. (Coastal had repudiated, or explicitly abandoned, its Garza lease, and therefore was no longer its lessee or operator.) While parties agreed that the hydraulic length of the fractures ("the distance the fracking fluid will travel" from the well) and the propped length (the distance proppant traveled) substantially exceeded 467 feet, their expert witnesses sharply disagreed about the effective length and direction of the fractures and about what volumes they would actually drain. The Court found itself wholly unable to scientifically or technically resolve these issues.[55] In a fractured 6 to 3 decision, the Court ruled that whatever drainage may have occurred was protected by the rule of capture, that no damages could be claimed, and that since no damages could be claimed, no trespass had occurred either.

The dissenting opinion, penned by Associate Justice Phil Johnson, was more focused. Johnson maintained, quoting *Peterson v. Grayce Oil Company* (1931), that the rule of capture applies only "whenever such flow occurs solely through the operation of natural agencies in a normal manner, as distinguished from artificial means applied to stimulate such a flow."[56] Since there is no evidence that oil or natural gas will flow through relatively impermeable tight sandstone or shale formations other than on geological time scales, the rule of capture was inapposite for drainage artificially induced by fracking.[57]

For the time being, in Texas, the rule of capture, which for the sixty years prior to 1961 had been the sword, shield, and, often, license to pilfer, for small-tract owners and operators, was now resurrected and stalks the land like a zombie. Five years after *Garza*, the Federal District Court for the Northern District of West Virginia, Wheeling, in *Stone v. Chesapeake Appalachia* (2013) found that plaintiffs had reasonable grounds to claim trespass and damages from Chesapeake's fracking operations in the Marcellus Shale. The Court explicitly rejected the reasoning of the majority opinion in *Garza*, which Chesapeake had cited, but looked favorably upon Johnson's dissenting argument. In terms that likely would have pleased the Oklahoma Supreme Court's C. J. Riley as well as Joseph C. Hutcheson Jr, circa 1931, and would have been familiar to A. W. Pollard or A. A. Nance, Federal District Judge John Preston Bailey wrote: "The *Garza* opinion gives oil and gas operators a blank check to steal from the small landowner. Under such a rule, the companies may tell a small landowner that either they sign a lease on the company's terms or the company will just hydraulically fracture under the property and take the oil and gas without compensation. In the alternative, a company may just take the gas without even contacting small landowners."[58] Memories are long in the country where Mr. Peabody's coal train done hauled it away.

Just as disputes about the rule of capture and the determination of what exactly a lessor held title to under his land and was entitled to produce vexed federal and Texas courts between 1880 and 1961, the legal status of fracking is likely to remain unsettled for some time. Where induced fractures go and what they actually drain will most likely not be resolved until 3D computer simulation models and visualization are accorded evidentiary status similar to that of heterogeneous flow theory (phase-flow) in combination with core analysis, laboratory experiment, and logging data: in Will Wilson's prophetic words from seventy years ago, until these new forms of knowledge "win the same acceptance in judicial thought that they are accorded in business and financial thinking."[59] Because the 1992 US Supreme Court case *Daubert v. Merrell Dow Pharmaceuticals, Inc.* sets a fairly high bar for the admissibility of substantive scientific evidence, this process may be protracted, although the central role simulation models play in disputes about climate change may hasten the process of legal clarification or accommodation.[60]

## SCIENCE, IDEOLOGY, AND OIL

As much as the oil fraternity's world has changed and continues to change, its alienated representation, a synthesis founded in the oil fraternity's original moral economy and that of petroleum engineering science, has been stable for eighty years. One of the clearest expressions of this encompassing representation came from Everette Lee DeGolyer on an über ritual occasion, the celebration in 1951 of the Fiftieth Anniversary of the discovery well at Spindletop. DeGolyer could wear any hat in the room—geologist, geophysicist, scientist, major company CEO, university professor, consultant, entrepreneur and innovator, wildcatter, fraternal spokesman—and had excelled in all of them. Much of what DeGolyer had to say Ernest O. Thompson also preached, and would preach, in his tireless, peripatetic lectures, talks, and testimonies over the thirty years he served on the Railroad Commission of Texas. DeGolyer began his address by appropriating Anthony Lucas and his expertise as the prototypical petroleum engineer. DeGolyer predicated Lucas's unique success on an unfettered free-enterprise system with private property (including minerals), secure property rights, a spirit of entrepreneurship, enlightened regulation, all unfettered by "the hampering hand of bureaucracy and crushed under a pyramid of strict regimentation."[61] DeGolyer concluded his canticle to Lucas with, "What a marvelous example of the working of our free economy. Every man free to dream a dream and every man free to pursue it to the end! Every man free to be wrong!"[62]

Thirty years later, Michel T. Halbouty voiced the same ideas, albeit more belligerently. Similarly, oil and gas (or energy) attorneys celebrate not only the oil fraternity but also the institutional structure it fabricated. Three Baker Botts lawyers in 2015, on the occasion of the firm's 175th anniversary, lauded the scientific and technological innovations that animated the shale booms as well as the factors in Texas law that made it possible, among them, private ownership of minerals in place, the legal severance of mineral and surface estates, the rule of capture, prevention of waste, the protection of correlative rights, consistent and uniform Railroad Commission of Texas regulation, strong property rights and limited zoning, and strong protection of intellectual property and trade secrets.[63] Sixty-five years after DeGolyer's address, in 2016, despite the myriad changes history had wrought, Bernard F. Clark Jr., in his *Oil Capital*, also recited this canonical representation. He characterized the radical transformations engineering science and technology engendered as perfect gales of Schumpeterian "creative destruction." Yet, "Through such innovation, the technology has changed . . . the spirit and drive remain constant. It is the same spirit that drove the early wildcatter to spend his—and his banker's—last dime in search of riches just waiting to be discovered."[64]

One hundred twenty years after the oil century erupted on the Big Hill south of Beaumont, everything has changed and nothing has. The moral economy of the oil fraternity in Texas perseveres, only slightly haunted by memories of the poverty, hardship, danger, swindling, waste, and pollution that were its manger mates. Scientific knowledge and technological capabilities progress in a mutually stimulating, positive feedback loop, supported by and supporting the oil fraternity in Texas. None go far without the others. Together, it's where opportunity comes from. Despite its costs—selection is neither pretty nor cheap—evolution happens, lagged by self-awareness. Schumpeterian gales continue to brew: The same attitudes and the

The oil fraternity in Texas: Edward W. Constant, Bear Creek No. 1, Matagorda County, Texas, 1975. "The usual dry hole." (Photo courtesy of Frank C. Nelms)

same conditions that made the oil fraternity in Texas have made Texas the largest producer of wind power in the United States (28 percent), and the second largest producer of solar power

(10 percent, behind only California).[65]

Half a century ago, Donald Campbell asked me a question to which I dared not venture an answer until now. I'd just returned to graduate school and was eyeball deep in "my" turbojet stuff.[66] I'd asked Campbell to supervise a research-related field in the psychology of invention. I prattled on and on about "my" turbojet inventors, and Campbell listened patiently and graciously, as was his custom. Then he asked me, "Suppose I break a pot of honey in the middle of this room, then break an ant farm at the door. Why should I care which ant finds the honey?" Random variation and selective retention. Looking at the wildcatters and innovators, roughnecks and petroleum engineering scientists who populated the oil fraternity in Texas, I think maybe the answer is, "Because it matters that there be a sufficient number of ants who believe in pots of honey." Or as my daddy used to say, in one of those metaphors in which the oil fraternity speaks, "You just never know when the top rail is going to be on the bottom, and the bottom rail is going to be on top."

# ACKNOWLEDGMENTS

This book is about the tribe I left. I am inside out. For all that, I am still a child of the oil fraternity in Texas.

Unlike Blanche DuBois, I have not so much depended on the kindness of strangers as upon the generosity and tolerance of people who had every reason to know better. From my first years in high school, I've been blessed with stimulating, encouraging, and indulgent teachers, colleagues, and friends: Louis Galambos, Frank Vandiver, Don Campbell, Bill Coleman, Carl Condit, Joel Tarr, David Miller, Don Sutton, Clark Glymour, Tom Hughes, Walter Vincenti, Erik Rau, Fred Quivik, Ron Overman, and Wiebe Bijker, among many others. Northwestern University, Carnegie Mellon University, and the Center for the Philosophy of Science at the University of Pittsburgh were places where the disciplinary silos were less jealously guarded than seems usually the case and I could wander about like a little kid in a candy store. After forty years of off-and-on fiddling with this topic, the people who have helped are legion, far too many to enumerate; many I fear I would fail to recall. Their anonymity might be my greatest gift to them. Some, nevertheless, are credited in the notes to the various chapters. I also owe a debt to the characters who people this book. For me, they have been a continuing source of engagement and entertainment.

Like a long cattle drive, the argument here covers a lot of ground, much of it already homesteaded by well-established communities: game theory, economics, social constructivism, behavioral and evolutionary economics, evolutionary biology, philosophy of science, cultural anthropology, law, petroleum engineering science, multiple varieties of history. I am expert in none. I hope these folks will forgive me my trespasses and find in my mistakes more to amuse than to offend.

The people at Texas Tech University Press have been uniformly helpful, congenial, and, most of all, patient: Travis Snyder, Christie Perlmutter, Hannah Gaskamp, and John Brock. I had no real idea how much they had to do.

As in any undertaking of this sort, I too have counted on the support, understanding, and forbearance of those closest to me, especially my wife, Susan C. Constant, and son, William S. Constant. Only they know how much they've had to put up with; I don't ask.

# NOTES

## EPIGRAPH

1. Donald T. Campbell, "Unjustified Variation and Selective Retention in Scientific Discovery," in *Studies in the Philosophy of Biology: Reduction and Related Problems*, ed. Francisco José Ayala and Theodosius Dobzhansky (Berkeley: University of California Press, 1974), 145.


## CHAPTER 1

1. Willie Nelson, "The Time of the Preacher," Lyrics © Sony/ATV Music Publishing LLC, lyrics licensed & provided by LyricFind, https://www.lyrics.com>lyric>13879608>Willie Nelson> Time of the Preacher, written by Willie Nelson.
2. For other stories in this vein, see Tasha Tudor, *The Tasha Tudor Book of Fairy Tales* (New York: Platt & Munk, 1961).
3. John Bunyan, *The Pilgrim's Progress: From This World to That Which Is to Come, Delivered Under the Similitude of a Dream* (1678; Bungay, UK: Brightly and Childs, 1817).
4. Which, of course, turned out to be a small boomtown in southwestern Arkansas.
5. June Rayfield Welch and J. Larry Nance, *The Texas Courthouse* (Dallas: G. L. A. Press, 1971), 1.
6. Edgar Eggleston Townes, interview by William A. Owens, June 26, 1952, Tape No. 5, 13, Oral History of the Texas Oil Industry (Dolph Briscoe Center for American History, University of Texas at Austin, hereinafter cited as "OH").
7. Townes interview, June 21, 1952, Tape No. 3, OH, 9.
8. Landon Hayes Collum, interview by Robert O. Stephens, January 31, 1959, Tape No. 199, OH, 4.
9. W. C. Gilbert, interview by W. A. Owens, July 22, 1953, Tape No. 116, OH, 17–18.
10. *The Texas Almanac, 1958–1959* (Dallas: Dallas Morning News, 1960): 387; Everett S. Lee, Ann Ratner, Carol P. Brainerd, and Richard Easterlin, under the direction of Simon Kuznets and Dorothy Swaine Thomas, *Population Redistribution and Economic Growth, United States, 1870–1950, l: Methodological Considerations and Reference Tables* (Philadelphia, PA: American Philosophical Society, 1957): 753. "Agricultural service income per worker" stood at $229 a year, exactly equal to the national average (755); "Non-agricultural service income per worker" stood at $597, slightly above the national average (756); "Property income per capita" was $20, only two-thirds the national average (757).
11. A conversation with Mindy and Fred Quivik about farming in eastern North Dakota and western Minnesota brought home to me just how exceptional the opportunity afforded the oil fraternity in Texas was.
12. Eugene G. Leonardon, "Logging, sampling, and testing," in American Petroleum Institute, Executive Committee on Drilling and Production Practice, Division of Production, *History of Petroleum Engineering* (Dallas: American Petroleum Institute, 1961, hereinafter cited as "*API History*"): 493–578, 500; R. E. Collom, "Prospecting and testing for oil and gas," *U.S. Bureau of Mines Bulletin 201* (1922): 52.
13. J. E. Brantly, *History of Oil Well Drilling* (Houston: Gulf Publishing Company, 1971), 216–20; Anthony Sampson, *The Seven Sisters: The Great Oil Companies and the World They Shaped* (New York: Viking, 1975): 44–48.
14. James F. Ross, interview by Dr. Mody C. Boatright, September 4, 1956, Tape No. 193, OH, 16.
15. E. P. (Matt) Matteson, interview by Mody C. Boatright, June 19, 1953, Tape No. 93, OH, 11.
16. Henrietta M. Larson and Kenneth Wiggins Porter, *History of Humble Oil and Refining Company: A Study in Industrial Growth* (New York: Harpers, 1959): 11.
17. Everette Lee DeGolyer, *Spindletop ... 1901–1951*, "Delivered at Spindletop Monument, Beaumont, Texas, January 10, 1951, in Celebration of the Fiftieth Anniversary of the Completion of the Lucas Gusher" (Dallas: DeGolyer and MacNaughton, 1951): 2–3.
18. DeGolyer, *Spindletop*, 4–5.
19. Brantly, *Oil Well Drilling*.

20. Early C. Deane, interview by W. A. Owens, July 10, 1953, Tape No. 106, OH, 21.

21. James A. Clark and Michel T. Halbouty, *Spindletop* (1952; Houston: Gulf Publishing Company, 1980), 72.

22. Clark and Halbouty, *Spindletop*, 75.

23. Clark and Halbouty, *Spindletop*, 82.

24. Clark and Halbouty, *Spindletop*, 88.

25. Clark and Halbouty, *Spindletop*, 108–9.

26. Deane interview, 21.

27. Deane interview, 92.

28. Deane interview, 77–79.

29. Michel Halbouty, September 27, 1978, quoted in Jack Donahue, *Wildcatter: The Story of Michel T. Halbouty and the Search for Oil* (Houston: Gulf Publishing Company, 1979): 251.

30. Bertram Wyatt-Brown, *Southern Honor: Ethics and Behavior in the Old South* (New York: Oxford University Press, 1982); Richard E. Nisbett and Dov Cohen, *Culture of Honor: The Psychology of Violence in the South* (Boulder, CO: Westview Press, 1996).

31. The first American case recognizing the rule of capture for wild game, or *ferae naturae*, is reputed to be *Pierson v. Post*, (1805) 3 Cal. R. 175, Am. Dec. 264 (Supreme Court of Judicature of NY). Cited in Bernard F. Clark Jr., *Oil Capital: The History of American Oil, Wildcatters, Independents, and Their Bankers* (Houston: Bernard F. Clark Jr., 2016): 56.

32. *Barnard v. Monongahela Gas Company*, 216 Pa. 362, at 65, Atl. 801 (1907), quoted in Rex G. Baker and Robert E. Hardwicke, "Conservation," in *API History*, 1119–20.

33. Constitution of the State of Texas, Article XIV, Section 7. The vicissitudes and nuances of mineral ownership, especially with regard to public lands, are discussed in Clark, *Oil Capital*, 17–39.

34. *Hail, etc. v. Reed, Etc.*, 15 Ben Monroe's Reports (Kentucky) 479, (Kentucky Supreme Court, 1854), at 490. James A. Veasey, "The Law of Oil and Gas," *Michigan Law Review* 18, no. 6 (April 1920): 451.

35. *Texas Co. v. Daugherty*, 107 Tex. 226, 176 S.W. 717 (Supreme Court of Texas, 1915).

36. *Houston &Texas Central Railroad v. East*, 98 Tex. 146, 81 S.W. 279 (Supreme Court of Texas, 1904). *Bender v. Brooks*, 103 Tex. 329 (Supreme Court of Texas, 1910). *Herman v. Thomas*, 143 S.W. 195 (Texas Court of Civil Appeals, 1911). Harry Grant Potter, III, "History and Evolution of the Rule of Capture," accessed March 12, 2019, https://www.twdb.texas.gov/publications/reports; Bruce M. Kramer, "The Nature of the Mineral Estate: A Guidebook for the Uninitiated," 2015, accessed September 16, 2017, www.cailaw.org/media/files/IEL/.../2013/TexasMineralTitle/BKramer-paper.pdf.

37. The orthodox view of oil formation is that ancient biota trapped in sedimentary layers, and subjected to heat and compression, produced oil (similar to coal formation). More recent theories, not widely credited, suggest that hydrocarbons are produced continuously in the earth's mantle, albeit slowly, migrate upward, and are trapped in sedimentary strata.

38. Jaqueline Lang Weaver, *Unitization of Oil and Gas Fields in Texas: A Study of Legislative, Administrative, and Judicial Politics* (Washington, DC: Resources for the Future, Baltimore, MD: Johns Hopkins University Press, 1986), 219–34.

39. *J. M. Guffey Petroleum Company v. Chaison Townsite Company* 48 Texas Civil Appeals 555, 107 S.W. 609, at 612 (Texas Court of Civil Appeals, February 17, 1908).

40. *Grubb v. McAfee* 109 Tex. 527, 212 S.W. 464, at 465 (Supreme Court of Texas, 1919). *Leonard v. Prater*, 18 S.W. 2d 681 (Court of Civil Appeals of Texas, 1929), cited in Clark, *Oil Capital*, 61n174. *Coastal Oil & Gas Corporation v. Garza Energy Trust* 268 S.W. 3d 1 (Supreme Court of Texas, 2008).

41. These include:
    A. Implied covenants to develop the leases.
        1. To drill an initial well.
        2. To reasonably develop the lease after production had been acquired.
    B. Implied covenants of protection.
        1. To protect against drainage.
        2. Not to depreciate the lessor's interest.
    C. Implied covenants relating to management and administration of the lease.
        1. To produce and market.
        2. To operate with reasonable care.
        3. To use successful modern methods of production and development.
        4. To seek favorable administration action.
    Dissenting opinion, Justice Johnson, joined by Chief Justice Jefferson and by Justice Medina, *Coastal Oil & Gas Corporation v. Garza Energy Trust* 268 S.W. 3d 1 (Supreme Court of Texas, 2008), note [5], citing R. Hemingway, *The Law of Oil and Gas* 8.1 (1971).

42. As Paul Davidson demonstrates, the marginal user costs deriving from an unrestrained rule of capture—the likelihood that one's oil would be drained away—swamps any positive marginal benefits accruing either from producing in such a way as to maximize ultimate physical recovery or to maximize long run revenue. Paul Davidson, "Public Policy Problems of the Domestic Crude Oil Industry," *American Economic Review* 53 (March 1963): 85–108. See also discussion of Present Value in chapter 3.

43. Recall the "wildcat banks" of the same character which blossomed after Andrew Jackson, in his most "populist" act, vetoed extension of the monopoly charter of the Second Bank of the United States in 1832. The wildcat banks' excesses are often credited with causing the Panic of 1837. Everette Lee DeGolyer Sr., "Problems of the Producing Branch of the Oil Industry" (typescript), May 24, 1939, The Papers of Everette Lee DeGolyer Sr., DeGolyer Library, Southern Methodist University, Dallas, Texas (hereafter cited as "DeGolyer Papers") Mss 60, Folder 3100, 8–11.

44. On the appropriateness of the use of present-day knowledge in historical reconstruction, David L. Hull, "In Defense of Presentism," in *The Metaphysics of Evolution* (Albany, NY: State University of New York Press, 1989), 205–20.

45. *MacMillan et al. v. Railroad Commission of Texas et al.* 51 F.2d 400 (Western District of Texas, July 28, 1931), National Archives, Federal Record Center, Fort Worth, RG 21 U.S. District Courts, Western District of Texas, Austin, Entry # 48 WO33A. Equity Case Files, EQ 390, Record of Testimony, Original Copy for Hon. Joseph C. Hutcheson, United States Circuit Judge, August 4, 1931; Extract of Testimony of E. A. Landreth, on behalf of the Railroad Commission (defendants) in first *Danciger* case, read into evidence, at 234–45, *Danciger Oil & Refining Co. v. Railroad Commission of Texas, et al.,* 49 S. W. 2d 837 (Court of Civil Appeals of Texas, 1932).

46. Texaco, Inc., "Texaco Canada announces promising tertiary recovery project in Alberta," in *Texaco Inc. Second Quarter Report, 1983* (White Plains, NY: Texaco Inc., 1983), 4.

47. Nathan Rosenberg, *Technology and American Economic Growth* (New York: Harpers, 1972), 144–45nn39–41.

48. *Amazon Petroleum Corporation, et al. v. Archie D. Ryan, et al.*, in Equity No. 652, National Archives, Federal Record Center, Fort Worth, RG 21, U.S. District Courts, Eastern District of Texas, Tyler, Entry # 48EO83A, Equity Cases EQ 652, Statement of Evidence Under Equity Rule No. 75. Filed April 2, 1934. Statement of Wirt Franklin, at 34–37; Statement of E. B. Reeser, at 59–60.

49. As Davidson argues: "Thus, when a large new field is discovered, the number of gushers and high-producing, free-flowing wells in the industry will increase and depress the supply curve. If the demand for crude is relatively inelastic, there may be a precipitous price decline. Expectations of further price declines could then develop (i.e., $dUm/dQ < 0$) and the supply curve would be further depressed. Under such circumstances there may not be any crude price above zero which would restore equilibrium." Davidson notes that something very close to this situation occurred in East Texas in 1931. See Paul Davidson, "Public Policy Problems of the Domestic Crude Oil Industry," *American Economic Review* 53 (March 1963): 85–108.

50. Clark and Halbouty, *Spindletop*, 102, 193.

51. Townes interview, June 26, 1952, Tape No. 5, OH, 6.

52. The phrase is Dillon Anderson's, from *I and Claudie* (Boston: Little, Brown, and Company, 1951).

53. Clark and Halbouty, *Spindletop*.

54. Clark and Halbouty, *Spindletop*, 101.

55. Clark and Halbouty, *Spindletop*, 87. Bryce Finley Ryan, "A Sociological Study of the Mail Order Oil promoter and His Methods" (MA thesis, University of Texas, 1933). Ryan is better known as the coauthor of the seminal article for virtually all modern diffusion of innovation studies: Bryce Ryan and Neal C. Gross (Iowa State College, Ames, Iowa), "The Diffusion of Hybrid Corn Seed in two Iowa Communities," *Rural Sociology* 8 (March 1943): 15–24, Journal Paper No. J-1092 of the Iowa Agricultural Experiment Station, Ames, Iowa. Project No. 776.

56. Clark and Halbouty, *Spindletop*, 132, 190–91.

57. Clark and Halbouty, *Spindletop*, 118–19.

58. When we dance together my world's in disguise
It's a fairyland tale that come true
And when you look at me with those stars in your eyes
I could waltz across Texas with you
—Ernest Tubb, "Waltz Across Texas"

59. Duncan J. Watts, *Small Worlds: The Dynamics of Networks between Order and Randomness* (Princeton, NJ: Princeton Studies in Complexity, Princeton University Press, 1999).

60. E. P. (Matt) Matteson, interview by Mody C. Boatright, June 19, 1953, Tape No. 93, OH, 13–14.

61. Walter Cline, interview by Mody C. Boatright, August 13, 1952, Tape No. 47, OH, at 23–24.

62. David L. Hull, *Science as a Process: An Evolutionary Account of the Social and Conceptual Development of Science* (Chicago: University of Chicago Press, 1988); David L. Hull, "A Mechanism and Its Metaphysics: An

Evolutionary Account of the Social and Conceptual Development of Science," *Biology and Philosophy* 3 (April 1988): 153–55.

63. Robert Boyd and Peter J. Richerson, *Culture and the Evolutionary Process* (Chicago: University of Chicago Press, 1985). Noemie Lamon, Christof Neumann, Thibaud Gruber, and Klaus Zuberbühler, "Kin-Based Cultural Transmission of Tool Use in Wild Chimpanzees," *Science Advances* 3 (26 April 2017): e1602750.

64. Stephen J. Shennan, Enrico R. Crema, and Tim Kerig, "Isolation-by-Distance, Homophily, and 'Core' vs. 'Package' Cultural Evolution Models in Neolithic Europe," *Evolution and Human Behavior* 3, no. 2 (2015): 103–19.

65. *Texas Almanac, 1941–1942* (Dallas: Dallas Morning News, 1943), 230–31.

66. Robert E. Hardwicke, "Conservation of Oil and Gas: Texas, 1938–1948," in American Bar Association, Section of Mineral Law, *Conservation of Oil and Gas: A Legal History, 1948,* ed. Blakely M. Murphy (Chicago: American Bar Association, 1949), 448–49.

67. Robert E. Sullivan, ed., *Conservation of Oil and Gas, A Legal History, 1958* (Chicago: American Bar Association, 1960), 218–19.

68. The best, and very nearly the only, source explicitly on the social origins of the oil fraternity is Roger M. and Diana Davids Olien, *Oil Booms: Social Change in Five Texas Towns* (Lincoln: University of Nebraska Press, 1982).

69. E. I. Thompson, interview by Dr. Boatright, September 3, 1952, Tape No. 67, OH, 2.

70. Again, the Oliens' work is the best available, but far from complete.

71. Terry G. Jordan, *German Seed in Texas Soil: Immigrant Farmers in Nineteenth Century Texas* (Austin: University of Texas Press, 1966). Terry G. Jordan, *Trails to Texas: Southern Roots of Western Cattle Ranching* (Lincoln: University of Nebraska Press, 1981).

72. Jordan, *German Seed*, 10–12, 27–29.

73. Jordan, *German Seed*, 17–19, 192–94.

74. Louis Galambos, *America at Middle Age: A New History of the U.S. in the Twentieth Century* (New York: McGraw Hill, 1981). Frederick Jackson Turner, "The Significance of the Frontier in American History," (1910), in *The Frontier in American History* (New York: Henry Holt and Company, 1921), 1–38.

75. Jordan, *German Seed*, 29.

76. Robert C. McMath Jr., "Sandy Land and Hogs in the Timber: (Agri)Cultural Origins of the Farmers' Alliance in Texas," in *The Countryside in the Age of Capitalist Transformation: Essays in the Social History of Rural America,* ed. Steven Hahn and Jonathan Prude (Chapel Hill: University of North Carolina Press, 1985).

77. Jordan, *Trails to Texas*, 155–57.

78. Jordan, *Trails to Texas*, 147–49.

79. Jordan, *Trails to Texas*, 31. In Louisiana and East Texas, the best of them persist as "Catahoula dogs."

80. Jordan, *Trails to Texas*, 34.

81. The links between populations, migration, technology, language, and culture that Jordan depicts appear at much larger scales over much longer periods—worldwide since 1000 BC: Diego Comin, William Easterly, and Erick Gong, "Was the Wealth of Nations Determined in 1000 BC?" *American Economic Journal: Macroeconomics* 2, no. 3 (July 2010): 65–97; Enrico Spolaore and Romain Wacziarg, "How Deep Are the Roots of Economic Development?" *Journal of Economic Literature* 51, no. 2 (June 2013): 325–69. As does Jordan, these sources use biological ancestor-descendant lineages, now measured more precisely as genetic distance between populations, as markers for cultural transmission and descent (Spolaore and Wacziarg, 341–48).

82. Jordan, *Trails to Texas*, 153.

83. Jordan, *Trails to Texas*, 155–56, 153.

84. Lawrence Goodwyn, *The Populist Moment: A Short History of the Agrarian Revolt in America* (Oxford, UK: Oxford University Press, 1978), 25; abridged edition of *Democratic Promise: The Populist Movement in America* (Oxford: Oxford University Press, 1976), 25.

85. John Bainbridge, *The Super-Americans: A Picture of Life in the United States, as Brought into Focus, Bigger than Life, in the Land of the Millionaires—Texas* (New York: Signet/New American Library, 1961).

86. Mody C. Boatright, "The Myth of Frontier Individualism," *Southwestern Social Science Quarterly* 22 (June 1941): 14–32.

87. Steven Hahn, *The Roots of Southern Populism: Yeoman Farmers and the Transformation of the Georgia Upcountry, 1850–1890* (New York: Oxford University Press, 1983), 52; Ruth Schwartz Cowan, *A Social History of American Technology* (New York: Oxford University Press, 1997), chapter 2, especially "Conclusion: The Myth of Self-Sufficiency," 39–43.

88. Hahn, *Roots of Southern Populism*, 58.

89. Hahn, *Roots of Southern Populism*, 58; Gary Kulik, "Dams, Fish, and Farmers: Defense of Public Rights in Eighteenth-Century Rhode Island," in *The Countryside in the Age of Capitalist Transformation*, 36–38.

90. Hahn, *Roots of Southern* Populism, 270.

91. Boatright, "Myth," 20–21.

92. Boatright, "Myth," 19.

93. William S. Link, "The Social Context of Southern Progressivism, 1880–1930," in *The Wilson Era: Essays in Honor of Arthur S. Link*, ed. John Milton Cooper Jr. and Charles E. Neu (Arlington Heights, IL: Harlan Davidson, 1991), 55–82. See also the highly colored account in John Evans Eubanks, *Ben Tillman's Baby: The Dispensary System of South Carolina, 1892–1915* (Augusta, GA: self-published, 1950), 84–104.

94. Link, "Social Context," 61, 64–68.

95. Hahn, *Roots of Southern Populism*, 52.

96. Hahn, *Roots of Southern Populism*, 73.

97. Boatright, "Myth," 18–19.

98. Hahn, *Roots of Southern Populism*, 81–84.

99. Texas declared her independence from Mexico March 2, 1836.

100. J. Evetts Haley, *Charles Goodnight: Cowman and Plainsman* (1936; Norman: University of Oklahoma Press, 1949), x. Goodnight's family virtually recapitulated Jordan's depiction of the northerly version of the trail to Texas. The family originally migrated from Germany in the mid-eighteenth century, settled briefly in Pennsylvania before moving to North Carolina. From North Carolina, they trailed to Kentucky in the late 1770s, then to southern Illinois in the 1830s, and finally to the Texas frontier (Haley, *Charles Goodnight*, 3–8).

101. Haley, *Charles Goodnight*, 364.

102. Haley, *Charles Goodnight*, 365. Boatright, "Myth," 25. Haley is perhaps better remembered now for his *A Texan Looks at Lyndon: A Study in Illegitimate Power* (Canyon, TX: The Palo Duro Press, 1964), which renders Robert Caro's tedious biography of LBJ both redundant and tame.

103. Haley, *Charles Goodnight*, 68–69.

104. Hahn, *Southern Populism*, chapter 7, 239–68. McMath, "Sandy Land and Hogs in the Timber," 212. For a contrary view, together with comparative costs of different kinds of fencing and an account of the coming of barbed wire to the Great Plains in the 1870s and 1880s, see Walter Prescott Webb, *The Great Plains* (1931; Lincoln: University of Nebraska Press, 1981), chapter 7, 280–319.

105. One of the best sources on this world is Blair Pittman, *The Stories of I. C. Eason, King of the Dog People* (Denton: University of North Texas Press, 1996).

106. Bruce Palmer, *"Man Over Money:" The Southern Populist Critique of American Capitalism* (Chapel Hill: University of North Carolina Press, 1980), xiv, note 5.

107. McMath, "Sandy Land and Hogs in the Timber," 215.

108. Goodwyn, *The Populist Moment*, 75; on the detailed workings of the Texas Exchange, at 74–78.

109. John Stricklin Spratt, *The Road to Spindletop: Economic Change in Texas, 1875–1901* (1955; Austin, University of Texas Press, 1983), 201–3. Goodwyn ignores these internal deficiencies in the Texas Exchange.

110. Texaco, adrift from its roots as Hogg's own Texas Company, is likely the most spectacular victim of Jim Hogg's law. Not only was Texaco assessed triple damages for its anti-competitive interference in Pennzoil's purchase of Getty Oil, but the $10.5 billion it was required to post in escrow was more than it had in petty cash. Bankruptcy and acquisition by Chevron followed. John Davidson, "The Man Who Crushed Texaco," *Texas Monthly* 16, March 1988, 92.

111. T. R. Fehrenbach, *Lone Star: A History of Texas and the Texans* (1968; New York: American Legacy Press, 1983), chapter 34.

112. Bunyan, *Pilgrim's Progress*; William Makepeace Thackeray, *Vanity Fair: Pen and Pencil Sketches of English Society or Vanity Fair: A Novel without a Hero* (London: *Punch* (serial); Bradbury and Evans (novel), 1847–1848).

113. Fehrenbach, *Lone Star*, 304.

114. Norman Pollock, *The Just Polity: Populism, Law, and Human Welfare* (Urbana: University of Illinois Press, 1987), 11, quoted in Robert C. McMath Jr., "Populist Base Communities: The Evangelical Roots of Farm Protest in Texas," *Locus: An Historical Journal of Regional Perspectives on National Topics* 1 (Fall 1988): 63.

115. McMath, "Populist Base Communities," 58.

116. McMath, "Populist Base Communities," 58.

117. Deane interview, 7.

118. Larson and Porter, *History of Humble Oil*.

119. Roger M. and Diana Davids Olien, *Wildcatters: Texas Independent Oilmen* (Austin: Texas Monthly Press, 1984), 24–25.

120. John H. Wynne, interview by Mrs. Maude Ross, January 26, 1960, Tape No. 210, OH, 24–25.

121. Landon Hayes Collum, interview by Robert O. Stephens, January 31, 1959, Tape No. 199, OH, 3–11.

122. Erich W. Zimmermann, *Conservation in the Production of Petroleum: A Study in Industrial Control* (New Haven, CT: Yale University Press, 1957), 143 chart 11 (data supplied by the American Petroleum Institute).

123. Texas Railroad Commission, "Crude Oil Production and Well Counts (since 1935): History of Texas Initial Crude Oil, Annual Production, and Producing Wells," accessed January 18, 2021, includes onshore and offshore wells within Texas's tidelands (up to ten miles out).

124. Michel T. Halbouty, September 27, 1978, quoted in Donahue, *Wildcatter*, 251.

125. Ernest O. Thompson, "The Application of the Market Demand Statute to Oil Production in Texas," *Journal of Petroleum Technology* 2 (March 1950): Section 1, 10.

126. Prudence Mackintosh, "The Soul of East Texas: Reflections and recollections of life among the Shadows of the Piney Woods," *Texas Monthly* 17 (October 1989), 116–29ff; Pittman, *The Stories of I. C. Eason*; Gordon A. Baxter, "Jenny in a Barn," *Flying* 116 (August 1989): 130–31.

## CHAPTER 2

1.    E. P. Thompson, "The Moral Economy of the English Crowd in the Eighteenth Century," *Past and Present* 50 (February 1971): 78–79.

2.    Edward Hallett Carr, *What Is History?* (London: Macmillan, 1961); Max Weber, "Objectivity in Social Science and Social Policy," in *The Methodology of the Social Sciences*, eds. and trans. E. A. Shils and H. A. Finch (1904; New York: Free Press, 1949).

3.    Thompson, "Moral Economy," 78–79.

4.    James C. Scott, "Hegemony and the Peasantry," *Politics and Society* 7, no. 3 (1977): 267–96; James C. Scott, *The Moral Economy of the Peasant: Rebellion and Subsistence in Southeast Asia* (New Haven, CT: Yale University Press, 1976).

5.    Scott, "Hegemony and the Peasantry," 279.

6.    Robert E. Kohler, "Drosophila and Evolutionary Genetics: The Moral Economy of Scientific Practice," *History of Science* 29, no. 4 (1991): 335–75; Robert E. Kohler, *Lords of the Fly: Drosophila Genetics and the Experimental Life* (Chicago: University of Chicago Press, 1994); David L. Hull, *Science as a Process: An Evolutionary Account of the Social and Conceptual Development of Science* (Chicago: University of Chicago Press, 1988); David L. Hull, "A Mechanism and Its Metaphysics: An Evolutionary Account of the Social and Conceptual Development of Science," *Biology and Philosophy* 3 (April 1988): 153–55; Peter Galison, *How Experiments End* (Chicago: University of Chicago Press, 1987); Peter Galison, *Image & Logic: A Material Culture of Microphysics* (Chicago: University of Chicago Press, 1997); Joan H. Fujimura, "Crafting Science: Standardized Packages, Boundary Objects, and 'Translation,'" in Andrew Pickering, ed., *Science as Practice and Culture* (Chicago: University of Chicago Press, 1992), 168–211; Harry M. Collins, "The TEA set: Tacit Knowledge and Scientific Networks," *Science Studies* 4 (1974): 165–86; reprinted in Barry Barnes and David Edge, eds., *Science in Context: Readings in the Sociology of Science* (Cambridge, MA: MIT Press, 1982).

7.    Norbert Götz traces the etymology of "moral economy" from ancient Greek philosophy through its wider usage in the eighteenth and nineteenth centuries to Thompson's very specific meaning, thence to its reappearance in a number of guises more recently. Norbert Götz, "'Moral Economy': its Conceptual History and Analytical Prospects," *Journal of Global Ethics* 11, no. 2 (29 July 2015): 147–62. Usage here denotes a local social system of altruistic cooperation in economic endeavor: moral economies plural.
Other sources include Samuel L. Plotkin, *The Rational Peasant: The Political Economy of Rural Society in Vietnam* (Berkeley: University of California Press, 1979); Charles F. Keyes, ed., "Peasant Strategies in Asian Societies: Moral and Rational Economic Approaches, A Symposium," *Journal of Asian Studies* 42 (August 1983): 753–868, especially Paul R. Greenough, "Indulgence and Abundance as Asian Peasant Values: A Bengali Case in Point," 828–29; Daniel Little, "Collective Action and the Traditional Village (undated manuscript—I am indebted to Professor Little for sharing with me an early draft of this paper). See also Daniel Little, *Varieties of Social Explanation: An Introduction to the Philosophy of Social Science* (Boulder, CO: Westview Press, 1991), and *Understanding Peasant China: Case Studies in the Philosophy of Social Science* (New Haven, CT: Yale University Press, 1989).
Samuel Bowles, *The Moral Economy: Why Good Incentives Are No Substitute for Good Citizens* (New Haven, CT: Yale University Press, 2016) uses the term more broadly, but not inconsistently. Other usages diverge more from moral economy's original meaning: Sara L. Ackerman, Katherine Weatherford Darling, Sandra Soo-Jin Lee, Robert A. Hiatt, and Janet K. Shim, "Accounting for Complexity: Gene-Environment Interaction Research and the Moral Economy of Quantification," *Science, Technology, & Human Values* 41 (March 2016): 194–218; Lorraine Datson, "The Moral Economy of Science," *Osiris* 10 (1995): 2–24.

8.    Francis Fukuyama, *Trust: Social Virtues and the Creation of Prosperity* (New York: Free Press, 1995); Philip Scranton, "Manufacturing Diversity: Production Systems, Markets, and an American Consumer Society, 1870–1930," *Technology and Culture* 35 (July 1994): 476–505. Walter G. Vincenti (personal communication);

J. Stephen Lansing, *Perfect Order: Recognizing Complexity in Bali* (Princeton, NJ: Princeton University Press, 2006); Peter T. Leeson, *The Invisible Hook: The Hidden Economics of Pirates* (Princeton, NJ: Princeton University Press, 2009); James Wilson, Liying Yan, and Carl Wilson, "The Precursors of Governance in the Maine Lobster Fishery," *Proceedings of the National Academy of Sciences U.S.A.* 104, no. 39 (September 25, 2007): 15212–17; Janet T. Landa, "The Bioeconomics of Homogeneous Middleman Groups as Adaptive Units: Theory and Empirical Evidence Viewed from a Group Selection Framework," *Journal of Bioeconomics* 10, no. 3 (2008): 259–78; Hywel Francis, "The Law, Oral Tradition and the Mining Community," *Journal of Law and Society* 12 (Winter 1985): 267–71; Kent V. Flannery, Joyce Marcus, Robert G. Reynolds, *The Flocks of the Wamani: A Study of Llama Herders on the Puntas of Ayacucho, Peru* (San Diego: Academic Press, 1989).

9.    E. P. Thompson, "The Moral Economy Reviewed," in *Customs in Common: Studies in Traditional Popular Culture* (London: Merlin Press, 1991), 259–351.

10.   *Amazon Petroleum Corporation, et al., v. Railroad Commission of Texas, et al.*, Eastern District of Texas, Tyler District Court, in Equity 652, National Archives, Federal Record Center, Fort Worth, Agency Box no. 17, Accession no. 90-5-18, FRC 61027; Trial Transcript Vol. I, Complainants' Exhibit 56, Committee of the Whole Senate, Forty-Second Legislature, Fourth Called Session, November 4, 1932, Testimony of C. V. Terrell, 300–379, at 310; Henry LaFayette Wilgus (University of Michigan School of Law), "State Regulations Affecting Interstate Commerce," *Michigan Law Review* 8 (1910): 656–60. "Stoppage" is the power of the Railroad Commission to "order the stoppage of trains if the company does not otherwise furnish proper and adequate accommodation to a particular locality, and in such cases the order may embrace a through *interstate* train actually running, and compel it to stop at the locality named."

11.   Thomas S. Kuhn, *The Structure of Scientific Revolutions*, 2nd ed. (1962; Chicago: University of Chicago Press, 1970); Thomas S. Kuhn, "The Road Since Structure," in *PSA 1990: Proceedings of the 1990 Biennial Meeting of the Philosophy of Science Association*, eds. Arthur Fine, Mickey Forbes, and Linda Wessels (East Lansing, MI: Philosophy of Science Association, 1991), 2: 3–13.

12.   Walter Cline, interview by Mody C. Boatright, August 13, 1952, Tape 46, OH, 1.

13.   Landon Hayes Cullum, interview by Robert O. Stephens, January 31, 1959, Tape No. 199, OH, 10.

14.   Cullum interview, 11–12.

15.   Bruce M. Kramer, "The Nature of the Mineral Estate: A Guidebook for the Uninitiated," 2015, accessed September 16, 2017, www.cailaw.org/media/files/IEL/.../2013/TexasMineralTitle/BKramer-paper.pdf, 5.

16.   Kramer, "Nature of the Mineral Estate," 22–24.

17.   John H. Wynne, interview by Mrs. Maude Ross, January 26, 1960, Tape No. 210, OH, 23–24.

18.   E. C. Laster, interview by R. M. Hays, 1956, Tape No. 174, OH, 3.

19.   Olien and Olien, *Wildcatters*, 70–71.

20.   Donahue, *Wildcatter*, 92–98.

21.   Cullum interview, 12.

22.   Clark and Halbouty, *Spindletop*, 208. On the many varieties overriding interests could take, see Kramer, "Nature of the Mineral Estate," 6–8.

23.   L. L. James, interview by R. M. Hayes, February 9, 1955, Tape No. 162, OH, 14–16.

24.   See the arrangements, trades, and promises recorded to get one well on 1.5 acres in the East Texas field drilled circa 1933: Brief of Appellant (Humble Oil), filed November 19, 1933, 127–30, *Humble Oil and Refining Company v. Railroad Commission of Texas et al.* 68 S.W. 2d 622 (Court of Civil Appeals of Texas, Austin, 1934), No. 6729, No. 7996. Texas State Archives, Austin.

25.   On the many forms production payments could take, see Kramer, "Nature of the Mineral Estate," 4.

26.   W. B. Hamilton, interview by M. C. Boatright and J. W. Williams, August 16, 1952, Tape No. 57, OH, 1.

27.   Clint Wood, interview by Mody C. Boatright, August 17, 1952, Tape No. 59, OH, 8.

28.   Townes interview, 5–6.

29.   Wynne, interview, 15.

30.   Wynne, interview, 15–16.

31.   Wynne, interview, 16–17.

32.   Quoted in Joseph A. Pratt, "The Petroleum Industry in Transition: Antitrust and the Decline of Monopoly Control in Oil," *Journal of Economic History* 40 (December 1980): 821.

33.   "Miss Ima," as she became known throughout Texas, never married. Her sisters, Youra and Theyrea, are apocryphal.

34.   Miss Ima Hogg, interview by Dr. Robert C. Cotner, July 31, 1954, Tape No. 158, OH, 10.

35.   Miss Ima Hogg, interview by W. A. Owens, March 24, 1955, Tape No. 165, OH, 7.

36.   John O. King, *Joseph Stephen Cullinan: A Study in Leadership in the Texas Petroleum Industry, 1897–1937* (Nashville, TN: published for the Gulf Coast Historical Association by Vanderbilt University Press, 1970), 184–95.

37. The Papers of James Lockhart Autry, Box 16, File No. 619, Woodson Research Center, Rice University Library, Houston, Texas (hereafter cited as "Autry Papers").

38. James L. Autry to Bryan T. Barry, December 3, 1913, MSS 69, Box 11/7, The Joseph S. Cullinan Papers, Houston Metropolitan Research Center, Houston Public Library, Houston, Texas.

39. F. C. Proctor to James L. Autry, November 26, 1913, Autry Papers. Compare Amartya K. Sen, "Rational Fools: A Critique of the Behavioral Foundations of Economic Theory," *Philosophy and Public Affairs* 6, no. 4 (Summer 1977): 335.

   The "implicit collusions" that have been observed in business behavior in oligopolies seem to work on a system of mutual trust and sense of responsibility that has well-defined limits, and attempts at "universalization" of the same kind of behavior in other spheres of action may not go with it at all. There is strictly a question of business ethics which is taken to apply within a fairly limited domain.

40. The Pew family controlled Sun Oil and associated companies: August W. Giebelhaus, *Business and Government in the Oil Business: A Case Study of Sun Oil Company, 1876–1945* (Greenwich, CT: JAI Press, 1980).

41. Townes, interview, 5–6.

42. Larson and Porter, *History of Humble Oil and Refining Company*, 45.

43. Clark and Halbouty, *Spindletop*, 119.

44. Diana Davids Olien and Roger M. Olien, *Oil in Texas: The Gusher Age, 1895–1945* (Austin: University of Texas Press, 2002), 9.

45. Gary D. Libecap, "Government Support of Private Claims to Public Minerals: Western Mineral Rights," *Business History Review* 53 (Autumn 1979): 364–85.

46. F. G. Swanson, interview by R. M. Hayes, February 20, 1955, Tape No. 167, OH, 47.

47. Inequality within the oil fraternity likely enhanced its rate of learning and discovery. Hisashi Ohtsuki and Shumpei Ujiyama model a group comprising subordinate and dominant phenotypes (better and worse off), who may learn either socially (via imitation or teaching) or individually (trial and error). In groups with both types, if the costs of individual learning are not exorbitant and the payoffs sufficient, subordinates do more learning: they have more proportionally to gain. In this model, the entire group is assumed to have acquired what is learned in the next generation. Strongly egalitarian groups learn little or nothing, and in competition with more unequal groups lose. Hisashi Ohtsuki and Shumpei Ujiyama, "Impact of Social Dominance on the Evolution of Individual Learning," *Journal of Theoretical Biology* 535 (21 February 2022): 110986.

48. E. P. (Matt) Matteson, interview by Mody C. Boatright, June 19, 1953, Tape No. 93, OH, 13.

49. Walter Cline, interview by Mody C. Boatright, August 13, 1952, Tape No. 48, OH, 6–8.

50. Matteson, interview, 15.

51. Joseph M. Weaver, interview by Mody C. Boatright, August 5, 1952, Tape No. 42, OH, 12.

52. Cline, interview, 12.

53. Clint Wood, interview by Mody C. Boatright, August 17, 1952, Tape No, 59, OH, 7–8.

54. Larson and Porter, *Humble Oil*, 95, 386–87.

55. Hamilton, interview, 11–12.

56. Hamilton, interview, 10.

57. Donahue, *Wildcatter*, 6–16.

58. Cullum, interview, 4–6.

59. Donahue, *Wildcatter*, 36–39.

60. Duncan J. Watts, *Small Worlds: The Dynamics of Networks between Order and Randomness* (Princeton, NJ: Princeton University Press, 1999), chapters 2–4. See also Duncan J. Watts and Steven H. Strogatz, "Collective Dynamics of 'Small-World' Networks," *Nature* 393 (14 June 1998): 440–42; Melvin M. Webber, "Order in Diversity: Community without Propinquity," in ed. L. Wirigo, *Cities and Space* (Baltimore, MD: Johns Hopkins University Press, 1963).

61. Wynne, interview, 18–19.

62. Townes, interview, 8–9.

63. Carl Angstadt, interview by Mody C. Boatright, August 4, 1952, Tape No. 41, OH, 28.

64. Michael E. Levine and Jennifer L. Forrence, "Regulatory Capture, Public Interest, and the Public Agenda: Toward a Synthesis," *Journal of Law and Economics* 6, special issue (1990): 167–98; papers from the Organization of Political Institutions Conference, April 1990.

65. Anthony Sampson, *The Seven Sisters: The Great Oil Companies and the World They Shaped* (New York: Viking, 1975).

66. Larson and Porter, *Humble Oil*; Joseph Nocera, "Humble Pie," *Texas Monthly* 14, January 1986, 68.

67. Olien and Olien, *Wildcatters*.

68. P. W. S. Andrews, "Competition in the Modern Economy" reprinted from *Competitive Aspects of Oil Operations* (London, UK: The Institute of Petroleum, 1958), 49 (paper read at the 1958 Summer Meeting of the Institute

of Petroleum, Scarborough, UK, June 1958). In contrast, outside the United States, 99 percent of production was controlled by major integrated producer-refiners. Unlike the United States, where landowners held fee title to mineral rights, most other sovereign states reserved mineral rights to themselves and either granted concessions to one or a small number of majors or created state owned and operated oil companies.

69. Alfred D. Chandler, *The Visible Hand: The Managerial Revolution in American Business* (Cambridge, MA: Belknap Press, 1977).

70. Alfred DuPont Chandler, "Organizational Capabilities and the Economic History of the Industrial Enterprise," *Journal of Economic Perspectives* 6, no. 3 (Summer 1992): 98–99.

71. In many cases, rather too literally. Bill Porterfield, "H. L. Hunt's Long Goodbye," *Texas Monthly* 3, March 1975, 63–69ff.

72. "Eulogy of H. L. Hunt by Sidney Latham, During Final Rites, Monday, December 2, 1974," plaque in the H. L. Hunt Memorial Room, East Texas Oil Museum, Kilgore, Texas. The presence of this room, which also contains a larger than life-sized bronze statue of Hunt, is likely related to the fact that Hunt's Placid Oil Company built the museum.

# CHAPTER 3

1. T. H. White, *The Book of Merlyn* (Austin: University of Texas Press, 1977), 102.

2. For example, Alex de Mendoza, "A Mammalian DNA Methylation Landscape: A Study of 348 Species Offers Clues into the Diversity of Mammalian Life Spans," *Science* 381, no. 6658 (11 August 2023): 602–3; and Amin Haghani et al., "DNA Methylation Networks underlying Mammalian Traits," *Science* 381, no. 6658 (11 August 2023): 647.

3. $rb > c$ Relatedness, $r$, is the proportion of genes shared by benefactor and recipient. W. D. Hamilton, "The Genetical Evolution of Social Behavior I," *Journal of Theoretical Biology* 7, no. 1 (1964): 1–16 and "The Genetical Evolution of Social Behavior II," *Journal of Theoretical Biology* 7, no. 1 (1964): 17–52; Harold P. de Vladar and Eörs Szathmáry, "Beyond Hamilton's Rule: A Broader View of How Relatedness Affects the Evolution of Altruism Is Emerging," *Science* 356 (5 May 2017): 485–86.

4. Quoted in Stephen Jay Gould, "Exaptation: A Crucial Tool for an Evolutionary Psychology," *Journal of Social Issues* 47 (Fall 1991): 50.

5. $pb > c$ The probability of reciprocation is $p$. Robert L. Trivers, "The Evolution of Reciprocal Altruism," *Quarterly Review of Biology* 46, no. 4 (1971): 35–57.

6. Plotkin, *The Rational Peasant*. Charles F. Keyes, ed., "Peasant Strategies in Asian Societies: Moral and Rational Economic Approaches, A Symposium," *Journal of Asian Studies* 42 (August 1983): 753–868, especially Paul R. Greenough, "Indulgence and Abundance as Asian Peasant Values: A Bengali Case in Point," 828–49. Greenough claims that in Bengali peasant societies, famine is accompanied not by assertions of subsistence rights by the poorest and meekest members of society, but rather by fatalistic acceptance of their own and their families' starvation for the ultimate good order of society. This variability in the content of moral economies, dependent upon specific historical and cultural factors, is why moral economies are treated here as spatiotemporal particulars with no "essences." Little, "Collective Action and the Traditional Village" (I am indebted to Professor Little for sharing with me an early draft of this paper); Daniel Little, *Varieties of Social Explanation: An Introduction to the Philosophy of Social Science* (Boulder, CO: Westview Press, 1991); Little, *Understanding Peasant China: Case Studies in the Philosophy of Social Science* (New Haven, CT: Yale University Press, 1989).

7. Elliott Sober and David Sloan Wilson, *Unto Others: The Evolution and Psychology of Unselfish Behavior* (Cambridge, MA: Harvard University Press, 1998); David Sloan Wilson and Elliott Sober, "Reintroducing Group Selection to the Human Behavioral Sciences, with Open Peer Commentary," *Behavioral and Brain Sciences* 17 (December 1994): 585–654; Richard McElreath and Robert Boyd, *Mathematical Models of Social Evolution: A Guide for the Perplexed* (Chicago: University of Chicago Press, 2007).

8. Michael T. Ghiselin, *The Economy of Nature and the Evolution of Sex* (Berkeley: University of California Press, 1974), 247, quoted in Sober and Wilson, *Unto Others*, 5.

9. Peter Hammerstein, "Why Is Reciprocity so Rare in Social Animals? A Protestant Appeal," in *The Genetic and Cultural Evolution of Cooperation*, ed. Peter Hammerstein (Cambridge, MA: MIT Press, 2003), 84; Elinor Ostrom, Roy Gardner, and James Walker, *Rules, Games, and Common-Pool Resources* (Ann Arbor, MI: University of Michigan Press, 1994), 147.

10. There is some empirical support for the "folk theorem." David Cooper and John Kagel found that 52 to 55 percent of teams (of two), and 37 to 39 percent of individuals converge on cooperation in indefinitely repeated prisoner's dilemma (IRPD) experiments. David J. Cooper and John H. Kagel, "Using Team Discussions to Understand Behavior in Indefinitely Repeated Prisoner's Dilemma Games," *American Economic Journal: Microeconomics* 15, no. 4 (2023): 114–45. John H. Kagel and Peter McGee, "Team versus Individual Play in

Finitely Repeated Prisoner's Dilemma Games," *American Economic Journal: Microeconomics* 8, no. 2 (2016): 253–76. Backward induction remains controversial: see Ken Binmore, "Rationality in the Centipede," accessed March 23, 2018, https://pdfs.semanticscholar.org/3ac3/4256852473048ff66e917e6f576db7836628.pdf; Robert J. Aumann, "Backward Induction and Common Knowledge of Rationality," *Games and Economic Behavior* 8, no. 1 (1995): 6–19.

11.  Garrett Hardin, "The Tragedy of the Commons," *Science* 162 (13 December 1968): 1243–48.

12.  Elinor Ostrom, *Governing the Commons: The Evolution of Institutions for Collective Action* (Cambridge, UK: Cambridge University Press, 1990); Elinor Ostrom, "A Behavioral Approach to the Rational Choice Theory of Collective Action," *American Political Science Review* 9 (March 1998): 1–22; Elinor Ostrom, T. Dietz, N. Dolsak, et al., eds., *The Drama of the Commons* (Washington, DC: National Academy Press, 2002); Elinor Ostrom, et al., *Rules, Games, and Common-Pool Resources*; Rajiv Sethi and E. Somanathan, "The Evolution of Social Norms in Common Property Resource Use," *American Economic Review* 86, no. 4 (September 1996): 766–88; Fenton Martin, *Common-Pool Resources and Collective Action: A Bibliography*, 2 vols. (Bloomington, IN: Workshop in Political Theory and Policy Analysis, 1989–1992).

13.  Herbert Gintis, *The Bounds of Reason: Game Theory and the Unification of the Behavioral Sciences* (Princeton, NJ: Princeton University Press, 2009), 132–35; Samuel Bowles and Herbert Gintis, *A Cooperative Species: Human Reciprocity and Its Evolution* (Princeton, NJ: Princeton University Press, 2011): 89–92; Trivers, "Reciprocal Altruism"; McElreath and Boyd, *Mathematical Models of Social Evolution*, 145–66.

14.  Bowles and Gintis offer a complex simulation for the evolution of reciprocity and sanctioning that depends upon a sufficiently large "quorum" of punishers (strong reciprocators) in some groups, and upon "economies of scale in punishing" (154) or "increasing returns to scale in punishing" (158, 166). This dynamic propagates through the global population by group selection. Bowles and Gintis use this model to support a gene-culture coevolution hypothesis for the secular emergence of altruistic cooperation in our species. GSD is agnostic in these matters, simply positing a minority of conditional altruistic cooperators (one-third) in its initial population, and does not require overt punishment at all. See Bowles and Gintis, *A Cooperative Species*. Sarah Mathew, in contrast, argues that innate moralistic aggression toward non-contributors and non-punishers in primitive societies might foster cultural traditions that solve the second-order free rider problem. Sarah Mathew, "How the Second-Order Free Rider Problem Is Solved in a Small-Scale," *American Economic Review* 107, no. 5 (2017): 578–81.

15.  Dan Jones, "A WEIRD View of Human Nature Skews Psychologists' Studies," *Science* 328 (25 June 2010): 162.

16.  Gintis, *Bounds of Reason*, 66–67; Ernst Fehr and Simon Gächter, "Cooperation and Punishment in Public Goods Experiments," *American Economic Review* 90 (September 2000): 980–94; Linnda R. Caporael, Robyn M. Dawes, John M. Orbell, and Alphons J. C. van de Kragt, "Selfishness Examined: Cooperation in the Absence of Egoistic Incentives," *Behavioral and Brain Sciences* 12, no. 4 (1989): 683–739; Joseph Henrich, et al., "Markets, Religion, Community Size, and the Evolution of Fairness and Punishment," *Science* 327 (19 March 2010): 1480–84; Elinor Ostrom, "Behavioral Approach to the Rational Choice Theory of Collective Action," *American Political Science Review* 92, no. 1 (1998): 1–22; Urs Fischbacher and Simon Gächter, "Social Preferences, Beliefs, and the Dynamics of Free Riding in Public Goods Experiments," *American Economic Review* 100, no. 1 (March 2010): 541–56.

17.  Fehr and Gächter, "Cooperation and Punishment," 980–94; Ernst Fehr, Urs Fischbacher, and Simon Gächter, "Strong Reciprocity, Human Cooperation, and the Enforcement of Social Norms," *Human Nature* 13 (2002): 1–25.

18.  Fehr and Gächter, "Cooperation and Punishment," 980–94.

19.  Özgür Gürerk, Bernd Irlenbusch, and Bettina Rockenbach, "The Competitive Advantage of Sanctioning Institutions," *Science* 312 (7 April 2006): 108–11; Benedikt Herrmann, Christian Thöni, and Simon Gächter, "Antisocial Punishment Across Societies," *Science* 319 (7 March 2008): 1362–67; Gintis, *Bounds of Reason*, 67; Simon Gächter, Elke Renner, and Martin Sefton, "The Long-Run Benefits of Punishment," *Science* 322 (5 December 2008): 1510.

20.  Ostrom, et al., *Rules, Games, and Common-Pool Resources*; based upon Elinor Ostrom, Roy Gardner, and James Walker, "Covenants with and without a Sword: Self-Governance Is Possible," *American Political Science Review* 86 (June 1992): 404–17.

21.  Ostrom, et al., *Rules, Games, and Common-Pool Resources*.

22.  Jones, "A WEIRD View," 1627.

23.  Herrmann, Thöni, and Gächter., "Antisocial Punishment." Herrmann, et al, used the same payoff function as Gächter, Renner, and Sefton, except $m = 0.40$ and $n = 4$. Groups stayed together for all twenty rounds of the experiment (the "Partner" treatment), ten rounds in the no-punishment condition (SFI) and ten with punishment (SI). See Box 3.5. Public Goods Games (PGG) and Sanctioning Institutions (SI).

24.  Robert Boyd and Peter J. Richerson, *Culture and the Evolutionary Process* (Chicago: University of Chicago

Press, 1985).

25.  Simon T. Powers, Daniel J. Taylor, and Joanna J. Bryson, "Punishment Can Promote Defection in Group-Structured Populations," *Journal of Theoretical Biology* 311 (21 October 2012): 107–16.

26.  "Particular participant pool" is all participants from a specific locale.

27.  Joseph Henrich, Jean Ensminger, Richard McElreath, Abigail Barr, Clark Barrett, Alexander Bolyanatz, Juan Camilo Cardenas, Michael Gurven, Edwins Gwako, Natalie Henrich, Carolyn Lesorogol, Frank Marlowe, David Tracer, and John Ziker, et al., "Markets, Religion, Community Size, and the Evolution of Fairness and Punishment," *Science* 327, no. 5972 (19 March 2010): 1480–84.

28.  In a DG, one subject receives an endowment (sum) that he can share, in any proportion or not at all, with another subject. A rational (utility maximizing) dictator would give the other nothing. Therefore, donations to the other are considered a measure of other-regarding preferences, or altruism. A TPG is simply a DG with a third party added who also receives an endowment (half that of the dictator). The third subject has the choice, ex ante, to punish the dictator, at a cost to herself. A rational dictator would offer nothing, and a rational third party would never pay anything to sanction. A third party's choice to punish is therefore altruistic and the threshold at which she punishes indicative of her sense of fairness.

29.  Samuel Bowles and Herbert Gintis, "The Moral Economy of Communities: Structured Populations and the Evolution of Pro-Social Norms," *Evolution and Human Behavior* 19, no. 1 (1998): 3–25; Christian Cordes, Peter J. Richerson, Richard McElreath, and Pontus Strimling, "A Naturalistic Approach to the Theory of the Firm: The Role of Cooperation and Cultural Evolution," *Journal of Economic Behavior and Organization* 68, no. 1 (2008): 125–39.

30.  Max Weber, *The Protestant Ethic and the Spirit of Capitalism*, trans. Talcott Parsons (1904–1905; London: George Allen & Unwin, 1930); Jonathan F. Schulz, et al., "The Church, Intensive Kinship, and Global Psychological Variation," *Science* 366, no. 6466 (8 November 2019): 707; Summary; full article: http://dx.doi.org/10.1126/science.aau5141 *Science* 366, eaau5141 (2019). Benjamin Enke, "Kinship, Cooperation, and the Evolution of Moral Systems," *Quarterly Journal of Economics* 134, no. 2 (2019): 953–1019.

31.  Bowles, *The Moral Economy*, 145.

32.  Joel A. C. Baum and Bill McKelvey, eds., *Variations in Organization Science: In Honor of Donald T. Campbell* (Thousand Oaks, CA.: Sage, 1999); Diana Crane, *Invisible Colleges: Diffusion of Knowledge in Scientific Communities* (Chicago: University of Chicago Press, 1972).

33.  Patrick Francois, Thomas Fujiwara, and Tanguy van Ypersele, "The Origins of Human Pro-Sociality: A Test of Cultural Group Selection on Economic Data and in the Laboratory," *Scientific Advances* 4 (19 September 2018): 1–10.

34.  Carla Handley and Sarah Mathew, "Human Large-Scale Cooperation as a Product of Competition between Cultural Groups," *Nature Communications* 11 (4 February 2020): 702; Joseph Henrich and Michael Muthukrishna, "The Origins and Psychology of Human Cooperation," *Annual Review of Psychology* 72 (2021): 207–40.

35.  Ostrom, "A Behavioral Approach to the Rational Choice Theory of Collective Action," 1–22; Binmore, "Rationality in the Centipede"; Herbert A. Simon, "Rational Choice and the Structure of the Environment," *Psychological Review* 63, no. 2 (1956): 129–38; Herbert A. Simon, *Sciences of the Artificial* (Cambridge, MA: MIT Press, 1969); Herbert A. Simon, "Altruism and Economics," *Eastern Economic Journal* 18 (Winter 1992): 73–83.

36.  John Maynard Smith and George R. Price, "The Logic of Animal Conflict," *Nature* 246, no. 5427 (1973): 15–18.

37.  Karl Sigmund, *The Calculus of Selfishness*, Princeton Series in Theoretical and Computational Biology (Princeton, NJ: Princeton University Press, 2010): 4–5, 49–50.

38.  Brian Skyrms, *The Stag Hunt and the Evolution of Social Structure* (Cambridge, UK: Cambridge University Press, 2004).

39.  The games discussed so far are "normal" or "strategic form" games: each player moves or chooses simultaneously. The centipede game is an "extensive form" game in which agents move sequentially. These games typically suffer the same backward induction affliction as normal form games. See Colin F. Camerer, *Behavioral Game Theory: Experiments in Strategic Interaction* (Princeton, NJ: Princeton University Press, 2003), 375–76. I am indebted to an anonymous referee for *The Journal of Evolutionary Economics* for suggesting these alternatives.

40.  In 2012, William Press and Freeman Dyson published their discovery of "zero determinant" strategies for iterated prisoner's dilemma games. ("Zero determinant" refers to vanishing terms in a matrix.) Although it is impossible for a player to determine her own payoff independent of the other player's choice, it is mathematically possible for a sophisticated agent to set her own payoff to any multiple of her opponent's. While a remarkable discovery after sixty years of research on PD games, so far there is no evidence that human players actually pursue such strategies. See William H. Press and Freeman J. Dyson, "Iterated Prisoner's Dilemma Contains Strategies that Dominate any Evolutionary Opponent," *Proceedings of the National Academy of Sciences* 109 (26 June 2012):

10409–14; Alexander Stewart and Joshua Plotkin, "Extortion and Cooperation in the Prisoner's Dilemma," *Proceedings of the National Academy of Sciences* 109, no. 26 (26 June 2012): 10134–35; Robert Axelrod and William D. Hamilton, "The Evolution of Cooperation," *Science* 211 (27 March 1981): 1390–96; Christian Hilbe, Bin Wu, Arne Traulsen, and Martin A. Nowak, "Evolutionary Performance of Zero-Determinant Strategies in Multiplayer Games," *Journal of Theoretical Biology* 374 (7 October 2015): 115–24; L. Pan, D. Hao, Z. Rong, and T. Zhou, "Zero-Determinant Strategies in the Iterated Public Goods Game," *Scientific Reports* 5 (21 August 2015): https://doi.org/10.1038/srep13096; Christian Hilbe, Bin Wu, Arne Traulsen, and Martin A. Nowak, "Cooperation and Control in Multiplayer Social Dilemmas," *Proceedings of the National Academy of Sciences USA* 111, no. 46 (2014): 16425–30; Genki Ichinose and Naoki Masuda, "Zero-Determinant Strategies in Finitely Repeated Games," *Journal of Theoretical Biology* 428 (7 February 2018): 61–77; Azumi Mamiya and Genki Ichinose, "Strategies that Enforce Linear Payoff Relationships under Observation Errors in Repeated Prisoner's Dilemma Game," *Journal of Theoretical Biology* 477 (21 September 2019): 63–76.

41.   Armin Falk and James J. Heckman, "Lab Experiments Are a Major Source of Knowledge in the Social Sciences," *Science* 326 (23 October 2009): 535–38.

42.   Compare Herbert Gintis, Samuel Bowles, Robert Boyd, and Ernst Fehr, eds., *Moral Sentiments and Material Interests: The Foundations of Cooperation in Economic Life* (Cambridge, MA: MIT Press, 2005). "The behavioral sciences have traditionally offered two contrasting explanations of cooperation. One, favored by sociologists and anthropologists, considers the willingness to subordinate self-interest to the needs of the social group to be part of human nature. Another, favored by economists and biologists, treats cooperation as the result of the interaction of selfish agents maximizing their long-term individual material interests" (xi).

To save the phenomena, a growing minority of microeconomists and game theorists are willing to forsake *Homo œconomicus* and replace her with something like *Homo moralis*, who, under the right social circumstances, incorporates social preferences into her (still individualistic) utility functions. Ingela Alger and Jörgen Weibull, "Homo Moralis—Preference Evolution under Incomplete Information and Assortative Matching," *Econometrica* 81, no. 6 (2013): 2269–302; Valerio Capraro, Joseph Y. Halpern, and Matjaž Perc, "From Outcome-Based to Language-Based Preferences," *Journal of Economic Literature* 62, no. 1 (2024): 115–54.

43.   David Sloan Wilson and Edward O. Wilson, "Rethinking the Theoretical Foundation of Sociobiology," *Quarterly Review of Biology* 82, no. 4 (December 2007): 327–48; Michael J. Wade, David Sloan Wilson, Charles Goodnight, Doug Taylor, Yaneer Bar-Yam, Marcus A. M. de Aguiar, Blake Stacey, et al., "Multilevel and Kin Selection in a Connected World," *Nature* 463 (18 February 2010): E8–E9; Theodore C. Bergstrom, "Evolution of Social Behavior: Individual and Group Selection," *Journal of Economic Perspectives* 16, no. 2 (2002): 67–88; Paul A. Samuelson, "Altruism as a Problem Involving Group versus Individual Selection in Economics and Biology," *American Economic Review* 83, no. 2 (1993): 143–48.

44.   Robert Boyd and Peter J. Richerson, "Culture and the Evolution of Human Cooperation," *Philosophical Transactions of the Royal Society London B. Biological Sciences* 364, no. 1533 (12 November 2009): 2381–88; Robert Boyd and Peter J. Richerson, "Transmission Coupling Mechanisms: Cultural Group Selection," *Philosophical Transactions of the Royal Society London B. Biological Sciences* 355 (1 November 2010): 3787–95; J.-B. André and O. Morin, "Questioning the Cultural Evolution of Altruism," *Journal of Evolutionary Biology* 24, no. 12 (December 2011): 2531–42.

45.   Julián García and Jeroen C. J. M. van den Bergh, "Evolution of Parochial Altruism by Multilevel Selection," *Evolution and Human Behavior* 32, no. 4 (2011): 277–87; Arne Traulsen and Martin A. Nowak, "Evolution of Cooperation by Multilevel Selection," *Proceedings of the National Academy of Sciences*, 103 (18 July 2006): 10952–55; Laurent Lehmann, Laurent Keller, Stuart West, and Denis Roze, "Group Selection and Kin Selection: Two Concepts but One Process," *Proceedings of the National Academy of Sciences* 104, no. 16 (2007): 6736–39; Laurent Lehmann and Laurent Keller, "The Evolution of Cooperation and Altruism—a General Framework and a Classification of Models," *Journal of Evolutionary Biology* 19, no. 46 (2006): 1365–76; Samuel Bowles, Jung-Kyoo Choi, and Astrid Hopfensitz, "The Co-Evolution of Individual Behaviors and Social Institutions," *Journal of Theoretical Biology* 223, no. 2 (2003): 135–47; Jung-Kyoo Choi and Samuel Bowles, "The Coevolution of Parochial Altruism and War," *Science* 318, no. 5850 (2007): 636–40; Jeroen C. J. M. van den Bergh and John M. Gowdy, "A Group Selection Perspective on Economic Behavior Institutions and Organizations," *Journal of Economic Behavior & Organization* 72, no. 1 (2009): 1–20; Joseph Henrich, "Cultural Group Selection, Coevolutionary Processes and Large-Scale Cooperation," *Journal of Economic Behavior and Organization* 53, no. 1 (2004): 3–35; Herbert A. Simon, "Altruism and Economics," *American Economic Review* 83, no. 2 (May 1993): 156–61; Jeffrey A. Fletcher and Michael Doebeli, "A Simple and General Explanation for the Evolution of Altruism," *Proceedings of the Royal Society B: Biological Sciences* 276 (7 January 2009): 13–19; Charles J. Goodnight, "Multilevel Selection: The Evolution of Cooperation in Non-Kin Groups," *Population Ecology* 47, no. 1 (2005): 3–12.

46.   Group selection selects for traits that would, were the actant to leave the group, "reduce rather than enhance

fitness." See J. E. Strassmann and David C. Queller, "The Social Organism: Congresses, Parties, and Committees," *Evolution* 64 (2010): 609; Deborah E. Shelton and Richard E. Michod, "Group and Individual Selection during Evolutionary Transitions in Individuality: Meanings and Partitions," *Philosophical Transactions of the Royal Society B* 375 (27 April 2020): 20190364.

47. Sober and Wilson, *Unto Others*, 24, fig. 1.1; Wilson and Sober, "Reintroducing Group Selection," 585–654; Richard C. Lewontin, "The Units of Selection," *Annual Review of Ecology and Systematics* 11 (1970): 1–18.

48. George R. Price, "Selection and Covariance," *Nature* 277, no. 5257 (1970): 520–21; George R. Price, "Extensions of Covariance Selection Mathematics," *Annals of Human Genetics* 35 (1972): 485–90; Steven A. Frank, "George Price's Contributions to Evolutionary Genetics," *Journal of Theoretical Biology* 175 (7 August 1995): 373–88; Lee Alan Dugatkin, *The Altruism Equation: Seven Scientists Search for the Origins of Goodness* (Princeton, NJ: Princeton University Press, 2006), chapter 6, 107–14. "Hamilton's rule" for inclusive fitness (kin selection) can be directly derived from the Price equation as a case of group selection. See McElreath and Boyd, *Mathematical Models*, 82–86. The same goes for reciprocity. See Sober and Wilson, *Unto Others*, 79–86.

49. Samir Okasha and Jun Otsuka, "The Price Equation and the Causal Analysis of Evolutionary Change," *Philosophical Transactions of the Royal Society B* 375 (27 April 2020): 20190365.

50. Joseph Henrich and Robert Boyd, "On Modeling Cognition and Culture—Why Cultural Evolution Does Not Require Replication of Representations," *Journal of Cognition and Culture* 2, no. 2 (2002): 93–94.

51. Henrich and Boyd, "On Modeling Cognition and Culture," 108; Inês S. Martins, Franziska Schrodt, Shane A. Blowes, et al., "Widespread Shifts in Body Size within Populations and Assemblages," *Science* 381 no. 6662 (8 September 2023): 1067–71; Andy Gardner and Joseph P. Conlon, "Cosmological Natural Selection and the Purpose of the Universe," *Complexity* 18 (May–June 2013): 48–56.

52. While the efficacy and prevalence of group selection remains controversial, it is much more widely understood and accepted now than it was even a generation ago: See *Philosophical Transactions of the Royal Society B* 375 (27 April 2020), "Fifty Years of the Price Equation," comprising sixteen contributions; Samuel Bowles and Herbert Gintis, *A Cooperative Species: Human Reciprocity and Its Evolution* (Princeton, NJ: Princeton University Press, 2011); Bowles, *The Moral Economy*, 145; Bergstrom, "Evolution of Social Behavior," 67–88. Steven Pinker, with responses from the Reality Group, "The False Allure of Group Selection," *Edge*, June 21 2012, http://edge.org/conversation/the-false-allure-of-group-selection; Burton Simon, "Continuous-Time Models of Group Selection, and the Dynamical Insufficiency of Kin Selection Models," *Journal of Theoretical Biology* 349 (21 May 2014): 22–31; David Sloan Wilson, "A Theory of Group Selection," *Proceedings of the National Academy of Sciences* 72 (January 1975): 143–46; Wilson and Sober, "Reintroducing Group Selection"; Sober and Wilson, *Unto Others*; Samir Okasha, "Biological Altruism," in *The Stanford Encyclopedia of Philosophy*, eds. Edward N. Zalta and Uri Nodelman, 2013 ed., accessed February 24, 2016, http://plato.stanford.edu/archives/fall2013/entries/altruism-biological/.

53. Hull, *Science as a Process*; Hull, "A Mechanism and Its Metaphysics," 123–55; Thorbjørn Knudsen, "General Selection Theory and Economic Evolution: The Price Equation and the Replicator/Interactor Distinction," *Journal of Economic Methodology* 11, no. 2 (June 2004): 147–73.

54. The phrase is attributed to CSA cavalry general Nathan Bedford Forrest.

55. Oliver E. Williamson, "Transaction Cost Economics: The Governance of Contractual Relations," *The Journal of Law and Economics* 22 (October 1979): 233–61; Oliver E. Williamson, *The Mechanisms of Governance* (New York: Oxford University Press, 1996). Completion costs for a well include production casing in the well itself, perforating and cementing, separators on the surface (to separate oil from natural gas and salt water), storage tanks, any necessary pumps or compressors, and, sometimes, pipeline connection.

56. Samuel Bowles, "Endogenous Preferences: The Cultural Consequences of Markets and other Economic Institutions," *Journal of Economic Literature* 36, no. 1 (March 1998): 75–111.
If competition is pairwise and individual relative fitness in the replicator equation is calculated one-on-one, rather than relative to the average fitness in the population as in the standard Price equation, spite, defined as harming oneself to harm the other more, can spread in the population, as Morgan and Steiglitz show. John Morgan and Ken Steiglitz, "Pairwise Competition and the Replicator Equation," *Bulletin of Mathematical Biology* 65, no. 6 (2003): 1163–72. In a typical oil or gas play, competition is among all deals in the play, and this situation does not arise.
If only two firms enter a field, spite is easy to come by, as in the Eugene Island natural gas field 125 miles out in the Gulf of Mexico, south of Morgan City, Louisiana. The first firm in the field assumed the usual twenty-year exploitation and installed very large, heavy, and expensive but highly efficient Cooper-Bessemer piston compressors to raise well-head pressure to pipeline delivery pressure. The second comer was desperate for gas to meet its delivery commitments, and installed light, small, and less expensive but relatively inefficient gas turbine powered centrifugal compressors to get gas out of the common reservoir fast. The result was an unseemly sucking contest as each firm festooned its production platforms with turbine compressors to get

more gas out and better the other.

57.  Lawrence Goodwyn, *Texas Oil, American Dreams: A Study of the Texas Independent Producers and Royalty Owners Association* (Austin: Texas State Historical Association, 1996); Jack Donahue, *Wildcatter: The Story of Michel T. Halbouty and the Search for Oil* (Houston: Gulf Publishing Company, 1979). The majors too suffered bouts of being "oil poor," especially the Texas Company, Humble, and Gulf. John O. King, *Joseph Stephen Cullinan: A Study in Leadership in the Texas Petroleum Industry, 1897–1937* (Nashville, TN: published for the Texas Gulf Coast Historical Association by Vanderbilt University Press, 1970); Larson and Porter, *History of Humble Oil and Refining Company*; Clark and Halbouty, *Spindletop*.

58.  When a well blew out and caught fire, it typically took drilling rig, boilers, and all with it, leaving behind only a good-sized, smoldering crater. Hence a well, a deal, a car, or a marriage might "crater."
An opportunist pursues a strategy analogous to what MacArthur and Wilson call "*r* selection": "favoring increased reproductive rates . . . where climates are rigorously seasonal and winter survivors recolonize each spring, in the presence of a bloom of foliage and food, we expect *r* selection favoring large productivity. . . . Among other things, *r* selection may favour shorter generation time, harvesting most food (even if wastefully), and greater mobility of the young or dispersal of seed." R. H. MacArthur and E. O. Wilson, *The Theory of Island Biogeography* (Princeton, NJ: Princeton University Press, 1967), discussed in Theodosius Dobzhansky, "Chance and Creativity in Evolution," in Ayala and Dobzhansky, *Studies in the Philosophy of Biology*, 324.

59.  Bergstrom, "Evolution of Social Behavior"; Robyn M. Dawes, "Social Dilemmas," *Annual Review of Psychology* 31 (1980): 169–93; Christoph Hauert, Franziska Michor, Martin A. Nowak, and Michael Doebeli, "Synergy and the Discounting of Cooperation in Social Dilemmas," *Journal of Theoretical Biology* 239 (21 March 2006): 195–202.

60.  Gintis, *Bounds of Reason*, 230.

61.  Peter J. Richerson and Robert Boyd, "Simple Models of Complex Phenomena: The Case of Cultural Evolution," in *The Latest on the Best: Essays on Evolution and Optimality*, ed. John Dupré (Cambridge, MA: MIT Press, 1987), 27–52, reproduced in Boyd and Richerson, *Origin and Evolution of Culture*, chapter 19, 397–419. Sean Rice notes a difficulty with iterating the Price equation over multiple generations if phenotypes vary. This problem does not arise in the GSD implementation since phenotypes—*a* or *o*—are assumed to "breed true" and do not vary from generation to generation. Sean H. Rice, "Universal Rules for the Interaction of Selection and Transmission in Evolution," *Philosophical Transactions of the Royal Society B* 375 (27 April 2020): 20190353; Alan Grafen, "The Price Equation and Reproductive Value," *Philosophical Transactions of the Royal Society B* 375 (2020): 20190356; theme issue "Fifty Years of the Price Equation."

62.  Gintis, *Bounds of Reason*, 66–67; Fehr and Gächter, "Cooperation and Punishment in Public Goods Experiments"; Caporael, et al., "Selfishness Examined"; Jones, "A WEIRD View of Human Nature"; Joseph Henrich et al., "Markets, Religion, Community Size, and the Evolution of Fairness and Punishment," *Science* 327 (19 March 2010): 1480–84; Ostrom, "Behavioral Approach to the Rational Choice Theory of Collective Action"; Urs Fischbacher and Simon Gächter, "Social Preferences, Beliefs, and the Dynamics of Free Riding in Public Goods Experiments," *American Economic Review* 100, no. 1 (March 2010): 541–56; Ernst Fehr and Urs Fischbacher, "Why Social Preferences Matter—the Impact of Non-Selfish Motives on Competition, Cooperation, and Incentives," *Economic Journal* 112, no. 478 (March 2002): C1–C33.

63.  This deal size is conservative, since a larger size makes the evolution of cooperation more difficult, but it accords well both with the anecdotal evidence in the voluminous literature on the oil business and with the stories told in the Oral History of the Oil Industry. The literature on the oil business is immense and of extraordinarily variable quality, ranging from paragons of corporate history, such as Larson and Porter, *History of Humble Oil and Refining Company*, to shamelessly hagiographic biographies of individual oilmen. The several works of Roger and Diana Olien on independent oilmen in Texas and Goodwyn, *Texas Oil, American Dreams*, however, are of especially high quality. Despite their diversity, what virtually all of these sources agree upon is the nature and importance of "the deal."

64.  Sigmund, *The Calculus of Selfishness*; Fehr and Gächter, "Cooperation and Punishment in Public Goods Experiments"; Geoffrey M. Hodgson and Kainan Huang, "Evolutionary Game Theory and Evolutionary Economics: Are They Different Species?" *Journal of Evolutionary Economics* 22, no. 2 (2012): 345–66. For a spirited and persuasive defense of the use and merit of explanatory models, see Richerson and Boyd, "Simple Models of Complex Phenomena," in Dupré, ed., *The Latest on the Best*, 27–52.

65.  Williamson, "Transaction Cost"; Williamson, *Mechanisms of Governance*.

66.  Partha Dasgupta, "Dark Matters: Exploitation as Cooperation," *Journal of Theoretical Biology* 299 (21 April 2012): 180–87; Jacob Rubæk Holm, Esben Sloth Andersen, and J. Stanley Metcalfe, "Confounded, Augmented and Constrained Replicator Dynamics: Complex selection processes and their measurement," *Journal of Evolutionary Economics* 26 (October 2016): 803–22; Hodgson and Huang, "Evolutionary Game Theory." Joseph B. Kadane and Patrick D. Larkey, "Subjective Probability and the Theory of Games," *Management*

*Science* 28, no. 2 (1982): 113–20; John Harsanyi, "Subjective Probability and the Theory of Games: Comments on Kadane and Larkey's Paper," *Management Science* 28, no. 2 (1982): 120–24; Asya Magazinnik, "Game Theory. Lecture 6: Static Games of Incomplete Information," accessed June 9, 2023, https://ocw.mit.edu. I am indebted to Professor Clark Glymour for directing my attention to Bayesian games.

67.  Karen M. Page and Martin A. Nowak, "Unifying Evolutionary Dynamics," *Journal of Theoretical Biology* 219 (7 November 2002): 93–98. Pascale Gerbault, et al., "Storytelling and Story Testing in Domestication," *National Academy of Sciences USA* 111, no. 17 (2014): 6159–64.

68.  Simon, "Rational Choice and the Structure of the Environment"; Simon, *Sciences of the Artificial*.

69.  The notion of fitness used here is simple-minded. As Richard Lewontin argues, actual calculation of relative "fitnesses" of different genotypes in natural populations is complex and can depend upon the differential effects of the types on their shared environment, whether they reproduce in discrete generations, their age distributions, their densities, and so on. For the actants with the specified behavioral propensities in the environment specified in our GSD model's parameters, fitness is as defined. Richard C. Lewontin, "Four Complications in Understanding the Evolutionary Process," in *Worlds Hidden in Plain Sight: The Evolving Idea of Complexity at the Santa Fe Institute, 1984–2019*, ed. David C. Krakauer (Santa Fe, NM: Santa Fe Institute Press, 2019), 105–12; originally published in *SFI Bulletin*, Winter 2003).

70.  The size of fields conforms to an inverse power law, with the probability of a field of size $n = 1/n^2$. Within fields, the distribution of well quality depends upon the specific geological character of the subsurface pool, while individual wells are idiosyncratic to the point of perversity. Both $L$ and $b$ could be varied either stochastically or with the number of iterations after a colonization event (reflecting the typically declining price of oil as a boom developed), but the model is already messy enough.

71.  Assuming the benefit provided to each group member is $b$, the cost to the benefactor $c$, and the size of the group $n$, if $b/n < c$, the game is a prisoner's dilemma and linear self-excluding and results in strong altruism; if $b/n > c$, it is a public goods game and produces weak altruism. Bergstrom, "Evolution of Social Behavior," 67–88, 75–76. See Yuriy Pichugin, Chaitanya S. Gokhale, Julián Garcia, Arne Traulsen, and Paul B. Rainey, "Modes of Migration and Multilevel Selection in Evolutionary Multiplayer Games," *Journal of Theoretical Biology* 387 (21 December 2015): 144–53; Hilbe, et al., "Evolutionary Performance of Zero-Determinant Strategies"; Hannelore De Silva, Christoph Hauert, Arne Traulsen, and Karl Sigmund, "Freedom, Enforcement, and the Social Dilemma of Strong Altruism," *Journal of Evolutionary Economics* 20, no. 3 (2010): 203–17; David Sloan Wilson, "Structured Demes and the Evolution of Group-Advantageous Traits," *American Naturalist* 111 (January–February 1977): 157–85; David Sloan Wilson, *The Natural Selection of Populations and Communities* (Menlo Park, CA: Benjamin Cummings, 1980).

72.  Christian Cordes, Peter J. Richerson, Richard McElreath, and Pontus Strimling, "A Naturalistic Approach to the Theory of the Firm: The Role of Cooperation and Cultural Evolution," *Journal of Economic Behavior and Organization* 68, no. 1 (2008): 131.

73.  Michael J. Wade, "An Experimental Study of Group Selection," *Evolution* 31 (1 March 1977): 137–43, and figure 1, 138; Michael J. Wade, Piter Bijma, Esther D. Ellen, and William Muir, "Group Selection and Social Evolution in Domesticated Animals," *Evolutionary Applications* 3, nos. 5–6 (2010): 453–65; Sober and Wilson, *Unto Others*, 113–15; John S. Chuang, Olivier Rivoire, and Stanislas Liebler, "Simpson's Paradox in a Synthetic Microbial System," *Science* 323 (9 January 2009): 272–75.

74.  B. Morsky and C.T. Bauch, "Truncation Selection and Payoff Distributions Applied to the Replicator Equation," *Journal of Theoretical Biology* 404 (7 September 2016): 383–90. Karthik Panchanathan speculates that George Price's realization that his equation's selection for individual altruism at one level entailed a "struggle for existence" at a higher level (groups) depressed him, so he sought a mystical solution in doing personal good, and that his failure to prove the eventual universal triumph of good may have led to his suicide. Karthik Panchanathan, "George Price, the Price Equation, and Cultural Group Selection," *Evolution and Human Behavior* 32, no. 5 (2011): 368–71; Review of Oren Harman, *The Price of Altruism: George Price and the Search for the Origins of Kindness*, (New York: Norton, 2010).

75.  Bergstrom, "Evolution of Social Behavior," 67–88.

76.  Richard Dawkins, *The Selfish Gene* (Oxford, UK: Oxford University Press, 1976); Richard Dawkins, *The Extended Phenotype* (Oxford, UK: Oxford University Press, 1982); Samir Okasha, *Evolution and the Levels of Selection* (Oxford, UK: Oxford University Press, 2006); Stuart A. West and Andy Gardner, "Altruism, Spite, and Greenbeards," *Science* 327 (12 March 2010): 1341–44; Julián García, Matthijs van Veelen, and Arne Traulsen, "Evil Green Beards: Tag Recognition Can Also Be Used to Withhold Cooperation in Structured Populations," *Journal of Theoretical Biology* 360 (7 November 2014): 181–86; Marco Campenni and Gabriele Schino, "Partner Choice Promotes Cooperation: The Two Faces of Testing with Agent-Based Models," *Journal of Theoretical Biology* 344 (7 March 2014): 49–55; Andrew M. Colman, Lindsay Browning, and Briony D. Pulford, "Spontaneous Similarity Discrimination in the Evolution of Cooperation," *Journal of Theoretical*

*Biology* 299 (21 April 2012): 162–71. Donald T. Campbell, "From Evolutionary Epistemology via Selection Theory to a Sociology of Scientific Validity," *Evolution and Cognition* 3, no. 1 (1997): 5–38. Trivers, "Reciprocal Altruism."

77.     More complex population processes similar to our assortative group formation are modeled in Brian Skyrms and Robin Pemantle, "A Dynamic Model of Social Network Formation," *Proceedings of the National Academy of Sciences* 97, no. 16 (August 2000): 9340–46. The internal, intragroup (intra-deal) manner in which actants assess the "trustworthiness" (expectation of cooperation) of other group members could be modeled as Bayesian updating: Each actant adjusts his priors and partner preferences even in the face of considerable noise or uncertainty, and conditional altruistic cooperators thereby come to recognize each other and stick together. Rajeev Bhattacharya, Timothy M. Devinney, and Madan M. Pillutla, "A Formal Model of Trust Based on Outcomes," *Academy of Management Review* 23, no. 3 (July 1998): 459–72.

78.     Gintis et al., *Moral Sentiments and Material Interests*; Amy R. Poteete, Marco A. Janssen, and Elinor Ostrom, *Working Together: Collective Action, the Commons, and Multiple Methods in Practice* (Princeton, NJ: Princeton University Press, 2010); Alessandro Tavoni, Maja Schlüter, and Simon Levin, "The Survival of the Conformist: Social Pressure and Renewable Resource Management," *Journal of Theoretical Biology* 299 (21 April 2012): 152–61; Dasgupta, "Dark Matters," 180–87; Oliver P. Hauser, Martin A. Nowak, and David G. Rand, "Punishment Does Not Promote Cooperation under Exploration Dynamics When Anti-Social Punishment Is Possible," *Journal of Theoretical Biology* 360 (7 November 2014): 163–71.

79.     Watts, *Small Worlds*, chapters 2 through 4; Duncan J. Watts and Steven H. Strogatz, "Collective Dynamics of 'Small-World' Networks," *Nature* 393 (14 June 1998): 440–42.

80.     Karl Sigmund, "Moral Assessment in Indirect Reciprocity," *Journal of Theoretical Biology* 299 (21 April 2012): 25–30; Motohide Seki and Mayuko Nakamaru, "A Model for Gossip-Mediated Evolution of Altruism with Various Types of False Information by Speakers and Assessment by Listeners," *Journal of Theoretical Biology* 407 (21 October 2016): 90–105; Robert C. Ellickson, *Order Without Law: How Neighbors Settle Disputes* (Cambridge, MA: Harvard University Press, 1991).

81.     Whitney R. Cross, *The Burned-Over District: The Social and Intellectual History of Enthusiastic Religion in Western New York, 1800–1850* (Ithaca, NY: Cornell University Press, 1950).

82.     In later iterations, to the extent the population is dominated by all-altruists groups (*har* deals), this model captures features of Boyd and Richerson's "propagule" model for the spread of altruism via differential colonization by altruist-rich groups: Robert Boyd and Peter J. Richerson, "Group Selection among Alternatively Evolutionarily Stable Strategies," *Journal of Theoretical Biology* 145 (9 August 1990): 338.

83.     David Sloan Wilson, "Altruism and Organization: Disentangling the Themes of Multilevel Selection Theory," *American Naturalist* 150, supplement 1 (July 1997): S122–S134.

84.     Sober and Wilson, *Unto Others*; McElreath and Boyd, *Mathematical Models of Social Evolution*; Frank, "George Price's Contributions to Evolutionary Genetics," 373–88; Matthijs van Veelen, "On the Use of the Price Equation," *Journal of Theoretical Biology* 237 (21 December 2005): 412–26.

85.     This is Ronald A. Fisher's "fundamental theorem of natural selection": "The rate of increase in fitness of any organism at any time is equal to its genetic variance in fitness at that time." See Ronald A. Fisher, *The Genetical Theory of Natural Selection* (Oxford, UK: Clarendon Press, Oxford University, 1929), 37; quoted in Richard R. Nelson and Sidney G. Winter, *An Evolutionary Theory of Economic Change* (Cambridge, MA: Harvard University Press, 1982), 243n3; George R. Price, "Fisher's 'Fundamental Theorem' Made Clear," *Annals of Human Genetics* 36 (1972): 129–40; Esben Sloth Andersen, "Population Thinking, Price's Equation and the Analysis of Economic Evolution," *Evolutionary and Institutional Economics Review* 1, no. 1 (2004): 127–48.

86.     A model with pure types—altruists who never defect and opportunists who never cooperate—and a much smaller proportion of initial altruists (five percent) as well as fewer altruists among new entrants in colonization events (also five percent), but with otherwise unchanged parameter values, yields similar results: Altruists constitute greater than 90 percent of the population by generation 74. For our historical exemplar, conditional altruistic cooperation seems more empirically plausible than unadulterated altruism.

87.     In our model, 70 percent of participants in each new boom are from within the oil fraternity, independent of the new play's location. More realistic models might make the oil fraternity's dispersal an inverse function of distances from established production in Texas. The proportion of entrants would correspondingly rise with distance, thus decreasing the proportion of altruists in the population and, therefore, the probability of encountering them in a deal. Such a dynamic likely underlay the oil fraternity's xenophobic belief that alien lands—Oklahoma, or Kansas, or Illinois—were seedbeds of mendacity and chicanery.

88.     Albert-László Barabási, "Scale-Free Networks: A Decade and Beyond," *Science* 325 (24 July 2009): 412–13; Melanie Mitchell, *Complexity: A Guided Tour* (New York: Oxford University Press, 2009), 236–45; Karen Heyman, "Making Connections," *Science* 313 (4 August 2006): 604–6; Michael Kearns, Siddharth Suri, and Nick Montfort, "An Experimental Study of the Coloring Problem on Human Subject Networks," *Science* 313

(11 August 2006): 824–27; Albert-László Barabási and Réka Albert, "The Emergence of Scaling in Random Networks," *Science* 286 (15 October 1999): 509–12.

For a critique of power laws and their enthusiasms, see Evelyn Fox Keller, "Revisiting 'Scale-Free' Networks," *BioEssays* 27 (14 September 2005): 1060–68. For an earlier exploration of power law properties, Herbert A. Simon, "On a Class of Skew Distribution Functions," *Biometrika* 42 (December 1955): 425–40. Duncan Watts cautions against putting too much weight on such distributions without detailed knowledge of the underlying social dynamics. See Watts, *Small Worlds*.

"Preferential attachment," of course, is the famous Matthew Effect: "For whosoever hath, to him shall be given, and he shall have more abundance" (Matthew 13:12). Robert K. Merton, "The Matthew Effect in Science," *Science* 159 (5 January 1968): L 56–63; reprinted in Robert King Merton, *The Sociology of Science* (Chicago, IL: University of Chicago Press, 1973).

89.    Carol M. Rose, "Common Property, Regulatory Property, and Environmental Protection: Comparing Community-Based Management to Tradable Environmental Allowances," in Ostrom et al., eds., *The Drama of the Commons*, 234–35.

90.    Dasgupta, "Dark Matters"; Holm, Andersen, and Metcalfe, "Confounded, Augmented and Constrained Replicator Dynamics"; Howard E. Aldrich, Geoffrey M. Hodgson, David L. Hull, Thorbjørn Knudsen, Joel Mokyr, and Viktor J. Vanberg, "In Defense of Generalized Darwinism," *Journal of Evolutionary Economics* 18, no. 5 (2008); 577–96; Hull, *Science as a Process*. Hull, "A Mechanism and Its Metaphysics"; Knudsen, "General Selection Theory and Economic Evolution: The Price Equation and the Replicator/Interactor Distinction."

91.    Donald T. Campbell, "Levels of Organization, Downward Causation, and the Selection-Theory Approach to Evolutionary Epistemology," in *Theories of the Evolution of Knowing*, eds. G. Greenberg and E. Tobach (Hillsdale, NJ: Lawrence Erlbaum, 1990), 1–17; Hodgson and Huang, "Evolutionary Game Theory and Evolutionary Economics"; Page and Nowak, "Unifying Evolutionary Dynamics," 93–98; David C. Queller, "Quantitative Genetics, Inclusive Fitness, and Group Selection," *The American Naturalist* 139, no. 3 (March 1992): 540–58.

In the GSD model, altruists and opportunists synchronously and asexually "breed true," that is, produce "offspring" interests that are also phenotypic altruists or opportunists. Even though interests may defect or cooperate conditional on the behavior of other interests in a deal, if they survive truncation they revert to type in their next deal with a high probability, 0.90, while the replicator dynamics of the model eliminate opportunistic phenotypic behaviors.

92.    Okasha, *Evolution and the Levels of Selection*, 59. See also Louis Galambos, *Competition and Cooperation: The Emergence of a National Trade Association* (Baltimore, MD: Johns Hopkins University Press, 1966).

93.    Jonathan Bendor and Dilip Mookherjee, "Institutional Structure and the Logic of Ongoing Collective Action," *American Political Science Review* 81, no. 1 (March 1987): 129–54; Mancur Olson, *The Logic of Collective Action* (Cambridge, MA: Harvard University Press, 1965).

94.    Rajiv Sethi and E. Somanathan, "The Evolution of Social Norms in Common Property Resource Use," *American Economic Review* 86, no. 4 (September 1996): 782.

95.    Christian Hilbe, Arne Traulsen, Torsten Röhl, and Manfred Milinski, "Democratic Decisions Establish Stable Authorities that Overcome the Paradox of Second-Order Punishment," *Proceedings of the National Academy of Sciences of the United States of America* 111, no. 2 (2014): 752–56. In these investigators' experiments, voting for institutional sanctioning of second-order free riders followed subjects' unhappy experiences with no sanctioning institution and no sanctioning of second-order free riders. Subjects were WEIRD undergraduates.

96.    W. Brian Arthur, "Complexity Economics: Why Does Economics Need This Different Approach?" in *Complexity Economics: Dialogues of the Applied Complexity Network*, eds. W. Brian Arthur, Eric D. Beihocker, and Allison Stranger (Santa Fe, NM: Santa Fe Institute Press, 2020), 133–35; Joshua M. Epstein and Robert Axtell, *Growing Artificial Societies: Social Science from the Bottom Up* (Washington, DC: Brookings Institution Press, 1996); Robert Axtell, "Agent-Based Models for Economics," in Arthur et al., *Complexity Economics*, 155–56; Matthew G. Burgess et al., "Opportunities for Agent-Based Modelling in Human Dimensions of Fisheries," *Fish and Fisheries* 21 (2020): 570–87; Fletcher and Doebeli, "A Simple and General Explanation for the Evolution of Altruism," 13–19; Emilie Lindkvist et al., "Navigating Complexities: Agent-Based Modeling to Support Research, Governance, and Management in Small-Scale Fisheries," *Frontiers in Marine Science* 6 (17 January 2020): https://doi.org/10.3389/fmars.2019.00733; James Wilson, Liying Yan, and Carl Wilson, "The Precursors of Governance in the Maine Lobster Fishery," *Proceedings of the National Academy of Sciences U.S.A.* 104, no. 39 (September 25, 2007): 15212–17.

97.    Matthew Hutson, "Guinea Pigbots," *Science* 381, no. 6654 (14 July 2023): 121–23.

98.    Jeffrey C. Schank, Paul E. Smaldino, and Matt L. Miller, "Evolution of Fairness in the Dictator Game by Multilevel Selection," *Journal of Theoretical Biology* 382 (7 October 2015): 69.

99.    Giangiacomo Bravo, "Agents' Beliefs and the Evolution of Institutions for Common-Pool Resource

Management," *Rationality and Society* 23, no. 1 (2011): 144; Schank, et al., "Evolution of Fairness in the Dictator Game"; Maja Schlüter, Alessandro Tavoni, Simon Levin, "Robustness of Norm-Driven Cooperation in the Commons," *Proceedings of the Royal Society B: Biological Sciences* 283 (13 January 2016): 1–22; Lindkvist, et al., "Navigating Complexities"; Bowles, Choi, and Hopfensitz, "The Co-Evolution of Individual Behaviors and Social Institutions," 135–47; Emilie Lindkvist, Xavier Basurto, and Maja Schlüter, "Micro-level Explanations for Emergent Patterns of Self-Governance Arrangements in Smallscale Fisheries—A Modeling Approach," *PLoS ONE*, 12 (13 April 2017): e0175532; Tavoni, Schlüter, and Levin, "The Survival of the Conformist," 152–61; Michele L. Barnes, John Lynham, Kolter Kalberg, and PingSun Leung, "Social Networks and Environmental Outcomes," *Proceedings of the National Academy of Sciences* 113, no. 23 (2016): 6466–71; John W. Pepper, "Simple Models of Assortment through Environmental Feedback," *Artificial Life* 13, no. 1 (2007): 6–7; Takuya Iwamura, Eric F. Lambin, Kirsten M. Silvius, Jeffrey B. Luzar, and José M. V. Fragoso, "Agent-Based Modeling of Hunting and Subsistence Agriculture on Indigenous Lands: Understanding Interactions between Social and Ecological Systems," *Environmental Modelling and Software* 58 (2014): 113; Schlüter, Tavoni, and Levin, "Robustness of Norm-Driven Cooperation in the Commons," 1–22. Princeton University Press has recently released a comprehensive guide to and comparison of mathematical and agent-based models: Paul E. Smaldino, *Modeling Social Behavior: Mathematical and Agent-Based Models of Social Dynamics and Cultural Evolution* (Princeton: Princeton University Press, 2023).

100. Daniel Nettle, "Selection, Adaptation, Inheritance, and Design in Human Culture: The View from the Price Equation," *Philosophical Transactions of the Royal Society B* 375 (27 April 2020): 20190358.

101. P. Steven Sangren, *History and Magical Power in a Chinese Community* (Stanford, CA: Stanford University Press, 1987).

102. P. Steven Sangren, "History and the Rhetoric of Legitimacy: The Ma Tsu Cult of Taiwan," *Comparative Studies of Society and History* 30 (October 1988): 674–97; P. Steven Sangren, "Dialectics of Alienation: Individuals and Collectivities in Chinese Religion," *Man* 26 (March 1991): 68–69; P. Steven Sangren, "Ma Tsu's Black Face: Individuals and Collectivities in Chinese Religion," and "Multilectics of Alienation: Worship and Testimony in the Ma Tsu Pilgrimages of Taiwan" (undated manuscripts). I am indebted to my colleague Donald Sutton for sharing preliminary drafts of these last two papers with me. See also P. Steven Sangren, "Ritual as Production, Reproduction, and Transformation: The Ma Tsu Cult of Taiwan," University of Pittsburgh, National Resource Center for East Asian Studies, March 16, 1989.

103. Sangren, "Dialectics of Alienation," 67–68.

104. Sangren, *History and Magical Power*, 212.

105. Sangren, "Dialectics of Alienation," 74–75.

106. Sangren, "Dialectics of Alienation," 79. Ronald Fischer and Dimitris Xygalatas, "Extreme Rituals as Social Technologies," *Journal of Cognition and Culture* 14, no. 5 (2014): 345–55. Martin Lang, Benjamin Grant Purzycki et al., "Moralizing Gods, Impartiality and Religious Parochialism across 15 Societies," *Proceedings of the Royal Society B: Biological Sciences* 286 (6 March 2019): 1–10.

107. Karl Marx, *Capital I*, "The Fetishism of Commodities and the Secret Thereof." Émile Durkheim, *The Elementary Forms of the Religious Life* (1915; New York: Free Press, 1965), vol. 2, chapter 6, and vol. 3, chapter 4. Jürgen Habermas, *The Theory of Communicative Action*, vol. 2, *Lifeworld and System: A Critique of Functionalist Reason,* trans. Thomas McCarthy (1981; Boston, MA: Beacon Press, 1987).

108. Habermas, *Theory of Communicative Action*, 2:19–25.

109. Hull, *Science as a Process*, 513.

110. Nicolas Claidière, Thomas C. Scott-Phillips, and Dan Sperber, "How Darwinian Is Cultural Evolution?" *Philosophical Transactions of the Royal Society B* 369 (19 May 2014): 20130368; Gary L. Downey, "Reproducing Cultural Identity in Negotiating Nuclear Power: The Union of Concerned Scientists and Emergency Core Cooling," *Social Studies of Science* 18 (May 1988): 256; Gary L. Downey, "Structure and Practice in the Cultural Identities of Scientists: Negotiating Nuclear Wastes in New Mexico," *Anthropological Quarterly* 61 (January 1988): 26–38.

111. John Keegan, *The Mask of Command* (New York: Viking, 1987).

112. Thomas Woodrow Wilson, *A History of the American People*, 5 vols. (New York: Harper and Brothers, 1902).

113. Alternatively, Purzycki, et al. stress supernatural representations. Their concern, however, is the universalizing effects of moralizing, punishing gods, not the intracommunity altruistic cooperation of interest here. Benjamin Grant Purzycki, et al., "Moralistic Gods, Supernatural Punishment and the Expansion of Human Sociality," *Nature* 530 (18 February 2016): 327–330.

114. Claidière et al., "How Darwinian Is Cultural Evolution?"

115. To perform or attest to (as in an experiment) in a culturally defined public space. Steven Shapin, "The House of Experiment in Seventeenth Century England," *Isis* 79 (September 1988): 373–404; Steven Shapin, *A Social History of Truth: Civility and Science in Seventeenth Century England* (Chicago, IL: University of Chicago

Press, 1994).

116. Henrich and Boyd, "On Modeling Cognition and Culture"; Gardner and Conlon, "Cosmological Natural Selection and the Purpose of the Universe," 48–56.

117. David Sloan Wilson, "Species of Thought: A Comment on Evolutionary Epistemology," *Biology and Philosophy* 5 (January 1990): 43–44.

118. Harold Vance, "Evaluation," in *API History*, 999–1080; David T. Day, "The Petroleum Resources of the U.S.," *U.S. Geological Survey Bulletin 394*, National Conservation Commission, February 1909; J. O. Lewis and Carl H. Beal, "Some New Methods for Estimating the Future Production of Oil Wells," *Transactions of the American Institute of Mining and Metallurgical Engineers* 59 (1918): 492; Carl H. Beal, "The Decline and Ultimate Production of Oil Wells," *U.S. Bureau of Mines Bulletin 177*, 1919. U.S. Treasury Department, Internal Revenue Service, *Manual for the Oil and Gas Industry Under the Revenue Act of 1918*, 1919.

119. "Decline Curve Analysis (DCA)," PetroWiki, accessed June 8, 2018.

120. William A. Kirkland, *Old Bank–New Bank: The First National Bank, Houston, 1866–1966* (Hinsdale, IL: Dryden Press, 1974); Walter L. Buenger and Joseph A. Pratt, *But Also Good Business: Texas Commerce Banks and the Financing of Houston and Texas, 1886–1986* (College Station: Texas A&M University Press, 1986); Vance, "Evaluation."

121. For a more realistic assessment, see Vance, "Evaluation."

122. Weber, "Objectivity in Social Science."

123. Poteete, Janssen, and Ostrom, *Working Together*, 195.

124. Donald T. Campbell, "A Phenomenology of the Other One: Corrigible, Hypothetical, and Critical," in *Human Action: Conceptual and Empirical Issues*, ed. Theodore Mischel (New York: Academic Press, 1969), 41–69.

125. Bruce Glymour and Chris French provide an analytical defense of causal group selection (Bruce Glymour and Chris French, "Causation, Equivalence, and Group-Selection," 15–19, Appendix 33–35; Draft, 2008; downloaded 11/16/22 from Bruce Glymour's web page). "Their Model 8 is structurally, dynamically, and causally nearly identical to our GSD model, and Glymour and French show that such a model 'predicts group reproductive success, type reproductive success, and next generation type frequencies without error.'" Jake M. Hofman, Amit Sharma, and Duncan J. Watts, "Prediction and Explanation in Social Systems," *Science* 355 (3 February 2017): 486–88.

126. Sigmund, *The Calculus of Selfishness*, 2–4.

127. Colin F. Camerer, *Behavioral Game Theory: Experiments in Strategic Interaction* (Princeton, NJ: Princeton University Press, 2003), 45.

128. Andy Gardner and Allen Grafen, "Capturing the Superorganism: A Formal Theory of Group Adaptation," *Journal of Evolutionary Biology* 22, no. 4 (2009): 660–61.
   Oil fields are finite: latecomers miss the boat, and each boom eventually peters out. This sad circumstance is formally equivalent to density dependent mortality. John Maynard Smith's original "haystack model" for the evolution of altruism imagined a species of mouse that lives in haystacks. Each season, exactly two mice drawn at random colonize each haystack. Two randomly paired altruistic mice produce altruistic offspring, which reproduce through the season. At the end of the season, the population remixes and new pairs form to colonize new haystacks next season. Since altruistic mice do better than selfish mice in each season, the proportion of altruistic mice increases in the total population. According to Theodore Bergstrom, if next season there are too many mice and not enough haystacks, the excess mice are presumed to be "consumed by predators," although that explanation is not in the original published paper. John Maynard Smith, "Group Selection and Kin Selection," *Nature* 201, no. 4924 (14 March 1964): 1145–47; Bergstrom, "Evolution of Social Behavior," 72.

129. Sober and Wilson, *Unto Others*, 35–54.

130. Sober and Wilson, *Unto Others*, 73–75. For derivation of the Price equation, see McElreath and Boyd, *Mathematical Models of Social Evolution*, 228–32; and Frank, "George Price's Contributions to Evolutionary Genetics."
   Samir Okasha claims that in some circumstances the Price equation is misleading, and that "contextual analysis" is superior. As Sober and Wilson point out, "Traits that increase the fitness of individuals without having any effect on other members of the group do not evolve by group selection," even though "these differences appear in the group component of the Price equation" (*Unto Others*, 343n6). For example, suppose that taller *individuals* were advantaged independent of context, groups varied in average height, and groups with higher average height produced more offspring. The apparent enhanced fitness of taller *groups* is the result purely of the fitness advantages of taller individuals: there are no group effects on "organismic" fitness. Okasha, *Evolution and the Levels of Selection*, 85–87, citing Elliott Sober, *The Nature of Selection: Evolutionary Theory in Philosophical Focus*, 2nd ed. (Chicago: University of Chicago Press, 1994). Samir Okasha, "Multilevel Selection and the Partitioning of Covariance: A Comparison of Three Approaches," *Evolution* 58, no. 3 (2004): 486–94.
   In our model, however, altruistic conditional cooperators not only benefit others in their group but also are

singularly disadvantaged compared to opportunists within the same group: following Sober and Wilson, altruism increases only as the result of group effects, via higher-level group selection. The conditions that concern Okasha do not arise, and his reservations regarding the Price equation are moot. See Okasha, *Evolution and the Levels of Selection*, 124n12, and 155–56; Sober and Wilson, *Unto Others*, 343n6. In any event, for equal group sizes, contextual analysis and the Price partition give equivalent results. Jussi Lehtonen, "The Price Equation and the Unity of Social Evolution Theory, *Philosophical Transactions of the Royal Society B* 375 (27 April 2020): 20190362.

Bruce Glymour criticizes both the Price equation and contextual analysis: Bruce Glymour, "Stable Models and Causal Explanation in Evolutionary Biology," *Philosophy of Science* 75 (December 2008): 571–83. Our model satisfies the causal Markov condition that exercises Glymour, namely, that "the state of the system at a [given] time render future states of the system statistically independent of past states of the system" (572n1, 580).

131. "Decline Curve Analysis (DCA)," PetroWiki, accessed 6/8/18.
132. "Decline Curve Analysis (DCA)," PetroWiki.
133. Erich W. Zimmermann, *Conservation in the Production of Petroleum: A Study in Industrial Control* (New Haven, CT: Yale University Press, 1957; published under the auspices of the Editorial Board of the Petroleum Monograph Series, established by a grant from the American Petroleum Institute), 55 Table I. Source: *Petroleum Productive Capacity, a Report of the National Petroleum Council* (Washington, DC: 1952), 64.

# CHAPTER 4

1. *H. F. Wilcox Oil & Gas Co. v. State*, 162 Okla. 89, Okl. 19 Pac. (2d) 347, at 353 (Supreme Court of Oklahoma, February 15, 1933).
2. The personal, ideological, and political affinities among conservationists are illuminated in Roger M. Olien and Diana Davids Olien, *Oil and Ideology: The Cultural Creation of the American Petroleum Industry* (Chapel Hill: University of North Carolina Press, 2000). The broader historical conservation movement is summarized in Samuel P. Hays, *Conservation and the Gospel of Efficiency: The Progressive Conservation Movement, 1890–1920* (Pittsburgh, PA: University of Pittsburgh Press, 1959); Michael Aaron Dennis, "Drilling for Dollars: The Making of US Petroleum Reserve Estimates, 1921–1925," *Social Studies of Science* 15 (May 1985): 241–65. R. H. Montgomery, *The Brimstone Game* (New York: Vanguard Press, 1940).
3. Olien and Olien, *Oil and Ideology*.
4. Bobby D. Weaver, "Doherty, Henry Latham," *The Encyclopedia of Oklahoma History and Culture*, accessed September 23, 2017, www.okhistory.org.
5. Thomas P. Hughes, "The Electrification of America: The System Builders," *Technology and Culture* 20 (January 1979): 124–61; Thomas P. Hughes, *Networks of Power: Electrification in Western Society, 1880–1930* (Baltimore, MD: Johns Hopkins University Press, 1983).
6. Thomas P. Hughes, *American Genesis: A Century of Invention and Technological Enthusiasm* (New York: Viking, 1989); Edwin T. Layton Jr., *The Revolt of the Engineers: Social Responsibility and the American Engineering Profession* (Cleveland, OH: Case Western Reserve University Press, 1971).
7. Charles A. Warner, "Sources of Men," in *API History*, 55. This pattern of focused educational innovation was later repeated by Texas Instruments' creation of the Graduate Research Center of the Southwest (1961), which became the University of Texas at Dallas (1969).
8. Norman E. Nordhauser, *The Quest for Stability: Domestic Oil Regulation, 1917–1935* (New York: Garland, 1979), 9–10. Nordhauser provides the most complete and even-handed account of Doherty's odyssey, as well as of federal oil policy disputes through 1935 (the demise of the National Industrial Recovery Act).
9. Nordhauser, *Quest for Stability*, 10; Clark Jr., *Oil Capital*, 71n206.
10. Robert (?) Briggs, *Proceedings, American Philosophical Society* 10 (1865): 136, cited in Everette Lee DeGolyer Sr., "Waste in Oil and Gas Production" (typescript), 1941, Mss 60, Box 25, Folder 2458, DeGolyer Papers, 6; L. G. Huntley, "Possible Causes of the Decline of Oil Wells and Suggested Methods of Prolonging Yield," *U.S. Bureau of Mines Technical Paper 51* (1913), 7; quoted in DeGolyer, "Waste in Oil and Gas Production," 7; William F. Murray and James O. Lewis, "Underground Wastes in Oil and Gas Fields and Methods of Prevention," *U.S. Bureau of Mines Technical Paper 130* (1916); James O. Lewis, "Methods for Increasing the Recovery from Oil Sands," *United States Bureau of Mines Bulletin 148* (1917), 7.
11. Murray and Lewis, "Underground Wastes in Oil and Gas Fields."
12. Doherty memorandum, n. d. but probably September 6, 1923, File 651, Records of the Bureau of Mines, RG 70, National Archives, Washington, DC (summarized and quoted in Nordhauser, *Quest for Stability*, 11–13).
13. Hughes, "Electrification of America"; Hughes, *Networks*.
14. The collapse of Insull's pyramid scheme during the Depression led to the Holding Company Act of 1935, under which Cities Service was forced to divest its utility holdings, which proceeded with all deliberate speed

from 1938 to 1958.

15.  Nordhauser, *Quest for Stability*, 13n21.

16.  Olien and Olien, *Oil and Ideology*; James C. Scott, *Seeing Like a State: How Certain Schemes to Improve the Human Condition Have Failed* (New Haven, CT: Yale University Press, 1998); Poteete, Janssen, and Ostrom, *Working Together*.

17.  Nelson and Winter, *Evolutionary Theory of Economic Change*; Chandler, "Organizational Capabilities and the Economic History of the Industrial Enterprise," 79–100; Chandler, *The Visible Hand*.

18.  Nicholas George Malavis, *Bless the Pure and Humble: Texas Lawyers and Oil Regulation, 1919–1936* Number Eight: Kenneth E. Montague Series in Oil and Business History (College Station: Texas A&M University Press, 1996), 19.

19.  Everette Lee DeGolyer, "Problems of the Producing Branch of the Oil Industry" (typescript), May 24, 1939, Mss 60, Box 25, Folder 3100, DeGolyer Papers, 19.

20.  Rodney P. Carlisle and August W. Giebelhaus, *Energy Center: The Federal Government in Petroleum Research, Bartlesville, 1918–1983* (Bartlesville, OK: U.S. Department of Energy, Bartlesville Energy Technology Center, CE-AC19-80BC10126, July 13, 1985), 18–19.

21.  Eric von Hippel, "Cooperation Between Rivals: Informal Know-How Trading," *Research Policy* 16, no. 6 (December 1987): 291–302.

22.  DeGolyer, "Waste in Oil and Gas Production," 30.

23.  C. W. Beecher and I. P. Parkhurst, "Effect of Dissolved Gas upon the Viscosity and Surface Tension of Crude Oil," *Petroleum Development and Technology in 1926, Transactions of the American Institute of Mining, Metallurgical, and Petroleum Engineers* (1926): 51–69. See Doherty's long letter to Hubert Work, Secretary of the Interior and chairman of the Federal Oil Commission (later FOCB), dated July 3, 1925, included in a comment by C. E. Beecher on Umpleby's 1932 paper: J. B. Umpleby (Petroleum Engineer), "Changing Concepts in the Petroleum Industry," *Transactions of the AIME* 98, no. 1 (December 1932): 45–49.

24.  Beecher and Parkhurst, "Effect of Dissolved Gas," 51.

25.  API oil gravity, Schlumberger Oil Field Glossary, accessed November 13, 2017: "A specific gravity scale developed by the American Petroleum Institute (API) for measuring the relative density of various petroleum liquids, expressed in degrees. API gravity is graduated in degrees on a hydrometer instrument and was designed so that most values would fall between 10 and 70 degrees API gravity. The arbitrary formula used to obtain this effect is: API gravity = (141.5/SG at 60 degrees Fahrenheit) - 131.5, where SG is the specific gravity of the fluid."

26.  Beecher and Parkhurst, "Effect of Dissolved Gas," 51–52.

27.  Lester C. Uren, "The 'Gas Factor' as a Measure of Oil-Production Efficiency," *Transactions of AIME* 77, no. 1 (December 1927): 156 (paper presented at the Fort Worth Meeting, Petroleum Division, October 1927).

28.  D. B. Dow and L. P. Calkin, "Solubility and Effects of Natural Gas and Air in Crude Oils," *U.S. Bureau of Mines RI 2732* (February 1926), 6.

29.  Theodore E. Swigart and C. R. Bopp, "Experiments in the Use of Backpressure on Oil Wells," *U.S. Bureau of Mines TP 322* (1924).

30.  Carl E. Reistle Jr., "Reservoir Engineering," *API History*, 822.

31.  C. E. Beecher and H. C. Fowler, "Production Techniques and Control," *API History*, 746.

32.  Reistle, "Reservoir Engineering," 812.

33.  D. B. Dow and Carl E. Reistle Jr., "Absorption of Natural Gas and Air in Crude Petroleum," *Mining and Metallurgy* 5, no. 5 (July 1924): 336–37.

34.  Beecher and Parkhurst, "Effect of Dissolved Gas," 55.

35.  Dow and Calkin, "Solubility and Effects," 2.

36.  Dow and Calkin, "Solubility and Effects," 5 and 5n6.

37.  C. Northcote Parkinson, *Parkinson's Law: The Pursuit of Progress* (London: John Murray, 1958).

38.  Robert K. Merton, "Multiple Discoveries as a Critical Research Site," in *The Sociology of Science: Theoretical and Empirical Investigations* (Chicago: University of Chicago Press, 1970).

39.  Olien and Olien, *Oil and Ideology*, 156–58.

40.  Uren, "'Gas Factor,'" 146.

41.  Uren, "'Gas Factor,'" 147.

42.  Umpleby, "Changing Concepts in the Petroleum Industry," 43.

43.  Morris Muskat, *Physical Principles of Oil Production* (New York: McGraw-Hill, 1949), 98–101. Morris Muskat and M. W. Meres, "The Flow of Heterogeneous Fluids Through Porous Media," *Journal of Applied Physics* 7 (1936): 346–63. Morris Muskat, R. D. Wyckoff, H.G. Botset, and M. W. Meres, "Flow of Gas-liquid Mixtures through Sands," *Transactions of AIME* 123 (1937): 69–96. Bruce H. Sage and William N. Lacey, *Volumetric and Phase Behavior of Hydrocarbons* (Stanford, CA: Stanford University Press, 1939). Bruce H. Sage, William N. Lacey, and Jan G. Schaafsma, "Phase Equilibria in Hydrocarbon Systems II. Methane-propane System,"

*Industrial and Engineering Chemistry* 26, no. 2 (1 February 1934): 214–17. Bruce H. Sage and William N. Lacey, "Phase Equilibria in Hydrocarbon Systems," *Industrial and Engineering Chemistry* 31, no. 12 (1 December 1939): 1497–1509.

44.  Muskat, *Physical Principles of Oil Production*; Derek J. de Solla Price, *Science since Babylon*, enlarged ed. (New Haven, CT: Yale University Press, 1975).

45.  J. B. Umpleby, "Production Engineering in 1927," *Transactions of the American Institute of Mining and Metallurgical Engineers: Petroleum Division* 77, no. 1 (December 1927): 1–18.

46.  Nordhauser, *Quest for Stability*, 22–25.

47.  Harold Carl Miller, *Function of Natural Gas in the Production of Oil: A Report of the U.S. Bureau of Mines, Department of Commerce, in Cooperation with the Division of Development and Production Engineering of the American Petroleum Institute, Based on Data Gathered and Reported by the Kansas and Oklahoma, Pacific Coast, Rocky Mountain, and Texas and Louisiana Regional Committees of the Gas Conservation Committee of the American Petroleum Institute* (New York: American Petroleum Institute, 1929): 2.

48.  Hugh S. Taylor, *The ABC's of Science in Oil Recovery* (New York: American Petroleum Institute, 1927).

49.  DeGolyer, "Waste in Oil and Gas Production," 10.

50.  Miller, *Function of Natural Gas*, 6.

51.  DeGolyer, "Concepts of Occurrence of Oil and Gas," *API History*, 31

52.  R. A. Cattell and H. C. Fowler, "Fluid-Energy Relations in Production of Petroleum and Natural Gas," in *Minerals Yearbook, 1934*, edited by O. E. Kiessling (Washington, DC: Government Printing Office, 1935): 717.

53.  Walter G. Vincenti, *What Engineers Know and How They Know It: Analytical Studies from Aeronautical History* (Baltimore, MD: Johns Hopkins University Press, 1990); Louis L. Bucciarelli, *Designing Engineers* (Cambridge, MA: MIT Press, 1994); Carliss Y. Baldwin and Kim B. Clark, *Design Rules: The Power of Modularity* (Cambridge, MA: MIT Press, 2000); John M. Ziman, *Technological Innovation as an Evolutionary Process* (New York: Cambridge University Press, 2000), 219–33; William Brian Arthur, *The Nature of Technology: What It Is and How It Evolves* (New York: The Free Press, 2009).

54.  Layton, *Revolt of the Engineers*; Edwin T. Layton Jr., "American Ideologies of Science and Engineering," *Technology and Culture* 17 (October 1976): 688–701.

55.  Umpleby, "Changing Concepts in the Petroleum Industry," 39. Gas and Related Papers for 1929 and 1930 includes 13 on unit operation.

56.  Miller, *Function of Natural Gas*, 7.

57.  Miller, *Function of Natural Gas*, 20–21, 47.

58.  Imre Lakatos, "Falsification and the Methodology of Scientific Research Programmes," in *Criticism and the Growth of Knowledge*, eds. Imre Lakatos and Alan Musgrave (Cambridge, UK: Cambridge University Press, 1970), 91–196. Examples abound: Ptolemaic astronomy and Copernicus (which also engaged the interest of the Church); the protracted squabble between Cartesians and Newtonians, which echoed again in disputes over relativity and acausal physics (quantum mechanics); the hard-won victory of uniformitarianism over catastrophism, which haunted advocates of continental drift.

59.  Umpleby, "Changing Concepts," 44.

60.  Donald MacKenzie, "The Trough of Certainty," in *Exploring Expertise: Issues and Perspectives*, eds. Robin Williams, Wendy Faulkner, and James Fleck (Basingstoke, UK: Macmillan, 1998), 325–29.

61.  Miller, *Function of Natural Gas*, 5, 72, 251, 256.

62.  Miller, *Function of Natural Gas*, 235, 240.

63.  John R. Suman, "Evolution by Companies," *API History*, 64. Theodore E. Swigart, "Handling Oil and Gas in the Field," *API History*, 908.

64.  C. E. Beecher, in "Appendix—Can Engineers Determine Extractable Acreage Content within Standard Set by Lawyers?" Earl Oliver, "Stabilizing Influences for the Petroleum Industry," presented at the meeting of the Petroleum Division, AIME, at Houston, Texas, in October,1931; Reprinted from the *Transactions of the American Institute of Mining and Metallurgical Engineers, Petroleum Division* (1932), 10.

65.  Warner, "Sources of Men," *API History*, 40, 59.

66.  B. E. Lindsly, comment in Joseph B. Umpleby, "A.I.M.E. Stabilization Forum, Article 8; Reservoir Energy: Source, Ownership, Use," *Oil & Gas Journal*, July 20, 1933, from a symposium conducted by Joseph B. Umpleby at the AIME meeting at Ponca City in September 1932.

67.  William R. Childs, *The Texas Railroad Commission: Understanding Regulation in America to the Mid-Twentieth Century* (College Station: Texas A&M University Press, 2005).

68.  Olien and Olien, *Oil in Texas*, 83–88.

69.  Childs, *Texas Railroad Commission*, 154–60.

70.  Childs, *Texas Railroad Commission*, 166. Samuel D. Myres, *The Permian Basin: Petroleum Basin of the Southwest—Era of Discovery* (El Paso, TX: Permian Basin Press, 1973).

71. As Ralph Waldo Emerson famously said in *Self-Reliance* (1841), "A foolish consistency is the hobgoblin of little minds."
72. Ross S. Sterling with Ed Kilman and Don Carleton, *Ross Sterling: Texan. A Memoir by the Founder of the Humble Oil and Refining Company* (Austin: University of Texas Press, 2007), 36–41.
73. Malavis, *Bless the Pure and Humble*, 27. The oil fraternity in Texas is hardly unique in its distrust of regulatory nostrums: Nathan James Bennett and Philip Dearden, "Why Local People Do Not Support Conservation: Community Perceptions of Marine Protected Area Livelihood Impacts, Governance and Management in Thailand," *Marine Policy* 44 (February 2014): 107–16.
74. Larson and Porter, *Humble Oil and Refining Company*, 315.
75. Malavis, *Bless the Pure and Humble*, 30.
76. Malavis, *Bless the Pure and Humble*, 211n5.
77. Malavis, *Bless the Pure and Humble*, 31–32.
78. Olien and Olien, *Wildcatters*, 46–47.
79. Suman, "Evolution by Companies," 116. Malavis, *Bless the Pure and Humble*, 39.
80. Olien and Olien, *Wildcatters*, 47–48. Malavis, *Bless the Pure and Humble*, 39.
81. Weaver, *Unitization of Oil and Gas Fields in Texas*, 46.
82. Hines Baker, quoted in Malavis, *Bless the Pure and Humble*, 26.
83. Olien and Olien, *Wildcatters*, 48–51. Childs, *Texas Railroad Commission*, 168.
84. Malavis, *Bless the Pure and Humble*, 42, 216n39.
85. Malavis, *Bless the Pure and Humble*, 41–43.
86. Railroad Commission of the State of Texas, *39th Annual Report of the Railroad Commission of Texas for the Year 1930* (Austin, 1930), 1.
87. *Ohio Oil Co. v. Indiana* 177 U.S. 190 (US Supreme Court, April 9, 1900).
88. *Stephens County, et al. v. Mid-Kansas Oil and Gas Co.* 254 S.W. 290 (Supreme Court of Texas, June 30, 1923).
89. Childs, *Texas Railroad Commission*.
90. Weaver, *Unitization of Oil and Gas Fields in Texas*, 43, 387n32; R. D. Parker, affidavit, *Champlin Refining Co. v. Corporation Commission of Oklahoma, et al.*, No. 1156 in Equity, U.S. District Court for the Western District of Oklahoma, R75-B Appellant's [Plaintiff's] Statement of the Testimony, National Archives, Federal Record Center, Fort Worth, RG 21 U.S. District Courts, Western District of Oklahoma, Entry 40W11, Equity Case Files EQ 1156.
91. *Danciger Oil & Refining Company v. Railroad Commission* 49 S.W. 2d 837, at 845 (Court of Civil Appeals of Texas, March 23, 1932); Weaver, *Unitization of Oil and Gas Fields in Texas*, 43; David F. Prindle, *Petroleum Politics and the Texas Railroad Commission* (Austin: University of Texas Press, 1981), 32–33.
92. *Danciger Oil & Refining Company v. Railroad Commission* 49 S.W. 2d 837, at 845.
93. Hardwicke, "Conservation of Oil and Gas: Texas, 1938–1948," 464.
94. Joe R. Greenhill, "Practice and Procedure before the Railroad Commission," Mineral Law Section, July 3, 1953, in Langdon Papers, Texas State Archives, 1980/164-9 Folder JCL, at 1. Greenhill provides considerable detail on how and by whom different hearings were conducted in the Oil and Gas Division of the Commission.
95. *Burford et al. v Sun Oil Co. et al., Sun Oil Co. et al. v. Burford et al.* Nos. 495, 496, 60 S. Ct. 1098, 319 U.S. 315 at 326 (US Supreme Court, 1943). *Texas Steel Co. v. Fort Worth and D.C. Railway Co.* 120 Texas 597, at 604,
96. Barbara Sue Thompson Day, "The Oil and Gas Industry and Texas Politics, 1930–1935" (PhD diss., Rice University, 1973).
97. *H. F. Wilcox Oil & Gas Co. v. State, et al.*, 162 Okla. 89, 19 Pac. 2d 347 (Oklahoma Supreme Court, 1933).

## CHAPTER 5

1. Robert E. Hardwicke, "Legal History of Conservation of Oil in Texas," in American Bar Association, Section of Mineral Law, *Legal History of Conservation of Oil and Gas: A Symposium* (Chicago: American Bar Association, 1939), 247.
2. The discovery and development of the East Texas field takes pride of place in any history of the oil business in Texas, and in any legal history of oil conservation. James A. Clark and Michel T. Halbouty, *The Last Boom* (New York: Random House, 1972), is the classic account; Olien and Olien, *Oil in Texas*, 167–92, the most insightful recent version.
3. Olien and Olien, *Oil in Texas*, 169–71. Clark and Halbouty, *Last Boom*, 107.
4. Clark and Halbouty, *Last Boom*, 108.
5. For example, John Steinbeck's *The Grapes of Wrath* (1939) and Pete Seeger's "My Oklahoma Home (It Blowed Away)," the latter also recorded by Bruce Springsteen.
6. Testimony of Neal G. Norris, superintendent of East Texas wells for Alfred MacMillan, 53–54, *MacMillan*

*et al., v. Railroad Commission of Texas et al.*, National Archives, Federal Record Center, Fort Worth, RG 21 U.S. District Court, Western District of Texas, Austin, Entry # 48 WO33A, Equity Case Files, EQ 390.

7.   Norris Testimony, *MacMillan et al., v. Railroad Commission of Texas et al.*, 87, 93.

8.   Statement of Ralph H. Kinsloe (President, Magnolia Petroleum Company), 52–53, Statement of Evidence Under Equity Rule No. 75. Filed April 2, 1934, *Amazon Petroleum Corporation, et al. v. Archie D. Ryan, et al.*, in Equity No. 652, National Archives, Federal Record Center, Fort Worth, RG 21, U.S. District Courts, Eastern District of Texas, Tyler, Entry # 48EO83A, Equity Cases EQ 652.

9.   Weaver, *Unitization of Oil and Gas Fields in Texas*, 49, 390–91n63. Weaver's published sums don't add up; they've been corrected here. She attributes the majors' relatively lower production percentages to prudence and wider well spacing. Smaller independents and new entrants were eager for cash flow. Integrated majors' strategies dictated long term control of reserves, but in 1931 none were desperate for new crude production.

10.  The circumstances and genesis of this first order are summarized in *Champlin Refining Co. v. Corporation Commission of Oklahoma, et al.*, U.S. District Court for the Western District of Oklahoma, R75-B Appellant's [Plaintiff's] Condensed Narrative Statement of the Testimony, Answer of Defendants to Second Amended and Supplemental Bill of Complaint, hearing for a temporary injunction, filed April 22, 1931, Affidavit of R. D. Parker. National Archives, Federal Record Center, Fort Worth, RG 21 U.S. District Courts, Western District of Oklahoma, Entry 40W11, Equity Case Files EQ 1156.

11.  *Danciger Oil and Refining Company v. Railroad Commission of Texas*, 49 S.W.2d 837, at 839 (Court of Civil Appeals of Texas, Austin, 1932).

12.  Malavis, *Bless the Pure and Humble*, 63.

13.  *Danciger Oil and Refining Company v. Railroad Commission of Texas*, 49 S.W.2d 837, at 839.

14.  *Danciger Oil and Refining Company v. Railroad Commission of Texas*, 49 S.W.2d 837, at 842.

15.  *Danciger Oil and Refining Company v. Railroad Commission of Texas*, 49 S.W.2d 837, at 838, 842–43.

16.  *Danciger Oil and Refining Company v. Smith et al.*, 4 F. Supp. 236, at 236n1 (Federal District Court for the Northern District of Texas, Amarillo Division, June 25, 1933):
     The entire area constitutes one gas reservoir. Scattered through nearly the length of that area are more than thirty pools known to be productive of oil, more or less separated by reason of their position in and on the structure, in the sense that there is no migration of oil from one pool to another, but all of the fields are indirectly connected with each other through the formations containing gas. These formations, besides oil and gas, also contain water. In a well drilled within any of the known oil pools, two or more producing formations are often encountered. In such cases the higher formation usually produces gas only, while both gas and oil are found in the lower formation. By reason of the fact that this field is one gas reservoir more or less intimately connected, production either of gas, or of oil and gas in any part of the reservoir theoretically affects the pressure all over it.

17.  *Danciger Oil and Refining Company v. Railroad Commission of Texas*, 49 S.W.2d 837, at 842.

18.  R. D. Parker Affidavit, *Champlin Refining Co. v. Corporation Commission of Oklahoma, et al.*

19.  Different cases and citations use different spellings for MacMillan's name: MacMillan (adopted here) or Macmillan.

20.  *MacMillan et al., v. Railroad Commission of Texas et al.*, 51 F. 2d 400, at 400 (Federal District Court for the Western District of Texas, Austin Division, 1931).

21.  Clark and Halbouty, *Last Boom*, 162.

22.  *MacMillan et al., v. Railroad Commission of Texas et al.*, 51 F. 2d 400, at 400.

23.  *MacMillan et al., v. Railroad Commission of Texas et al.*, 51 F. 2d 400, at 400–401.

24.  *Swift v. Tyson*, 16 Pet. 1 (1842); Charles Zelden, "Regional Growth and the Federal District Courts: The Impact of Judge Joseph C. Hutcheson Jr., on Southeast Texas, 1918–1931," *Houston Review* 11, no. 2 (1989): 68–69; Malavis, *Bless the Pure and Humble*, 236n33. The *Swift* doctrine was overturned in 1938 in *Erie R. R. Co. v. Tompkins*, 504 U.S. 65 (US Supreme Court, 1938).

25.  *MacMillan et al., v. Railroad Commission of Texas et al.*, 51 F. 2d 400, Western District of Texas, Austin Division, July 28, 1931, National Archives, Federal Record Center, Fort Worth, RG 21 U.S. District Courts, Western District of Texas, Austin, Entry # 48 WO33A, Equity Case Files, EQ 390, Record of Testimony, Original Copy for Hon. Joseph C. Hutcheson, United States Circuit Judge, August 4, 1931.

26.  Testimony of E. V. Foran, Railroad Commission witness, *MacMillan et al., v. Railroad Commission of Texas et al.*, National Archives, Federal Record Center, Fort Worth, Transcript, 182.

27.  Testimony of E. V. Foran, *MacMillan et al., v. Railroad Commission of Texas et al.*, National Archives, Federal Record Center, Fort Worth, Transcript, 154–80.

28.  Testimony incorporated from first *Danciger* case, *MacMillan et al., v. Railroad Commission of Texas et al.*, National Archives, Federal Record Center, Fort Worth, Transcript, 219–71.

29.  Testimony of Robert R. Penn, Railroad Commission witness, *MacMillan et al., v. Railroad Commission of Texas et al.*, Transcript, 121–24; Penn, under cross-examination by Hutcheson, *MacMillan et al., v. Railroad*

*Commission of Texas et al.*, 138–39. Testimony of E. V. Foran, Railroad Commission witness, *MacMillan et al., v. Railroad Commission of Texas et al.*, Transcript, 188, *MacMillan et al., v. Railroad Commission of Texas et al.*, National Archives, Federal Record Center, Fort Worth.

30.  Testimony of Robert R. Penn, Railroad Commission witness, under cross-examination by Hutcheson, *MacMillan et al., v. Railroad Commission of Texas et al.*, 138–39 at 143, *MacMillan et al., v. Railroad Commission of Texas et al.*, National Archives, Federal Record Center, Fort Worth.

31.  Testimony of Wallace Pratt, Railroad Commission rebuttal witness, *MacMillan et al., v. Railroad Commission of Texas et al.* Transcript, 371–74, *MacMillan et al., v. Railroad Commission of Texas et al.*, National Archives, Federal Record Center, Fort Worth.

32.  Testimony of Wallace Pratt, *MacMillan et al., v. Railroad Commission of Texas et al.* Transcript, 388, *MacMillan et al., v. Railroad Commission of Texas et al.*, National Archives, Federal Record Center, Fort Worth.

33.  Testimony of J. S. Hutnall, MacMillan witness, *MacMillan et al., v. Railroad Commission of Texas et al.* Transcript, 270–88, *MacMillan et al., v. Railroad Commission of Texas et al.*, National Archives, Federal Record Center, Fort Worth.

34.  "A Guide to the R. D. Parker Family Papers, 1924–1959," Briscoe Center for American History, The University of Texas at Austin. Full name Richard Denny Parker.

35.  *MacMillan et al., v. Railroad Commission of Texas et al.*, 51 F. 2d 400 (Western District of Texas, Austin Division, July 28, 1931), National Archives, Federal Record Center, Fort Worth, RG 21 U.S. District Court, Western District of Texas, Austin, Entry # 48 WO33A, Equity Case Files, EQ 390. Plaintiff's Exhibit "F", at 36–37. This letter is reproduced as Note 1 in *MacMillan et al., v. Railroad Commission of Texas et al.*, 51 F. 2d 400, at 401–5, Federal District Court for the Western District of Texas, Austin Division, July 28, 1931.

36.  Suman, "Evolution by Companies," *API History*, 144–45.

37.  *Amazon Petroleum Corporation, et al., v. Railroad Commission of Texas, et al.*, Eastern District of Texas, Tyler District Court, in Equity 652, National Archives, Federal Record Center, Fort Worth, Agency Box no. 17, Accession no. 90-5-18, FRC 61027.

38.  *MacMillan et al., v. Railroad Commission of Texas et al.*, National Archives, Federal Record Center, Fort Worth, Transcript, 201, 312, 313.

39.  *MacMillan et al., v. Railroad Commission of Texas et al.*, National Archives, Federal Record Center, Fort Worth, Transcript, 201.

40.  *MacMillan et al., v. Railroad Commission of Texas et al.*, National Archives, Federal Record Center, Fort Worth, Transcript, 404, 405.

41.  *MacMillan et al., v. Railroad Commission of Texas et al.* (1931), Findings of Fact and Conclusions of Law. National Archives, Federal Record Center, Fort Worth, Texas.

42.  *MacMillan et al., v. Railroad Commission of Texas et al.* (1931), Findings of Fact and Conclusions of Law. National Archives, Federal Record Center, 404, 405.

43.  Article 6014, as amended, *General and Special Laws of the State of Texas Passed by the Forty-second Legislature, First-called Session*, 1931, Chapter 26, Section 1, 46.

44.  Article 6014, as amended, *General and Special Laws of the State of Texas Passed by the Forty-first Legislature, Regular Session*, 1929, Chapter 313, 694:
     Neither natural gas nor crude petroleum shall be produced, transported, stored or used in such manner or under such conditions as to constitute waste; provided, however, that this shall not be construed to mean economic waste. The term "waste" in addition to its ordinary meaning, shall include permitting (a) escape into the open air of natural gas except as may be necessary in the drilling or operation of a well; (b) drowning with water of any stratum capable of producing oil or gas or both oil and gas in paying quantities; (c) underground waste; (d) any natural gas well to wastefully burn; (e) the wasteful utilization of natural gas; (f) the creation of unnecessary fire hazards. Hardwicke, "Legal History," (1939), 220n9.

45.  Article 6014, Rev. St. 1925 (as amended Acts of 1931, 1st Called Sess. Chapter 26 [Vernon's Ann. Civ. St. art. 6014]).

46.  *People's Petroleum Producers v. Sterling*, 60 F 2d 1041, at 1046, Federal District Court, Eastern District of Texas, Tyler, July 19, 1932.

47.  *MacMillan et al., v. Railroad Commission of Texas et al.*, 51 F. 2d 400, at 402.

48.  Malavis, *Bless the Pure and Humble*, 86.

49.  Hardwicke, "Legal History of Conservation of Oil in Texas," 232; Malavis, *Bless the Pure and Humble*, 91–92.

50.  Kenny A. Franks, "Hot Oil Controversy," *The Encyclopedia of Oklahoma History and Culture*, accessed December 30, 2017, www.okhistory.org.

51.  The genesis of Oklahoma's prorationing statute is summarized in *Champlin Refining Co. v. Corporation Commission of Oklahoma, et al.*, U.S. District Court for the Western District of Oklahoma, R75-B Appellant's [Plaintiff's] Condensed Narrative Statement of the Testimony, Answer of Defendants to Second Amended and

Supplemental Bill of Complaint, hearing for a temporary injunction, filed April 22, 1931, Affidavit of Ray M. Collins. National Archives, Federal Record Center, Fort Worth, RG 21, U.S. District Courts, Western District of Oklahoma, Entry 40W11, Equity Case Files EQ 1156.

52. *Julian Oil and Royalties Company v. Capshaw, et al.*, 145 Okla. 237 (Supreme Court of Oklahoma, 1930); W. P. Z. German, "Legal History of Conservation of Oil and Gas in Oklahoma," in Section of Mineral Law, American Bar Association, *Legal History of Conservation of Oil and Gas: A Symposium* (Chicago: American Bar Association, 1939), 110–213.

53. The majority opinion includes a lengthy discussion of correlative rights precedents and the reasonableness thereof. *Julian Oil and Royalties Company v. Capshaw, et al.*, 145 Okla. 237 (Supreme Court of Oklahoma, 1930), majority opinion, at 15–23.

54. *Julian Oil and Royalties Company v. Capshaw, et al.* (1930), majority opinion, at 24. *Julian Oil and Royalties Company v. Capshaw, et al.* (1930), supplemental opinion, at 17.

55. Affidavit of H. H. Champlin, 3, *Champlin Refining Co. v. Corporation Commission of Oklahoma, et al.*, U.S. District Court for the Western District of Oklahoma, R75-B Appellant's [Plaintiff's] Condensed Narrative Statement of the Testimony, hearing for a temporary injunction, filed April 22, 1931, National Archives, Federal Record Center, Fort Worth, RG 21, U.S. District Courts, Western District of Oklahoma, Entry 40W11, Equity Case Files EQ 1156; *Champlin Refining Co. v. Corporation Commission of Oklahoma, et al.*, U.S. District Court for the Western District of Oklahoma, Special Findings of Fact, September 11, 1931, at 5–7, National Archives, Federal Record Center, Fort Worth, RG 21, U.S. District Courts, Western District of Oklahoma, Entry 40W11, Equity Case Files EQ 1156.

56. *Champlin Refining Co. v. Corporation Commission of Oklahoma, et al.*, U.S. District Court for the Western District of Oklahoma, Special Findings of Fact, September 11, 1931, at 46, National Archives, Federal Record Center, Fort Worth, RG 21 U.S. District Courts, Western District of Oklahoma, Entry 40W11, Equity Case Files EQ 1156.

57. Dr. W. P. Haseman, plaintiff's witness, re-direct examination by Harry O. Glasser, *Champlin Refining Co. v. Corporation Commission of Oklahoma, et al.*, U.S. District Court for the Western District of Oklahoma, Transcript of Proceedings, July 20, 1931, at 364–65, National Archives, Federal Record Center, Fort Worth, RG 21, U.S. District Courts, Western District of Oklahoma, Entry 40W11, Equity Case Files EQ 1156.

58. Testimony of Ralph Arnold, Plaintiff's witness, 86–88, *Champlin Refining Co. v. Corporation Commission of Oklahoma, et al.*, U.S. District Court for the Western District of Oklahoma, R75-B Appellant's [Plaintiff's] Condensed Narrative Statement of the Testimony, Final Hearing, filed October 17, 1931, National Archives, Federal Record Center, Fort Worth, RG 21, U.S. District Courts, Western District of Oklahoma, Entry 40W11, Equity Case Files EQ 1156.

59. Testimony of J. R. McWilliams, Defendant's witness, at 105, *Champlin Refining Co. v. Corporation Commission of Oklahoma, et al.*, U.S. District Court for the Western District of Oklahoma, R75-B Condensed Narrative Statement of the Testimony, Final Hearing, filed October 17, 1931, National Archives, Federal Record Center, Fort Worth, RG 21, U.S. District Courts, Western District of Oklahoma, Entry 40W11, Equity Case Files EQ 1156.

60. Affidavit of Ray M. Collins, *Champlin Refining Co. v. Corporation Commission of Oklahoma, et al.*, U.S. District Court for the Western District of Oklahoma, R75-B Appellant's [Plaintiff's] Condensed Narrative Statement of the Testimony, Answer of Defendants to Second Amended and Supplemental Bill of Complaint, hearing for a temporary injunction, filed April 22, 1931, National Archives, Federal Record Center, Fort Worth, RG 21, U.S. District Courts, Western District of Oklahoma, Entry 40W11, Equity Case Files EQ 1156.

61. *Champlin Refining Co. v. Corporation Commission of Oklahoma, et al.*, 286 U.S. 210, at 232 (US Supreme Court, 1932).

62. *Champlin Refining Co. v. Corporation Commission of Oklahoma, et al.*, U.S. District Court for the Western District of Oklahoma, Special Findings of Fact, September 11, 1931, at 34–35, National Archives, Federal Record Center, Fort Worth, RG 21, U.S. District Courts, Western District of Oklahoma, Entry 40W11, Equity Case Files EQ 1156.

63. German, "Legal History of Conservation of Oil and Gas in Oklahoma," 181.

64. *Champlin Refining Co. v. Corporation Commission of Oklahoma, et al.*, 51 F. 2d 823, at 826 (US District Court for the Western District of Oklahoma, 11, 1931); *Chicago, Burlington & Quincy Railway Co. v. Illinois*, 200 U.S. 561 (US Supreme Court, argued December 14, 1905, decided March 5, 1906).

65. *Champlin Refining Co. v. Corporation Commission of Oklahoma et al.*, 286 U.S. 210, at 228 (US Supreme Court, argued March 23, 1932; Decided May 16, 1932).

66. *Champlin Refining Co. v. Corporation Commission of Oklahoma et al.*, 286 U.S. 210, at 233.

67. *Champlin Refining Co. v. Corporation Commission of Oklahoma et al.*, 286 U.S. 210, at 232.

68. *Champlin Refining Co. v. Corporation Commission of Oklahoma et al.*, 286 U.S. 210, at 233–34.

69. Delightfully recounted in Clark and Halbouty, *The Last Boom*, 167–87.

70. *Amazon Petroleum Corporation, et al., v. Railroad Commission of Texas, et al.,* Eastern District of Texas, Tyler District Court, in Equity 652, National Archives, Federal Record Center, Fort Worth, Agency Box no. 17, Accession no. 90-5-18, FRC 61027, *Amazon* Complainants' Trial Brief, J. N. Saye, H. P. Smead, W. T. Saye, Solicitors for Complainants, Longview, Texas. W. C. Spooner, Consulting Geologist, Technical Appendix, 281–85.

71. Clark and Halbouty, *Last Boom*, 173.

72. *Constantin et al. v. Smith et al.,* 57 F.2d 227, at 234 (Federal District Court for the Eastern District of Texas, Tyler Division, February 18, 1932).

73. *Sterling v. Constantin*, 287 U.S. 378, at 402 (US Supreme Court, argued November 15, 16, 1932).

74. *Alfred MacMillan, et al. v. Ross S. Sterling, Governor of Texas, et al.,* Texas Railroad Commission Answer, No. 395 in Equity, Eastern District of Texas, Tyler District Court, National Archives, Fort Worth, Texas, RG 21, FRC 61010; Railroad Commission of Texas, Oil and Gas Division, Austin, Texas, Oil and Gas Docket No. 120, Special Order Adopting Rules and Regulations Pursuant to Hearing February 12, 1932 (Order attached to Answer).

75. *Amazon Petroleum Corporation, et al., v. Railroad Commission of Texas, et al.,* Eastern District of Texas, Tyler District Court, in Equity 652, National Archives, Federal Record Center, Fort Worth, Agency Box no. 17, Accession no. 90-5-18, FRC 61027, *Amazon* Complainants' Trial Brief, J. N. Saye, H. P. Smead, W. T Saye, Solicitors for Complainants, Longview, Texas. W. C. Spooner, Consulting Geologist, Technical Appendix, at 290–94.

76. *Bill & Dave Oil Corporation, et al. v. Lon Smith, et al.,* No. 392 in Equity, Amended Bill of Complaint. Eastern District of Texas, Tyler District Court, National Archives, Fort Worth, Texas, RG 21, FRC 61010.

77. John Donnelly (*JPT* editor), "Membership Rolls Have Contained Many Notable Pioneers," *Journal of Petroleum Technology, Special Commemorative Issue Celebrating 50 Years of the Society of Petroleum Engineers* 59, no. 10 (October 2007): 79.

78. Lester C. Uren (University of California), "The 'Gas Factor,' as a Measure of Oil-production Efficiency," *Transactions of the AIME* 77, no. 1 (December 1927): 1–51, Fort Worth Meeting, Petroleum Division, October 1927.

79. Uren, "'Gas Factor,'" 151.

80. All of plaintiffs' expert witnesses' testimony from *People's* was incorporated into the *Amazon* transcript verbatim: *Amazon Petroleum Corporation, et al., v. Railroad Commission of Texas, et al.,* Eastern District of Texas, Tyler District Court, in Equity 652, National Archives, Federal Record Center, Fort Worth, Agency Box no. 17, Accession no. 90-5-18, FRC 61027, Trial Transcript, Vol. 1, at 553–900.

81. *Alfred MacMillan, et al. v. Ross S. Sterling, Governor of Texas, et al.,* No. 395 in Equity, Texas Railroad Commission Answer. Eastern District of Texas, Tyler District Court, National Archives, Fort Worth, Texas, RG 21, FRC 61010, Railroad Commission of Texas, Oil and Gas Division, Austin, Texas, Oil and Gas Docket No. 120, Special Order Adopting Rules and Regulations Pursuant to Hearing February 12, 1932 (Order attached to Answer).

82. *Arthur F. Graf, et al. v. Lon A. Smith, et al.,* No. 394 in Equity, Application to make tests, and list of test rates of production. Eastern District of Texas, Tyler District Court, National Archives, Fort Worth, Texas, RG 21, FRC 61010.

83. Deposition of A. H. Tarver, May 14, 1932, *A. S. Palmer, J. G. Palmer and A. H. Tarver v. Ross S. Sterling, Governor of the State of Texas, et al.,* No. 393 in Equity, Eastern District of Texas, Tyler District Court, National Archives, Fort Worth, Texas, RG 21, FRC 61010.

84. Malavis, *Bless the Pure and Humble*, 116, 240n20.

85. *People's Petroleum Producers v. Sterling*, 60 F 2d 1041, at 1046 (Federal District Court, Eastern District of Texas, Tyler, 1932).

86. *People's Petroleum Producers v. Sterling*, 60 F 2d 1041, at 1046.

87. E. O. Buck, interview by Louis J. Marchiafava, April 16, 1981, Archive Number: OH 281, Houston Metropolitan Research Center of the Houston Public Library, Houston, Texas.

88. Buck interview.

89. Malavis, *Bless the Pure and Humble*, 128.

90. *People's Petroleum Producers, Inc., v. Smith et al., and six other cases,* 1 F. Supp. 361, 361–62, Federal District Court, Eastern District of Texas, Tyler, October 24, 1932; *Amazon Petroleum Corporation, et al., v. Railroad Commission of Texas, et al.,* Eastern District of Texas, Tyler District Court, in Equity 652, National Archives, Federal Record Center, Fort Worth, Agency Box no. 17, Accession no. 90-5-18, FRC 61027. Trial Transcript, Vol. 1, at 839, 841, 869–70. All of plaintiffs' witnesses' testimony in *People's* was incorporated into the trial transcript in *Amazon*.

91. *Amazon Petroleum Corporation, et al., v. Railroad Commission of Texas, et al.,* Eastern District of Texas, Tyler District Court, in Equity 652, National Archives, Federal Record Center, Fort Worth, Agency Box no. 17, Accession no. 90-5-18, FRC 61027. Trial Transcript, Vol. 1, at 819.

92. *Amazon Petroleum Corporation, et al., v. Railroad Commission of Texas, et al.,* Eastern District of Texas, Tyler District Court, in Equity 652, National Archives, Federal Record Center, Fort Worth, Agency Box no. 17, Accession no. 90-5-18, FRC 61027. Trial Transcript, Vol. 1, at 799.

93. *People's Petroleum Producers, Inc., v. Smith et al., and six other cases,* 1 F. Supp. 361, at 365 (Federal District Court, Eastern District of Texas, Tyler, October 24, 1932).

94. *People's Petroleum Producers, Inc., v. Smith,* 1 F. Supp. 361, 363.

95. *People's Petroleum Producers, Inc., v. Smith,* 1 F. Supp. 361, 363.

96. *People's Petroleum Producers, Inc., v. Smith,* 1 F. Supp. 361, 364.

97. *People's Petroleum Producers, Inc., v. Smith,* 1 F. Supp. 361, 364.

98. Malavis, *Bless the Pure and Humble,* 128–29, 245–46n72.

99. Clark and Halbouty, *Last Boom,* 190–95.

100. Malavis, *Bless the Pure and Humble,* 248n6.

101. Malavis, *Bless the Pure and Humble,* 138.

102. *People's Petroleum Producers, Inc. v. Lon A. Smith, et al., and Conjoined Cases,* No. 386 in Equity, March 17, 1933, Federal Record Center, Fort Worth, RG 21, U.S. District Courts, District Court for the Eastern District of Texas, Tyler Division, FRC 61013, Equity Case Files, EQ 386.

103. Malavis, *Bless the Pure and Humble,* 140–41.

104. Clark and Halbouty, *Last Boom,* 211.

105. Malavis, *Bless the Pure and Humble,* 143; Clark and Halbouty, *Last Boom,* 211–12.

106. Malavis, *Bless the Pure and Humble,* 145–46, 254–55n66; *Hunt Production Company v. Lon A. Smith, et al.* (Unpublished Opinion, Northern District of Texas, May 26, 1933).

107. *Danciger Oil and Refining Company of Texas v. Smith,* 4 F. Supp. 236, U.S. District Court for the Northern District of Texas, Amarillo, 1933, Complaint, National Archives, Federal Record Center, Fort Worth, RG 21, U.S. District Courts, Northern District of Texas, Amarillo Division, FRC 152787, Equity Case Files EQ 391.

108. *Danciger Oil and Refining Company of Texas v. Smith,* 4 F. Supp. 236, at 239 (US District Court for the Northern District of Texas, Amarillo Division, June 25, 1933).

109. *Danciger Oil and Refining Company of Texas v. Smith,* 4 F. Supp. 236, at 239.

110. *Danciger Oil and Refining Company of Texas v. Smith,* 4 F. Supp. 236, at 239–40.

111. Hardwicke, "Legal History of Conservation of Oil in Texas," 1939, 234; Childs, *Texas Railroad Commission,* 211–12.

112. Malavis, *Bless the Pure and Humble,* 134, 248–49n11, citing Transcript of the Railroad Commission Hearing in Tyler, Texas, January 12, 1933, 6–41.

113. *Amazon Petroleum Corporation, et al. v. Archie D. Ryan, et al.,* in Equity No. 652, National Archives, Federal Record Center, Fort Worth, RG 21, U.S. District Courts, Eastern District of Texas, Tyler, Entry # 48EO83A, Equity Cases EQ 65. Statement of Evidence Under Equity Rule No. 75. Filed April 2, 1934. Statement of Avory H. Alcorn (Department of the Interior, Division of Investigations), at 80.

114. See the operators' statements summarized in *Amazon Petroleum Corporation, et al. v. Archie D. Ryan, et al.,* in Equity No. 652, National Archives, Federal Record Center, Fort Worth, RG 21, U.S. District Courts, Eastern District of Texas, Tyler, Entry # 48EO83A, Equity Cases EQ 652; Statement of Evidence Under Equity Rule No. 75, Filed April 2, 1934, tattling on themselves, and especially on operators of offsetting leases. Invariably, an operator's excess production was justified as defensive.

115. Olien and Olien, *Oil in Texas,* 195.

116. Rajiv Sethi and E. Somanathan, "The Evolution of Social Norms in Common Property Resource Use," *American Economic Review* 86, no. 4 (September 1996): 782. Jonathan Bendor and Dilip Mookherjee, "Institutional Structure and the Logic of Ongoing Collective Action," *American Political Science Review* 81, no. 1 (March 1987): 129–54.

117. Buck interview.

118. German, "Legal History of Conservation of Oil and Gas in Oklahoma," 196.

119. James A. Clark, *Three Stars for the Colonel* (New York: Random House, 1954), 101.

120. Ross S. Sterling, with Ed Kilman and Don Carleton, *Ross Sterling: Texan. A Memoir by the Founder of the Humble Oil and Refining Company* (Austin: University of Texas Press, 2007), 179.

# CHAPTER 6

1. *Amazon Petroleum Corporation v. Railroad Commission of Texas et al.,* 5 F. Supp. 633, at 639 (Federal District

Court, Eastern District of Texas, Tyler, February 12, 1934).

2.  Zelden, "Regional Growth and the Federal District Courts: The Impact of Judge Joseph C. Hutcheson Jr., on Southeast Texas, 1918–1931," *Houston Review* 11, no. 2 (1989): 90, quoting Walter P. Armstrong, "Joseph C. Hutcheson Jr.: Chief Judge, Fifth Circuit Court of Appeals," *American Bar Association Journal* 35 (June 1949): 548.

3.  Zelden, "Regional Growth and the Federal District Courts."

4.  DuVal West to Joseph C. Hutcheson Jr., May 23, 1933, National Archives, RG 21, Northern District of Texas, Amarillo District Court, FRC 152787, Federal Records Center, Fort Worth, Texas.

5.  Joseph C. Hutcheson Jr., "The Judgement Intuitive: The Function of the 'Hunch' in Judicial Decision," *Cornell Law Review* 14, no. 3 (April 1929): 274–88.

6.  Hutcheson, "Judgement Intuitive," 278.

7.  Hutcheson, "Judgement Intuitive," 286.

8.  Hutcheson, "Judgement Intuitive," 274.

9.  Hutcheson, "Judgement Intuitive," 279–81.

10. Henri Poincaré, *Science and Method*, "Chapter III: Mathematical Creation," reprinted in *The Foundations of Science: Science and Hypothesis; The Value of Science; Science and Method*, trans. George Bruce Halsted (1908; New York and Garrison, NY: The Science Press, 1913), 389.

11. Poincaré, "Mathematical Creation," 389.

12. Poincaré, "Mathematical Creation," 390–92.

13. Poincaré, "Mathematical Creation," 389–90. Poincaré's analysis was reprised, elaborated, and affirmed by another French mathematician, Jacques Hadamard, in his *An Essay on the Psychology of Invention in the Mathematical Field* (Princeton, NJ: Princeton University Press, 1945).

14. Hutcheson, "Judgement Intuitive," 274.

15. Zelden, "Regional Growth and the Federal District Courts," 69–70.

16. *People v. Associated Oil Co.*, 211 Cal. 93, 294 Pacific 717 (Supreme Court of California, December 3, 1930). J. Howard Marshall, "Legal History of Conservation of Oil in California," in American Bar Association, Section of Mineral Law, *Legal History of Conservation of Oil and Gas: A Symposium* (Chicago: American Bar Association, 1938), 33.

17. *Bandini Petroleum Company v. Superior Court of California ex Rel. Los Angeles County*, 284 U.S. 8, 52 Sup. Ct. 10 (US Supreme Court, November 23, 1931).

18. *Chicago, Burlington & Quincy Railway Co. v. Illinois*, 200 U.S. 561 (US Supreme Court, March 5, 1906) and the voluminous precedents cited therein.

19. *Ohio Oil Co. v. Indiana* (1900), 177 U.S. 190 (US Supreme Court, April 9,1900); *Champlin Refining Co. v. Corporation Commission of Oklahoma et al.*, 286 U.S. 210 (US Supreme Court, 1932) and the cases cited therein.

20. *Amazon Petroleum Corporation v. Railroad Commission of Texas et al.*, 5 F. Supp. 633, at 634 (Federal District Court for the Eastern District of Texas, Tyler, February 12, 1934).

21. *Amazon Petroleum Corporation v. Railroad Commission of Texas et al.*, 5 F. Supp. 633, at 634.

22. *Panama Refining Co. et al. v. Ryan et al.*, 5 F. Supp. 699 (Eastern District of Texas, 1934); *Panama Refining Co. et al. v. Ryan et al.*, 55 S.Ct. 241, 293 U.S. 388 (US Supreme Court, argued December 10,11, 1934, decided January 7, 1935); *A. L. A. Schechter Poultry Corp. v. United States* 295 U.S. 495 (US Supreme Court, argued May 23, 1935, Decided May 27, 1935).

23. Malavis, *Bless the Pure and Humble*, 164–65.

24. With apologies to Walter Vincenti and his magisterial *What Engineers Know and How They Know It* (1990).

25. Railroad Commission Orders for the East Texas field, September 2, 1931, Rule 6; summarized in W. C. Spooner, Appendix to Complainants' Trial Brief, at 285, *Amazon Petroleum Corporation, et al., v. Railroad Commission of Texas, et al.*, Eastern District of Texas, Tyler District Court, in Equity 652. National Archives, Federal Record Center, Fort Worth, Agency Box no. 17, Accession no. 90-5-18, FRC 61027.

26. DeGolyer Sr., "Waste in Oil and Gas Production," typescript, 1941, 15, Papers of Everette Lee DeGolyer Sr., DeGolyer Papers Library, SMU, Dallas, Box 25, Folder 2458; Kenneth C. Sclater and B. R. Stephenson (Marland Production Co., Ponca City, OK), "Measurements of Original Pressure, Temperature, and Gas-Oil Ratio in Oil Sands," *Transactions of the American Institute of Mining and Metallurgical Engineers* 82 (1928–1929): 119–36, presented at the American Institute of Mining Engineers, Petroleum Division, Tulsa Meeting, October 1928. Marland merged with Continental Oil (Conoco) in 1929.

27. Sclater and Stephenson, "Measurements of Original Pressure, Temperature, and Gas-oil Ratio," 126, fig. 6, 127, table 1.

28. Sclater and Stephenson, "Measurements of Original Pressure, Temperature, and Gas-oil Ratio," 119, 128, fig. 7.

29. C. V. Millikan and Carroll V. Sidwell (Amerada Petroleum Co.), "Bottom-Hole Pressure in Oil Wells," *Transactions of the American Institute of Mining and Metallurgical Engineers*: *Petroleum Development and*

*Technology* 92, no. 1 (December 1931): 194; presented at the Tulsa Meeting of the Petroleum Division, American Institute of Mining and Metallurgical Engineers, October 1930.

30.   D.G. Hawthorn (Barnsdall Oil Co.), "Subsurface Pressures in Oil Wells and Their Field of Application," *Transactions of the American Institute of Mining Engineers* 103, no. 1 (December 1933): 148 (paper presented at the Ponca City Meeting of the Petroleum Division, American Institute of Mining and Metallurgical Engineers, October 1932). See also Buck, interview.

31.   E. V. Foran, "Interpretation of Bottom-Hole pressure in East Texas Oil Field," *American Association of Petroleum Geologists Bulletin* 16, no. 9 (1 September 1932): 907–8 (paper presented at the spring meeting of the Division of Production of the American Petroleum Institute, Tyler, Texas, April 14, 1932).

32.   C. E. Beecher, in "Appendix—Can Engineers Determine Extractable Acreage Content within Standard Set by Lawyers?" Earl Oliver, "Stabilizing Influences for the Petroleum Industry," Reprinted from the *Transactions of the American Institute of Mining and Metallurgical Engineers, Petroleum Division* (1932): 10, presented at the meeting of the Petroleum Division, AIME, at Houston, Texas, October 1931.

33.   Foran, "Bottom-Hole Pressure in East Texas," 908, 911.

34.   Donald T. Campbell, H. W. Riecken, Robert F. Boruch, N. Kaplan, T.K. Glennan, J. Pratt, A. Rees, and W. Williams, "Quasi-Experimental Designs," in *Social Experimentation: A Method for Planning and Evaluating Social Interventions*, ed. H. W. Riecken and Robert F. Boruch (New York: Academic Press, 1974), 87–116; Donald T. Campbell and H. Laurence Ross, "The Connecticut Crackdown on Speeding: Time-Series Data in Quasi-Experimental Analysis," *Law and Society Review* 3, no. 1 (1968): 33–53; reproduced in Donald T. Campbell, *Methodology and Epistemology for Social Science: Selected Papers*, ed. E. Samuel Overman (Chicago: University of Chicago Press, 1988), 222–37.

35.   Foran, "Bottom-Hole Pressure in East Texas," 911; C. V. Millikan (Amerada Petroleum), "Geological Application of Bottom-Hole Pressures," *American Association of Petroleum Geologists Bulletin* 16, no. 9 (1 September 1932): 891–906. Read before the AAPG meeting, Oklahoma City, March 25, 1932.

36.   Dale R. Snow (Barnsdall Oil), "Water Encroachment in Bartlesville Sand Pools of Northeastern Oklahoma and its Bearing on East Texas Recovery Problem," *American Association of Petroleum Geologists Bulletin* 16, no. 9 (1 September 1932): 881–90. Paper read before the American Association of Petroleum Geologists meeting in Oklahoma City, March 25, 1932.

37.   Millikan (Amerada Petroleum), "Bottom-Hole Pressures," 891–906.

38.   Hawthorn, "Subsurface Pressures," 159. H. D. Wilde Jr., "Why Measure Bottom Hole Pressures?" API Mid-year Meeting, Tulsa, Oklahoma, June 3, 1932.

39.   Hawthorn, "Subsurface Pressures," 156–59. The firms were Humble Oil, Gulf, Amerada, Tidal Oil Co., Atlantic Oil, Barnsdall, Sun Oil, Shell Petroleum, Mid-Kansas Oil & Gas, Stanolind Oil &Gas, Sinclair-Prairie Oil, and Arkansas Fuel Oil.

40.   Steven Shapin and Simon Schaffer, *Leviathan and the Air-Pump: Hobbes, Boyle, and the Experimental Life* (Princeton, NJ: Princeton University Press, 1985).

41.   Ralph D. Wyckoff, "Discussion" of Ralph J. Schilthuis and William Hurst, "Variations in Reservoir Pressure in the East Texas Field," *Transactions of the American Institute of Mining and Metallurgical Engineers: Petroleum Development and Technology* 114, no. 1 (December 1935): 173–74. Paper read at the Tulsa meeting of the AIME Petroleum Division, October 1934.

42.   H. D. Wilde Jr. and T. V. Moore (Humble Oil and Refining Co., Houston), "Hydrodynamics of Reservoir Drainage and Its Relation to Well Spacing," *Proceedings of the American Petroleum Institute, Production Bulletin 210* (1932): 83–90; Bas C. van Fraassen, *The Scientific Image* (Oxford, UK: Oxford University Press, 1980).

43.   Buck, interview.

44.   Deposition of Gordon Griffin, Railroad Commission petroleum engineer in charge of the Engineering Department of the East Texas Field, December 14, 1933, Trial transcript, Vol. III, 1294–98, *Amazon Petroleum Corporation, et al., v. Railroad Commission of Texas, et al.*, Eastern District of Texas, Tyler District Court, in Equity 652, National Archives, Federal Record Center, Fort Worth, Agency Box no. 17, Accession no. 90-5-18, FRC 61027; Schilthuis and Hurst, "Variations in Reservoir Pressure in the East Texas Field," 164.

45.   Ben E. Lindsly, "A Study of 'Bottom-Hole' Samples of East Texas Crude Oil," *United States Bureau of Mines Report of Investigations RI 3212*, May 1933, read before the American Petroleum Institute, Division of Production, Tulsa, Oklahoma, May 19, 1933). Transcribed verbatim and in its entirety in the transcript of *Amazon Petroleum Corporation v. Railroad Commission of Texas et al.*, Trial transcript, at 1334–80, plus 5 unnumbered plates. See also Carl E. Reistle Jr. and E. P. Hayes, "A Study of Subsurface Pressures and Temperatures in Flowing Wells in the East Texas Field and the Application of These Data to Reservoir and Vertical flow problems," *United States Bureau of Mines RI 3211*, May 1933, also read before the American Petroleum Institute, Division of Production, Tulsa, Okla., May 19, 1933. The University of California Library, Berkeley, catalogued its copy of Reistle and Hayes' *RI* November 18, 1933, which gives some indication of

diffusion time.

46. Ben E. Lindsly to H. C. Fowler, July 27, 1931, Location # C2300946, Accession No. 65-A-855, RG 70, National Archives, Federal Records Center, Fort Worth, Texas.

47. Lindsly, "Study of 'Bottom-hole' Samples of East Texas Crude Oil," Trial transcript, at 1336, *Amazon Petroleum Corporation v. Railroad Commission of Texas et al.*

48. Lindsly, "Study of 'Bottom-hole' Samples of East Texas Crude Oil," Trial transcript at 1338–39, *Amazon Petroleum Corporation v. Railroad Commission of Texas et al.* Carter Oil was the Oklahoma operating subsidiary of Standard Oil of New Jersey.

49. Lindsly, "Study of 'Bottom-hole' Samples of East Texas Crude Oil," Trial transcript at 1341, *Amazon Petroleum Corporation v. Railroad Commission of Texas et al.*

50. Reistle and Hayes, "Study of Subsurface Pressures and Temperatures in Flowing Wells," 1.

51. Lindsly, "Study of 'Bottom-hole' Samples of East Texas Crude Oil," Trial transcript at 1368, *Amazon Petroleum Corporation v. Railroad Commission of Texas et al.*

52. Percy W. Bridgman, "Thermodynamic Properties of Liquid Water to 80° (C) and 12,000 KG N (Kg/Cm$^2$)," *Proceedings of the American Academy of Arts and Sciences* 48, no. 9 (September 1912): 357.

53. Lindsly, "Study of 'Bottom-hole' Samples of East Texas Crude Oil," Trial transcript at 1373–74, *Amazon Petroleum Corporation v. Railroad Commission of Texas et al.*

54. Lindsly, "Study of 'Bottom-hole' Samples of East Texas Crude Oil," Trial transcript at 1375, *Amazon Petroleum Corporation v. Railroad Commission of Texas et al.*; Schilthuis and Hurst, "Variations in Reservoir Pressure in the East Texas Field," *Transactions of the American Institute of Mining and Metallurgical Engineers: Petroleum Development and Technology* 114, no. 1 (December 1935): 165, presented at the Tulsa meeting of the Petroleum Division, American Institute of Mining and Metallurgical Engineers, October 1934 (Received at the Institute September 27, 1934).

55. Lindsly, "Study of 'Bottom-hole' Samples of East Texas Crude Oil," Trial transcript at 1380, *Amazon Petroleum Corporation v. Railroad Commission of Texas et al.*

56. Deposition of T. V. Moore and C. E. Reistle Jr., December 12, 1933, Trial transcript at 1381–97, *Amazon Petroleum Corporation v. Railroad Commission of Texas et al.*

57. Wyckoff, discussion of Schilthuis and Hurst, "Variations in Reservoir Pressure in the East Texas Field," 174.

58. Lester C. Uren and E. J. Bradshaw, "Experimental Study of Pressure Conditions within the Oil Reservoir Rock in the Vicinity of a High-Pressure Producing Well," *Transactions of the American Institute of Mining Engineers: Petroleum Development and Technology, 1932*: 438–60, presented at the Houston Meeting of the Petroleum Division of the American Institute of Mining Engineers, October 1931.

59. Reistle and Hayes, "Study of Subsurface Pressures and Temperatures in Flowing Wells," 6–7.

60. Reistle and Hayes, "Study of Subsurface Pressures and Temperatures in Flowing Wells," 6–7.

61. Reistle and Hayes, "Study of Subsurface Pressures and Temperatures in Flowing Wells," 8–11.

62. The forty-one contributing operators were Amerada Petroleum, American-Liberty Oil, Arkansas Fuel Oil, Atlantic Oil Producing, Barnsdall Oil, Bass and Dillard, Jim Cloud, Inc., Columbia Oil & Gas, Conservative Oil Company, Cranfill-Reynolds, Deep Rock Oil, Devonian Oil, Empire Oil & Refining, Gulf Production Company, Haynes Drilling Company, Humble Oil & Refining, Hunt Production Company, Kirby Petroleum, Louisiana Oil Refining, Lucey Petroleum, Mid-Kansas Oil & Gas, Navarro Oil, Perren & Fell, Pilot Oil Company, Republic Production, Roeser & Pendleton, Rowan & Nichols, Selby Oil & Gas, Shell Petroleum, Showers & Moncrief, Simms Oil Company, Sinclair-Prairie Oil, Snowden & McSweeney, Stanolind Oil & Gas, Stroube & Stroube, Sun Oil, Superior Oil, the Texas Company, Tidal Oil, Turman Oil, and Yount-Lee Oil Company. Lindsly, "Study of 'Bottom-hole' Samples of East Texas Crude Oil," Trial transcript at 1339–40, *Amazon Petroleum Corporation v. Railroad Commission of Texas et al.*

63. The operators were Amerada, Arkansas Natural Gas, Atlantic Oil Producing, Barnsdall Oil, Empire Oil & Gas, Shell Petroleum, Sinclair-Prairie Oil, Stanolind, Sun, the Texas Company, and Tidal Oil. Lindsly, "Study of 'Bottom-hole' Samples of East Texas Crude Oil," Trial transcript at 1340–41, *Amazon Petroleum Corporation v. Railroad Commission of Texas et al.*

64. Ben E. Lindsly, "Solubility and Liberation of Gas from Natural Oil-Gas Solutions," *U.S. Bureau of Mines Technical Paper 554* (1933); Ben E. Lindsly, "Effect of Gas Withdrawal upon Reservoir Fluids," *Transactions of the American Institute of Mining and Metallurgical Engineers: Petroleum Development and Technology* 107, no. 1 (December 1934): 94 (paper presented at the Dallas Meeting of the Petroleum Division, American Institute of Mining and Metallurgical Engineers, October 1933).

65. Lindsly, "Effect of Gas Withdrawal upon Reservoir Fluids," 96.

66. C. E. Reistle Jr., discussion of Lindsly, "Effect of Gas Withdrawal upon Reservoir Fluids," 97.

67. *Amazon* Complainants' Trial Brief, J. N. Saye, H. P. Smead, W. T Saye, Solicitors for Complainants, Longview, Texas; W. C. Spooner, Consulting Geologist, technical appendix, at 299–300, *Amazon Petroleum Corporation,*

*et al., v. Railroad Commission of Texas, et al.,* Eastern District of Texas, Tyler District Court, in Equity 652, National Archives, Federal Record Center, Fort Worth, Agency Box no. 17, Accession no. 90-5-18, FRC 61027.

68.    G. L. Nye and Carl E. Reistle Jr. (The Tide Water Oil Co.), "Recent Changes in Reservoir Pressure Conditions in the East Texas Field," *Transactions of the American Institute of Mining Engineers* 107, no. 1 (December 1934): 77–78 presented at the Dallas Meeting of the Petroleum Division, American Institute of Mining and Metallurgical Engineers, October 1933).

69.    Nye and Reistle, "Recent Changes," 79–80.

70.    Nye and Reistle, "Recent Changes," 82.

71.    Stewart Coleman (Manager of Development Department), H. D. Wilde Jr. (Director of Production Research), and Thomas W. Moore (Production Research Engineer), Humble Oil and Refining Co., Houston, "Quantitative Effect of Gas-Oil Ratios on Decline of Average Rock Pressure," *Transactions of the American Institute of Mining and Metallurgical Engineers* 86, no. 1 (December 1930): 174–84, presented at the Tulsa Meeting of the Petroleum Division, American Institute of Mining and Metallurgical Engineers, October 1929.

72.    Coleman, Wilde, and Moore, "Quantitative Effect," 175–77.

73.    R. J. Schilthuis, "Active Oil and Reservoir Energy," *Transactions of the American Institute of Mining and Metallurgical Engineers: Petroleum Development and Technology* 118, no. 1 (December 1936): 33.

74.    Henry Darcy, *Les fontaines publiques de la ville de Dijon* (Paris: Dalmont, 1856).

75.    In Morris Muskat's later notation, applicable to single-phase flow in a petroleum reservoir:

$$Q = \frac{-kA(p_b - p_a)}{\mu L}$$

where:

$Q$ = volume of flow
$k$ = permeability of medium
$A$ = cross sectional area of pipe
$p_a$ = initial (higher) pressure
$p_b$ = final (lower) pressure
$\mu$ = viscosity of fluid
$L$ = length of pipe through which fluid flows

76.    F. H. King, "Conditions and Movements of Underground Waters," in *Nineteenth Annual Report of the United States: Papers Chiefly of a Theoretic Nature* (Washington, DC: Government Printing Office, 1899): 2: 59–294. C. S. Slichter, "Theoretical Investigation of the Motion of Ground Waters," *U.S. Geological Survey 19th Annual Report, 1897–1898, Part II—Papers Chiefly of a Theoretic Nature* (1899), 2: 295–384; W. Heber Green and G. A. Ampt, "Studies on Soil Physics," *Journal of Agricultural Science* 4 Part I (May 1911): 1–24.

77.    Jan Versluys, "The Origin of Artesian Pressure," *Proceedings of the Royal Netherlands Academy of Arts and Sciences* 33, no. 3 (1930): 217 (communicated at the meeting of March 29, 1930). Similarly, Steven Frank maintains that the Price equation itself (discussed in chapter 3) shares a similar structure with other conservation laws in multiple disciplines. See Steven A. Frank, "Simple Unity among the Fundamental Equations of Science," *Philosophical Transactions of the Royal Society B* 375 (2020): 20190351; theme issue "Fifty Years of the Price Equation"), while John Campbell argues even more broadly that the Price equation is subsumed under a "universal Darwinism" that in all instantiations constitutes Bayesian inference. See John O. Campbell, "Hypothesis Theory: Universal Darwinism as a Process of Bayesian Inference," *Frontiers in Systems Neuroscience* 10 (7 June 2016): 49.

78.    Jan Versluys, "De capillaire in den bodem (Capillarity in Soils)," thesis (in Dutch), Amsterdam, 1916; Jan Versluys, "Die Kapillarität im Boden," *Internationale Mitteilungen für Bodenkunde* (1917): 117–40.

79.    Jan Versluys (Bataafsche Petroleum Maatschappij), "Hydraulics in Flowing Wells: Mathematical Development of the Theory of Flowing Oil Wells," Technical Publication 213 (1929), and *Transactions of the American Institute of Mining Engineers, Petroleum Division* (1930): 192 (paper presented at the Tulsa Meeting of the Petroleum Division, October 1929).

80.    Versluys, "Hydraulics in Flowing Wells," 194–200; Reistle and Hayes, "Study of Subsurface Pressures and Temperatures in Flowing Wells."

81.    Jan Versluys, "Applied Geology: The Equation of Flow of Oil and Gas to a Well after Dynamic Equilibrium Has Been Established," *Akademie van Weenschappen, Amsterdam: Proceedings* 33-3 nos. 6–10, Proceedings of the Section of Science (1930): 578–86. Communicated at the meeting of June 28, 1930,

82.    Morris Muskat and Holbrook G. Botset, "Flow of Gases through Porous Media," *Physics* 1, no. 1 (July 1931): 27–47, received May 1, 1931.

83.    Muskat and Botset, "Flow of Gases through Porous Media," 27, 31ff.

84.    Muskat and Botset, "Flow of Gases through Porous Media," 35.

85.    $\delta p^2 = k\,(\rho v)^n$, where $p$ = pressure, $k$ = "permeability constant of the sand and gas,"

$\rho$ = fluid density, $v$ = gas velocity, and $n$ = [1,2]. Muskat and Botset, "Flow of Gases through Porous Media," 27–28, 35–36.

86. Muskat and Botset, "Flow of Gases through Porous Media," 33–34.

87. Muskat and Botset, "Flow of Gases through Porous Media," 46–47. H. R. Pierce (Consulting natural-gas engineer BOM), and E. L. Rawlins (Associate natural-gas engineer, BOM Petroleum Experiment Station, Bartlesville), "The Study of a Fundamental Basis for Controlling and Gauging Natural-Gas Wells: Part 1. Computing the Pressure at the Sand in a Gas Well," *U.S. Bureau of Mines Report of Investigations 2929* (May 1929), 46–47; H. R. Pierce (Consulting natural-gas engineer BOM), and E. L. Rawlins (Associate natural-gas engineer, BOM Petroleum Experiment Station, Bartlesville), "The Study of a Fundamental Basis for Controlling and Gauging Natural-Gas Wells: Part 2. A Fundamental Relation for Gauging Gas-Well Capacities," *U.S. Bureau of Mines Report of Investigations 2930* (May 1929).

88. Morris Muskat, "Potential Distributions in Large Cylindrical Disks with Partially Penetrating Electrodes," *Physics* 2, no. 5 (May 1932): 329–64, received March 3, 1932.

89. Muskat, "Potential Distributions in Large Cylindrical Disks," 362–63.

90. H. C. Fowler to Joseph Chalmers, December 18, 1931, Accession No. 65-A-855, Box 224-316 #7, RG 70, National Archives, Federal Record Center, Fort Worth, Texas.

91. John Donnelly, *Journal of Petroleum Technology* (*JPT*) Editor, *Journal of Petroleum Technology, Special Commemorative Issue Celebrating 50 Years of the Society of Petroleum Engineers* 59, no. 10 (October 2007): 37, citing DeGolyer's fellow geologist and close friend, Wallace Pratt.

92. Ralph D. Wyckoff, Holbrook G. Botset, Morris Muskat, and D. W. Reed (Gulf Production Company), "Measurement of Permeability of Porous Media," *Bulletin of the American Association of Petroleum Geologists* 18, no. 2 (February 1934): 161–90, manuscript received August 12, 1932.

93. Wyckoff, Botset, Muskat, and Reed, "Measurement of Permeability of Porous Media," 163.

94. Wyckoff, Botset, Muskat, and Reed, "Measurement of Permeability of Porous Media," 167.

95. Wyckoff, Botset, Muskat, and Reed, "Measurement of Permeability of Porous Media," 181.

96. Wyckoff, Botset, Muskat, and Reed, "Measurement of Permeability of Porous Media."

97. Wilde and Moore, "Hydrodynamics of Reservoir Drainage and Its Relation to Well Spacing," 84.

98. Wilde and Moore, "Hydrodynamics of Reservoir Drainage and Its Relation to Well Spacing," 84–86.

99. Wilde and Moore, "Hydrodynamics of Reservoir Drainage and Its Relation to Well Spacing," 88.

100. Wilde and Moore, "Hydrodynamics of Reservoir Drainage and Its Relation to Well Spacing," 89.

101. Reistle Jr., "Reservoir Engineering," 831.

102. Thomas V. Moore, Ralph J. Schilthuis, and William Hurst (Humble Oil and Refining Co., Houston), "The Determination of Permeability from Field Data," *Proceedings of the American Petroleum Institute, Production Bulletin 211* (1933): 11, 5–10; also published in *Oil Weekly* 69 (May 21, 1933).

103. William Hurst, "Unsteady Flow of Fluids in Oil Reservoirs," *Physics* (*Journal of Applied Physics*) 5, no. 1 (January 1934): 20–30.

104. Morris Muskat, "The Flow of Compressible Fluids through Porous Media and Some Problems in Heat Conduction," *Physics* 5 no. 1 (January 1934): 82n9.

105. Muskat, "The Flow of Compressible Fluids through Porous Media and Some Problems in Heat Conduction," 77–84.

106. Muskat, "The Flow of Compressible Fluids through Porous Media and Some Problems in Heat Conduction," 78.

107. Muskat, "The Flow of Compressible Fluids through Porous Media and Some Problems in Heat Conduction," 84n10.

108. Muskat, "The Flow of Compressible Fluids through Porous Media and Some Problems in Heat Conduction," 84.

109. Muskat, "The Flow of Compressible Fluids through Porous Media and Some Problems in Heat Conduction," 90–92.

110. Muskat, "The Flow of Compressible Fluids through Porous Media and Some Problems in Heat Conduction," 71.

111. Schilthuis and Hurst, "Variations in Reservoir Pressure in the East Texas Field," 164–76.

112. Schilthuis and Hurst, "Variations in Reservoir Pressure in the East Texas Field," 167–68.

113. Schilthuis and Hurst, "Variations in Reservoir Pressure in the East Texas Field," 168–69.

114. Schilthuis and Hurst, "Variations in Reservoir Pressure in the East Texas Field," 169–70.

115. Bridgman, "Thermodynamic Properties of Liquid Water," 357.

116. Wyckoff, discussion of Schilthuis and Hurst, "Variations in Reservoir Pressure in the East Texas Field," 174–75, 175, fig. 5.

117. Wyckoff, discussion of Schilthuis and Hurst, "Variations in Reservoir Pressure in the East Texas Field," 174.

118. E. A. Stephenson, discussion of Schilthuis and Hurst, "Variations in Reservoir Pressure in the East Texas Field," 176.

119. Donald T. Campbell and Donald W. Fiske, "Convergent and Discriminant Validation by the

Multitrait-Multimethod Matrix," *Psychological Bulletin* 56 (March 1959): 81–105, reprinted in Donald T. Campbell, *Methodology and Epistemology for Social Science: Selected Papers*, ed. E. Samuel Overman (Chicago, IL: University of Chicago Press, 1988), 37–61.

120. Ralph D. Wyckoff, Gulf Research & Development Corporation, Pittsburgh, discussion of Nye and Reistle, "Recent Changes," 82.

121. D. C. Barton and J. B. Umpleby, discussion of Nye and Reistle, "Recent Changes," 83.

122. C. B. Carpenter and G. B. Spencer, "Measurement of Compressibility of Consolidated Oil-Bearing Sandstones," *Bureau of Mines RI 3540* (October 1940), 6n17; J. S. Hundall, Unpublished paper read before the American Association of Petroleum Geologists annual meeting, Dallas, March 1934; J. O. Lewis, "Rock Pressure," *Petroleum Engineer* (May 1934): 40–44.

123. Carpenter and Spencer, "Measurement of Compressibility," 7.

124. Carpenter and Spencer, "Measurement of Compressibility," 2, reiterated at 18 and 19.

125. Joseph Bertram Umpleby, "Changing Concepts in the Petroleum Industry," *Transactions of the AIME* 98, no. 1 (December 1932). Paper read before AIME Petroleum Division, New York, February 1932; *Petroleum Development and Technology 1932* (New York, Petroleum Division, American Institute of Mining and Metallurgical Engineers, 1933), 43.

126. Schilthuis and Hurst, "Variations in Reservoir Pressure in the East Texas Field," 173.

127. Everette Lee DeGolyer Sr., "Production Under Effective Water Drive as a Standard for Conservation Practice," 2, MSS 60 DeGolyer, Box 18, Folder 2295, DeGolyer Papers.

128. *Panama Refining Company, et al. v. Ryan, et al.* 5 F. Supp. 639 (Eastern District of Texas, Tyler, February 12, 1934); *Ryan v. Amazon Petroleum Corporation*, 293 U.S. 388 (US Supreme Court, January 7, 1935).

129. *Amazon Petroleum Corporation v. Railroad Commission of Texas et al.*, 5 F. Supp. 633, at 638 (Federal District Court, Eastern District of Texas, Tyler, February 12, 1934).

130. Trial Transcript, Vol. I, at 500, *Amazon Petroleum Corporation, et al., v. Railroad Commission of Texas, et al.*, Eastern District of Texas, Tyler District Court, in Equity 652, National Archives, Federal Record Center, Fort Worth, Agency Box no. 17, Accession no. 90-5-18, FRC 61027.

131. Trial Transcript, Vol. I, at 553–900, *Amazon Petroleum Corporation, et al., v. Railroad Commission of Texas, et al.*

132. Complainants' Trial Brief, at 26–29, *Amazon Petroleum Corporation, et al., v. Railroad Commission of Texas, et al.*

133. Complainants' Trial Brief, Appendix, at 233, *Amazon Petroleum Corporation, et al., v. Railroad Commission of Texas, et al.*

134. Complainants' Trial Brief, Appendix, at 234–37, *Amazon Petroleum Corporation, et al., v. Railroad Commission of Texas, et al.*

135. Trial Transcript, Vol. 1, at 500. Affidavit of Stanley Gill, Trial Transcript, Vol. I, at 516, *Amazon Petroleum Corporation, et al., v. Railroad Commission of Texas, et al.*

136. Petroleum Investigation, House of Representatives, Subcommittee of the Committee on Interstate and Foreign Commerce [the "Cole Committee"], Dallas, Texas, November 20,1934, "Statement of Stanley Gill, Consulting Engineer, Houston, Tex." at 2113, Affidavit of Stanley Gill, Trial Transcript, Vol. I, at 505, *Amazon Petroleum Corporation, et al., v. Railroad Commission of Texas, et al.*

137. Trial Transcript, Vol. 1, at 500, Affidavit of Stanley Gill, Trial Transcript, Vol. I, at 512–13, *Amazon Petroleum Corporation, et al., v. Railroad Commission of Texas, et al.*

138. Complainants' Trial Brief, at 18, *Amazon Petroleum Corporation, et al., v. Railroad Commission of Texas, et al.*

139. Trial Transcript, Vol. III, at 1294–1383, *Amazon Petroleum Corporation, et al., v. Railroad Commission of Texas, et al.*

140. Deposition of T. V. Moore and C. E. Reistle Jr., December 12, 1933, Trial Transcript Vol. III, at 1381–88, *Amazon Petroleum Corporation, et al., v. Railroad Commission of Texas, et al.*

141. Deposition of T. V. Moore and C. E. Reistle Jr., December 12, 1933, Trial Transcript Vol. III, at 1381–83, *Amazon Petroleum Corporation, et al., v. Railroad Commission of Texas, et al.*

142. *Amazon Petroleum Corporation v. Railroad Commission of Texas et al.*, 5 F. Supp. 633, at 638–39.

143. *Amazon Petroleum Corporation v. Railroad Commission of Texas et al.*, 5 F. Supp. 633, at 638.

144. *Rubáiyát of Omar Khayyám*, rendered into English verse by Edward Fitzgerald (New York: Garden City Books, 1952), Verse LXI, 109.

145. Faced with near-schism in the Church as monastic orders across Europe gleefully digested and debated the corpus of Greek learning acquired during the Crusades (largely known by Latin retranslations of Arabic texts translated from Greek or Latin), Saint Thomas Aquinas invented natural theology: It was okay to study nature, natural philosophy, or even philosophy as (further) evidence of God's wisdom and design. Science, Greek and Islamic, survived and prospered in Western Europe. But Aquinas, immersed in a tradition of textual exegesis as the one ordained way to absolute theological truth, conflated "Truth" with well-founded scientific belief. He, and his

successors, therefore placed the imprimatur of Holy Truth on specific scientific hypotheses, especially those of Aristotle. Hence the travails of Copernicus, Galileo, and Darwin. Aquinas's bit of metaphysical legerdemain wasn't really rectified until the Kuhnian revolution, although a variety of scholars, among them John Locke, Ernst Mach, Henri Poincaré, Karl Popper, Donald Campbell, Bas van Fraassen, Wesley Salmon, Wilfried Sieg, and a whole cadre of Bayesians, had reached similar conclusions by dissimilar routes.

146. Complainants' Trial Brief, at 29, *Amazon Petroleum Corporation, et al., v. Railroad Commission of Texas, et al.*

147. Complainants' Trial Brief, at 223, *Amazon Petroleum Corporation, et al., v. Railroad Commission of Texas, et al.*

148. *Amazon Petroleum Corporation v. Railroad Commission of Texas et al.*, 5 F. Supp. 633, at 639 (Federal District Court, Eastern District of Texas, Tyler, February 12, 1934).

149. Hardwicke, "Legal History of Conservation of Oil in Texas," 214–68; Hardwicke, "Conservation of Oil and Gas: Texas, 1938–1948," 447–512.

150. Hardwicke, "Legal History of Conservation of Oil in Texas," 1938, 215.

151. Hardwicke, "Legal History of Conservation of Oil in Texas," 1938, 245–47.

152. Weaver, *Unitization of Oil and Gas Fields in Texas*; Malavis, *Pure and Humble*.

# CHAPTER 7

1. Reistle Jr., "Reservoir Engineering," 830.

2. Morris Muskat, Ralph Dewey Wyckoff, Holbrook Gorham Botset, and Milan W. Meres, "Flow of Gas-Liquid Mixtures through Sands," *Transactions of the American Institute of Mining and Metallurgical Engineers: Petroleum Development and Technology* 123, no. 1 (December 1937): 69–70; New York meeting of the Petroleum Division, American Institute of Mining and Metallurgical Engineers, February 1937; received at the Institute February 15, 1937. This paper is a fusion and abbreviation of two separate papers published in *Physics* by Muskat and Meres, and Wyckoff and Botset.

3. Morris Muskat, *The Flow of Homogeneous Fluids Through Porous Media* (New York: McGraw-Hall, 1937; Ann Arbor, MI: J. W. Edwards, 1946): vii–x.

4. Muskat, Wyckoff, Botset, and Meres, "Flow of Gas-Liquid Mixtures through Sands," 70.

5. H. H. Evinger and Morris Muskat, "Calculation of Theoretical Productivity Factor," *Transactions of the American Institute of Mining and Metallurgical Engineers: Petroleum Development and Technology* 146, no. 1 (December 1942): 126–27; New York Meeting of the Petroleum Division, American Institute of Mining and Metallurgical Engineers, February 1941, received February 7, 1941, Issued as T.P. 1352 *Petroleum Technology*, September 1941.

6. Muskat, Wyckoff, Botset, and Meres, "Flow of Gas-Liquid Mixtures through Sands," 70–71.

7. Betty Jo Teeter Dobbs, *The Foundations of Newton's Alchemy, or, "the Hunting of the Greene Lyon"* (Cambridge, UK: Cambridge University Press, 1975).

8. Peter Galison, *Einstein's Clocks, Poincaré's Maps: Empires of Time* (New York: W. W. Norton, 2004).

9. Stuart A. Kauffman, J. Lobo, and W. G. Macready, "Optimal Search on a Technological Landscape," *Journal of Economic Behavior and Organization* 43 (October 2000): 141–66. Stuart A. Kauffman, *The Origins of Order: Self-Organization and Selection in Evolution* (Oxford, UK: Oxford University Press, 1993).

10. Muskat, *Physical Principles of Oil Production*, 273.

11. Howard C. Pyle and John E. Sherborne (Union Oil Company of California), "Core Analysis," *Transactions of the American Institute of Mining and Metallurgical Engineers: Petroleum Development and Technology* 132, no. 1 (December 1939): 33–61; AIME Petroleum Division San Antonio Meeting and Los Angles Meeting, October 1938, received at Institute October 14, 1938. Issued as T.P. 1024 in *Petroleum Technology*, February 1939.

12. Ralph D. Wyckoff and Holbrook G. Botset, "The Flow of Gas-Liquid Mixtures through Unconsolidated Sands," *Physics* 7, no. 9 (September 1936): 325–45, Received July 22, 1936; Morris Muskat and Milan W. Meres, "The Flow of Heterogeneous Fluids through Porous Media," *Physics* 7, no. 9 (September 1936): 346–63. Received July 22, 1936.

13. Muskat, Wyckoff, Botset, and Meres, "Flow of Gas-Liquid Mixtures through Sands," 69–96.

14. Wyckoff and Botset, "Flow of Gas-Liquid Mixtures through Unconsolidated Sands," 328.

15. Muskat, Wyckoff, Botset, and Meres, "Flow of Gas-Liquid Mixtures through Sands," 75–77.

16. Wyckoff and Botset, "Flow of Gas-Liquid Mixtures through Unconsolidated Sands," 330–31.

17. Wyckoff and Botset, "Flow of Gas-Liquid Mixtures through Unconsolidated Sands."

18. Wyckoff and Botset, "Flow of Gas-Liquid Mixtures through Unconsolidated Sands," 331–33.

19. Wyckoff and Botset, "Flow of Gas-Liquid Mixtures through Unconsolidated Sands," 332.

20. Muskat, Wyckoff, Botset, and Meres, "Flow of Gas-Liquid Mixtures through Sands," 76, fig. 4.

21. Wyckoff and Botset, "Flow of Gas-Liquid Mixtures through Unconsolidated Sands," 332.

22. Muskat, Wyckoff, Botset, and Meres, "Flow of Gas-Liquid Mixtures through Sands," 77.

23.  Muskat and Meres, "The Flow of Heterogeneous Fluids through Porous Media."

24.  Willard W. Cutler Jr., "Estimation of Underground Oil Reserves by Oil-Well Production Curves," *U.S. Bureau of Mines Bulletin 228* (1924), 85–90.

25.  John R. Suman, "The Well Spacing Problem—Low Well Density Increases Ultimate Recovery," *API Drilling and Production Practice* 158 (1934). John R. Suman, "Evolution by Companies," *API History*, 117.

26.  Muskat and Meres, "The Flow of Heterogeneous Fluids through Porous Media," 358–60.

27.  Muskat and Meres, "The Flow of Heterogeneous Fluids through Porous Media," 360.

28.  Muskat, Wyckoff, Botset, and Meres, "Flow of Gas-Liquid Mixtures through Sands," 94.

29.  Muskat and Meres, "The Flow of Heterogeneous Fluids through Porous Media," 361–62.

30.  Muskat and Meres, "The Flow of Heterogeneous Fluids through Porous Media," 361–62.

31.  Hull, *Science as a Process*.

32.  L. S. Reid (Phillips Petroleum) and R. L. Huntington (Petroleum Engineering School, University of Oklahoma), "Flow of Oil-Gas Mixtures through Unconsolidated Sands," *Transactions of the American Institute of Mining and Metallurgical Engineers* 127, no. 01 (December 1938): 228, 232–33; presented at the AIME Petroleum Division Oklahoma City Meeting, October 1937, received at the Institute August 18, 1937; "Paper presented by L. S. Reid in partial fulfillment of the requirements of the Master of Science Degree in Petroleum Engineering, University of Oklahoma."

33.  M.C. Leverett, "Flow of Oil-water Mixtures through Unconsolidated Sands," *Transactions of the American Institute of Mining and Metallurgical Engineers* 132, no. 1 (December 1939): 149–71; San Antonio Meeting, October 1938; received at the Institute September 23, 1938; "Part of a thesis submitted to the Massachusetts Institute of Technology in partial fulfillment for the degree of Doctor of Science in Chemical Engineering." Issued as T.P. 1003 in *Petroleum Technology*, November 1938.

34.  Holbrook G. Botset, "Flow of Gas-Liquid Mixtures through Consolidated Sand," *Transactions of the American Institute of Mining Engineers* 136, no. 1 (December 1940): 91–105; Galveston Meeting, October 1939, received at the institute April 28, 1939, Issued as T.P. 1111 in *Petroleum Technology*, November 1939. See Holbrook G. Botset, comment, October 1938, on Leverett, "Flow of Oil-Water Mixtures through Unconsolidated Sands," 170.

35.  Stuart E. Buckley, comment on Botset, "Flow of Gas-Liquid Mixtures through Consolidated Sand," 104.

36.  Botset, "Flow of Gas-Liquid Mixtures through Consolidated Sand," 94, fig. 3.

37.  M.C. Leverett and W. B. Lewis, "Steady Flow of Gas-Oil-Water Mixtures through Unconsolidated Sands," *Transactions of the American Institute of Mining and Metallurgical Engineers* 142 no. 1 (December 1941): 101–16; New York Meeting, February 1940; Received at the Institute January 30, 1940, Issued as T.P. 1206 in *Petroleum Technology*, May 1940.

38.  Leverett and Lewis, "Steady Flow of Gas-Oil-Water Mixtures," 113–14.

39.  Leverett and Lewis, "Steady Flow of Gas-Oil-Water Mixtures," 115.

40.  Leverett and Lewis, "Steady Flow of Gas-Oil-Water Mixtures," 114, fig. 9.

41.  Stuart E. Buckley and M. C. Leverett (Humble Oil & Refining), "Mechanism of Fluid Displacement in Sands," *Transactions of the American Institute of Mining and Metallurgical Engineers: Petroleum Development and Technology* 146, no. 1 (December 1942): 107–16; AIME New York Meeting, February 1941, received at the Institute January 6, 1941. Issued as T.P. 1337 in *Petroleum Technology*, May 1941.

42.  Buckley and Leverett, "Mechanism of Fluid Displacement in Sands," 112–13.

43.  H. H. Evinger and Morris Muskat, "Calculation of Theoretical Productivity Factor," *Transactions of the American Institute of Mining and Metallurgical Engineers: Petroleum Development and Technology* 146, no. 1 (December 1942): 126–39; New York Meeting, February 1941, received at the Institute February 7, 1941. Issued as T.P. 1352 in *Petroleum Technology*, September 1941. H. H. Evinger and Morris Muskat, "Calculation of Productivity Factors for Oil-Gas-Water Systems in the Steady State," *Transactions of the American Institute of Mining and Metallurgical Engineers: Petroleum Development and Technology* 146, no. 1 (December 1942): 194–203; New York Meeting, February 1942, received at the Institute July 30, 1941. Issued as T.P. 1416 in *Petroleum Technology*, January 1942.

44.  R. E. Old Jr. (Chemical Engineer, Core Laboratories Inc., Dallas), "Analysis of Reservoir Performance," *Transactions of the American Institute of Mining and Metallurgical Engineers* 151, no. 1 (December 1943): 86–98; Austin Meeting of the Petroleum Division of AIME, October 1942, received at the Institute August 6, 1942, Issued in *Petroleum Technology*, November 1942; Rex W. Woods (Gulf Oil Co., Tulsa) and Morris Muskat (Gulf Research and Development Co., Pittsburgh), "An Analysis of Material-Balance Calculations," *Transactions of the Institute of Mining Engineers* 160, no. 1 (December 1945): 124–39; Houston Meeting of the Petroleum Division, AIME, May 1944, received at the Institute May 24, 1944, Issued as T.P. 1780 in *Petroleum Technology*, January 1945.

45.  James A. Lewis, William L. Horner, and Marion Stekoll (Core Laboratories), "Productivity Index and

Measurable Reservoir Characteristics," American Institute of Mining and Metallurgical Engineers, Technical Publication No. 1467 (Class G. Petroleum Division, No. 160), 1942. Evinger and Muskat, "Calculation of Productivity Factors for Oil-Gas-Water Systems in the Steady State," 202.

46.    T. M. Geffen, D. R. Parrish, G. W. Haynes, and R. A. Morse (Stanolind Oil and Gas, Tulsa), "Efficiency of Gas Displacement from Porous Media by Liquid Flooding," *Transactions of the American Institute of Mining and Metallurgical Engineers: Petroleum Development and Technology* 195 (1952): 29; Alternative citation: *Journal of Petroleum Technology* 4, no. 2 (February 1952); presented at the Fall Meeting of the Petroleum Branch of AIME, Oklahoma City, October 1951, Issued as T. P. 3279.

47.    Evinger and Muskat, "Calculation of Productivity Factors for Oil-Gas-Water Systems in the Steady State," 194.

48.    Thomas P. Hughes, *American Genesis: A Century of Invention and Technological Enthusiasm* (New York: Viking, 1989): 52.

49.    Muskat, *Physical Principles of Oil Production*, vii–viii.

50.    Dennis Denney (*JPT* Technology Editor), "Technical Papers Have Been the Backbone of SPE's Mission," *Journal of Petroleum Technology, Special Commemorative Issue Celebrating 50 Years of the Society of Petroleum Engineers* 59, no. 10 (October 2007): 103.

51.    A. F. van Everdingen (Shell Oil) and William Hurst (Petroleum Consultant), "The Application of the La Place Transformation to Flow Problems in Reservoirs," *Journal of Petroleum Technology* 1, no. 12 (December 1949): 305; presented at the AIME Annual Meeting, San Francisco, February 13–17, 1949, received at the office of the Petroleum Branch, January 12, 1949, Issued as T.P. 2732; G. H. Bruce, D. W. Paceman, H. H. Rachford Jr. (all Humble Oil & Refining), and J. D. Rice (Rice Institute), "Calculations of Unsteady-State Gas Flow through Porous Media," *Transactions of the American Institute of Mining and Metallurgical Engineers: Petroleum Development and Technology* 198 (1953): 79; *Journal of Petroleum Technology* 5, no. 3 (March 1953): 79–92; Petroleum Branch Fall Meeting, Houston, October 1–3, 1952; received August 12, 1952, Issued as T.P. 3518. J. S. Aronofsky and R. Jenkins, "Unsteady Flow of Gas Through Porous Media—One Dimensional Case," in *Proceedings of the First U.S. National Congress of Applied Mechanics* (Ann Arbor, MI: J. Edwards, 1952). R. Jenkins and J. S. Aronofsky, "Unsteady Radial Flow of Gas Through Porous Media," *Journal of Applied Mechanics*, A.S.M.E. 20, no. 2 (June 1953); Annual Meeting of the American Society of Mechanical Engineers, December 1952. W. A. Bruce (Carter Oil Co.), "An Electrical Device for Analyzing Oil-Reservoir Behavior," *Transactions of the American Institute of Mining and Metallurgical Engineers* 151, no. 1 (December 1943): 112–24; Austin Meeting of the Petroleum Division of AIME, October 1942, received at the Institute September 28, 1942; Issued in *Petroleum Technology*, T.P. 1550, January 1943.

52.    Henry J. Welge (Carter Oil Co., Tulsa), "A Simplified Method for Computing Oil Recovery by Gas or Water Drive," *Transactions of the American Institute of Mining and Metallurgical Engineers* 195 (1952): 92–93, 95–96; Alternative citation: *Journal of Petroleum Technology* 4, no. 14 (April 1952): 91–98; presented at the Petroleum Branch Fall Meeting, Oklahoma City, October 1951, received at the Petroleum Branch September 25, 1951.

53.    J. S. Aronofsky and R. Jenkins (Magnolia Petroleum, Dallas), "A Simplified Analysis of Unsteady Radial Gas Flow," *Journal of Petroleum Technology* 6, no. 7 (July 1954): 24; Received at the Petroleum Branch Office, February 13, 1953. T.P. 3802.

54.    PetroMehras, accessed 2/19/20.

55.    Derek J. De Solla Price, *Science Since Babylon*, enlarged edition (New Haven, CT: Yale University Press, 1975), 128.

56.    Known among Francophones as Mariotte's Law for Edme Mariotte, who in 1676 noted that Boyle's relationship held only for constant temperature. From Mariotte comes the more common relation P'V'/T' = PV/T.

57.    Li Fan (College Station, Texas), Billy W. Harris (Wagner & Brown, Ltd., Midland), A. (Jamal) Jamaluddin (Rosharon, Texas), Jairam Kamath (Chevron Energy Technology Co., San Ramon, California), Robert Mott (Independent Consultant, Dorchester, UK), Gary A. Pope (University of Texas, Austin), Alexander Shandrygin (Moscow, Russia), Curtis Hays Whitson (Norwegian University of Science and Technology and PERA, A/S, Trondheim, Norway), "Understanding Gas-Condensate Reservoirs," *Oil Field Review* 17, no. 4 (Winter 2005–2006): 14n3 (published by Schlumberger Technology Corporation); Donald L. Katz and Frederick Kurata (University of Michigan), "Retrograde Condensation," *Industrial and Engineering Chemistry* 32, no. 6 (1 June 1940): 817.

58.    J. P. Kuenen, "On Retrograde Condensation and the Critical Phenomena of Mixtures of Two Substances," Communication, Physics Laboratory, University of Leiden, The Netherlands (1892) No. 4B, 7–14; Katz and Kurata, "Retrograde Condensation," 817; Donald L. Katz (University of Michigan), "Overview of Phase Behavior in Oil and Gas Production," *Journal of Petroleum Technology* 35, no. 6 (June 1983): 1205.

59.    Ben Edwin Lindsly, "Solubility and Liberation of Gas from Natural Oil-Gas Solutions," *U.S. Bureau of Mines Technical Paper 554* (1933).

60.    Ben E. Lindsly to H. C. Fowler, "Monthly letter," December 31, 1931, Accession No. 65-A-855, RG 70,

National Archives, Federal Record Center, Fort Worth, Texas.

61.   Bruce H. Sage, Jan G. Schaafsma, and William N. Lacey, *American Petroleum Institute Bulletin 212* (1933).

62.   Ben E. Lindsly to H. C. Fowler, September 30, 1931, Accession No. 65-A-855, RG 70, National Archives, Federal Record Center, Fort Worth, Texas.

63.   Fan, et al., "Understanding Gas-Condensate Reservoirs," 14–27; Katz and Kurata, "Retrograde Condensation," 817–27; Katz, "Overview of Phase Behavior in Oil and Gas Production," 1205–14; Stuart E. Buckley and J. H. Lightfoot (Humble Oil & Refining), "Effects of Pressure and Temperature on Condensation of Distillate from Natural Gas," *Transactions of the American Institute of Mining Engineers* 142, no. 01 (December 1941): 232–45; E. V. Foran (Consulting Engineer, Austin), "Development and Production Problems in High-pressure Distillate Pools," *Transactions of the American Institute of Mining Engineers* 132, no. 1 (December 1939): 22–32; San Antonio Meeting, October 1938, received at the Institute October 18, 1938, issued as T.P. 1023 in *Petroleum Technology*, February 1939.

64.   *Clymore Production Company v. Thompson*, 11 F. Supp. 791 (W.D. Tex. 1935; prelim. inj.) 13 F. Supp. 469 (W.D. Tex. 1936; final hearing). Clymore was expanding gas containing condensate through a nozzle (thereby cooling it), blowing it against baffles, and claiming the distillate that condensed out made the well an oil well. Therefore, they were free to "pop" (release) or flare any residual dry gas. The Railroad Commission argued, successfully, that the effluent from the well was all gas at reservoir conditions, that Clymore's process, crude as it was, artificially altered the gas, and that flaring the residual constituted illegal waste of natural gas from a gas well.

65.   Muskat, *Physical Principles of Oil Production*, 787–88.

66.   Li Fan, et al., "Understanding Gas-Condensate Reservoirs," 20.

67.   PERM Inc., TIPM Laboratory, "Fundamentals of Fluid Flow in Porous Media: Chapter 2. Porosity: Laboratory Porosity Measurement," accessed April 24, 2020; Muskat, *Physical Principles of Oil Production*, 148–68; Muskat, *Flow of Homogeneous Fluids Through Porous Media*, 113–20.

68.   Charles R. Fettke (Carnegie Institute of Technology, Pittsburgh), "Core Studies of the Second Sand of the Venango Group from Oil City, Pennsylvania," *Transactions of the American Society of Mining and Metallurgical Engineers: Petroleum Development and Technology in 1926* G26 no. 1 (December 1926): 229, 231 table 6; Reistle, "Reservoir Engineering," *API History*, 827.

69.   N. T. Lindtrop and V. M. Nicolaef, "Oil and Water Content of Oil Sands: Grozny, Russia," *American Association of Petroleum Geologists Bulletin* 13, no. 7 (July 1929): 811–22.

70.   Jan Versluys, "Factors Involved in Segregation of Oil and Gas from Subterranean Water," *American Association of Petroleum Geologists Bulletin* 16, no. 9 (September 1932): 924–42; Reistle, "Reservoir Engineering," *API History*, 827.

71.   R. J. Schilthuis, "Connate Water in Oil and Gas Sands," *Transactions of the American Institute of Mining and Metallurgical Engineers: Petroleum Development and Technology* 127 (December 1938): 199n2–8. Allen D. Garrison (Rice Institute), "Selective Wetting of Reservoir Rocks and its Relation to Oil Production," *API Drilling and Production Practice* (1 January 1935): 330–41.

72.   Schilthuis, "Connate Water," 199–201.

73.   Schilthuis, "Connate Water," 201–4.

74.   Schilthuis, "Connate Water," 204–7.

75.   Schilthuis, "Connate Water," 211–12.

76.   Schilthuis, "Connate Water," 214.

77.   Leverett and Lewis, "Steady Flow of Gas-Oil-Water Mixtures," 114.

78.   M.C. Leverett, "Capillary Behavior in Porous Solids," *Transactions of the American Institute of Mining and Metallurgical Engineers* 142, no. 1 (December 1941): 152–74 (paper presented Petroleum Division of AIME, Tulsa Meeting, October 1940).

79.   E. DeGolyer (Professor of Geology, University of Texas), "The Development of Some of Our Concepts of Fundamentals Governing Petroleum Production Practice" (typescript), March 12, 1940, MSS 60, Box 26, Folder 2537, DeGolyer Papers, 14–15.

80.   DeGolyer, "Waste in Oil and Gas Production," 36.

81.   Startling revision is not unique to petroleum engineering science. Recent investigations using whole-genome sequencing of mitochondrial DNA in baleen whale trios (sire, dam, offspring) indicate that actual populations of humpback whales before extensive whaling were 86 percent lower than widely accepted estimates based upon phylogenetic analyses. See Marco Suárez-Menéndez, Martine Bérube, et al., "Wild Pedigrees Inform Mutation Rates and Historic Abundance in Baleen Whales," *Science* 381 no. 6661 (1 September 2023): 990–95; A. Rus Hoetzel and Michael Lynch, "The Raw Material of Evolution: Estimates of Whale Mutation Rates Contribute to Understanding Evolutionary Processes," *Science* 381 no. 6661 (1 September 2023): 942–43.

82.   Resit Kenan, "The Mathematical Analysis of the Flow of Compressible Fluids through Porous Media, and

the Application to the East Texas Field Pressure Decline" (MS thesis, University of Texas at Austin, 1941), 27; Stuart E. Buckley (Humble Oil and Refining Co.), "The Pressure-Production Relationship in the East Texas Field," *API Drilling and Production Practice* (1938): 140.

83.  Kenan, "Mathematical Analysis," 27.

84.  Kenan, "Mathematical Analysis," 31.

85.  Kenan, "Mathematical Analysis," 31–35; Kuhn, *The Structure of Scientific Revolutions*.

86.  R. C. Rumble, H. H. Spain, and H. E. Stamm III (Humble Oil and Refining Co.), "A Reservoir Analyzer Study of the Woodbine Basin," *Journal of Petroleum Technology* 3, no. 12 (December 1951): 331–40. Alternative citation: *Petroleum Transactions, AIME* 192 (1951): 331–40, at 331; Petroleum Branch Fall Meeting, American Institute of Mining Engineers, Oklahoma City, October 3–5, 1951, paper received August 28, 1951; T.P. 3219; Percy W. Bridgman, "Thermodynamic Properties of Liquid Water to 80° (C) and 12,000 KG N (Kg/Cm²)," *Proceedings of the American Academy of Arts and Sciences* 48, no. 9 (September 1912): 357.

87.  Rumble, Spain, and Stamm, "Reservoir Analyzer Study," 331–32.

88.  John S. Bell and J. M. Shepherd, "Pressure Behavior in the Woodbine Sand," *Journal of Petroleum Technology* 3, no. 1 (January 1951): 20–22; alternative citation: *Transactions of the American Institute of Mining Engineers, Petroleum Transactions* 192 (1951): 20–22; paper presented at the Fall Meeting of the Petroleum Branch of the American Institute of Mining Engineers, New Orleans, October 1950, T.P. 3000.

89.  Edward Allen Wendlandt, Thomas H. Shelby, and John Smith Bell, "Hawkins Field, Wood County, Texas," *Bulletin of the American Association of Petroleum Geologists* 30, no. 11 (November 1946): 1830–56.

90.  Bell and Shepherd, "Pressure Behavior in the Woodbine Sand," 20–22.

91.  Bell and Shepherd, "Pressure Behavior in the Woodbine Sand," 22.

92.  Bell and Shepherd, "Pressure Behavior in the Woodbine Sand," 23, 22, fig. 3.

93.  Bell and Shepherd, "Pressure Behavior in the Woodbine Sand," 23–26.

94.  Bell and Shepherd, "Pressure Behavior in the Woodbine Sand," 27.

95.  Bell and Shepherd, "Pressure Behavior in the Woodbine Sand," 28.

96.  Rumble, Spain, and Stamm, "Reservoir Analyzer Study," 331–32R. C. Rumble, H. H. Spain, and H. E. Stamm III (all Humble Oil and Refining Co., Houston), "A Reservoir Analyzer Study of the Woodbine Basin," *Petroleum Transactions, AIME* 192 (1951): 331–40. Alternative citation: *Journal of Petroleum Technology* 3 no. 12 (December, 1951): SPE-122-G; Petroleum Branch Fall Meeting, American Institute of Mining Engineers, Oklahoma City, October 3–5, 1951 (paper received August 28, 1951); also published as T.P. 3219.

97.  W. A. Bruce (Carter Oil Co.), "An Electrical Device for Analyzing Oil-Reservoir Behavior," *Transactions of the American Institute of Mining and Metallurgical Engineers* 151, no. 01 (December 1943): 112–24. Rumble, Spain, and Stamm, "Reservoir Analyzer Study," 332.

98.  Derek J. de Solla Price, "Gears from the Greeks. The Antikythera Mechanism: A Calendar Computer from ca. 80 BC," *Transactions of the American Philosophical Society* 64, no. 7 (1974): 1–70; Price, *Science since Babylon*; David N. McKnight, *The Rittenhouse Orrery: A Marvel of History, Science and Craftsmanship* (Philadelphia, PA: University of Pennsylvania, n. d.).

99.  Rumble, Spain, and Stamm, "Reservoir Analyzer Study," 331–40.

100.  C. B. Carpenter and G. B. Spencer, "Measurement of Compressibility of Consolidated Oil-Bearing Sandstones," *Bureau of Mines RI 3540*, October 1940.

101.  Rumble, Spain, and Stamm, "Reservoir Analyzer Study," 336.

102.  George W. Barineau, "Supplement, 1962: The East Texas Field," in Rumble, Spain, and Stamm, "Reservoir Analyzer Study," 331–40.

103.  Barineau, "Supplement, 1962: The East Texas Field," 331–40.

104.  Hull, *Science as a Process*, 520.

105.  Most of the early research on these topics was supported by the National Science Foundation, History and Philosophy of Science Program, Grants Nos. SOC-7912933 and SES-8618890.

106.  Robert Boyd and Peter J. Richerson, *Culture and the Evolutionary Process* (Chicago: University of Chicago Press, 1985); Robert Boyd and Peter J. Richerson, *The Origin and Evolution of Cultures* (Oxford, UK: Oxford University Press, 2005).

107.  Bryan Pfaffenberger, "The Harsh Facts of Hydraulics: Technology and Society in Sri Lanka's Colonization Schemes," *Technology and Culture* 31 (July 1990): 361–97; Ostrom, *Governing the Commons*; Poteete, Janssen, and Ostrom, *Working Together*; James C. Scott, *Seeing Like a State: How Certain Schemes to Improve the Human Condition Have Failed* (New Haven, CT: Yale University Press, 1998).

108.  American Petroleum Institute, *Summary of API Program of Fundamental Research in Petroleum 1926–1951* (unpublished, API Library, Washington, DC), 4–5. I am indebted to Ms. Lois Schuermann, API Librarian, for her help in using these materials.

109.  Suman, "Evolution by Companies," 63–132.

110. Warner, "Sources of Men," 44.

111. Erica Shillings, "SPE: A 1957 Snapshot," in *Journal of Petroleum Technology, Special Commemorative Issue Celebrating 50 Years of the Society of Petroleum Engineers* 59, no. 10 (October 2007): 49–51.

112. Based on a random sample (N=201) drawn from the membership records for both active and inactive members of the Society of Petroleum Engineers, Dallas, Texas, in 1980. Criteria for inclusion in the sample were that the member not be a student member or foreign and be a member or reinstated member of the society prior to 1960. To preserve anonymity, no names were recorded, just sociometric data. Within the sample, 180 members provided degree data; 21 either had no degree or no data indicated. I am indebted to Mr. Dan K. Adamson, Executive Director of the society, for granting me access to its membership records. This data, and that discussed below regarding the geographical origins, parents' occupations, religious preferences, and employment are presented in greater detail in Edward W. Constant II, "Science in Society: Petroleum Engineers and the Oil Fraternity in Texas, 1925–65," *Social Studies of Science* 19 (1989): 439–72.

113. Education Committee, Society of Petroleum Engineers of American Institute of Mining Engineers, *Petroleum Engineering Schools of the United States—1965* (Dallas, TX: Office of the Executive Secretary, Society of Petroleum Engineers, 1965).

114. Handwritten notes of Dr. H. Y. Bennedict, president of the University of Texas at Austin, undated, in a file with "Recommendations of the Committee to Consider the Future of the College of Engineering," January 14, 1928, and which were apparently made in preparation for the first committee meeting because the committee members are listed at the bottom of the page (by last name only). University of Texas, President's Office Records, Box VF 1/A, University of Texas Archives, Austin, Texas.

115. F. B. Plummer, "Petroleum Engineering Education—Present Curricula and Future Possibilities," *Mining and Metallurgy* (American Institute of Mining and Metallurgical Engineers), Vol. 17 (October 1936): 485–87.

116. Plummer, "Petroleum Engineering Education."

117. W. R. Woolrich, et al., *Men of Ingenuity: From Beneath the Orange Tower, 1884–1964* (Austin: College of Engineering of the University of Texas, 1969).

118. "College of Engineering," *University of Oklahoma Bulletin, 1924*, 94.

119. Agricultural and Mechanical College of Texas, *Bulletin, 1941–42*, 208.

120. Agricultural and Mechanical College of Texas, *Bulletin, 1958–1959*, 338.

121. Yearly data on number of engineering graduates at UT, and number of petroleum engineering graduates, is from W. R. Woolrich, et al., *Men of Ingenuity*.

122. The sample of UT petroleum engineering graduates was constructed in 1984 in the following way. The University of Texas *Bulletin* for these years lists graduates by major. Within each decade, names were selected at random. Based on that list, Ms. Martha F. Wingren, of the Registrar's Office at the University of Texas, then searched the original students' transcripts for information on students' home addresses and occupations of parents; religious data was not available from these records at UT. I am indebted to Ms. Wingren for collecting and recording the data, to Mr. Albert K. Meerzo, Registrar of the University of Texas, for permitting access to these records; and to Dr. Ralph Elder and Dr. Larry Landis, at the University of Texas Archives, Eugene C. Barker Texas History Center, for helping to arrange access.

    The sample for A&M was defined as for UT. I am indebted to Mr. Robert A. Lacey, Registrar at Texas A&M, for granting me direct access to matriculation records. While names as published in the *Bulletin* were used to identify petroleum engineering graduates, only data on geographical origin, occupations of parents, religious preference, and parents' education was recorded. I am also indebted to Dr. Arthur H. Blair, Assistant to the President; Dr. Frank E. Vandiver, President of Texas A&M; and Dr. David Chapman and Dr. Terry Anderson of the Texas A&M Archives, for arranging access to records. The people in the Registrar's Office, especially Ms. Lorrain Kitchner, assistant registrar, were extraordinarily accommodating, since they had literally to climb over the author during an especially hectic time for them—registration, 1983.

    For OU sample population constructed as for UT and A&M. I am indebted to the Registrar of the University of Oklahoma for access to records, and to Dr. John S. Ezell and Mr. Jack Haley, of the Western History Collections at the University of Oklahoma, for their help.

123. Society of Petroleum Engineers, Dallas, Texas, random sample drawn in 1980.

124. O. G. Lawson, interview by Mr. Mody Boatright, July 29, 1952, Tape No. 315, OH.

# CHAPTER 8

1. *Railroad Commission of Texas v. Rowan & Nichols Oil Company*, 311 U.S. 570, at 573 (US Supreme Court, January 6, 1941).

2. *Marbury v. Madison*, 5 U.S. 137, 2 L. Ed. 60, 1803 U.S. LEXIS 352 (US Supreme Court, 1803). *Munn v.*

*Illinois*, 94 US 113, 24 L. Ed. 77, 1876 U.S. LEXIS 1842 (US Supreme Court, 1877).

3.   E. Blythe Stason (Dean, University of Michigan Law School), "'Substantial Evidence' in Administrative Law," *University of Pennsylvania Law Review* 89 (June 1941): 1026, 1029, 1035, 1040–41; *Interstate Commerce Commission v. Illinois Central*, 215 U.S. 452 (US Supreme Court, 1910); *Interstate Commerce Commission v. Union Pacific Railroad*, 222 U.S. 541 (US Supreme Court, 1911).

4.   Chief Justice Charles Evans Hughes, Majority Opinion, *Consolidated Edison Co. v. National Labor Relations Board*, 305 U.S. 197, at 229 (US Supreme Court, 1938), quoted in Whitney R. Harris, "A Reappraisal of the Substantial Evidence Rule in Texas Administrative Law," *Southwestern Law Journal* 3, no. 4 (1949): 419–20.

5.   *Rowan & Nichols Oil Co. v. Railroad Commission of Texas*, 28 F. Supp. 131, at 134 (Western District of Texas, June 12, 1939).

6.   *Rowan & Nichols Oil Co. v. Railroad Commission of Texas*, 28 F. Supp. 131, at 134.

7.   *Railroad Commission of Texas et al. v. Rowan & Nichols Oil Co.*, 107 F. 2d 70 (United States Court of Appeals, Fifth Circuit, November 3, 1939).

8.   *Railroad Commission of Texas et al. v. Rowan & Nichols Oil Co.*, 60 S. Ct. 1021, 310 U.S. 573 (U.S. Supreme Court, June 3, 1940); As Modified on Denial of Petition for Rehearing, October 21, 1940.

9.   This deference to the findings of regulatory agencies is again in play. On June 28, 2024, in a pair of companion cases, *Loper Bright Enterprises v. Raimondo* and *Relentless, Inc. v. Department of Commerce*, the U.S. Supreme Court overturned its *Chevron U. S. A., Inc. v. Natural Resources Defense Council* 467 U.S. 837 (U.S. Supreme Court, 1984), which had become perhaps the ultimate extension of the rationale in *Amazon* and *Rowan & Nichols*. The Court found that *Chevron* violated the Administrative Practices Act of 1946 in presuming that statutory ambiguities are implicit delegations to agencies. Writing for the 6–3 majority, Chief Justice John Roberts held that, "Courts must exercise their independent judgment in deciding whether an agency has acted within its statutory authority," reasoning that "*Chevron*'s presumption is misguided because agencies have no special competence in resolving statutory ambiguities. Courts do." https://www.supremecourt.gov > opinions.

10.  *Railroad Commission of Texas et al. v. Rowan & Nichols Oil Co.*, 311 U.S. 570 (US Supreme Court, January 6, 1941).

11.  *Railroad Commission of Texas et al. v. Rowan & Nichols Oil Co.*, 311 U.S. 570, at 573, 574.

12.  *Railroad Commission of Texas et al. v. Rowan & Nichols Oil Co.*, 311 U.S. 570, at 573.

13.  *Burford et al. v. Sun Oil Co. et al., Sun Oil Co. et al. v. Burford et al.*, 60 S. Ct. 1098, 319 U.S. 315 (US Supreme Court, May 24, 1943).

14.  *Gulf Land Co. v. Atlantic Refining Co.*, 134 Tex. 74, 131 S. W. (2d) 73, at 82 (Supreme Court of Texas, July 26, 1939).

15.  *Railroad Commission v. Shell Oil Co.*, 139 Tex. 66, 161 S. W. (2d) 1022, at 1028–30 (Supreme Court of Texas, 1942). *Marrs v. Railroad Commission*, 142 Tex. 293, 177 S. W. (2d) 941 (Supreme Court of Texas, January 5, 1944). Hardwicke, "Conservation of Oil and Gas," 509; *Trapp v. Shell Oil Co.*, 145 Tex. 323, 198 S.W. 2d 424 (Supreme Court of Texas, May 15, 1946); *Thomas v. Stanolind Oil and Gas Company*, 145 Tex. 270, 198 S.W. (2d) 420 (Supreme Court of Texas, May 15, 1946); *Hawkins v. Texas Co.*, 209 S. W. (2d) 338, 146 Tex. 511, at 513–14 (Supreme Court of Texas, February 4, 1948).

16.  Railroad Commission of Texas, Oil and Gas Division, Oil and Gas Docket No. 120 #6-1456, "In RE: Conservation and Prevention of Waste of Crude Petroleum and Natural Gas in the East Texas Oil Field," Austin, Texas, March 29, 1940.

17.  Staff of the East Texas Salt Water Disposal Company, *Salt Water Disposal: East Texas Field* (Austin: Petroleum Extension Service, The University of Texas–Division of Extension, 1953), 5–8.

18.  East Texas Salt Water Disposal Company, *Salt Water Disposal*, 5–11.

19.  East Texas Salt Water Disposal Company, *Salt Water Disposal*, 8–14.

20.  Bryan W. Payne, foreword to East Texas Salt Water Disposal Company, *Salt Water Disposal*.

21.  I remember as a child traveling across old US Highway 90 between Houston and Lake Charles, Louisiana, at night, which we did frequently to visit family. Oil field gas flares lit up stretches of the road "bright as day." Headlights were unnecessary and, had I been able to read, I could have easily read the proverbial newspaper.

22.  Prindle, *Petroleum Politics*, 63–66; Christopher James Castaneda, *Regulated Enterprise: Natural Gas Pipelines and Northeastern Markets, 1938-1954* (Columbus, OH: Ohio State University Press, 1993), 23–33.

23.  "Big and Little Inch Pipelines," Wikipedia, accessed June 1, 2020; Castaneda, *Regulated Enterprise*, 34–65.

24.  "Big and Little Inch Pipelines," Wikipedia.

25.  Castaneda offers a detailed account of these machinations. Castaneda, *Regulated Enterprise*, 66–93.

26.  Richard W. Hooley, *Financing the Natural Gas Industry: The Role of Life Insurance Investment Policies* (New York: Columbia University Press, 1961), 33–35.

27.  Oliver E. Williamson, "Transaction Cost Economics: The Governance of Contractual Relations," *The Journal of Law and Economics* 22 (1979): 233–61; Oliver E. Williamson, "The Economics of Organization: The

Transaction Cost Approach," *American Journal of Sociology* 87 (1981): 548–77.

28.  Prindle, *Petroleum Politics*, 65.

29.  Thomas P Hughes, *Networks of Power: Electrification in Western Society, 1880–1930* (Baltimore, MD: Johns Hopkins University Press, 1983). On the actual conversion process and its pitfalls and discontents, Joel A. Tarr and Bill C. Lamperes, "Changing Fuel Use Behavior and Energy Transitions: The Pittsburgh Smoke Control Movement, 1940–1950," *Journal of Social History* 14 (Summer 1981): 561–88; and Castaneda, *Regulated Enterprise*, 94–166.

30.  El Paso Natural Gas Company, *Fourth Supplemental Indenture* (March 1, 1949), 46–47; quoted in Hooley, *Financing the Natural Gas Industry*, 64.

31.  *Railroad Commission et al. v. Shell Oil Co., Inc. et al.*, 206 S.W. 2d 235, at 237 (Supreme Court of Texas, November 16, 1947).

32.  Brief for Appellants, in the Supreme Court of Texas, *Railroad Commission et al. v. Shell Oil Co., Inc. et al.*, Case No. A-1255, Filed May 21, 1947, at 2–4, Railroad Commission of Texas, Oil and Gas Division, Oil and Gas Docket No. 129, #4-10,351, Special Order Restricting the Production of Gas from Oil and Gas Wells Completed in the Various Reservoirs Underlying the Seeligson Field, Jim Wells and Kleberg Counties, Texas, March 17, 1947.

33.  Statement of Facts, *Shell Oil Company, Incorporated, et al. v. Railroad Commission of Texas, et al.*, in the District Court of Travis County, Texas, for the 126th Judicial District of Texas, Roy C. Archer, Judge, filed in the Supreme Court of Texas, April 19,1947, Testimony of Col. Ernest O. Thompson, at 47–55; Brief for Appellants, in the Supreme Court of Texas, *Railroad Commission et al. v. Shell Oil Co., Inc. et al.*, Case No. A-1255, Filed May 21, 1947, at 58–60.

34.  "Inasmuch as there is presently not available any additional beneficial use to which such casinghead gas may be devoted, it has no present market value in the field." Finding of Fact and Conclusions of Law, *Shell Oil Company, Incorporated, et al. v. Railroad Commission of Texas, et al.*, in the District Court of Travis County, Texas, in *Railroad Commission et al. v. Shell Oil Co., Inc. et al.*, Transcript at 88–89.

35.  *Railroad Commission et al. v. Shell Oil Co., Inc. et al.*, 206 S.W. 2d 235 (Supreme Court of Texas, November 16,1947).

36.  United States Constitution, Article I, Section 8, Clause 18: Congress shall have the authority, "To make all Laws which shall be necessary and proper for carrying into Execution the foregoing Powers [powers of Congress enumerated in other parts of the Constitution]."

37.  Statement of Facts, *Shell Oil Company, Incorporated, et al. v. Railroad Commission of Texas, et al.*, in the District Court of Travis County, Texas, Testimony of Col. Ernest O. Thompson, at 73.

38.  Railroad Commission of Texas, Oil and Gas Division, Oil and Gas Docket No. 129 #2-12,541, in RE: Conservation and Prevention of Waste of Crude Petroleum and Natural Gas in the Heyser Field, Calhoun County, Texas, Austin, Texas, November 22, 1948; *Railroad Commission v. Sterling Oil and Refining*, 147 Tex. 547, 218 S.W. 2d 415 (Supreme Court of Texas, 1949); Weaver, *Unitization of Oil and Gas Fields in Texas*, 145; *Railroad Commission et al. v. Flour Bluff Oil Corporation et al.*, 219 S.W. 2d 506 (Court of Civil Appeals of Texas, Austin, March 23, 1949); Railroad Commission of Texas, Oil and Gas Division, Oil and Gas Docket No. 129 #4-13,551, "In RE: Conservation and Prevention of Waste of Crude Petroleum and Natural Gas in the Flour Bluff Field, Nueces County, Texas," Austin, Texas, November 22, 1948.

39.  "If the prevention of waste of natural resources such as gas is to await the time when direct and immediate profits can be realized from the operation, there would have been little need for the people of Texas to have amended their Constitution by declaring that the preservation and conservation of natural resources of the State are public rights and duties and directing that the Legislature pass such laws as may be appropriate thereto (Sec. 59a, Art. 16, Tex. Constitution, Vernon's Ann. St.), for private enterprise would not need the compulsion of law to conserve these resources if the practice were financially profitable. . . .
     "Many of our laws, such as health and safety regulations, are upheld under police powers of the government although obedience results in financial loss. Conservation laws are similar in nature and noncompliance therewith cannot be justified on economic grounds." *Railroad Commission et al. v. Flour Bluff Oil Corporation et al.*, 219 S.W. 2d 506, at 509–10 (Court of Civil Appeals of Texas, Austin, March 23, 1949).

40.  A section is 640 acres, one square mile.

41.  Bernard F. Clark Jr., *Oil Capital: The History of American Oil, Wildcatters, Independents, and Their Bankers* (Houston: Bernard F. Clark Jr., 2016), 270–71.

42.  H. Harold Wineburgh, *The Texas Banker: The Life and Times of Fred Farrell Florence* (Dallas: H. Harold Wineburgh, 1981), 129–30.

43.  Clark, *Oil Capital*, 100–101.

44.  Clark, *Oil Capital*, 110.

45.  Walter L. Buenger and Joseph A. Pratt, *But Also Good Business: Texas Commerce Banks and the Financing of*

*Houston and Texas, 1886–1986* (College Station: Texas A&M University Press, 1986), 215.

46.  William A. Kirkland, *Old Bank–New Bank: The First National Bank, Houston, 1866–1966* (Hinsdale, IL: Dryden Press, 1974), 85–88.

47.  Wineburgh, *The Texas Banker*, 131.

48.  Clark, *Oil Capital*, 107–11; Wineburgh, *The Texas Banker*, 131–33, quoting W. L. Perryman Jr. (President, General American Oil Company of Texas), address to the Newcomen Society in North America, March 16, 1961.

49.  Clark, *Oil Capital*, 108n340.

50.  Clark, *Oil Capital*, 113–15; Buenger and Pratt, *But Also Good Business*, 213.

51.  Wineburgh, *Texas Banker*, 131–33, quoting W. L. Perryman Jr. (President, General American Oil Company of Texas), address to the Newcomen Society in North America, March 16, 1961.

52.  U.S. Energy Information Administration, "U.S. Crude Oil Imports." 1968: 472,323,000 bbls., 1972: 1,183,996,000 bbls.

53.  Clark, *Oil Capital*, 162, quoting Gerald E. Sherrod (Vice President, First National City Bank, New York), "What Makes Those Bank Engineers So Conservative," *Journal of the Society of Petroleum Evaluation Engineers* 1 (1968–1969): 18.

54.  Clark, *Oil Capital*, 162–96.

55.  Clark, *Oil Capital*, 120–59.

56.  E. O. Buck, interview; Buenger and Pratt, *But Also Good Business*, 149, 151–55; Harold Vance, "Completion Methods," in *API History*, 579–616.

57.  Lon Tinkle, *Mr. De: A Biography of Everette Lee DeGolyer* (Boston, MA: Little, Brown, 1970), 224–28.

58.  Harold Vance, "Evaluation," in *API History*, 1073.

59.  *Consolidated Gas Utilities Corporation v. Thompson, et al., Texoma Natural Gas Co. v. Same* Nos. 539, 550, 14 F. Supp. 318, at 319–21 note 3 (Western District of Texas, March 30, 1936), which reproduces the Railroad Commission order of December 10, 1935, and describes the character and history of the field. *Thompson v. Consolidated Gas Utilities Corporation et al.,* 57 S. Ct. 364, 300 U.S. 55, at 59–61, 66 note 12 (US Supreme Court, February 1, 1937). Maurice Cheek, "Legal History of Conservation of Gas in Texas," in Section of Mineral Law, American Bar Association, *Legal History of Conservation of Oil and Gas: A Symposium* (Chicago: American Bar Association, 1939): 269–70.

60.  *F. C. Henderson, Inc. v. Railroad Commission,* 56 F. (2d) 218 (US District Court for the Western District of Texas, Austin Division, February 10, 1932); *Texoma Natural Gas Co. v. Railroad Commission of Texas et al. and three other cases,* 59 F. 2d 750 (Federal District Court, Western District of Texas, Austin, June 8, 1932); *Texoma Natural Gas Company v. Terrell,* 2 F. Supp. 168 (Western District of Texas, December 23, 1932); *Canadian River Gas Co. v. Terrell,* 4 F. Supp. 222, at 226 (Western District of Texas, Austin, June 22, 1933).

61.  *Consolidated Gas Utilities Corporation v. Thompson, et al.,* 14 F. Supp. 318 (Federal District Court, Western District of Texas, Austin, 1936); *Thompson v. Consolidated Gas Utilities Corporation et al.,* 57 S. Ct. 364, 300 U.S. 55, at 79 (US Supreme Court, February 1, 1937).

62.  "Paradise," track 5 on John Prine, *John Prine*, Atlantic, 1971 ("Won't you take me back to Muhlenberg County?").

63.  Wallace F. Lovejoy and Paul T. Homan, *Economic Aspects of Oil Conservation Regulation* (Baltimore, MD: Johns Hopkins University Press, 1967), 154–56.

64.  David A. Hounshell, *From the American System to Mass Production, 1800–1932: The Development of Manufacturing Technology in the United States* (Baltimore, MD: Johns Hopkins University Press, 1984).

65.  Lovejoy and Homan, *Economic Aspects of Oil Conservation*, 141–45, 206–10. Lovejoy and Homan reproduce both the 1947 and 1965 Yardstick schedules (144–45) and offer calculations of effects on spacing (142–43).

66.  H. D. Wilde Jr., and T. V. Moore (Humble Oil and Refining Co., Houston), "Hydrodynamics of Reservoir Drainage and Its Relation to Well Spacing," *Proceedings of the American Petroleum Institute, Production Bulletin 210* (1932): 83–90.

67.  "Katie Mae Blues" by Lightnin' Hopkins, recorded November 9, 1946, Aladdin Records.

68.  Eleanor Andrews and James McCarthy, "Scale, Shale, and the State: Political Ecologies and Legal Geographies of Shale Gas Development in Pennsylvania," *Journal of Environmental Studies and Sciences* 4, no. 1 (2013): 7–16, and the sources in legal geographies and political ecology cited therein.

69.  Railroad Commission of Texas, Oil and Gas Division, Oil and Gas Docket No. 128, #20-550, "In RE: Conservation and Prevention of Waste of Crude Petroleum and Natural Gas in the State of Texas, Texas: Old Ocean Field, Brazoria County, Texas," Austin, Texas, October 28, 1937. Transcript in the Everette Lee DeGolyer Papers, SMU, MSS 60, Box 165, Folder 6080.

70.  Railroad Commission of Texas, Oil and Gas Division, Oil and Gas Docket No. 128, #20-550, "Old Ocean Field, Brazoria County, Texas," Transcript, 50–52.

71.  Railroad Commission of Texas, Oil and Gas Division, Oil and Gas Docket No. 128, #20-550, "Old Ocean Field, Brazoria County, Texas," Transcript, 53–54.

72.  *Oxford Oil Co. v. Atlantic Oil & Producing Co.*, 16 F.2d 639, at 641 (Northern District of Texas, Dallas, December 27, 1926); *Oxford Oil Co. v. Atlantic Oil & Producing Co.*, 22 F.2d 597 (Federal Fifth Circuit Court of Appeals, 1927). In 1934, the wording was amended to read, "to prevent waste or to prevent confiscation of property": Zimmermann, *Conservation in the Production of Petroleum*, 338.

73.  *Texas Company v. W. H. Daugherty*, 107 Tex. 234, 176 S.W. 717 (Supreme Court of Texas, May 21, 1915). *Stephens County v. Mid-Kansas Oil & Gas Co.*, 254 S.W. 290, 113 Tex. 160 (Supreme Court of Texas, June 30, 1923).

74.  York Young Willbern, "Administrative Control of Petroleum Production in Texas," in *Public Administration and Policy Formation: Studies in Oil, Gas, Banking, River Development, and Corporate Investigations*, ed. Emmette S. Redford (Austin: University of Texas Press, 1956), 37.

75.  *Smith v. Stewart*, 68 S.W. 2d 627 (Court of Civil Appeals of Texas, Austin, 1934); *Smith v. Stewart*, 83 S.W. 2d 945 (Supreme Court of Texas, 1935); *Railroad Commission v. Magnolia Petroleum Co.*, 105 S.W. 2d 787 (Court of Civil Appeals of Texas, 1937); *Railroad Commission v. Magnolia Petroleum Co.*, 109 S.W. 2d 967 (Supreme Court of Texas, 1937). *Nash v. Shell Petroleum Corp.*, 120 S W. 2d 522 (Court of Civil Appeals of Texas, Austin, May 4, 1938). *Humble Oil & Refining Company v. Lasseter* 120 S.W. 2d 541 (Court of Civil Appeals of Texas, Austin, 1938).

76.  *Sun Oil Co. v. Railroad Commission et al.*, 68 S.W. 2d 609 (Court of Civil Appeals, Austin, December 22, 1933).

77.  W. W. Cutler, *Estimation of Underground Oil Reserves by Oil-Well Production Curves, U.S. Bureau of Mines Bulletin 228* (1924).

78.  *Stanolind Oil & Gas Co. v. Midas Oil Co.* 123 S.W. 2d 911 (Court of Civil Appeals of Texas, Austin, May 25, 1938), Justice Blair dissent, at 914, citing *Railroad Commission v. Marathon Oil Company* 89 S.W. 2d 517, Court of Civil Appeals of Texas; Justice Blair dissent, at 915, citing *Magnolia Petroleum Co. v. Railroad Commission of Texas* 93 S.W. 2d 587, Court of Civil Appeals of Texas, at 588.

79.  *C. H. Brown et al. v. Humble Oil & Refining Co.*, 83 S.W. 2d 935 (Supreme Court of Texas, June 12, 1935). *C. H. Brown et al. v. Humble Oil & Refining Co.*, 87 S.W 2d 1069 (Supreme Court of Texas, November 27, 1935); motion for rehearing denied. *Humble Oil & Refining Co. v. Railroad Commission of Texas, et al.*, 68 S.W. 2d 622 (Court of Civil Appeals of Texas, Austin, 1934), Brief of Appellant (Humble Oil), November 19, 1933, Texas State Library and Archives, Austin, Texas.

80.  *Humble Oil & Refining Co. v. Railroad Commission of Texas, et al.*, 68 S.W. 2d 622 (Court of Civil Appeals of Texas, Austin, 1934), Brief of Appellant (Humble Oil), November 19, 1933; Statement of Facts, filed in Court of Civil Appeals of Texas, Austin, June 19, 1933, in appeal of *Humble Oil & Refining Co. v. Railroad Commission of Texas, et al.*, District Court of Travis County, June 1, 1933, Texas State Library and Archives, Austin, Texas.

81.  *Humble Oil & Refining Co. v. Railroad Commission of Texas, et al.* 68 S.W. 2d 622, Brief of Appellant (Humble Oil), November 19, 1933; Statement of Facts, filed in Court of Civil Appeals of Texas, Austin, June 19, 1933, in appeal of *Humble Oil & Refining Co. v. Railroad Commission of Texas, et al.*, District Court of Travis County, June 1, 1933.

82.  *C. H. Brown et al. v. Humble Oil & Refining Co.*, in the Supreme Court of Texas, J. W. Wheeler, Argument for Plaintiffs in Error (Brown), April 19, 1935, at 1–2, Texas State Library and Archives, Austin, Texas.

83.  *C. H. Brown et al. v. Humble Oil & Refining Co.*, in the Supreme Court of Texas, J. W. Wheeler, Argument for Plaintiffs in Error (Brown), April 19, 1935, at 1–2.

84.  *C. H. Brown et al. v. Humble Oil & Refining Co.*, in the Supreme Court of Texas, Hines H. Baker, Oral Argument for Defendants in Error (Humble), April 24, 1935, at 22.

85.  *C. H. Brown et al. v. Humble Oil & Refining Co.*, in the Supreme Court of Texas, Hines H. Baker, Oral Argument for Defendants in Error (Humble), April 24, 1935, at 8–9.

86.  *C. H. Brown et al. v. Humble Oil & Refining Co.*, No. 6729, in the Supreme Court of Texas, Hines H. Baker, Oral Argument for Defendants in Error (Humble), April 24, 1935, at 8; Ralph D. Wyckoff, " Ralph J. Schilthuis and William Hurst, "Variations in Reservoir Pressure in the East Texas Field," *Transactions of the American Institute of Mining and Metallurgical Engineers: Petroleum Development and Technology* 114, no. 1 (December 1935): 173–74, Tulsa meeting of the American Institute of Mining Engineers Petroleum Division, October 1934.

87.  *C. H. Brown et al. v. Humble Oil & Refining Co.* 83 S.W. 2d 935, at 940 (Supreme Court of Texas, June 12, 1935).

88.  *C. H. Brown et al. v. Humble Oil & Refining Co.* 83 S.W. 2d 935, at 944 (Supreme Court of Texas, June 12, 1935).

89.  John E. Kilgore, *Amicus Curiae* in the Supreme Court of Texas, *C. H. Brown et al. v. Humble Oil and Refining Company*, No. 6729, October 7, 1935, at 2, Texas State Library and Archives, Austin, Texas.

90.  *C. H. Brown et al. v. Humble Oil & Refining Co.* 87 S.W. 2d 1069, at 1069–70 (Supreme Court of Texas, November 27, 1935).

91.  *Humble Oil & Refining Co. v. Railroad Commission* 112 S.W. 2d 222 (Court of Civil Appeals of Texas, 1937); Weaver, *Unitization of Oil and Gas Fields*, 424n114.
92.  *Railroad Commission v. Magnolia Petroleum Co.* 109 S.W. 2d 967 (Supreme Court of Texas, 1937). *Nash v. Shell Petroleum Corp.*, 120 S.W. 2d 522 (Court of Civil Appeals of Texas, Austin, 1938); *Humble Oil & Refining Company v. Lasseter*, 120 S.W. 2d 541 (Court of Civil Appeals of Texas, Austin, 1938); *Stanolind Oil & Gas Co. v. Midas Oil Co.* 123 S.W. 2d 911 (Court of Civil Appeals of Texas, Austin, May 25, 1938); *Magnolia Petroleum Co. v. Railroad Commission et al.* 120 S.W. 2d 548 (Court of Civil Appeals of Texas, Austin, June 1, 1938); *Gulf Land Co. v. Atlantic Refining Co.* 122 S.W. 2d 197, aff'd, 131 S.W.2d 73 (Supreme Court of Texas, July 26, 1939); *Gulf Land Co. v. Atlantic Refining Co.* 113 F. 2d 902 (Federal Circuit Court of Appeals, Fifth Circuit, September 16, 1940).
93.  *Railroad Commission v. Shell Oil Co.* 154 S.W. 2d 507, at 513 (Court of Civil Appeals of Texas, Austin, June 25, 1941). *Railroad Commission v. Shell Oil Co.* 161 S.W.2d 1022, 139 Tex. 66, at 72–74 (Supreme Court of Texas, March 11, 1942).
94.  *Railroad Commission v. Humble Oil & Refining Co.* 193 S.W. 2d 824, at 832 (Court of Civil Appeals of Texas, Austin, March 6, 1946). The "Hawkins Case," not to be confused with *Hawkins v. Texas Company*.
95.  *Ryan Consolidated Petroleum Corp. v. W. L. Pickens* 285 S.W. 2d 201, at 210 (Supreme Court of Texas, November 23, 1955).
96.  Frank E. Vandiver, *Their Tattered Flags: The Epic of the Confederacy* (New York: Harper Collins, 1970).

# CHAPTER 9

1.  "My Love for the Rose," track 1 from Willie Nelson, *Tougher than Leather*, Columbia, 1983. Written by Willie Nelson.
2.  The "Hawkins (townsite) case" is *Railroad Commission, et al. v. Humble Oil & Refining Co., et al.*, 193 S. W. 2d 824 (Court of Civil Appeals of Texas, 1946), and is not to be confused with *Hawkins v. Texas Company*, 203 S.W.2d 1003 (Court of Civil Appeals of Texas, 1947) or *Hawkins v. Texas Company*, 209 S.W.2d 338 (Supreme Court of Texas, 1948).
3.  *Atlantic Refining Company et al., Appellants v. Railroad Commission of Texas et al. Appellees*, 346 S.W. 2d 801 (Supreme Court of Texas, No. A-7355, March 8, 1961), "Normanna"; *Michel T. Halbouty et al., Appellants v. Railroad Commission of Texas et al., Appellees*, 347 S.W. 2d 364 (Supreme Court of Texas, No. A-8200, February 14, 1962), "Port Acres"; *Railroad Commission of Texas et al., Petitioners, v. Shell Oil Company et al., Respondents*, 380 S. W. 2d 556 (Supreme Court of Texas, No. A-9759, May 27, 1964), "Quitman."
4.  Majority opinion, *Atlantic Refining Company et al. v. Railroad Commission of Texas et al.*, 346 S. W. 2d 80 (Supreme Court of Texas, March 8, 1961).
5.  Railroad Commission of Texas, Oil and Gas Division, Oil and Gas Docket No. 129 #2-36,177, In Re: Conservation and Prevention of Waste of Crude Petroleum and Natural Gas in the Normanna Field, Bee County, Texas, Hearing Held in Austin, Texas, on September 18 and 19, 1957, 14–19; in *Atlantic Refining Company, et al. v. Railroad Commission of Texas, et al.*, A-7355, Supreme Court of Texas, Texas State Library and Archives, Austin, Texas.
6.  S. W. Wilcox, discussion of Gustave E. "Gus" Archie (Shell Oil), "The Electrical Resistivity Log as an Aid in Determining Some Reservoir Characteristics," *Transactions of the American Institute of Mining Engineers* 146, no. 1 (December 1942): 61–62 (American Institute of Mining Engineers Petroleum Division, Dallas Meeting, October 1941, received at the Institute September 27, 1941, revised December 8, 1941, Issued as T.P. 1422 January 1942. Wilcox had been a Geophysical Engineer for the Minnesota Department of Highways, 1933–1936.
7.  Alexander Deussen and Eugene C. Leonardon (Consulting Geologist, Houston; Schlumberger Well Survey Corporation of Texas, Houston), "Electrical Exploration of Drill holes," in Central Committee on Drilling and Production Practice, Division of Production, American Petroleum Institute, *Drilling and Production Practice 1935* (New York: API, 1936): 47–59; Eugene G. Leonardon, "Logging, Sampling, and Testing," in *API History*, 493–578; Geoffrey C. Bowker, *Science on the Run: Information Management and Industrial Geophysics at Schlumberger, 1920–1940* (Cambridge, MA: MIT Press, 1994).
8.  *Railroad Commission v. Humble Oil & Refining Company*, 193 S.W. 2d 824, at 830 (Court of Civil Appeals of Texas, Austin, 1946).
9.  Gustave E. "Gus" Archie (Shell Oil), "The Electrical Resistivity Log as an Aid in Determining Some Reservoir Characteristics," *Transactions of the American Institute of Mining Engineers* 146, no. 1 (December 1942): 54–57; American Institute of Mining Engineers, Petroleum Division, Dallas Meeting, October 1941, received at the Institute September 27, 1941, revised December 8, 1941, issued as T.P. 1422 January 1942.
10.  Deussen and Leonardon, "Electrical Exploration of Drill Holes," 47–59. Leonardon, "Logging, Sampling, and

Testing," 493–578. Bowker, *Science on the Run*.

11.  Dissenting opinion, *Atlantic Refining Company, et al. v. Railroad Commission of Texas, et al.*, 346 S. W. 2d 801, at 813 (Supreme Court of Texas, March 8, 1961).

12.  Railroad Commission of Texas, Oil and Gas Division, Oil and Gas Docket No. 129 #2-36,177, 18–38, 71, 83–105; in *Atlantic Refining Company, et al. v. Railroad Commission of Texas, et al.*, A-7355, Supreme Court of Texas, Texas State Library and Archives, Austin, Texas.

13.  Railroad Commission of Texas, Oil and Gas Division, Oil and Gas Docket No. 129 #2-36,177, 66–67; in *Atlantic Refining Company, et al. v. Railroad Commission of Texas, et al.*, A-7355.

14.  Railroad Commission of Texas, Oil and Gas Division, Oil and Gas Docket No. 129 #2-36,177, 69; in *Atlantic Refining Company, et al. v. Railroad Commission of Texas, et al.*, A-7355.

15.  Statement of Facts, Testimony of William J. Murray, at 711–13, in *Atlantic Refining Company, et al. v. Railroad Commission of Texas, et al.*, A-7355. Judge Jim C. Langdon, "The Influence of Court Decisions upon Railroad Commission Policy in Rule 37 Cases and the Allocation of Allowables to the Small Tract Well," n.d. (probably late 1963), 1, Langdon Papers, 1980/164-10 Folder 9, Oil and Gas Division Proceedings, Texas State Library and Archives, Austin, Texas.

16.  *Atlantic Refining Company v. Railroad Commission of Texas*, 330 S.W. 2d 494 (1960). Langdon, "Rule 37 Cases," 5–6; Dissenting opinion, *Atlantic Refining Company, et al. v. Railroad Commission of Texas, et al.*, 346 S. W. 2d 801 (Supreme Court of Texas, March 8, 1961).

17.  Majority opinion, *Atlantic Refining Company et al. v. Railroad Commission of Texas, et al.*, 346 S. W. 2d 801 (Supreme Court of Texas, March 8, 1961); Brief for Appellants (Atlantic Refining Company, et al.), in the Supreme Court of Texas, No. A 7355, Filed July 28, 1959, *Atlantic Refining Company, et al. v. Railroad Commission of Texas, et al.*

18.  Majority opinion, *Atlantic Refining Company, et al. v. Railroad Commission of Texas, et al.*, 346 S. W. 2d 801 (Supreme Court of Texas, March 8, 1961).

19.  Brief for Appellants (Atlantic Refining Company et al.), in the Supreme Court of Texas, No. A 7355, Filed July 28, 1959, at 15–18, *Atlantic Refining Company et al. v. Railroad Commission of Texas et al.*

20.  Statement of Facts, Volume II, Testimony of John W. Crutchfield, at 499–502, in *Atlantic Refining Company et al. v. Railroad Commission of Texas et al.*, Supreme Court of Texas, A-7355.

21.  Deposition of Caroll D. Hudson, February 5, 1959, at 14–16, 53–56, *Atlantic Refining Company et al. v. Railroad Commission of Texas*, in the 98th District Court of Travis County, Texas, No. 112,966, Texas State Library and Archives, Austin, Texas; *Bright and Schiff v. Atlantic Refining Company et al.*, 156th District Court of Bee County, Texas, No. 9,159.

22.  Brief for Appellants (Atlantic Refining Company et al.), at 18–19, in the Supreme Court of Texas, No. A 7355, Filed July 28, 1959, *Atlantic Refining Company et al. v. Railroad Commission of Texas et al.*; Statement of Facts, Volume II, Testimony of John W. Crutchfield, at 499–510, in *Atlantic Refining Company et al. v. Railroad Commission of Texas et al.*, Supreme Court of Texas, A-7355.

23.  Statement of Facts, Volume II, Testimony of John W. Crutchfield, at 506–32, in *Atlantic Refining Company et al. v. Railroad Commission of Texas et al.*, Supreme Court of Texas, A-7355.

24.  Statement of Facts, Volume II, Testimony of John W. Crutchfield, at 537–43, in *Atlantic Refining Company et al. v. Railroad Commission of Texas et al.*, Supreme Court of Texas, A-7355.

25.  Statement of Facts, Volume II, Testimony of John W. Crutchfield, at 543–45, in *Atlantic Refining Company et al. v. Railroad Commission of Texas et al.*, Supreme Court of Texas, A-7355; Majority opinion, *Atlantic Refining Company et al. v. Railroad Commission of Texas et al.*, 346 S.W. 2d 801 (Supreme Court of Texas, March 8, 1961).

26.  Statement of Facts, Volume II, Testimony of William H. Price, at 762–63, in *Atlantic Refining Company et al. v. Railroad Commission of Texas et al.*, Supreme Court of Texas, A-7355.

27.  Statement of Facts, Volume III, Testimony of Joseph Ballanfonte, at 897–1022, in *Atlantic Refining Company et al. v. Railroad Commission of Texas et al.*, Supreme Court of Texas, A-7355.

28.  Dissenting opinion, *Atlantic Refining Company et al. v. Railroad Commission of Texas et al.*, 346 S. W. 2d 801 (Supreme Court of Texas, March 8, 1961); *Atlantic Refining Company et al. v. Bright & Schiff* 321 S.W. 2d 167 (Court of Civil Appeals of Texas).

29.  Statement of Facts, Volumes II and III, Testimony of William H. Price and Joseph Ballanfonte, in *Atlantic Refining Company et al. v. Railroad Commission of Texas et al.*, Supreme Court of Texas, A-7355.

30.  Statement of Facts, Volumes I and II, in *Atlantic Refining Company et al. v. Railroad Commission of Texas et al.*, Supreme Court of Texas, A-7355.

31.  Langdon, "Rule 37 Cases," 3.

32.  Majority opinion, *Atlantic Refining Company et al. v. Railroad Commission of Texas et al.*, 346 S. W. 2d 801, at 804 (Supreme Court of Texas, March 8, 1961).

33.  Brief for Appellants (Atlantic Refining Company et al.), in the Supreme Court of Texas, No. A 7355, Filed July 28, 1959, at 50, *Atlantic Refining Company et al. v. Railroad Commission of Texas et al.*, Texas State Archives, citing Statement of Facts, Volume III, at 1072–74, *Atlantic Refining Company et al. v. Railroad Commission of Texas et al.*, Supreme Court of Texas, A-7355.

34.  Rejoinder of Appellee Bright & Schiff, in the Supreme Court of Texas, No. A 7355, Filed December 14, 1959, *Atlantic Refining Company et al. v. Railroad Commission of Texas et al.*; Brief of Appellees Bright & Schiff, et al., in Support of Motion for Rehearing, in the Supreme Court of Texas, No. A 7355.

35.  Brief for Appellants (Atlantic Refining Company et al.), in the Supreme Court of Texas, No. A 7355, Filed July 28, 1959, *Atlantic Refining Company et al. v. Railroad Commission of Texas et al.*, Texas State Library and Archives; Supplemental Brief for Appellants, in the Supreme Court of Texas, No. A 7355, Filed July 28, 1959, *Atlantic Refining Company et al. v. Railroad Commission of Texas et al.*; Appellants' Rejoinder, in the Supreme Court of Texas, No. A 7355, Filed December 22, 1959, *Atlantic Refining Company et al. v. Railroad Commission of Texas et al.*

36.  *Atlantic Refining Company et al. v. Railroad Commission of Texas et al.*, 346 S. W. 2d 801, at 804–9 (Supreme Court of Texas, March 8, 1961). Cases discussed include *Corzelius v. Harrell*, 143 Tex. 509, 186 S.W. 2d 961 (Supreme Court of Texas, April 4, 1945); *Henderson v. Terrell*, 24 F. Supp. 147 (Federal Western District of Texas, July 23, 1938); *Marrs v. Railroad Commission*, 142 Tex. 293, 177 S.W. 2d 941 (Supreme Court of Texas, January 5,1944); *Gulf Land Co. v. Atlantic Refining Co.* 134 Tex. 59, 131 S.W. 2d 254 (Supreme Court of Texas, July 26, 1939); *Railroad Commission v. Gulf Production Company*, 134 Tex. 122, 132 S.W. 2d 254 (1939); *Brown v. Humble Oil & Refining Co.* 126 Tex. 296 and 314; 83 S.W. (2d) 935 (Supreme Court of Texas, June 12, 1935); *Magnolia Petroleum Company v. Railroad Commission* 93 S.W. 2d 587, 588 (Court of Civil Appeals of Texas); *Atlantic Oil Production Company v. Railroad Commission* 85 S.W. 2d 655 (Court of Civil Appeals of Texas); *Atlantic Refining Company et al. v. Railroad Commission of Texas et al.*, 346 S. W. 2d 801, at 809–11; *Ryan Consolidated Petroleum Corporation v. W. L. Pickens, et al.*, 285 S.W. 2d 201 (Supreme Court of Texas, November 23, 1955); *Railroad Commission v. Humble Oil & Refining Co.*, 193 S.W. 2d 824 (Court of Civil Appeals of Texas, Austin, March 6, 1946); *Standard Oil Company v. Railroad Commission*, 215 S.W. 2d 633 (Court of Civil Appeals of Texas, December 8, 1948).

37.  *Atlantic Refining Company et al. v. Railroad Commission of Texas et al.*, 346 S.W. 2d 801, at 811.

38.  *Atlantic Refining Company et al. v. Railroad Commission of Texas et al.*, 346 S.W. 2d 801, at 811.

39.  Dissenting opinion, *Atlantic Refining Company et al. v. Railroad Commission of Texas et al.*, 346 S. W. 2d 801, at 818–19, 823.

40.  Dissenting opinion, *Atlantic Refining Company et al. v. Railroad Commission of Texas et al.*, 346 S. W. 2d 801, at 815.

41.  Railroad Commission of Texas, Special Order No. 2-46,673, October 1, 1961, Adopting Operating Rules for the Normanna Field. This order is reproduced *verbatim* in Clyde E. Smith, dissenting opinion, *Michel T. Halbouty et al. v. Railroad Commission of Texas et al.*, 347 S.W. 2d 364, at 377n3 (Supreme Court of Texas, February 14, 1962).

42.  Railroad Commission of Texas, Special Order No. 2-46,673, October 1, 1961, Adopting Operating Rules for the Normanna Field, in Clyde E. Smith, Dissenting Opinion, *Michel T. Halbouty et al. v. Railroad Commission of Texas et al.*, 347 S.W. 2d 364, at 377–78n4.

43.  Fred Young, "Memorandum to Commissioner Thompson," September 8, 1961, Railroad Commission of Texas, Oil and Gas Division, Texas State Library and Archives, Austin, Texas.

44.  Smith, dissenting opinion, *Michel T. Halbouty et al., Appellants v. Railroad Commission of Texas et al., Appellees*, 347 S.W. 2d 364, at 382 (Supreme Court of Texas, No. A-8200, February 14, 1962).

45.  *Michel T. Halbouty et al., Appellants v. Railroad Commission of Texas et al., Appellees*, 347 S.W. 2d 364.

46.  *Halbouty v. Railroad Commission*, 347 S.W. 2d 364.

47.  Langdon, "Rule 37 Cases," 6–8.

48.  Brief for Appellees (H. L. Dillon et al), map attached, in the Supreme Court of Texas, *Michel T. Halbouty et al. v. Railroad Commission of Texas et al., Halbouty v. Railroad Commission*, 347 S. W. 2d 364, at 371.

49.  *Halbouty v. Railroad Commission*, 347 S. W. 2d 364, at 371–72.

50.  *Halbouty v. Railroad Commission*, 347 S. W. 2d 364, at 371–72, quoting Article 6008, § 12, Vernon's Annotated Civil Statutes.

51.  *Halbouty v. Railroad Commission*, 347 S. W. 2d 364, at 376.

52.  "Port Acres Group Hires Price Daniel," *Dallas Morning News*, February 16, 1963, Section 10-4. Langdon, "Rule 37 Cases," 8–10.

53.  Testimony of Granville Dutton (Proration Engineer, Sun Oil Company, Dallas), *Shell Oil Company et al. v. Railroad Commission of Texas et al.*, 53rd Judicial District Court of Travis County, Texas, Statement of Facts, Volume I, at 37–50, filed in the Court of Civil Appeals, Austin, Texas, March 1, 1963, filed in the Supreme

Court of Texas, November 6, 1963, *Railroad Commission of Texas et al., Petitioners, v. Shell Oil Company et al., Respondents*, Supreme Court of Texas, No. A-9759; Texas State Library and Archives, Austin, Texas.

54. *Railroad Commission of Texas et al., Petitioners, v. Shell Oil Company et al., Respondents*, 380 S.W. 2d 556, at 559 (Supreme Court of Texas, No. A-9759, May 27, 1964).

55. *Atlantic Refining Company et al. v. Railroad Commission of Texas et al.*, 346 S. W. 2d 801, at 811 (Supreme Court of Texas, March 8, 1961).

56. *Railroad Commission of Texas et al., Petitioners, v. Aluminum Company of America, et al., Respondents*, 380 S.W. 2d 558 at 602 (Supreme Court of Texas, No. A-9760, May 27, 1964). *Standard Oil Company v. Railroad Commission*, 215 S.W. 2d 633 (Court of Civil Appeals of Texas, 1948).

57. Weaver, *Unitization of Oil and Gas Fields in Texas*, 127–31.

58. John Kenneth Galbraith, *American Capitalism: The Concept of Countervailing Power* (Boston: Houghton Mifflin, 1952).

59. R. E. Allen (Assistant Oil Umpire of California), "Theory and Practice of Directed Drilling," *Transactions of the American Institute of Mining Engineers* 107, no. 1 (December 1934).

60. Prindle, *Petroleum*, 81.

61. James D. Hughes, "Value of Oil-Well Surveying and Applications of Controlled Directional Drilling," in Central Committee on Drilling and Production Practice, Division of Production, American Petroleum Institute, *Drilling and Production Practice 1935* (New York: API, 1936), 60–70 (Spring Meeting, Mid-Continent District, Amarillo, Texas, April 1935).

62. D. K. Weaver (Wilshire Oil Co.), "Practical Aspects of Directional Drilling," *API Drilling and Production Practice* 1 (January 1946).

63. Trevor Pinch and Nelly Oudshoorn, eds., *How Users Matter: The Co-Construction of Users and Technologies* (Cambridge, MA: MIT Press, 2003). Joy Parr, "What Makes Washday Less Blue? Gender, Nation, and Technology Choice in Postwar Canada," *Technology and Culture* 38 (January 1997): 153–86.

64. From Falstaff's apologia in *Henry IV, Part I*: "The better part of Valour, is Discretion; in the which better part, I haue saued my life" (spelling and punctuation from the *First Folio*, Act 5, Scene 3, lines 3085–3086). The website of Prof. Paul Brians, Washington State University, accessed 1/3/21.

65. Sir Arthur Conan Doyle, "The Adventure of the Silver Blaze," *The Memoirs of Sherlock Holmes*.

66. Unless otherwise noted, this section is drawn from Prindle, *Petroleum Politics*, 81–94, and Clark and Halbouty, *The Last Boom*, 273–85.

67. Ed Cocke, "Dallas Firm Plans to File Oil Theft Charges Saturday," *Dallas Morning News*, June 9, 1962, Section 1-5.

68. Richard M. Morehead, "State Order Bans Oil Well Plugging," *Dallas Morning News*, June 2, 1962, Section 1-7. Fred Pass, "Rangers Called to Aid in Drilling Probe," *Dallas Morning News*, June 3, 1962, 1-6; Dawson Duncan, "32 Wells Checked, 28 Found Slanted," *Dallas Morning News*, June 21, 1962, Section 1-10.

69. "Group to Tell ETex Row 'Other Side,'" *Dallas Morning News*, June 7, 1962, Section 4-3; Fred Pass, "Independent Oilmen Call for Mediation in Probe," *Dallas Morning News*, June 20, 1962 Section 1-4; Fred Pass, "4 East Texas Operators Sue Majors for 10 Million Dollars," *Dallas Morning News*, June 23, 1962, Section 1-5, "Conoco Attorneys Tell of Vandalism on Lease," June 30, 1962, Section 3-6.

70. Of Welsh coal miners during the 1984 strike, Hywel Francis wrote, "Shot through the above testimony is a controlled and reasoned anger over the injustice of generations." Francis, "The Law, Oral Tradition and the Mining Community," 268.

71. *Railroad Commission of Texas, et al. v. Rowan & Nichols Oil Co.* 311 U.S. 570, at 574 (US Supreme Court, Argued December 12, 13, 1940, Decided January 6, 1941). Zimmermann, *Conservation in the Production of Petroleum*, 335.

72. Fred Pass, "Oilman Long Silent in Slant-Well Suit," *Dallas Morning News*, June 30, 1962, Section 3-6.

73. Jim C. Langdon, Commissioner, Railroad Commission of Texas, "The Role of the Petroleum Engineer as an Expert Witness," February 22, 1965, Address Before the Victoria Chapter—API (American Petroleum Institute), Victoria, Texas, 4–5, Langdon Papers, 1980/164-10 Folder 36, Oil and Gas Division Proceedings, Texas State Library and Archives, Austin, Texas.

74. "Oil Operator No Stranger to Probes," *Dallas Morning News*, September 21, 1962, Section 4-3. Fred Pass, "Situation Deplored by Jury," *Dallas Morning News*, December 12, 1962, Section 1-1.

75. Prindle, *Petroleum Politics*, 91; *Dallas Morning News*, August 28, 1963, July 11, 1963.

76. Clark and Halbouty, *Last Boom*, 273–85; Prindle, *Petroleum Politics*, 81–94.

77. Clark and Halbouty, *Last Boom*, 273–85.

78. Prindle, *Petroleum Politics*, 81–94.

79. E. P. Thompson, "The Moral Economy of the English Crowd in the Eighteenth Century," *Past & Present* 50, no. 1 (February 1971): 76–136.

80. Michael Muthukrishna observes that whether an observed instance of cooperation is viewed as corruption or prosocial cooperation is a matter of perspective and scale: altruistic cooperation at one level is often seen as corruption from a higher level. Michael Muthukrishna, "Corruption, Cooperation, and the Evolution of Prosocial Institutions," *SSRN Journal* (2017), https://dx.doi.org/10.2139/ssrn.3082315.

81. East Texas oil operators were not alone. Welsh coal miners in the mid-1980s shared many of the same feelings. See Francis, "The Law, Oral Tradition and the Mining Community," *Journal of Law and Society* 12 (1985): 267–71.

## CHAPTER 10

1. Waylon Jennings, "My Heroes Have Always Been Cowboys," track 1 of Waylon Jennings, Willie Nelson, Jessi Colter, and Tompall Glaser, *Wanted: The Outlaws*, RCA Victor, 1976. Written by Sharon Vaughn.

2. Lawrence Goodwyn, *Texas Oil, American Dreams: A Study of the Texas Independent Producers and Royalty Owners Association* (Austin: Published for the Center for American History, University of Texas, by the Texas State Historical Association, 1996), 187.

3. Quoted in Gregory Zuckerman, *The Frackers: The Outrageous Inside Story of the New Billionaire Wildcatters* (New York: Portfolio/Penguin, 2013), 87.

4. Mark DeCambre, "About 150 Years of Oil-Price History: This One Chart Illustrates Crude's Spectacular Plunge Below $0 a Barrel," from *Deutsche Bank, Global Financial Data*, in *Market Watch*, April 29, 2020, accessed January 18, 2021, https://www.marketwatch.com/story/about-150-years-of-oil-price-history-in-one-chart-illustrates-crudes-spectacular-plunge-below-0-a-barrel-2020-04-22.

5. International Energy Agency, "World Oil Production by Region, 1971–2019." U. S. Energy Information Administration, "U. S. Field Production of Crude Oil, 1870-2020." Texas Railroad Commission, "Crude Oil Production and Well Counts (since 1935)." World Production is calculated from IEA data reported in tonnes (metric tons, 2,204 pounds each), assuming 7.3 barrels of crude oil per tonne. Reported Texas Production for 2019 is for November 2018 to October 2019. Accessed January 18–21, 2021.

6. US Energy Information Administration, *Monthly Energy Review*, Table 3.1, March 2020, accessed January 21, 2021.

7. John Donnelly, ed., *Journal of Petroleum Technology, Special Commemorative Issue: Celebrating 50 Years of the Society of Petroleum Engineers* 59, no. 10 (October 2007): 49–50, 63–64, 134–35.

8. Donnelly, *Journal of Petroleum Technology*, 65, 19.

9. Donnelly, *Journal of Petroleum Technology*, 64–65, 89, 129–31.

10. Oscar Parkes, *British Battleships: "Warrior" 1860 to "Vanguard" 1950; A History of Design, Construction and Armament* (Hamden, CT: Archon Books, The Shoe String Press, 1970). William H. Garzke Jr. and Robert O. Dullin Jr., *Battleships in World War II: Allied Battleships* (Annapolis, MD: Naval Institute Press, 1980). R. A. Burt, *British Battleships of World War One* (Annapolis, MD: Naval Institute Press, 1986). William H. McNeill, *The Pursuit of Power: Technology, Armed Force, and Society Since A. D. 1000* (Chicago: University of Chicago Press, 1982).

11. Vic Rao, "1984 and Beyond: The Advent of Horizontal Wells," *Journal of Petroleum Technology Special Commemorative Issue: Celebrating 50 Years of the Society of Petroleum Engineers* 59, no. 10 (October 2007): 120.

12. Rao, "1984 and Beyond," 120. Nathan Rosenberg, "Technological Change in the Machine Tool Industry, 1840–1910," *Journal of Economic History* 23 (1963): 414–43.

13. Rao, "1984 and Beyond," 120.

14. Rosenberg, "Technological Change in the Machine Tool Industry."

15. Rao, "1984 and Beyond," 120.

16. The "log jam" theory of technological innovation proposes that as technological systems evolve, certain problems emerge which block further progress: "log jams." The classic example is getting cotton seeds out of cotton bolls, solved by Eli Whitney's cotton gin, which thereby enabled the economical, large-scale manufacture of cotton textiles, and indirectly prolonged slavery in the United States. Thomas P. Hughes's "reverse salients in an advancing technological front" and resulting "critical problems" are better articulated portrayals of similar ideas. See Hughes, *Networks of Power*. Wikipedia, "CGG (company)," accessed January 28, 2021.

17. Ron Lord, "Technological Breakthroughs Advanced Upstream E&P's Evolution," *Journal of Petroleum Technology* 59, no. 10 (October 2007): 111–16.

18. Penn State College of Earth and Mineral Sciences, John A. Dutton Institute for Teaching and Learning Excellence, PNG 301 Introduction to Petroleum and Natural Gas Engineering; licensed by Creative Commons CC BY-NC-SA 4.0, accessed April 27, 2023.

19. Bruce M. Kramer (Thompson Visiting Professor, University of Colorado Law School; Maddox Professor of Law Emeritus, Texas Tech University School of Law), "Horizontal Drilling and Trespass: A Challenge to the Norms of Property and Tort Law," *Colorado Natural Resources, Energy & Environmental Law Review* 25, no. 2 (2014): 292–340.

20.  "Drilling Mud Motor Components," accessed January 31, 2021, www.drillingmanual.com. Jens Gravesen, "The Geometry of the Moineau Pump," Technical University of Denmark, Department of Mathematics, March 26, 2008, accessed February 4, 2021.

21.  John Thorogood (BP, and SPE Director, Drilling and Completions, 2001–2004), "Drilling in 2032—Back to the Future," *Journal of Petroleum Technology, Special Commemorative Issue: Celebrating 50 Years of the Society of Petroleum Engineers* 59, no. 10 (October 2007): 110.

22.  "Equations for Calculating Inclination, Toolface and Azimuth" are provided in "MWD—Directional Tools," December 12, 2017, accessed January 31, 2021, www.drillingmanual.com/2017.

23.  Iain Dowell (Halliburton Energy Services) and Andrew A. Mills (Esso Australia Ltd.), "Chapter 4—Measurement-While-Drilling [MWD], Logging-While-Drilling [LWD] and Geosteering," in Michael J. Economides (Texas A&M), Larry T. Watters (Halliburton Energy Services), and Shari Dunn-Norman (University of Missouri–Rolla), *Petroleum Well Construction* (New York: John Wiley, 1998), 155–211; "Content provided has been authored or coauthored by Halliburton employees to be used for educational purposes." Wiki, "Formation Evaluation," accessed January 31, 2021.

24.  Schlumberger Oilfield Glossary, "Density Measurement," accessed February 11, 2021; Schlumberger Oilfield Glossary, "Electrical Log," accessed January 31, 2021; Wiki, "Formation Evaluation," accessed January 31, 2021; Dowell and Mills, "Measurement-While-Drilling [MWD], Logging-While-Drilling [LWD] and Geosteering," 155–211.

25.  J. J. Arps (British-American Oil Producing Company, Dallas) and J. L. Arps (Arps Corporation, Garland, Texas), "The Subsurface Telemetry Problem—A Practical Solution," *Journal of Petroleum Technology* 16, no. 5 (May 1964): 487–93. Manuscript received in SPE office August 8, 1963 (presented at the SPE Annual Fall Meeting, New Orleans, October 6–8, 1963). This paper includes considerable detail on the history of alternative well telemetry systems.

26.  Rao, "1984 and Beyond," 117.

27.  Rao, "1984 and Beyond," 117.

28.  Rao, "1984 and Beyond," 120.

29.  "Acidizing Treatment in Oil and Gas Operators," API Digital Media (DM2014-113 | 06.14), 2014, accessed February 15, 2021.

30.  J. B. Clark (Stanolind Oil & Gas Co., Tulsa), "A Hydraulic Process for Increasing the Productivity of Wells, *Journal of Petroleum Technology* 1, no. 1 (1 January 1949): 1–8, T.P. 2510; paper presented at Petroleum Division Fall Meeting, Dallas, Texas, October 4–6, 1948. Carl T. Montgomery and Michael B. Smith (NSI Technologies), "Fracturing: History of an Enduring Technology," *Journal of Petroleum Technology* 63 (December 2010): 26ff; Wiki, "Hydraulic Fracturing," accessed January 19, 2021.

31.  Montgomery and Smith, "Fracturing," 26ff; Wiki, "Hydraulic Fracturing," accessed February 19, 2021.

32.  Montgomery and Smith, "Fracturing," 26ff; Wiki, "Hydraulic Fracturing," accessed February 19, 2021; M. King Hubbert and D. G. Willis (Shell Development Co., Houston), "Mechanics of Hydraulic Fracturing," *Petroleum Transactions, American Institute of Mining Engineers* 210, no. 1 (January 1957): 153 (paper presented at Petroleum Branch Fall Meeting in Los Angeles, October 14–17, 1956). T. P. 459.

33.  Hubbert and Willis, "Mechanics of Hydraulic Fracturing," 153–68.

34.  "Austin Chalk Revival," accessed February 5, 2021; Laurentian Research, "Austin Chalk Revived: An Emerging Unconventional Play," March 7, 2019, accessed February 5, 2021.

35.  Harry Hurt III, "New Oil: The Giddings Gamble," *Texas Monthly*, February 1981, 106 ff.

36.  "Austin Chalk Revived: An Emerging Unconventional Play," accessed February 5, 2021.

37.  Wiki, "Shale," accessed February 18, 2021.

38.  Clark Jr., *Oil Capital*, 321–29.

39.  Robert A. Gardner Jr. (Geologist, Christie, Mitchell and Mitchell Company), "The Boonsville Bend Conglomerate Gas Field, Wise County, Texas," *Abilene Geological Society: Geological Contributions 1960* (1960): 7–18. Clark, *Oil Capital*, 321–29. M. Y. Tanakov and M. Kelkar, "Integrated Reservoir Description for Boonsville, Texas Field Using 3D Seismic Well and Production Data" (paper presented at the SPE/DOE Improved Oil Recovery Symposium, Tulsa, Oklahoma, April 2000); Published: April 3, 2000: SPE-59349-MS. Aamer Ali Alhakeem, "3D Seismic Data Interpretation of Boonsville Field, Texas" (MS thesis, Missouri University of Science and Technology, 2013).

40.  Clark, *Oil Capital*, 321–29.

41.  Montgomery and Smith, "Fracturing," 26ff. Wiki, "Hydraulic Fracturing," accessed February 19, 2021; Clark, *Oil Capital*, 321–29; Loren Steffy, *George P. Mitchell: Fracking, Sustainability, and an Unorthodox Quest to Save the Planet* (College Station: Texas A&M University Press, 2019).

42.  Wiki, "Barnett Shale," accessed February 17, 2021; Alfred R. Jennings Jr., "Fracturing Fluids—Then and Now," *Journal of Petroleum Technology* 48, no. 7 (1 July 1996): 604–10; Wiki, "Hydraulic Fracturing," accessed

February 19, 2021; Clark, *Oil Capital*, 321–29.

43. Montgomery and Smith, "Fracturing," 26ff; Wiki, "Hydraulic Fracturing," accessed February 19, 2021. Clark, *Oil Capital*, 321–29; Steffy, *George P. Mitchell*.

44. Montgomery and Smith, "Fracturing," 26ff; Clark, *Oil Capital*, 321–29; Wiki, "Hydraulic Fracturing," accessed February 19, 2021.

45. Montgomery and Smith, "Fracturing," 26ffl; Clark, *Oil Capital*, 321–29.

46. Wiki, "Barnett Shale," accessed February 17, 2021.

47. Lynn Helms, "Horizontal Drilling, *DMR Newsletter* 35, no. 1 (n.d.): 1–3. Wiki, "Hydraulic Fracturing," accessed February 19, 2021.

48. Hughes, "Electrification of America," 124–61; Thomas P. Hughes, *Networks of Power*.

49. Texas Administrative Code, Title 16 ECONOMIC REGULATION, Part 1 RAILROAD COMMISSION OF TEXAS, Chapter 3 OIL AND GAS DIVISION, Rule §3.29 Hydraulic Fracturing Chemical Disclosure Requirements.

50. Texas Administrative Code, Title 16 ECONOMIC REGULATION, Part 1 RAILROAD COMMISSION OF TEXAS, Chapter 3 OIL AND GAS DIVISION, Rule §3.86 Horizontal Drainhole Wells.

51. Texas Administrative Code, Title 16, Rule §3.86 Horizontal Drainhole Wells. Bill Kroger, Jason Newman, Ben Sweet, and Justin Lipe (Baker Botts, Houston), "How Texas Law Promoted Shale Play Development," Baker Botts 175th Anniversary, 2015.

52. *Gregg v. Delhi-Taylor Oil Corporation,* 344 S.W. 2d 411 (Supreme Court of Texas, February 22, 1961). *Delhi-Taylor Oil Corporation v. Holmes,* 344 S.W. 2d 317 (Supreme Court of Texas, February 22, 1961). *Geo Viking, Inc. v. Tex-Lee Operating Company,* 817 S.W. 2d 357 (Court of Civil Appeals of Texas, Texarkana, July 16, 1991). *Geo Viking, Inc. v. Tex-Lee Operating Company,* No. D-1678, 1992, WL 80263 (Supreme Court of Texas, April 22, 1992). *Gifford Operating Company v. Indrex, Inc.,* No. 2:89-CV-0189, 1992 U.S. Dist. LEXIS 22505, at 16–17 (Federal Northern District of Texas, Aug. 7, 1992); Christopher S. Kulander (Director and Professor, Harry L. Reed Oil & Gas Law Institute; Of Counsel, Haynes and Boone LLP, BS Geology, MS Geophysics, Wright State University, PhD Petroleum Seismology, Texas A&M University, JD University of Oklahoma), and R. Jordan Shaw (BA Indiana University of Pennsylvania, JD Texas Tech University), "Comparing Subsurface Trespass Jurisprudence—Geophysical Surveying and Hydraulic Fracturing," *New Mexico Law Review* 46 (Winter 2016): 98–99. Bruce M. Kramer (Thompson Visiting Professor, University of Colorado Law School; Maddox Professor of Law Emeritus, Texas Tech University School of Law), "Horizontal Drilling and Trespass: A Challenge to the Norms of Property and Tort Law," *Colorado Natural Resources, Energy & Environmental Law Review* 25, no. 2 (2014): 303–4.

53. *Coastal Oil & Gas Corporation v. Garza Energy Trust,* 268 S.W. 3d 1 (Supreme Court of Texas, August 29, 2008).

54. Paul Burka, "Power Politics: How One Company's Wheeling and Dealing Brought the Energy Crisis into Your Life," *Texas Monthly* 3 (May 1975): 68–78+. *Coastal Oil & Gas Corporation v. Garza Energy Trust* 268 S.W. 3d 1 (2008).

55. *Coastal Oil & Gas Corporation v. Garza Energy Trust* 268 S.W. 3d 1 (2008).

56. *Peterson v. Grayce Oil Company,* 37 S.W. 2d 367, at 370–71 (Court of Civil Appeals, Fort Worth, 1931), affirmed, 128 Tex. 550, 98 S.W.2d 781 (Supreme Court of Texas, 1936).

57. Associate Justice Phil Johnson, dissenting opinion, *Coastal Oil & Gas Corporation v. Garza Energy Trust,* 268 S.W. 3d 1.

58. *Marion Stone and Brian Corwin, Plaintiffs, v. Chesapeake Appalachia, LLC, Statoil USA Onshore Properties, Inc., and Jamestown Resources, Inc.,* No. 5: 12-CV-102, 2013 U.S. Dist. Lexis 71121, at Page ID #: 303 (United States District Court for the Northern District of West Virginia, Wheeling, WV, April 10, 2013). Barclay Nicholson and Brian Albrecht (Fulbright & Jaworski, LLP), "Hydraulic Fracturing as a Subsurface Trespass: A New Ruling in West Virginia Provides Protection for the Landowner," *The Energy Law Advisor,* The Center for American and International Law, Plano, Texas, accessed March 15, 2021.

59. *Ryan Consolidated Petroleum Corp. v. W. L. Pickens* 285 S.W. 2d 201, at 210 (Supreme Court of Texas, 1955).

60. *Daubert v. Merrell Dow Pharmaceuticals, Inc.* 509 U.S. 579, 113 S. Ct. 2786 (U.S. Supreme Court, 1992); Caroline D. Walker, "Using Computer-Generated Animation and Simulation Evidence at Trial: What You Should Know," American Bar Association, PRACTICE POINTS, January 11, 2018; accessed April 20, 2023.

61. Everette Lee DeGolyer, *Spindletop ... 1901-1951,* "Delivered at Spindletop Monument, Beaumont, Texas, January 10, 1951, In Celebration of the Fiftieth Anniversary of the Completion of the Lucas Gusher" (Dallas: DeGolyer and MacNaughton, 1951), 6–7; Michel T. Halbouty, September 27, 1978, quoted in Jack Donahue, *Wildcatter: The Story of Michel T. Halbouty and the Search for Oil* (Houston: Gulf Publishing Company, 1979), 251.

62. DeGolyer, *Spindletop,* 5.

63. Kroger, Newman, Sweet, and Lipe, "How Texas Law Promoted Shale Play Development." Baker Botts 175th

Anniversary, 2015.

64. Clark, *Oil Capital*, 385–88.
65. Choose Energy/Data Center/Wind Generation by State; Choose Energy/Solar Energy/Solar Energy by State; accessed October 9, 2021.
66. Edward W. Constant II, *The Origins of the Turbojet Revolution* (Baltimore, MD: Johns Hopkins University Press, 1980).

# BIBLIOGRAPHY

T his bibliography is divided into three sections: Methodological and Historical Sources; Technical Articles, Papers, and Publications; and Cases Cited. Each section is in alphabetical order.

## METHODOLOGICAL AND HISTORICAL SOURCES

Note: Individual Interviews in the Oral History of the Texas Oil Industry are not listed separately but are individually cited in the endnotes to the various chapters. This invaluable collection, housed in the Eugene C. Barker Texas History Center (later the Eugene C. Barker American History Center, now the Dolph Briscoe Center for American History) at the University of Texas at Austin, comprises transcribed interviews, mostly conducted in the 1950s.

Akerlof, George A. "The Market 'Lemons': Quality Uncertainty and the Market Mechanism." *Quarterly Journal of Economics* 84, no. 3 (1970): 488–500.
Ackerman, Sara L., Katherine Weatherford Darling, Sandra Soo-Jin Lee, Robert A. Hiatt, and Janet K. Shim. "Accounting for Complexity: Gene-Environment Interaction Research and the Moral Economy of Quantification." *Science, Technology, & Human Values* 41 (March 2016): 194–218.
Adams, Nathan. *The First National in Dallas*. Dallas: First National Bank, 1942.
Alchian, A. "Uncertainty, Evolution, and Economic Theory." *Journal of Political Economy* 58, no. 2 (1950): 211–22.
Aldrich, Howard E., Geoffrey M. Hodgson, David L. Hull, Thorbjørn Knudsen, Joel Mokyr, and Viktor J. Vanberg. "In Defense of Generalized Darwinism." *Journal of Evolutionary Economics* 18, no. 5 (2008): 577–96.
Alger, Ingela, and Jörgen Weibull. "Homo Moralis—Preference Evolution under Incomplete Information and Assortative Matching." *Econometrica* 81, no. 6 (2013): 2269–302.
Allaby, Robin G., Chris J. Stevens, and Dorian Q. Fuller. "A Novel Cost Framework Reveals Evidence for Competitive Selection in the Evolution of Complex Traits during Plant Domestication." *Journal of Theoretical Biology* 537 (21 March 2022): https://doi.org/10.1016/j.jtbi.2022.111004.
Allen, Robert C. "Collective Invention." *Journal of Economic Behavior and Organization* 4, no. 1 (March 1983): 1–24.
American Bar Association, Section of Mineral and Natural Resources Law. *Conservation of Oil and Gas: A Legal History, 1958*. Edited by Robert E. Sullivan. Chicago: American Bar Association, 1960.
American Bar Association, Section of Mineral Law. *Conservation of Oil and Gas: A Legal History, 1948*. Edited by Blakely M. Murphy. Chicago: American Bar Association, 1949.
American Oil & Gas Historical Society. "The Legend of Technology and the 'Conroe Crater.'" URL: https://aoghs.org/technology/directional-drilling, June 1, 2005. Accessed October 26, 2020.
American Petroleum Institute. *History of Petroleum Engineering*. Dallas: American Petroleum Institute, 1961
American Petroleum Institute. *Summary of API Program of Fundamental Research in Petroleum 1926–1951*. Washington, DC: API Library, unpublished.
Andersen, Esben Sloth. "Population Thinking, Price's Equation and the Analysis of Economic Evolution." *Evolutionary and Institutional Economics Review* 1, no. 1 (2004): 127–48.
Andersen, Esben Sloth, and Jacob Rubæk Holm. "The Signs of Change in Economic Evolution." *Journal of Evolutionary Economics* 24 (27 March 2014): 291–316.
Andersen, Esben Sloth, and Jacob Rubæk. "The Signs of Change in Economic Evolution: An Analysis of Directional, Stabilizing and Diversifying Selection Based on Price's Equation." (2016) Corpus ID: 210721833.

Anderson, Dillon. *I and Claudie*. Boston: Little, Brown, and Company, 1951.

André, J.-B., and O. Morin. "Questioning the Cultural Evolution of Altruism." *Journal of Evolutionary Biology* 24, no. 12 (December 2011): 2531–42.

Andrews, Eleanor, and James McCarthy. "Scale, Shale, and the State: Political Ecologies and Legal Geographies of Shale Gas Development in Pennsylvania." *Journal of Environmental Studies and Sciences* 4, no. 1 (2013): 7–16.

Andrews, P. W. S. "Competition in the Modern Economy." Reprinted from *Competitive Aspects of Oil Operations*. Paper presented at the 1958 Summer Meeting of the Institute of Petroleum, Scarborough, UK, June 1958. London: The Institute of Petroleum, 1958.

Ariew, André, and Richard C. Lewontin. "The Confusions of Fitness." *British Journal for the Philosophy of Science* 55, no. 2 (June 2004): 347–63.

Arthur, William Brian. "All Systems Will be Gamed: Exploitative Behavior in Economic and Social Systems." Santa Fe Institute (SFI) Working Paper 2014-06-016 (2014).

———. "Complexity Economics: Why Does Economics Need This Different Approach?" In *Complexity Economics: Dialogues of the Applied Complexity Network*, edited by W. Brian Arthur, Eric D. Beihocker, and Allison Stranger, 121–43. Santa Fe, NM: Santa Fe Institute Press, 2020.

———. "Economics in Nouns and Verbs." PREPRINT at arXiv https://arxiv.org/abs/2104.01868 2021 (5 April 2021).

———. "Foundations of Complexity Economics." *Nature Review Physics* 3 (29 January 2021): 136–45.

———. *The Nature of Technology: What It Is and How It Evolves*. New York: The Free Press, 2009.

Aumann, Robert J. "Agreeing to Disagree." *Annals of Statistics* 4 (November 1976): 1236–39.

———. "Backward Induction and Common Knowledge of Rationality." *Games and Economic Behavior* 8, no. 1 (1995): 6–19.

———. "Correlated Equilibrium as an Expression of Bayesian Rationality." *Econometrica* 55, no. 1 (1987): 1–18.

Axelrod, Robert, and William D. Hamilton. "The Evolution of Cooperation," *Science* 211 (27 March 1981): 1390–96.

Axtell, Robert. "Agent-Based Models for Economics." In *Complexity Economics: Dialogues of the Applied Complexity Network*, edited by W. Brian Arthur, Eric D. Beihocker, and Allison Stranger. Santa Fe, NM: Santa Fe Institute Press, 2020.

Ayala, Francisco José, and Theodosius Dobzhansky. *Studies in the Philosophy of Biology: Reduction and Related Problems*. Berkeley: University of California Press, 1974.

Bailey, R. M., E. Carrella, R. L. Axtell, M. G. Burgess, R. B. Cabral, M. Drexler, Chris Dorsett, Jens Koed Madsen, Andreas Merkl, and Steven Saul. "A Computational Approach to Managing Coupled Human-Environmental Systems: The POSEIDON Model of Ocean Fisheries." *Sustainability Science* 14, no. 2 (2019): 259–75.

Baker, Rex G., and Robert E. Hardwicke. "Conservation." In American Petroleum Institute, *History of Petroleum Engineering*, 1115–65.

Baldwin, Carliss Y., and Kim B. Clark. *Design Rules: The Power of Modularity*. Cambridge, MA: MIT Press, 2000.

Baldwin, James S., Peter M. Allen, Belinda Winder, and Keith Ridgway. "Simulating the Cladistic Evolution of Manufacturing." *Innovation: Management, Policy and Practice* 5, no. 3 (2003): 144–56.

Barabási, Albert-László. "Scale-Free Networks: A Decade and Beyond." *Science* 325 (24 July 2009): 412–13.

Barabási, Albert-László, and Réka Albert. "The Emergence of Scaling in Random Networks." *Science* 286 (15 October 1999): 509–12.

Barnes, Michele L., John Lynham, Kolter Kalberg, and Ping Sun Leung. "Social Networks and Environmental Outcomes." *Proceedings of the National Academy of Sciences* 113, no. 23 (2016): 6466–71.

Baum, Joel A. C., and Bill McKelvey, eds. *Variations in Organization Science: In Honor of Donald T. Campbell*. Thousand Oaks, CA: Sage, 1999.

Baxter, Gordon A. "Jenny in a Barn." *Flying* 116 (August 1989): 130–31.

Beard, Charles Austin. *An Economic Interpretation of the Constitution of the United States*. New York: Macmillan, 1913.

Beecher, C. E., and H. C. Fowler. "Production Techniques and Control." In American Petroleum Institute, *History of Petroleum Engineering*, 745–810.

Beinhocker, E. D. "Evolution as Computation: Integrating Self-Organization with Generalized Darwinism." *Journal of Institutional Economics* 7 (2011): 393–423.

Bell, A. V., Peter J. Richerson, and Richard McElreath. "Culture Rather Than Genes Provides Greater Scope for the Evolution of Large-Scale Human Prosociality." *Proceedings of the National Academy of Sciences USA* 106, no. 42 (2009): 17671–74.

Ben-David, Eyal, Alejandro Burga, and Leonid Kruglyak. "A Maternal-Effect Selfish Genetic Element in *Caenorhabditis elegans*." *Science* 366 (9 June 2017): 1051–55.

Bendor, Jonathan, and Dilip Mookherjee. "Institutional Structure and the Logic of Ongoing Collective Action." *American Political Science Review* 81, no. 1 (March 1987): 129–54.

Bennett, Nathan James, and Philip Dearden. "Why Local People Do Not Support Conservation: Community

Perceptions of Marine Protected Area Livelihood Impacts, Governance and Management in Thailand." *Marine Policy* 44 (February 2014): 107–16.

Bergstrom, Theodore C. "Evolution of Social Behavior: Individual and Group Selection." *Journal of Economic Perspectives* 16, no. 2 (2002): 67–88.

Berlin, Isaiah. *The Hedgehog and the Fox: An Essay on Today's View of History*. New York: Simon and Schuster, 1953.

Bhattacharya, Rajeev, Timothy M. Devinney, and Madan M. Pillutla. "A Formal Model of Trust Based on Outcomes." *Academy of Management Review* 23 (July 1998): 459–72.

Bijker, Wiebe. *Of Bikes, Bakelites, and Bulbs: Towards a Theory of Socio-Technical Change*. Cambridge, MA: MIT Press, 1997.

Bijker, Wiebe, Thomas P. Hughes, and Trevor Pinch, eds. *The Social Construction of Technological Systems: New Directions in the Sociology and History of Technology*. Cambridge, MA: MIT Press, 1987.

Bijma, P. "The Price Equation as a Bridge Between Animal Breeding and Evolutionary Biology." *Philosophical Transactions of the Royal Society B* 375 (27 April 2020): 1–10.

Binmore, Ken. "Rationality in the Centipede." URL: https://pdfs.semanticscholar.org/3ac3/4256852473048f-f66e917e6f576db7836628.pdf. Accessed June 23, 2018.

Blair, John M. *The Control of Oil*. New York: Pantheon Books, 1976.

Boatright, Mody C. "The Myth of Frontier Individualism." *Southwestern Social Science Quarterly* 22 (June 1941): 14–32.

Bookstaber, Richard M. "Using Agent-Based Models for Analyzing Threats to Financial Stability." Office of Financial Research Working Paper No. 0003, December 21, 2012.

Bowker, Geoffrey C. *Science on the Run: Information Management and Industrial Geophysics at Schlumberger, 1920–1940*. Cambridge, MA: MIT Press, 1994.

Bowles, Samuel. "Endogenous Preferences: The Cultural Consequences of Markets and Other Economic Institutions." *Journal of Economic Literature* 36, no. 1 (March 1998): 75–111.

Bowles, Samuel, Jung-Kyoo Choi, and Astrid Hopfensitz. "The Coevolution of Individual Behaviors and Group Level Institutions." *Journal of Theoretical Biology* 223, no. 2 (2003): 135–47.

Bowles, Samuel, and Herbert Gintis. "The Moral Economy of Communities: Structured Populations and the Evolution of Pro-Social Norms." *Evolution and Human Behavior* 19 (1998): 3–25.

Boyd, Robert. "Mistakes Allow Evolutionary Stability in the Repeated Prisoner's Dilemma Game." *Journal of Theoretical Biology* 136 (1989): 47–56.

Boyd, Robert, and Peter J. Richerson. "Collective Actions, Cultural Norms." *Science* 362, no. 6420 (14 November 2018): 1236–37.

———. "Culture and the Evolution of Human Cooperation." *Philosophical Transactions of the Royal Society London B. Biological Sciences* 364, no. 1533 (12 November 2009): 2381–88.

———. *Culture and the Evolutionary Process*. Chicago: University of Chicago Press, 1985.

———. "Group Selection Among Alternative Evolutionarily Stable Strategies." *Journal of Theoretical Biology* 145 (9 August 1990): 331–42.

———*The Origin and Evolution of Cultures*. Oxford, UK: Oxford University Press, 2005.

———. "Transmission Coupling Mechanisms: Cultural Group Selection." *Philosophical Transactions of the Royal Society B* 365 (2010): 3787–95.

Brantly, J. E. *History of Oil Well Drilling*. Houston: Gulf Publishing Company, 1971.

Bravo, Giangiacomo. "Agents' Beliefs and the Evolution of Institutions for Common-Pool Resource Management." *Rationality and Society* 23, no. 1 (2011): 117–52.

Brenner, T. "Can Evolutionary Algorithms Describe Learning Processes?" *Journal of Evolutionary Economics* 8, no. 3 (1998): 271–83.

Brown, C. L. "Bank Financing of Secondary Recovery Projects." *Journal of Petroleum Technology* 9 (March 1957): 22–25.

Bruderer, Erhard, and Jitendra V. Singh. "Organization Evolution, Learning, and Selection: A Genetic-Algorithm Based Model." *Academy of Management Journal* 39, no. 5 (October 1996): 1322–29.

Brush, Eleanor R., David C. Krakauer, and Jessica C. Flack. "Conflicts of Interest Improve Collective Computation of Adaptive Social Structures." *Science Advances* 4 (17 January 2018): https://www.science.org/doi/10.1126/sciadv.1603311.

———. "A Family of Algorithms for Computing Consensus about Node State from Network Data." *PLOS Computational Biology* 9 (18 July 2013): https://doi.org/10.1371/journal.pcbi.1003109.

Bucciarelli, Louis L. *Designing Engineers*. Cambridge, MA: MIT Press, 1994.

Buck, E. O., interview by Louis J. Marchiafava, April 16, 1981. Houston Metropolitan Research Center of the Houston Public Library. Archive Number: OH 281.

Buenger, Walter L., and Joseph A. Pratt. *But Also Good Business: Texas Commerce Banks and the Financing of Houston and Texas, 1886–1986.* College Station: Texas A&M University Press, 1986.

Bunyan, John. *The Pilgrim's Progress: From This World to That Which Is to Come, Delivered Under the Similitude of a Dream.* 1678. Reprint, Bungay, UK: Brightly and Childs, 1817.

Burgess, Matthew G., Ernesto Carrella, Michael Drexler, Robert L. Axtell, Richard M. Bailey, James R. Watson, Reniel B. Cabral, Michaela Clemence, et al. "Opportunities for Agent-Based Modelling in Human Dimensions of Fisheries." *Fish and Fisheries* 21 (2020): 570–87.

Burka, Paul. "Power Politics: How One Company's Wheeling and Dealing Brought the Energy Crisis into Your Life." *Texas Monthly*, May 1975, 68–78+.

Burt, R. A. *British Battleships of World War One.* Annapolis, MD: Naval Institute Press, 1986.

Camerer, Colin F. *Behavioral Game Theory: Experiments in Strategic Interaction.* Princeton, NJ: Princeton University Press, 2003.

Campbell, Donald T. "'Downward Causation' in Hierarchically Organized Biological Systems." In *Studies in the Philosophy of Biology: Reduction and Related Problems*, edited by Francisco José Ayala and Theodosius Dobzhansky, 179–86. Berkeley: University of California Press, 1974.

———. "Ethnocentrism of Disciplines and the Fish-Scale Model of Omniscience." In *Interdisciplinary Relationships in the Social Sciences*, edited by M. Sherif and C. W. Sherif, 328–48. Chicago: Aldine, 1969.

———. "Evolutionary Epistemology." In *The Philosophy of Karl Popper*, Vol. 14, bk. 1, of *The Library of Living Philosophers,* edited by Paul A. Schlipp, 413–63. La Salle, IL: Open Court, 1974.

———. "From Evolutionary Epistemology Via Selection Theory to a Sociology of Scientific Validity." *Evolution and Cognition* 3, no. 1 (2001): 5–38.

———. "How Individual and Face-to-Face-Group Selection Undermine Firm Selection in Organizational Evolution." In *Evolutionary Dynamics of Organizations*, edited by J. A. C. Baum and J. V. Singh, 23–38. New York: Oxford University Press, 1994.

———. "Levels of Organization, Downward Causation, and the Selection-Theory Approach to Evolutionary Epistemology." In *Theories of the Evolution of Knowing*, edited by G. Greenberg and E. Tobach, 1–17. Hillsdale, NJ: Lawrence Erlbaum, 1990.

———. "Neurological Embodiments of Belief and the Gaps in the Fit of Phenomena to Noumena." In *Naturalistic Epistemology*, edited by Abner Shimony and Debra Nails, 165–92. Dordrecht, Holland, and Boston, MA: D. Reidel, 1987.

———. "Objectivity and the Social Locus of Scientific Knowledge." Presidential Address to the Division of Social and Personality Psychology of the American Psychological Association, 1969.

———. "On the Conflicts Between Biological and Social Evolution and Between Psychology and Moral Tradition." *American Psychologist* (December 1975): 1103–26.

———. "A Phenomenology of the Other One: Corrigible, Hypothetical, and Critical." In *Human Action: Conceptual and Empirical Issues*, edited by Theodore Mischel, 41–69. New York: Academic Press, 1969.

———. "Selection Theory and the Sociology of Scientific Validity." In *Evolutionary Epistemology: A Multiparadigm Program*, edited by Werner Calebaut and R. Pinxten, 139–58. Dordrecht, Holland, and Boston, MA: D. Reidel, 1987.

———. "The Social Psychology of Scientific Validity: An Epistemological Perspective and a Personalized History." In *The Social Psychology of Science*, edited by William R. Shadish and Steve Fuller, 124–61. New York: Guilford Press, 1994.

———. "Unjustified Variation and Selective Retention in Scientific Discovery." In *Studies in the Philosophy of Biology: Reduction and Related Problems*, edited by Francisco José Ayala and Theodosius Dobzhansky, 139–61. Berkeley: University of California Press, 1974.

Campbell, Donald T., and Donald W. Fiske. "Convergent and Discriminant Validation by the Multitrait-Multimethod Matrix." *Psychological Bulletin* 56 (March 1959): 81–105, reprinted in Donald T. Campbell, *Methodology and Epistemology for Social Science: Selected Papers*, edited by E. Samuel Overman, 37–61. Chicago, IL: University of Chicago Press, 1988.

Campbell, Donald T., and Bonnie T. Paller. "Extending Evolutionary Epistemology to 'Justifying' Scientific Beliefs (A Sociological Rapprochement with a Fallibilist Perceptual Foundationalism?)." In *Issues in Evolutionary Epistemology*, edited by E. Halweg and C. A. Hookers. Albany, NY: State University of New York Press, 1989.

Campbell, Donald T., H. W. Riecken, Robert F. Boruch, N. Kaplan, T. K. Glennan, J. Pratt, A. Rees, and W. Williams. "Quasi-Experimental Designs." In *Social Experimentation: A Method for Planning and Evaluating Social Interventions*, edited by H. W. Riecken and Robert F. Boruch, 87–116. New York: Academic Press, 1974.

Campbell, Donald T., and H. Laurence Ross. "The Connecticut Crackdown on Speeding: Time-Series Data in Quasi-Experimental Analysis." *Law and Society Review* 3, no. 1 (1968): 33–53; reproduced in Donald T. Campbell, *Methodology and Epistemology for Social Science: Selected Papers*, edited by E. Samuel Overman, 222–37. Chicago: University of Chicago Press, 1988.

Campbell, Donald T., and J. Stanley. *Experimental and Quasi-Experimental Design for Research*. Boston, MA: Houghton Mifflin, 1963.

Campbell, John O. "Hypothesis Theory: Universal Darwinism as a Process of Bayesian Inference." *Frontiers in Systems Neuroscience* 10 (7 June 2016): 49.

Campenni, Marco, and Gabriele Schino. "Partner Choice Promotes Cooperation: The Two Faces of Testing with Agent-Based Models." *Journal of Theoretical Biology* 344 (7 March 2014): 49–55.

Caporael, Linnda R., Robyn M. Dawes, John M. Orbell, and Alphons J. C. van de Kragt. "Selfishness Examined: Cooperation in the Absence of Egoistic Incentives." *Behavioral and Brain Sciences* 12, no. 4 (1989): 683–739.

Capraro, Valerio, Joseph Y. Halpern, and Matjaž Perc. "From Outcome-Based to Language-Based Preferences." *Journal of Economic Literature* 62, no. 1 (2024): 115–54.

Carlisle, Rodney P., and August W. Giebelhaus. *Energy Center: The Federal Government in Petroleum Research, Bartlesville, 1918–1983*. Bartlesville, OK: US Department of Energy, Bartlesville Energy Technology Center, 1985.

Carr, Edward Hallett. *What Is History?* London: Macmillan, 1961.

Carrella, E., R. M. Bailey, and J. K. Madsen. "Repeated Discrete Choices in Geographical Agent Based Models with an Application to Fisheries." *Environmental Modelling and Software* 111 (2019): 204–30.

Carrella, Ernesto, Steven Saul, Kristin Marshall, Matthew G. Burgess, Reniel B. Cabral, R. M. Bailey, C. Dorsett, M. Drexler, J. K. Madsen, and Andreas Merkl. "Simple Adaptive Rules Describe Fishing Behaviour Better Than Perfect Rationality in the U.S. West Coast Groundfish Fishery." *Ecological Economics* 16 (2020): 106449.

Cartwright, Nancy. *How the Laws of Physics Lie*. Oxford, UK: Clarendon Press, 1983.

Castaneda, Christopher James. *Regulated Enterprise: Natural Gas Pipelines and Northeastern Markets, 1938–1954*. Columbus: Ohio State University Press, 1993.

Castaneda, Christopher James, and Joseph A. Pratt. *From Texas to the East: A Strategic History of Texas Eastern Corporation*. College Station: Texas A&M University Press, 1993.

Chandler, Alfred DuPont, Jr. "Organizational Capabilities and the Economic History of the Industrial Enterprise." *Journal of Economic Perspectives* 6, no. 3 (1992): 79–100.

———. *The Visible Hand: The Managerial Revolution in American Business*. Cambridge, MA: Belknap Press, 1977.

Chandler, Alfred DuPont, Jr., with Takashi Hikino. *Scale and Scope: The Dynamics of Industrial Capitalism*. Cambridge, MA: Harvard University Press, 1990.

Channel, David F. "The Emergence of the Engineering Sciences: A Historical Analysis." In *Handbook of the Philosophy of Technology and the Engineering Sciences*, edited by Anthonie Meijers, 117–54. Amsterdam: North Holland, 2009.

Cheek, Maurice. "Legal History of Conservation of Gas in Texas." In *Legal History of Conservation of Oil and Gas: A Symposium*, Section of Mineral Law, American Bar Association, 269–70. Chicago: American Bar Association, 1939.

Chen, Jing, and Aleksey Zinger. "The Robustness of Zero-Determinant Strategies in Iterated Prisoner's Dilemma Games." *Journal of Theoretical Biology* 357 (21 September 2014): 46–54.

Childs, William R. *The Texas Railroad Commission: Understanding Regulation in America to the Mid-Twentieth Century*. College Station: Texas A&M University Press, 2005.

Cho, Adrian. "Quantum Darwinism Seen in Diamond Traps: Concept of Survival of the Fittest Could Explain How Reality Emerges from Quantum Haze." *Science* 365, no. 6458 (13 September 2019): 1070.

Choi, Jung-Kyoo, and Samuel Bowles. "The Coevolution of Parochial Altruism and War." *Science* 318, no. 5850 (2007): 636–40.

Chuang, John S., Olivier Rivoire, and Stanislas Liebler. "Simpson's Paradox in a Synthetic Microbial System." *Science* 323 (9 January 2009): 272–75.

Claidière, Nicolas, Thomas C. Scott-Phillips, and Dan Sperber. "How Darwinian Is Cultural Evolution?" *Philosophical Transactions of the Royal Society B* 369 (19 May 2014): https://doi.org/10.1098/rstb.2013.0368.

Clark, Bernard F., Jr. *Oil Capital: The History of American Oil, Wildcatters, Independents, and Their Bankers*. Houston: Bernard F. Clark Jr., 2016.

Clark, James A. *Three Stars for the Colonel*. New York: Random House, 1954.

Clark, James A., and Michel T. Halbouty. *The Last Boom*. New York: Random House, 1972.

———. *Spindletop*. 1952. Reprint, Houston: Gulf Publishing Company, 1980.

Cohen, Wes M., and D. A. Levinthal. "Adsorptive Capacity: A New Perspective on Learning and Innovation." *Administrative Science Quarterly* 35 (1990): 128–52.

Collins, Harry M. "The TEA Set: Tacit Knowledge and Scientific Networks." *Science Studies* 4 (1974): 165–86; reprinted in Barry Barnes and David Edge, eds., *Science in Context: Readings in the Sociology of Science* (Cambridge, MA: MIT Press, 1982).

Colman, Andrew M., Lindsay Browning, and Briony D. Pulford. "Spontaneous Similarity Discrimination in the Evolution of Cooperation." *Journal of Theoretical Biology* 299 (21 April 2012): 162–71.

Comin, Diego, William Easterly, and Erick Gong. "Was the Wealth of Nations Determined in 1000 BC?" *American Economic Journal: Macroeconomics* 2, no. 3 (July 2010): 65–97.

Conrod, Robert Lucas. "Production Techniques and the Supply of Oil and Oil Products." PhD diss., University of Texas at Austin, 1934.

Constant, Edward W., II. "Cause or Consequence: Science, Technology, and Regulatory Change in the Oil Business in Texas, 1930–1975." *Technology and Culture* 30 (April 1989): 426–55.

———. "Limited Reach: The Visible Hand and the 'Oil Bidness' in Texas." In *Proceedings of the Conference on Business History*, edited by M. Davids, F. de Goey, and D. de Wit, 44–60. Rotterdam: Erasmus University, 1994.

———. *The Origins of the Turbojet Revolution*. Baltimore, MD: Johns Hopkins University Press, 1980.

———. "Science *in* Society: Petroleum Engineers and the Oil Fraternity in Texas, 1925–65." *Social Studies of Science* 19 (August 1989): 439–72.

———. "Why Evolution Is a Theory About Stability: Constraint, Causation, and Ecology on Technological Change." *Research Policy* 31 (2002): 1241–56.

Cooke, S. J., C. D. Suski, R. Arlinghaus, and A. J. Danylchuk. "Voluntary Institutions and Behaviours as Alternatives to Formal Regulations in Recreational Fisheries Management." *Fish and Fisheries* 14, no. 4 (2013): 439–57.

Cordes, Christian, Peter J. Richerson, Richard McElreath, and Pontus Strimling. "A Naturalistic Approach to the Theory of the Firm: The Role of Cooperation and Cultural Evolution." *Journal of Economic Behavior and Organization* 68, no. 1 (2008): 125–39.

Cowan, Ruth Schwartz. *A Social History of American Technology*. New York: Oxford University Press, 1997.

Craig, J. V., and W. M. Muir. "Group Selection for Adaptation to Multiple-Hen Cages: Beak-Related Mortality, Feathering, and Body Weight Responses." *Poultry Science* 75 (1995): 294–302.

Crane, Diana. *Invisible Colleges: Diffusion of Knowledge in Scientific Communities*. Chicago: University of Chicago Press, 1972.

Cross, Whitney R. *The Burned-Over District: The Social and Intellectual History of Enthusiastic Religion in Western New York, 1800–1850*. Ithaca, NY: Cornell University Press, 1950.

Cyert, Richard M., and Morris DeGroot. "Rational Expectations and Bayesian Analysis." *Journal of Political Economy* 82 (1974): 521–36.

Cyert, Richard M., and James G. March. *A Behavioral Theory of the Firm*. Englewood Cliffs, NJ: Prentice-Hall, 1963.

Daniels, George H., Jr. "The Process of Professionalization in American Science: The Emergent Period, 1820–1860." *Isis* 58 (Summer 1967): 151–66.

Dasgupta, Partha. "Dark Matters: Exploitation as Cooperation." *Journal of Theoretical Biology* 299 (21 April 2012): 180–87.

Datson, Lorraine. "The Moral Economy of Science." *Osiris* 10 (1995): 2–24.

David, Paul A. "Why Are Institutions the 'Carriers of History'?: Path Dependence and the Evolution of Conventions, Organizations and Institutions." *Structural Change and Economic Dynamics* 5, no. 2 (1994): 205–20.

Davidson, John. "The Man Who Crushed Texaco." *Texas Monthly*, March 1988, 92ff.

Davidson, Paul. "Public Policy Problems of the Domestic Crude Oil Industry." *American Economic Review* 53 (March 1963): 85–108.

Dawes, Robyn M. "Social Dilemmas." *Annual Review of Psychology* 31 (1980): 169–93.

Dawes, Robyn M., Alphons J. C. van de Kragt, and John M. Orbell. "Not Me or Thee but We: The Importance of Group Identity in Eliciting Cooperation in Dilemma Situations: Experimental Manipulations." *Acta Psychologica* 68 (1988): 83–97.

Dawkins, Richard. *The Extended Phenotype*. Oxford, UK: Oxford University Press, 1982.

———. *The Selfish Gene*. Oxford, UK: Oxford University Press, 1976.

Day, Barbara Sue Thompson. "The Oil and Gas Industry and Texas Politics, 1930–1935." PhD diss., Rice University, 1973.

Day, Troy, Todd Parsons, Anaury Lambert, and Sylvain Gandon. "The Price Equation and Evolutionary Epidemiology." *Philosophical Transactions of the Royal Society B* 375 (27 April 2020): https://doi.org/10.1098/rstb.2019.0357.

Deacon, R. T., D. P. Parker, and C. Costello. "Reforming Fisheries: Lessons from a Self-Selected Cooperative." *Journal of Law and Economics* 56, no. 1 (2013): 83–125.

DeCambre, Mark. "About 150 Years of Oil-Price History: This One Chart Illustrates Crude's Spectacular Plunge Below $0 a Barrel. Deutsche Bank, Global Financial Data." *Market Watch*, April 29, 2020. URL: https://www.marketwatch.com/story/about-150-years-of-oil-price-history-in-one-chart-illustrates-crudes-spectacular-plunge-below-0-a-barrel-2020-04-22. Accessed January 18, 2021.

"Decline Curve Analysis (DCA)." PetroWiki. URL: https://petrowiki.spe.org/Production_forecasting_decline_curve_analysis#::text=Decline%20curve%20analysis%20(DCA)%20is,fluids%2C%20are%20usually%20the%20cause. Accessed June 8, 2018.

DeGolyer, Everette Lee. "Concepts of Occurrence of Oil and Gas." In *History of Petroleum Engineering*, Executive Committee on Drilling and Production Practice, Division of Production, 15–33. Dallas: American Petroleum Institute, 1961.

———. "The Development of Some of Our Concepts of Fundamentals Governing Petroleum Production Practice." Typescript, March 12, 1940. The Papers of Everette Lee DeGolyer Sr., Mss. 60, Box 26, Folder 2537. DeGolyer Library, Southern Methodist University, Dallas, Texas.

———. "Problems of the Producing Branch of the Oil Industry." Typescript, May 24, 1939. The Papers of Everette Lee DeGolyer Sr., Mss. 60, Folder 3100. DeGolyer Library, Southern Methodist University, Dallas, Texas.

———. "Production Under Effective Water Drive as a Standard for Conservation Practice." The Papers of Everette Lee DeGolyer Sr., Mss. 60, Box 18, Folder 2295. DeGolyer Library, Southern Methodist University, Dallas, Texas.

———. *Spindletop . . . 1901–1951.* "Delivered at Spindletop Monument, Beaumont, Texas, January 10, 1951, in Celebration of the Fiftieth Anniversary of the Completion of the Lucas Gusher." Dallas: DeGolyer and MacNaughton, 1951.

———. "Waste in Oil and Gas Production." Typescript, 1941, The Papers of Everette Lee DeGolyer Sr., Mss. 60, Box 25, Folder 2458, DeGolyer Library, Southern Methodist University, Dallas, Texas.

de Mendoza, Alex. "A Mammalian DNA Methylation Landscape: A Study of 348 Species Offers Clues into the Diversity of Mammalian Life Spans." *Science* 381, no. 6658 (11 August 2023): 602–3.

De Moor, Tine. "Revealing Historical Resilience." *Science* 362, no. 6420 (14 November 2018): 1238.

Denney, Dennis. "Technical Papers Have Been the Backbone of SPE's Mission." *Journal of Petroleum Technology, Special Commemorative Issue Celebrating 50 Years of the Society of Petroleum Engineers* 59, no. 10 (October 2007): 103.

Dennis, Michael Aaron. "Drilling for Dollars: The Making of US Petroleum Reserve Estimates, 1921–25." *Social Studies of Science* 15 (May 1985): 241–65.

De Silva, Hannelore, Christoph Hauert, Arne Traulsen, and Karl Sigmund. "Freedom, Enforcement, and the Social Dilemma of Strong Altruism." *Journal of Evolutionary Economics* 20, no. 3 (2010): 203–17.

de Vladar, Harold P., and Eörs Szathmáry. "Beyond Hamilton's Rule: A Broader View of How Relatedness Affects the Evolution of Altruism Is Emerging." *Science* 356 (5 May 2017): 485–86.

de Waal, Frans B. M. "The Chimpanzee's Service Economy: Food for Grooming." *Evolution and Human Behavior* 18, no. 6 (November 1997): 375–86.

Dobbs, Betty Jo Teeter. *The Foundations of Newton's Alchemy, or, "The Hunting of the Greene Lyon."* Cambridge, UK: Cambridge University Press, 1975.

Dobzhansky, Theodosius. "Chance and Creativity in Evolution." In *Studies in the Philosophy of Biology: Reduction and Related Problems,* edited by Francisco José Ayala and Theodosius Dobzhansky, 307–38. Berkeley, CA: University of California Press, 1974.

Donahue, Jack. *Wildcatter: The Story of Michel T. Halbouty and the Search for Oil.* Houston: Gulf Publishing Company, 1979.

Donnelly, John. "1957–1969: The Early Years." *Journal of Petroleum Technology, Special Commemorative Issue: Celebrating 50 Years of the Society of Petroleum Engineers* 59, no. 10 (October 2007): 36–44.

Dopfer, Kurt. "The Economic Agent as Rule Maker and Rule User: Homo Sapiens Oeconomicus." *Journal of Evolutionary Economics* 14 (2004): 177–95.

Dopfer, Kurt, John Foster, and Jason Potts. "Micro-meso-macro." *Journal of Evolutionary Economics* 14, no. 3 (July 2004): 263–80.

Dosi, Giovanni, and Richard R. Nelson. "An Introduction to Evolutionary Theories in Economics." *Journal of Evolutionary Economics* 4 (1994): 153–72.

Dosi, Giovanni, and David J. Teece. "Introduction to 'Four Unpublished Manuscripts by Herbert A. Simon.'" *Industrial and Corporate Change* 11, no. 3 (1 June 2002): 591.

Downey, Gary L. "Reproducing Cultural Identity in Negotiating Nuclear Power: The Union of Concerned Scientists and Emergency Core Cooling." *Social Studies of Science* 18 (May 1988).

———. "Structure and Practice in the Cultural Identities of Scientists: Negotiating Nuclear Wastes in New Mexico." *Anthropological Quarterly* 61 (January 1988): 26–38.

Doyle, Arthur Conan, Sir. "The Adventure of the Silver Blaze." *The Memoirs of Sherlock Holmes.*

D'Souza, Raissa M. "Unlocking the Science of Success." *Science* 362, no. 6420 (14 December 2018): 1253.

Dugatkin, Lee Alan. *The Altruism Equation: Seven Scientists Search for the Origins of Goodness.* Princeton, NJ: Princeton University Press, 2006.

———. "*N*-person Games and the Evolution of Cooperation: A Model Based on Predator Inspection Behavior in Fish." *Journal of Theoretical Biology* (9 January 1990): 123–35.

Dugatkin, Lee Alan, Michael Mesterton-Gibbons, and Alasdair I. Houston. "Beyond the Prisoner's Dilemma: Towards Models to Discriminate Among Mechanisms of Cooperation in Nature." *Trends in Ecology and*

*Evolution* 7, no. 6 (June 1992): 202–5.

Durex, Maxime, Charles Perreault, and Robert Boyd. "Divide and Conquer: Intermediate Levels of Population Fragmentation Maximize Cultural Accumulation." *Philosophical Transactions of the Royal Society B—Biological Sciences* 373, no. 1743 (April 2018): 10.1098/rstb.2017.0062.

Durkheim, Émile. *The Elementary Forms of the Religious Life*. 1915. Reprint, New York: Free Press, 1965.

Durlauf, Steven N. "Complexity and Empirical Economics." *The Economic Journal* 115 (2005): F225–F243.

Eakens, Robert Henry Seale. "The Development of Proration in the East Texas Oil Field." MA thesis, University of Texas of Austin 1937.

East Texas Salt Water Disposal Company. *Salt Water Disposal: East Texas Field*. Austin: Petroleum Extension Service, University of Texas–Division of Extension, 1953.

Eerkens, J. W., and C. P. Lipo. "Cultural Transmission, Copying Errors, and the Generation of Variation in Material Culture and the Archaeological Record." *Journal of Anthropological Archaeology* 24 (2005): 316–24.

Elkins, L. E. "Research." In *History of Petroleum Engineering*, Executive Committee on Drilling and Production Practice, Division of Production, 1081–113. Dallas: American Petroleum Institute, 1961.

Ellickson, Robert C. *Order without Law: How Neighbors Settle Disputes*. Cambridge, MA: Harvard University Press, 1991.

El Mouden, C., J.-B. André, O. Morin, and D. Nettle. "Cultural Transmission and the Evolution of Human Behaviour: A General Approach Based on the Price Equation." *Journal of Evolutionary Biology* 27 (2014): 231–41.

Epstein, Joshua M., and Robert Axtell. *Growing Artificial Societies: Social Science from the Bottom Up*. Washington, DC: Brookings Institution Press, 1996.

Erard, Michael, and Catherine Matacic. "Did Kindness Prime Our Species for Language?" *Science* 361, no. 6401 (August 3, 2018): 436–37.

Eubanks, John Evans. *Ben Tillman's Baby: The Dispensary System of South Carolina, 1892–1915*. Augusta, GA: self-published, 1950.

Eyal, Gil. *The Crisis of Expertise*. New York: John Wiley & Sons, 2019.

Falk, Armin, Ernst Fehr, and Urs Fischbacher. "Approaching the Commons: A Theoretical Explanation." In *The Drama of the Commons*, edited by Elinor Ostrom, Thomas Dietz, Nives Dolšak, Paul C. Stern, Susan Stonich, and Elke U. Weber, 157–91. Washington, DC: National Academy Press, 2002.

Falk, Armin, and James J. Heckman. "Lab Experiments Are a Major Source of Knowledge in the Social Sciences." *Science* 326 (23 October 2009): 535–38.

Fehr, Ernst. "On the Economics and Biology of Trust." *Journal of the European Economic Association* 7 (April–May 2009): 235–66.

Fehr, Ernst, and Urs Fischbacher. "Why Social Preferences Matter—The Impact of Non-Selfish Motives on Competition, Cooperation, and Incentives." *The Economic Journal* 112, no. 478 (March 2002): C1–C33.

Fehr, Ernst, Urs Fischbacher, and Simon Gächter. "Strong Reciprocity, Human Cooperation and the Enforcement of Social Norms." *Human Nature* 13 (2002): 1–25.

Fehr, Ernst, and Simon Gächter. "Cooperation and Punishment in Public Goods Experiments." *American Economic Review* 90, no. 4 (September 2000): 980–94.

———. "Reciprocity and Economics: The Economic Implications of *Homo reciprocans*." *European Economic Review* 42 (1998): 845–59.

Fehr, Ernst, and Klaus M. Schmidt. "A Theory of Fairness, Competition, and Cooperation." *Quarterly Journal of Economics* 114 no. 3 (August 1999): 817–68.

Fehrenbach, T. R. "The Americanization of Texas." *Texas Monthly*, January 1975, 60–62ff.

———. *Lone Star: A History of Texas and the Texans*. 1968. Reprint, New York: American Legacy Press, 1983.

Feldman, Marcus W., and Ewart A. C. Thomas. "Behavior-Dependent Contexts for Repeated Plays of the Prisoner's Dilemma II: Dynamical Aspects of the Evolution of Cooperation." *Journal of Theoretical Biology* 128 (1987): 297–315.

Fischbacher, Urs, and Simon Gächter. "Social Preferences, Beliefs, and the Dynamics of Free Riding in Public Goods Experiments." *American Economic Review* 100, no. 1 (March 2010): 541–56.

Fischbacher, Urs, Simon Gächter, and Ernst Fehr. "Are People Conditionally Cooperative? Evidence from a Public Goods Experiment." *Economic Letters* 71, no. 3 (2001): 397–404.

Fisher, Ronald A. *The Genetical Theory of Natural Selection*. 1930. Reprint, New York: Dover, 1958.

Flannery, Kent V., Joyce Marcus, and Robert G. Reynolds. *The Flocks of the Wamani: A Study of Llama Herders on the Puntas of Ayacucho, Peru*. San Diego, CA: Academic Press, 1989.

Fletcher, J., and M. Doebeli. "A Simple and General Explanation for the Evolution of Altruism." *Proceedings of the Royal Society B: Biological Sciences* 276 (7 January 2009): 13–19.

Flierl, G., D. Grünbaum, S. Levins, and D. Olson. "From Individuals to Aggregations: The Interplay Between Behavior and Physics." *Journal of Theoretical Biology* 196, no. 4 (1999): 397–454.

Fogarty, Laurel. "Cultural Complexity and Evolution in Fluctuating Environments." *Transactions of the Royal Society B—Biological Sciences* 373, no. 1743 (April 2018): 10.1098/rstb.2017.0063.

Francis, Hywel. "The Law, Oral Tradition and the Mining Community." *Journal of Law and Society* 12 (Winter 1985): 267–71.

Francois, Patrick, Thomas Fujiwara, and Tanguy van Ypersele. "The Origins of Human Prosociality: Cultural Group Selection in the Workplace and Laboratory." *Science Advances* 4 (September 19, 2018): https://doi.org/10.1126/sciadv.aat2201.

Frank, R. H., T. Gilovich, and D. T. Regan. "The Evolution of One-Shot Cooperation." *Ethology and Sociobiology* 14 (1993): 247–56.

Frank, Steven A. "George Price's Contributions to Evolutionary Genetics." *Journal of Theoretical Biology* 175 (7 August 1995): 373–88.

———. "The Price Equation, Fisher's Fundamental Theorem, Kin Selection, and Causal Analysis." *Evolution* 51 (1995): 1712–29.

———. "Simple Unity Among the Fundamental Equations of Science." *Philosophical Transactions of the Royal Society B* 375 (27 April 2020): https://doi.org/10.1098/rstb.2019.0351.

Franks, Kenny A. "Hot Oil Controversy." *Encyclopedia of Oklahoma History and Culture.* URL: www.okhistory.org, accessed December 30, 2017.

Frenken, Koen. "Technological Innovation and Complexity Theory." *Economics of Innovation and New Technology* 15, no. 2 (2006): 137–55.

Frenken, Koen, and A. Nuvolari. "The Early History of Steam Engine Technology: An Evolutionary Interpretation Using Complexity Theory." *Industrial and Corporate Change* 13, no. 2 (2004): 419–50.

Fu, Feng, Christoph Hauert, Martin A. Nowak, and Long Wang. "Reputation-based Partner Choice Promotes Cooperation in Social Networks." *Physical Review E* 78, no. 2 (August 22, 2008): 02617.

Fudenberg, Drew, and Eric Maskin. "The Folk Theorem in Repeated Games with Discounting or with Incomplete Information." *Econometrica* 54, no. 3 (May 1986): 533–54.

Fujimura, Joan H. "Crafting Science: Standardized Packages, Boundary Objects, and 'Translation.'" In *Science as Practice and Culture,* edited by Andrew Pickering, 168–211. Chicago: University of Chicago Press, 1992.

Fukuyama, Francis. *Trust: Social Virtues and the Creation of Prosperity.* New York: Free Press, 1995.

Gächter, Simon, and Benedikt Herrmann. "The Limits of Self-Governance When Cooperators Get Punished—Experimental Evidence from Urban and Rural Russia." *European Economic Review* 55, no. 2 (February 2011): 193–210.

———. "Reciprocity, Culture, and Human Cooperation: Previous Insights and a New Cross-Cultural Experiment." *Philosophical Transactions of the Royal Society B* 364 (2009): 791–806.

Gächter, Simon, Benedikt Herrmann, and Christian Thöni. "Culture and Cooperation." *Philosophical Transactions of the Royal Society B* 365, no. 1553 (August 2, 2010): 2651–61.

Gächter, Simon, Elke Renner, and Martin Sefton. "The Long-Run Benefits of Punishment." *Science* 322 (5 December 2008): 1510.

Galambos, Louis. *America at Middle Age: A New History of the United States in the Twentieth Century.* New York: McGraw Hill, 1981.

———. *Competition and Cooperation: The Emergence of a National Trade Association.* Baltimore, MD: Johns Hopkins University Press, 1966.

Galbraith, John Kenneth. *American Capitalism: The Concept of Countervailing Power.* Boston: Houghton Mifflin, 1952.

Galison, Peter. *Einstein's Clocks, Poincaré's Maps: Empires of Time.* New York: W. W. Norton, 2004.

———. *How Experiments End.* Chicago: University of Chicago Press, 1987.

———. *Image & Logic: A Material Culture of Microphysics.* Chicago: University of Chicago Press, 1997.

García, Julián, and Jeroen C. J. M. van den Bergh. "Evolution of Parochial Altruism by Multilevel Selection." *Evolution and Human Behavior* 32, no. 4 (2011): 277–87.

García, Julián, Matthijs van Veelen, and Arne Traulsen. "Evil Green Beards: Tag Recognition Can Also Be Used to Withhold Cooperation in Structured Populations." *Journal of Theoretical Biology* 360 (7 November 2014): 181–86.

Gardner, Andy. "Price's Equation Made Clear." *Philosophical Transactions of the Royal Society B* 375 (27 April 2020): https://doi.org/10.1098/rstb.2019.0361.

Gardner, Andy, and Joseph P. Conlon. "Cosmological Natural Selection and the Purpose of the Universe." *Complexity* (May–June 2013): https://doi.org/10.1002/cplx.21446.

Gardner, Andy, and Alan Grafen. "Capturing the Superorganism: A Formal Theory of Group Adaptation." *Journal of Evolutionary Biology* 22, no. 4 (2009): 659–71.

Garzke, William H., Jr., and Robert O. Dullin Jr. *Battleships in World War II: Allied Battleships.* Annapolis, MD:

Naval Institute Press, 1980.

Gatti, Roberto Cazzolla, Roger Koppl, Brian D. Fath, Stuart Kauffman, Wim Hordijk, Robert E. Ulanowicz. "On the Emergence of Ecological and Economic Niches." *Journal of Bioeconomics* 22, no. 2 (2020): 10.1007/s10818-020-09295-4.

Geels, F. W. "Processes and Patterns in Transitions and System Innovations: Refining the Co-Evolutionary Multi-Level Perspective." *Technological Forecasting and Social Change* 72 (July 2005): 681–96.

———. "Technological Transitions as Evolutionary Reconfiguration Processes: A Multi-Level Perspective and a Case-Study." *Research Policy* 31 (December 2002): 1257–74.

Gelfand, Michele J. "Explaining the Puzzle of Human Diversity: Centuries of Church Exposure Promote More Individualistic and Less Conforming Psychology." *Science* 366, no. 6466 (8 November 2019): 686–87.

Gell-Mann, Murray. "Nature Conformable to Herself." In *Worlds Hidden in Plain Sight: The Evolving Idea of Complexity at the Santa Fe Institute, 1984–2019*, edited by David C. Krakauer, 31–40. Santa Fe, NM: The Santa Fe Institute Press, 2019.

Gerbault, Pascale, Robin G. Allaby, Nicole Boivin, Anna Rudzinski, Ilaria M. Grimaldi, J. Chris Pires, Cynthia Climer Vigueira, Keith Dobney, et al. "Storytelling and Story Testing in Domestication." *National Academy of Sciences USA* 111, no. 17 (2014): 6159–64.

German, W. P. Z. "Compulsory Unit Operation of Oil Pools." *Transactions of the American Institute of Mining and Metallurgical Engineers: Petroleum Development and Technology* 92, no. 1 (December 1931): 11–37.

———. "Legal History of Conservation of Oil and Gas in Oklahoma." In *Legal History of Conservation of Oil and Gas: A Symposium*, Mineral Law, American Bar Association, 110–213. Chicago, American Bar Association, 1939.

Ghang, Whan, and Martin A. Nowak. "Indirect Reciprocity with Optional Interactions." *Journal of Theoretical Biology* 365 (2015): 1–11.

Ghiselin, Michael T. *The Economy of Nature and the Evolution of Sex*. Berkeley, CA: University of California Press, 1974.

Giebelhaus, August W. *Business and Government in the Oil Business: A Case Study of Sun Oil Company, 1876–1945*. Greenwich, CT: JAI Press, 1980.

———. "The Influence of Engineering Methods in the Petroleum Industry, 1919–1930." Paper Presented to the Organization of American Historians, Detroit, Michigan, April 3, 1981.

Gilbert, Chester, and Joseph E. Pogue. *America's Power Resources: The Economic Significance of Coal, Oil, and Water Power*. New York: Century, 1921.

Gintis, Herbert. *The Bounds of Reason: Game Theory and the Unification of the Behavioral Sciences*. Princeton, NJ: Princeton University Press, 2009.

———. "The Hitchhiker's Guide to Altruism: Gene-Culture Coevolution and the Internalization of Norms." *Journal of Theoretical Biology* 220 (2003): 407–18.

———. "Punishment and Cooperation." *Science* 319 (7 March 2008): 1345–46.

———. "Strong Reciprocity and Human Sociality." *Journal of Theoretical Biology* 206 (2000): 169–79.

Gintis, Herbert, Samuel Bowles, Robert Boyd, and Ernst Fehr, eds. *Moral Sentiments and Material Interests: The Foundations of Cooperation in Economic Life*. Cambridge, MA: MIT Press, 2005.

Gintis, Herbert, Eric Alden Smith, and Samuel Bowles. "Costly Signaling and Cooperation." *Journal of Theoretical Biology* 213 (2001): 103–19.

Glymour, Bruce. "Modeling Environments: Interactive Causation and Adaptations to Environmental Conditions." *Philosophy of Science* 78, no. 3 (July 2011): 448–71.

———. "Stable Models and Causal Explanation in Evolutionary Biology." *Philosophy of Science* 75 (December 2008): 571–83.

———. "Wayward Modeling: Population Genetics and Natural Selection." *Philosophy of Science* 73 (2006): 369–89.

Glymour, Bruce, and Chris French. "Causation, Equivalence, and Group-Selection." Draft, 2008. downloaded November 16, 2022, from personal web page.

Gold, Russell. *The Boom: How Fracking Ignited the American Energy Revolution and Changed the World*. New York: Simon & Schuster, 2014.

Goodnight, Charles J. "Heritability at the Ecosystem Level." *Proceedings of the National Academy of Sciences USA* 97, no. 17 (2000): 9365–66.

———. "Multilevel Selection: The Evolution of Cooperation in Non-Kin Groups." *Population Ecology* 47, no. 1 (2005): 3–12.

Goodnight, Charles J., James M. Schwartz, and Lori Stevens. "Contextual Analysis of Models of Group Selection, Soft Selection, Hard Selection, and the Evolution of Altruism." *American Naturalist* 140 (1992): 743–61.

Goodwyn, Lawrence. *The Populist Moment: A Short History of the Agrarian Revolt in America*. New York: Oxford University Press, 1978.

———. *Texas Oil, American Dreams: A Study of the Texas Independent Producers and Royalty Owners Association*.

Austin: Texas State Historical Association, 1996.

Götz, Norbert. "'Moral Economy': Its Conceptual History and Analytical Prospects." *Journal of Global Ethics* 11, no. 2 (29 July 2015): 147–62.

Gould, Stephen Jay. "Exaptation: A Crucial Tool for an Evolutionary Psychology." *Journal of Social Issues* 47 (Fall 1991): 43–65.

Grafen, Alan. "The Price Equation and Reproductive Value." *Philosophical Transactions of the Royal Society B* 375 (27 April 2020): https://doi.org/10.1098/rstb.2019.0356.

Granovetter, Mark S. "Economic Action and Social Structure: The Problem of Embeddedness." *American Journal of Sociology* 91, no. 3 (November 1985): 481–510.

———. "The Strength of Weak Ties." *American Journal of Sociology* 78 (May 1973): 1360–80.

Grant, B. Rosemary, and Peter R. Grant. "Watching Speciation in Action: An Interplay Between Environmental and Genetic Change Is Responsible for the Emergence of New Species." *Science* 355 (3 March 2017): 910–11.

Grant, Verne, and Karen A. Grant. *Flower Pollination in the Phlox Family*. New York: Columbia University Press, 1965.

Greenhill, Joe R. "Practice and Procedure before the Railroad Commission." Mineral Law Section, July 3, 1953; in Langdon Papers, 1980/164-9, Folder JCL. Texas State Library and Archives, Austin, Texas.

Greenough, Paul R. "Indulgence and Abundance as Asian Peasant Values: A Bengali Case in Point." *Journal of Asian Studies* 42 (August 1983): 753–868.

Grozier, David F. "The Brooklyn Union Natural Gas Conversion: Biggest Changeover in the World." *Gas Age* 111 (1 January 1953): 29–74.

Gürerk, Özgür, Bernd Irlenbusch, and Bettina Rockenbach. "The Competitive Advantage of Sanctioning Institutions." *Science* 312 (April 7, 2006): 108–11

Habermas, Jürgen. *Lifeworld and System: A Critique of Functionalist Reason*, Vol. 2 of *The Theory of Communicative Action*. Translated by Thomas McCarthy. 1981. Reprint, Boston, MA: Beacon Press, 1987.

Hackett, Steven, Edella Schlager, and James M. Walker. "The Role of Communication in Resolving Commons Dilemmas: Experimental Evidence with Heterogeneous Appropriators." *Journal of Environmental Economics and Management* 27, no. 2 (September 1994): 99–126.

Hacking, Ian. *Representing and Intervening*. Cambridge, UK: Cambridge University Press, 1983.

Hadamard, Jacques. *An Essay on the Psychology of Invention in the Mathematical Field*. Princeton, NJ: Princeton University Press, 1945.

Haghani, Amin, Caesar Z. Li, Todd R. Robeck, Joshua Zhang, Ake T. Lu, Julia Ablaeva, Victoria A. Acosta-Rodríguez et al. "DNA Methylation Networks Underlying Mammalian Traits." *Science* 381, no. 6658 (11 August 2023): 647.

Hahlweg, Kai. "Popper versus Lorenz: An Exploration into the Nature of Evolutionary Epistemology." In *Contributed Papers*, Vol. 1 of *PSA 1986: Proceedings of the 1986 Biennial Meeting of the Philosophy of Science Association*, edited by Arthur Fine and Peter Machamer, 172–82. East Lansing, MI: Philosophy of Science Association, 1986.

Hahn, Steven. *The Roots of Southern Populism: Yeoman Farmers and the Transformation of the Georgia Upcountry, 1850-1890*. New York: Oxford University Press, 1983.

Hajduk, G. K., C. A. Walling, A. Cockburn, L. E. B. Kruuk. "The 'Algebra of Evolution': The Robertson–Price Identity and Viability Selection for Body Mass in a Wild Bird Population." *Philosophical Transactions of the Royal Society B* 375 (27 April 2020): https://doi.org/10.1098/rstb.2019.0359.

Haley, J. Evetts. *Charles Goodnight: Cowman and Plainsman*. 1936. Reprint, Norman: University of Oklahoma Press, 1949.

———. *A Texan Looks at Lyndon: A Study in Illegitimate Power*. Canyon, TX: Palo Duro Press, 1964.

Hamilton, W. D. "The Genetical Evolution of Social Behavior I." *Journal of Theoretical Biology* 7, no. 1 (1964): 1–16.

———. The Genetical Evolution of Social Behavior II." *Journal of Theoretical Biology* 7, no. 1 (1964): 17–52.

———. "Innate Social Aptitudes in Man: An Approach from Evolutionary Genetics." In *Biosocial Anthropology*, edited by Robin Fox, 143–69. New York: Wiley, 1975.

———. *Narrow Roads of Gene Land: The Collected Papers of W. D. Hamilton, Volume 1: Evolution of Social Behaviour*. New York: W. H. Freeman, 1996.

Hammerstein, Peter. "Why Is Reciprocity So Rare in Social Animals? A Protestant Appeal." In *The Genetic and Cultural Evolution of Cooperation*, edited by Peter Hammerstein. Cambridge, MA: MIT Press, 2003.

Handley, Carla, and Sarah Mathew. "Human Large-Scale Cooperation as a Product of Competition between Cultural Groups." *Nature Communications* 11 (4 February 2020): 702.

Hardin, Garrett. "The Tragedy of the Commons." *Science* 162 (13 December 1968): 1243–48.

Hardwicke, Robert E. *Antitrust Laws, et al., v. Unit Operation of Oil or Gas Pools*. New York: American Institute of Mining and Metallurgical Engineers, 1948.

———. "Conservation of Oil and Gas: Texas, 1938-1948." In *Conservation of Oil & Gas: A Legal History, 1948*, edited by Blakely M. Murphy, 447–512. Chicago: Section of Mineral Law, American Bar Association, 1949.

———. "Legal History of Conservation of Oil in Texas." In *Legal History of Conservation of Oil and Gas: A Symposium*, 214–68. Chicago: Section of Mineral Law, American Bar Association, 1938.

———. "Report on Progress in Conservation, 1948–1958." In *Conservation of Oil and Gas, A Legal History, 1958*, edited by Robert E. Sullivan, 4–5. Chicago: Section of Mineral and Natural Resources Law, American Bar Association, 1960.

———. "The Rule of Capture and Its Implications as Applied to Oil and Gas." *Texas Law Review* 13 (June 1935): 391.

Harman, Oren. "When Science Mirrors Life: On the Origins of the Price Equation." *Philosophical Transactions of the Royal Society B* 375 (27 April 2020): https://doi.org/10.1098/rstb.2019.0352.

Harris, Jared, and Philip Bromiley. "Incentives to Cheat: The Influence of Executive Compensation and Firm Performance on Financial Misrepresentation." *Organization Science* 18, no. 3 (May–June 2007): 350–67.

Harris, Whitney R. "A Reappraisal of the Substantial Evidence Rule in Texas Administrative Law." *Southwestern Law Journal* 3, no. 4 (1949): 416–36.

Harsanyi, John. "Subjective Probability and the Theory of Games: Comments on Kadane and Larkey's Paper." *Management Science* 28, no. 2 (1982): 120–24.

Haskell, Thomas L. *The Authority of Experts: Studies in History and Theory*. Bloomington: Indiana University Press, 1984.

Hauert, Christoph, Franziska Michor, Martin A. Nowak, and Michael Doebeli. "Synergy and the Discounting of Cooperation in Social Dilemmas." *Journal of Theoretical Biology* 239 (21 March 2006): 195–202.

Hawley, Ellis W. "Herbert Hoover, the Commerce Secretariat, and the Vision of the 'Associative State,' 1921–1928." *Journal of American History* 61 (June 1974): 116–40.

Hays, Samuel P. *Conservation and the Gospel of Efficiency: The Progressive Conservation Movement, 1890–1920*. Pittsburgh: University of Pittsburgh Press, 1959.

Hefner, Robert A., III. "The United States of Gas, Why the Shale Revolution Could Have Happened Only in America." *Foreign Affairs* 93 no. 3 (May–June 2014).

Heisler, I. L., and John Damuth. "A Method for Analyzing Selection in Hierarchically Structured Populations." *American Naturalist* 130 (1 October 1987): 582–602.

Helanterä, Heikki, and Tobias Uller. "Different Perspectives on Non-Genetic Inheritance Illustrate the Versatile Utility of the Price Equation in Evolutionary Biology." *Philosophical Transactions of the Royal Society B* 375 (27 April 2020): https://doi.org/10.1098/rstb.2019.0366.

Henrich, Joseph, "Cooperation, Punishment, and the Evolution of Human Institutions." *Science* 312 (7 April 2006): 60–61.

———. "Cultural Group Selection, Coevolutionary Processes and Large-Scale Cooperation." *Journal of Economic Behavior and Organization* 53, no. 1 (2004): 3–35.

———. "Cultural Transmission and the Diffusion of Innovations: Adoption Dynamics Indicate That Biased Cultural Transmission Is the Predominate Force in Behavioral Change." *American Anthropologist* 103, no. 4 (2001): 992–1013.

Henrich, Joseph, and Robert Boyd. "Division of Labor, Economic Specialization, and the Evolution of Social Stratification." *Current Anthropology* 49, no. 4 (August 2008): 715–24.

———. "The Evolution of Conformist Transmission and the Emergence of Between-Group Differences." *Evolution and Human Behavior* 19 (July 1998): 1–28.

———. "On Modeling Cognition and Culture: Why Cultural Evolution Does Not Require Replication of Representation." *Journal of Cognition and Culture* 2, no. 2 (2002): 87–112.

Henrich, Joseph, Robert Boyd, Samuel Bowles, Colin Camerer, Ernst Fehr, Herbert Gintis, Richard McElreath et al. "'Economic Man' in Cross-Cultural Perspective: Behavioral Experiments in 15 Small-scale Societies." *Behavioral and Brain Sciences* 28, no. 6 (2005): 795–815.

Henrich, Joseph, Jean Emsminger, Richard McElreath, Abigail Barr, Clark Barrett, Alexander Bolyanatz, Juan Camilo Cardenas et al. "Markets, Religion, Community Size, and the Evolution of Fairness and Punishment." *Science* 327, no. 5972 (19 March 2010): 1480–84.

Henrich, Joseph, and F. Gil-White. "The Evolution of Prestige." *Evolution and Human Behavior* 19 (June 2001): 165–96.

Henrich, Joseph, and Michael Muthukrishna. "The Origins and Psychology of Human Cooperation." *Annual Review of Psychology* 72 (2021): 207–40.

Herrmann, Benedikt, Christian Thöni, and Simon Gächter. "Antisocial Punishment Across Societies." *Science* 319 (7 March 2008): 1362–67.

Hess, Aimee. "Subsurface Trespass by Hydraulic Fracturing? Yes, Says Pennsylvania Court." Texas Oil and Gas Attorney Blog. URL: https://www.texasoilandgasattorneyblog.com/subsurface-trespass-by-hydrauli c-fracturing-yes-says-pennsylvania-court, August 24, 2018. Accessed March 20, 2021.

Heyes, Cecilia, and David Lee Hull, eds. *Selection Theory and Social Construction: The Evolutionary Naturalistic Epistemology of Donald T. Campbell*. Albany: State University of New York Press, 2001.

Heyman, Karen. "Making Connections." *Science* 313 (4 August 2006): 604–6.

Hilbe, Christian, Bin Wu, Arne Traulsen, and Martin A. Nowak. "Cooperation and Control in Multiplayer Social Dilemmas." *Proceedings of the National Academy of Sciences USA* 111, no. 46 (2014): 16425–30.

———. "Evolutionary Performance of Zero-Determinant Strategies in Multiplayer Games." *Journal of Theoretical Biology* 374 (October 7, 2015): 115–12.

Hinton, Diana Davids. "Introduction for a Special Issue on the Oil Industry." *Business History Review* 84 (Summer 2010): 195–201.

Hodgson, Geoffrey M., and Kainan Huang. "Evolutionary Game Theory and Evolutionary Economics: Are They Different Species?" *Journal of Evolutionary Economics* 22 no. 2 (2012): 345–66.

Hoetzel, A. Rus, and Michael Lynch. "The Raw Material of Evolution: Estimates of Whale Mutation Rates Contribute to Understanding Evolutionary Processes." *Science* 381 no. 6661 (1 September 2023): 942–43.

Hofman, Jake M., Amit Sharma, and Duncan J. Watts. "Prediction and Explanation in Social Systems." *Science* 355 (3 February 2017): 486–88.

Holm, Jacob Rubæk, Esben Sloth Andersen, and J. Stanley Metcalfe. "Confounded, Augmented and Constrained Replicator Dynamics." *Journal of Evolutionary Economics* 26 (October 2016): 803–22.

Hooley, Richard W. *Financing the Natural Gas Industry: The Role of Life Insurance Investment Policies*. New York: Columbia University Press, 1961.

Hounshell, David A. *From the American System to Mass Production, 1800–1932: The Development of Manufacturing Technology in the United States*. Baltimore, MD: Johns Hopkins University Press, 1984.

Hruschka, D. J., and Joseph Henrich. "Friendship, Cliquishness, and the Emergence of Cooperation." *Journal of Theoretical Biology* 239, no. 1 (2006): 1–15.

Hughes, Thomas P. *American Genesis: A Century of Invention and Technological Enthusiasm*. New York: Viking, 1989.

———. "The Electrification of America: The System Builders." *Technology and Culture* 20 (January 1979): 124–61.

———. *Elmer Sperry: Inventor and Engineer*. Baltimore, MD: Johns Hopkins University Press, 1971.

———. *Networks of Power: Electrification in Western Society, 1880–1930*. Baltimore, MD: Johns Hopkins University Press, 1983.

Hull, David Lee, "In Defense of Presentism." In *The Metaphysics of Evolution*, 205–20. Albany: State University of New York Press, 1989.

———. "In Search of Epistemological Warrant." In *Selection Theory and Social Construction: The Evolutionary Naturalistic Epistemology of Donald T. Campbell*, edited by C. Heyes and D. L. Hull, 155–67. Albany: State University of New York Press, 2001.

———. "A Mechanism and Its Metaphysics: An Evolutionary Account of the Social and Conceptual Development of Science." *Biology and Philosophy* 3, no. 2 (April 1988): 123–55.

———. *Science and Selection: Essays on Biological Evolution and the Philosophy of Science*. New York: Cambridge University Press, 2001.

———. *Science as a Process: An Evolutionary Account of the Social and Conceptual Development of Science*. Chicago: University of Chicago Press, 1988.

Hull, David Lee, R. E. Langman, and S. S. Glenn. "A General Account of Selection: Biology, Immunology and Behavior." *Behavioral and Brain Sciences* 24, no. 3 (2001): 511–28.

Hunt, Haroldson Lafayette, Jr. *Alpaca*. Dallas: H. L. Hunt Press, 1960.

Hurt, Harry, III. "The Most Powerful Texans." *Texas Monthly* 4, no. 4 (April 1976): 73–76ff.

———. "New Oil: The Giddings Gamble." *Texas Monthly*, February 1981, 106ff.

Hutcheson, Joseph C., Jr. "Judgement Intuitive: The Function of the 'Hunch' in Judicial Decision." *Cornell Law Review* 14, no. 3 (April 1929): 274–88.

Hutson, Matthew. "Guinea Pigbots." *Science* 381, no. 6654 (14 July 2023): 121–23.

Ichinose, Genki, and Naoki Masuda. "Zero-Determinant Strategies in Finitely Repeated Games." *Journal of Theoretical Biology* 428 (7 February 2018): 61–77.

Inglehart, R., and W. E. Baker. "Modernization, Cultural Change, and the Presence of Traditional Values." *American Sociological Review* 65 (2010): 19–51.

International Energy Agency. "World Oil Production by Region, 1971–2019." https://www.iea.org/data-and-statistics/charts/world-crude-oil-production-by-region-1971-2019.

Iwamura, Takuya, Eric F. Lambin, Kirsten M. Silvius, Jeffrey B. Luzar, and José M. V. Fragoso. "Agent-based Modeling of Hunting and Subsistence Agriculture on Indigenous Lands: Understanding Interactions Between Social and Ecological Systems." *Environmental Modelling and Software* 58 (August2014): 109–27.

Jensen, Keith. "Punishment and Spite, the Dark Side of Cooperation." *Philosophical Transactions of the Royal Society B* 365, no. 1553 (12 September 2010): 2635–50.

Johns, David Merritt, and Gerald M. Oppenheimer. "Was There Ever Really a 'Sugar Conspiracy'?" *Science* 359, no. 6377 (16 February 2018): 747–50.

Johnson, H. Thomas, and Robert S. Kaplan. *Relevance Lost: The Rise and Fall of Management Accounting*. Boston, MA: Harvard Business School Press, 1987.

Jones, Dan. "A WEIRD View of Human Nature Skews Psychologists' Studies." *Science* 328 (25 June 2010): 162.

Jones, Darrell K. "Factors that Contribute to the Drilling of Unnecessary Oil Wells in Texas." MBA thesis, University of Texas at Austin, 1964.

Jordan, Terry G. *German Seed in Texas Soil: Immigrant Farmers in Nineteenth Century Texas*. Austin: University of Texas Press, 1966.

———. *Trails to Texas: Southern Roots of Western Cattle Ranching*. Lincoln: University of Nebraska Press, 1981.

Kadane, Joseph B., and Patrick D. Larkey. "Subjective Probability and the Theory of Games." *Management Science* 28, no. 2 (1982): 113–20.

Kahneman, Daniel. *Thinking, Fast and Slow*. New York: Farrar, Straus and Giroux, 2011.

Kandler, Anne, and Adam Powell. "Generative Inference for Cultural Evolution." *Transactions of the Royal Society B—Biological Sciences* 373, no. 1743 (April 2018): 10.1098/rstb.2017.0056.

Kauffman, Stuart A. *The Origins of Order: Self-Organization and Selection in Evolution*. Oxford, UK: Oxford University Press, 1993.

Kauffman, Stuart A., J. Lobo, and W. G. Macready. "Optimal Search on a Technological Landscape." *Journal of Economic Behavior and Organization* 43 (October 2000): 141–66.

Kauffman, Stuart A., and Dean Radin. "Quantum Aspects of the Brain-Mind Relationship: A Hypothesis with Supporting Evidence." *BioSystems* 223 (January 2023): https://doi.org/10.1016/j.biosystems.2022.104820.

Kearns, Michael, Siddharth Suri, and Nick Montfort. "An Experimental Study of the Coloring Problem on Human Subject Networks." *Science* 313 (11 August 2006): 824–27.

Keegan, John. *The Mask of Command*. New York: Viking, 1987.

Keller, Evelyn Fox. "Revisiting 'Scale-Free' Networks." *BioEssays* 27 (14 September 2005): 1060–68.

Kenan, Resit. "The Mathematical Analysis of the Flow of Compressible Fluids through Porous Media, and the Application to the East Texas Field Pressure Decline." MS thesis, University of Texas at Austin, 1941.

Kerr, B., and Peter Godfrey-Smith. "Individualist and Multi-Level Perspectives on Selection in Structured Populations." *Biology and Philosophy* 17 (September 2002): 477–517.

Keyes, Charles F., ed. "Peasant Strategies in Asian Societies: Moral and Rational Economic Approaches, A Symposium." *Journal of Asian Studies* 42 (August 1983): 753–868.

Khayyám, Omar. *Rubáiyát of Omar Khayyám*. Rendered into English verse by Edward Fitzgerald. Garden City, NY: Garden City Books, 1952.

Kiessling, O. E., H. O. Rogers, G. R. Hopkins, N. Yaworski, R. L. Kiessling, J. Brian Eby, Lew Suverkrop, et al. *Technology, Employment, and Output Per Man in Petroleum and Natural-Gas Production*. Philadelphia, PA: Work Projects Administration, National Research Project, 1939.

King, John O. *Joseph Stephen Cullinan: A Study in Leadership in the Texas Petroleum Industry, 1897–1937*. Nashville, TN: Vanderbilt University Press, 1970.

Kingsolver, Joel G., and David W. Pfennig. "Patterns and Power of Phenotypic Selection in Nature." *Bioscience* 57, no. 7 (2007): 561–72.

Kipling, Rudyard. *Just So Stories for Little Children*. London: Macmillan, 1902.

Kirk, P. D. W., A. C. Babtie, and M. P. H. Stumpf. "Systems Biology (*Un*)certainties." *Science* 350 (23 October 2015): 386–88.

Kirkland, William A. *Old Bank–New Bank: The First National Bank, Houston, 1866–1966*. Hinsdale, IL: Dryden Press, 1974.

Kline, Ronald R. "Mathematical Models of Technological and Social Complexity." In *Technology and Mathematics: Philosophical and Historical Investigations*, edited by Sven Ove Hansson, 285–304. Berlin: Springer International, 2018.

Knudsen, Thorbjørn. "General Selection Theory and Economic Evolution: The Price Equation and the Replicator/Interactor Distinction." *Journal of Economic Methodology* 11, no. 2 (June 2004): 147–73.

Kohler, Robert E. "*Drosophila* and Evolutionary Genetics: The Moral Economy of Scientific Practice." *History of Science* 29, no. 4 (1991): 335–75.

———. *Lords of the Fly: Drosophila Genetics and the Experimental Life*. Chicago: University of Chicago Press, 1994.

Kollock, Peter. "An Eye for an Eye Leaves Everyone Blind: Cooperation and Accounting Systems." *American Sociological Review* 58, no. 6 (1993): 768–86.

Kramer, Bruce M. "Horizontal Drilling and Trespass: A Challenge to the Norms of Property and Tort Law." *Colorado Natural Resources, Energy & Environmental Law Review* 25, no. 2 (2014): 292–340.

———. "The Nature of the Mineral Estate: A Guidebook for the Uninitiated," 2015. URL: www.cailaw.org/media/

files/IEL/.../2013/TexasMineralTitle/BKramer-paper.pdf, accessed September 16, 2017.

Kraus, Edgar. "'MER'—A History." *Oil and Gas Journal* (15 November 1947): 269–72.

Kroger, Bill, Jason Newman, Ben Sweet, and Justin Lipe. "How Texas Law Promoted Shale Play Development." Baker Botts 175th Anniversary, Houston, 2015.

Kuchin, Joseph W. *How Mitchell Energy and Development Corp. Got Its Start and How It Grew: An Oral History and Narrative*. (N.p: University Publishers, 1998).

Kuhn, Thomas S. *The Copernican Revolution: Planetary Astronomy in the Development of Western Thought*. Cambridge, MA: Harvard University Press, 1957.

———. "The Road Since Structure." *PSA 1990: Proceedings of the 1990 Biennial Meeting of the Philosophy of Science Association*, vol. 2, edited by Arthur Fine, Mickey Forbes, and Linda Wessels, 3–13. East Lansing, MI: Philosophy of Science Association, 1991.

———. *The Structure of Scientific Revolutions*. 2nd ed. Chicago: University of Chicago Press, 1970.

Kulander, Christopher S., and R. Jordan Shaw. "Comparing Subsurface Trespass Jurisprudence—Geophysical Surveying and Hydraulic Fracturing." *New Mexico Law Review* 46 (Winter 2016): 67–114.

Kulik, Gary. "Dams, Fish, and Farmers: Defense of Public Rights in Eighteenth-Century Rhode Island." In *The Countryside in the Age of Capitalist Transformation: Essays in the Social History of Rural America*, edited by Steven Hahn and Jonathan Prude, 25–50. Chapel Hill: University of North Carolina Press, 1985.

Kunz, J. "Group-level Exploration and Exploitation: A Computer Simulation-based Analysis." *Journal of Artificial Societies and Social Simulation* 14 no. 4 (2011): 18.

Lade, Stephen J., Alessandro Tavoni, Simon A. Levin, and Maja Schlüter. "Regime Shifts in a Social Ecological System." *Theoretical Ecology* 6, no. 3 (2013): 359–72.

Lakatos, Imre. "Falsification and the Methodology of Scientific Research Programmes." In *Criticism and the Growth of Knowledge*, edited by Imre Lakatos and Alan Musgrave, 91–196. Cambridge, UK: Cambridge University Press, 1970.

Lamon, Noemie, Christof Neumann, Thibaud Gruber, and Klaus Zuberbühler. "Kin-based Cultural Transmission of Tool Use in Wild Chimpanzees." *Science Advances* 3 (26 April 2017): https://doi.org/10.1126/sciadv.1602750.

Landa, Janet T. "The Bioeconomics of Homogeneous Middleman Groups as Adaptive Units: Theory and Empirical Evidence Viewed from a Group Selection Framework." *Journal of Bioeconomics* 10, no. 3 (2008): 259–78.

Langdon, Jim C. "Current Engineering Problems Facing the Railroad Commission of Texas." Address Delivered Before the Houston Chapter, American Institute of Mining Engineers, January 21, 1964. Langdon Papers, 1980/164-10 Folder 11, Oil and Gas Division Proceedings. Texas State Archives, Austin, Texas.

———. "The Influence of Court Decisions upon Railroad Commission Policy in Rule 37 Cases and the Allocation of Allowables to the Small Tract Well," c. 1963. Langdon Papers, 1980/164-10, Folder 9, Oil and Gas Division Proceedings. Texas State Archives, Austin, Texas.

———. "The Role of the Petroleum Engineer as an Expert Witness." Address Before the Victoria Chapter of the American Petroleum Institute, Victoria, Texas, February 22, 1965. Langdon Papers, 1980/164-10 Folder 36, Oil and Gas Division Proceedings. Texas State Archives, Austin, Texas.

Lansing, J. Stephen. *Perfect Order: Recognizing Complexity in Bali*. Princeton, NJ: Princeton University Press, 2006.

Larson, Henrietta M., and Kenneth Wiggins Porter. *History of Humble Oil and Refining Company: A Study in Industrial Growth*. New York: Harpers, 1959.

Larson, Lennart V. "The Substantial Evidence Rule: Texas Version." *Southwestern Law Journal* 5, no. 2 (1951): 152–69.

Layton, Edwin T., Jr. "American Ideologies of Science and Engineering." *Technology and Culture* 17 (October 1976): 688–701.

———. *The Revolt of the Engineers: Social Responsibility and the American Engineering Profession*. Cleveland, OH: Case Western Reserve University Press, 1971.

Lee, Everett S., Ann Ratner, Carol P. Brainerd, and Richard Easterlin. *Population Redistribution and Economic Growth, United States, 1870–1950, l: Methodological Considerations and Reference Tables*. Under the direction of Simon Kuznets and Dorothy Swaine Thomas. Philadelphia, PA: American Philosophical Society, 1957.

Leeson, Peter T. *The Invisible Hook: The Hidden Economics of Pirates*. Princeton, NJ: Princeton University Press, 2009.

Leeston, Alfred M., John A. Crichton, and John C. Jacobs. *The Dynamic Natural Gas Industry: The Description of an American Industry from the Historical, Technical, Legal, Financial, and Economic Standpoints*. Norman: University of Oklahoma Press, 1963.

Lehmann, Laurent, and Laurent Keller. "The Evolution of Cooperation and Altruism—A General Framework and a Classification of Models." *Journal of Evolutionary Biology* 19, no. 5 (2006): 1365–76.

Lehmann, Laurent, Laurent Keller, Stuart West, and Denis Roze. "Group Selection and Kin Selection: Two Concepts but One Process." *Proceedings of the National Academy of Sciences* 104, no. 16 (2007): 6736–39.

Lehtonen, Jussi. "The Price Equation and the Unity of Social Evolution Theory." *Philosophical Transactions of the Royal Society B* 375 (27 April 2020): https://doi.org/10.1098/rstb.2019.0362.

Lehtonen, Jussi, Samir Okasha, and Heikki Helanterä. "Fifty Years of the Price Equation." *Philosophical Transactions of the Royal Society B* 375 (27 April 2020): https://doi.org/10.1098/rstb.2019.0350.

Leonardon, Eugene G. "Logging, Sampling, and Testing." In *History of Petroleum Engineering*, Executive Committee on Drilling and Production Practice, Division of Production, 493–578. Dallas: American Petroleum Institute, 1961.

Levine, David. "Modeling Altruism and Spitefulness in Experiments." *Review of Economic Dynamics* 1 (July 1998): 593–622.

Levine, Michael E., and Jennifer L. Forrence. "Regulatory Capture, Public Interest, and the Public Agenda: Toward a Synthesis." *Journal of Law and Economics* 6, Special Issue (1990): 167–98.

Levinson, Sanford V. "The Democratic Faith of Felix Frankfurter." *Stanford Law Review* 25 (February 1973): 430–48.

Lewontin, Richard C. "Four Complications in Understanding the Evolutionary Process." In *Worlds Hidden in Plain Sight: The Evolving Idea of Complexity at the Santa Fe Institute, 1984–2019*, edited by David C. Krakauer, 97–113. Santa Fe, NM: Santa Fe Institute Press, 2019.

———. "The Units of Selection." *Annual Review of Ecology and Systematics* 11 (1970): 1–18.

Libecap, Gary D. "Government Support of Private Claims to Public Minerals: Western Mineral Rights." *Business History Review* 53 (Autumn 1979): 364–85.

———. "The Political Allocation of Mineral Rights: A Re-evaluation of Teapot Dome." *Journal of Economic History* 44, no. 2 (June 1984): 381–91.

———. "The Political Economy of Crude Oil Cartelization in the United States, 1933–1972." *Journal of Economic History* 49, no. 4 (December 1989): 833–55.

———. "The Self-Enforcing Provisions of Oil and Gas Unit Operating Agreements: Theory and Evidence." *Journal of Law, Economics, and Organization* 15, no. 2 (1999): 526–48.

Libecap, Gary D., and James L. Smith. "The Economic Evolution of Petroleum Property Rights in the United States." A conference sponsored by the Searle Fund and Northwestern University School of Law, *The Journal of Legal Studies* 31, no. 2, part 2 (June 2002): S589–S608.

———. "Regulatory Remedies to the Common Pool: The Limits to Oil Field Unitization." *The Energy Journal* 22, no. 1 (2001): 1–26.

Libecap, Gary D., and Steven N. Wiggins. "The Influence of Private Contractual Failure on Regulation: The Case of Oil Field Unitization." *Journal of Political Economy* 93, no. 4 (1985): 690–714.

Libre, S. V. D., G. A. K. van Voorn, G. A. ten Broeke, M. Bailey, P. Berentsen, and S. R. Bush. "Effects of Social Factors on Fishing Effort: The Case of the Philippine Tuna Purse Seine Fishery." *Fisheries Research* 172 (December 2015): 250–60.

Lile, Stephen E. "The Religious Economy of Texas: An Historical Perspective." *Essays in Economic History: The Journal of the Economic and Business Historical Society* 18 (2000): 113–22.

Lindkvist, Emilie, Xavier Basurto, and Maja Schlüter. "Micro-level Explanations for Emergent Patterns of Self-Governance Arrangements in Small-Scale Fisheries—A Modeling Approach." *PLoS ONE* 12 (13 April 2017): https://doi.org/10.1371/journal.pone.0175532.

Lindkvist, Emilie, Nanda Wijermans, Tim M. Daw, Blanca Gonzalez-Mon, Alfredo Giron- Nava, Andrew F. Johnson, Ingrid van Putten et al. "Navigating Complexities: Agent-Based Modeling to Support Research, Governance, and Management in Small-Scale Fisheries." *Frontiers in Marine Science* 6 (17 January 2020): https://doi.org/10.3389/fmars.2019.00733.

Link, William S. "The Social Context of Southern Progressivism, 1880–1930." In *The Wilson Era: Essays in Honor of Arthur S. Link*, edited by John Milton Cooper Jr. and Charles E. Neu. Arlington Heights, IL: Harlan Davidson, 1991.

Little, Daniel. "Collective Action and the Traditional Village." manuscript, n.d.

———. *Understanding Peasant China: Case Studies in the Philosophy of Social Science*. New Haven, CT: Yale University Press, 1989.

———. *Varieties of Social Explanation: An Introduction to the Philosophy of Social Science*. Boulder, CO: Westview Press, 1991.

Lloyd, Elisabeth A. "Evaluation of Evidence in Group Selection Debates." *Contributed Papers*, Vol. 1 of *PSA 1986: Proceedings of the 1986 Biennial Meeting of the Philosophy of Science Association*, edited by Arthur Fine and Peter Machamer, 483–93. East Lansing, MI: Philosophy of Science Association, 1986.

Logan, Leonard Marion. *Stabilization of the Petroleum Industry*. Norman: University of Oklahoma Press, 1930.

Longo, Giuseppe, Maël Montévil, and Arnaud Pocheville. "From Bottom-up Approaches to Levels of Organization and Extended Critical Transitions." *Frontiers in Physiology* 3 (17 July 2012): doi:10.3389/fphys.2012.00232.

Lord, Ron. "Technological Breakthroughs Advanced Upstream E&P's Evolution." *Journal of Petroleum Technology* 59, no. 10 (October 2007): 111–16.

Lorenz, Konrad. "Kant's Lehre vom apriorischen im Lichte gegenwartiger Biologie." *Blatter fur Deutsche Philosophie*

15 (1941): 94–125. Reprinted as "Kant's Doctrine of the A Priori in the Light of Contemporary Biology," in *General Systems: Yearbook of the Society for General Systems Research*, Vol. 7, edited by Ludwig von Bertalanffy and Anatol Rapoport, 23–35, trans. Charlotte Ghurye. Ann Arbor, MI: Society for General Systems Research, 1962.

Lovejoy, Wallace F., and Paul T. Homan. *Economic Aspects of Oil Conservation Regulation*. Baltimore, MD: Johns Hopkins University Press, 1967.

Lowery, Velma. *The History of Hackberry, Louisiana*. Sulphur, LA: Wise Publications, 1991.

Lowood, Henry E. "The Calculating Forester: Quantification, Cameral Science, and the Emergence of Scientific Forestry Management in Germany." In *The Quantifying Spirit in the 18th Century*, edited by Tore Frängsmyr, J. L. Heilbron, and Robin E. Rider, 315–42. Berkeley: University of California Press, 1990.

Lucier, Paul. "Comstock Capitalism: The Law, the Lode, and the Science." *Osiris* 33, no. 1 (2018): 210–31.

MacArthur, R. H., and E. O. Wilson. *The Theory of Island Biogeography*. Princeton, NJ: Princeton University Press, 1967.

Macdonald, Stephen. *Petroleum Conservation in the United States: An Economic Analysis*. Baltimore, MD: Johns Hopkins University Press, 1971.

MacKenzie, Donald. "The Trough of Certainty." In *Exploring Expertise: Issues and Perspectives*, edited by Robin Williams, Wendy Faulkner, and James Fleck, 325–29. Basingstoke, UK: Macmillan, 1998.

Mackintosh, Prudence. "The Soul of East Texas: Reflections and Recollections of Life among the Shadows of the Piney Woods." *Texas Monthly* 17 (October 1989): 116–29ff.

Malavis, Nicholas George. *Bless the Pure and Humble: Texas Lawyers and Oil Regulation, 1919–1936*. College Station: Texas A&M University Press, 1996.

Malthus, Thomas Robert. *An Essay on the Principle of Population as It Affects the Future Improvement of Society, with Remarks on the Speculations of Mr. Godwin, M. Condorcet, and Other Writers*, 1798; *An Essay on the Principle of Population; or, a View of its Past and Present Effects on Human Happiness*, in two volumes. London: J. Johnson, 1806.

Mamiya, Azumi, and Genki Ichinose. "Strategies That Enforce Linear Payoff Relationships Under Observation Errors in Repeated Prisoner's Dilemma Game." *Journal of Theoretical Biology* 477 (21 September 2019): 63–76.

Mandelbaum, David. "Rule of Capture Is Back for Pennsylvania Oil and Gas Wells … Sort Of." URL: https://www.gtlaw-environmentalandenergy.com/2020/01/articles/court-cases/rule-of-capture-is-back-for-pennsylvania-oil-and-gas-wells-sort-of/, January 24, 2020, accessed March 1, 2021.

March, James G., and Herbert A. Simon. *Organizations*. New York: Wiley, 1967 (originally published 1958).

Marshall, J. Howard. "Legal History of Conservation of Oil in California." In *Legal History of Conservation of Oil and Gas: A Symposium*. Chicago: Section of Mineral Law, American Bar Association, 1938.

Martin, Fenton. *Common-Pool Resources and Collective Action: A Bibliography*. Vols. 1 and 2. Bloomington, IN: Workshop in Political Theory and Policy Analysis, 1989–92.

Marx, Karl. *Capital l*, "The Fetishism of Commodities and the Secret Thereof."

Masuda, Naoki. "Evolution via Imitation Among Like-Minded Individuals." *Journal of Theoretical Biology* 349 (21 May 2014): 100–108.

Mathew, Sarah, "How the Second-Order Free Rider Problem is Solved in a Small-Scale." *American Economic Review* 107, no. 5 (2017): 578–81.

Mathews, Don. "Wealth and Its Distribution in the Antebellum South: Where Do We Stand and Why Does It Matter." *Economic and Business History* 15 (1997): 109–20.

Maynard Smith, John. "Group Selection and Kin Selection." *Nature* 201, no. 4924 (14 March 1964): 1145–47.

———. "Honest Signaling: The Philip Sidney Game." *Animal Behaviour* 42 (December 1991): 1034–35.

Maynard Smith, John, and George R. Price. "The Logic of Animal Conflict." *Nature* 146, no. 5427 (1973): 15–18.

McDonald, Stephen L. *Federal Tax Treatment of Income from Oil and Gas*. Washington, DC: Brookings Institution, Studies of Government Finance, 1963.

———. *Petroleum Conservation in the United States: An Economic Analysis*. Baltimore, MD: Johns Hopkins University Press, 1971.

McDowell, A. "Real Property, Spontaneous Order, and Norms in the Gold Mines." *Law and Social Inquiry* 29, no. 4 (2004): 771–818.

McElreath, Richard. "Reputation and the Evolution of Conflict." *Journal of Theoretical Biology* 220, no. 3 (7 February 2003): 345–57.

McElreath, Richard, Adrian V. Bell, Charles Efferson, Mark D. Lubell, Peter J. Richerson, and Timothy Waring. "Beyond Existence and Aiming Outside the Laboratory: Estimating Frequency-Dependent and Payoff-Based Social Learning Strategies." *Philosophical Transactions of the Royal Society London B. Biological Sciences* 363, no. 1509 (12 November 2008): 3515–28.

McElreath, Richard, and Robert Boyd. *Mathematical Models of Social Evolution: A Guide for the Perplexed*. Chicago: University of Chicago Press, 2007.

McKelvey, B. "Toward a Cambellian Realist Organization Science." In *Variations in Organization Science: In Honor of Donald T. Campbell*, edited by J. A. C. Baum and B. McKelvey. Thousand Oaks, CA: Sage Publications, 1999.

McKnight, David N. *The Rittenhouse Orrery: A Marvel of History, Science and Craftsmanship*. Philadelphia: University of Pennsylvania, n.d.

McMath, Robert C., Jr. "Sandy Land and Hogs in the Timber: (Agri)Cultural Origins of the Farmers' Alliance in Texas." In *The Countryside in the Age of Capitalist Transformation: Essays in the Social History of Rural America*, edited by Steven Hahn and Jonathan Prude. Chapel Hill: University of North Carolina Press, 1985.

McNeill, William H. *Keeping Together in Time: Dance and Drill in Human History*. Cambridge, MA: Harvard University Press, 1995.

———. *The Pursuit of Power: Technology, Armed Force, and Society Since A.D. 1000*. Chicago: University of Chicago Press, 1982.

Merrill, Thomas W., and David M. Schizer. "The Shale Oil and Gas Revolution, Hydraulic Fracturing, and Water Contamination: A Regulatory Strategy." *Minnesota Law Review* 98 (2013): 145–265.

Merton, Robert King. "The Matthew Effect in Science." *Science* 159 (5 January 1968): 56–63.

———. "Multiple Discoveries as a Critical Research Site." In *The Sociology of Science: Theoretical and Empirical Investigations*. Chicago: University of Chicago Press, 1970.

Mervis, Jeffrey. "Legal Challenge Could Weaken Science's Role in U.S. Regulation." *Science* 380, no. 6647 (26 May 2023): 782–83.

Metcalfe, John Stanley. "Accounting for Economic Evolution: Fitnesses and the Population Method." *Journal of Bioeconomics* 10, no. 1 (April 2008): 23–49.

Miller, Bill. "From Theory to Application." In *Complexity Economics: Dialogues of the Applied Complexity Network*, Proceedings of the Santa Fe Institute's 2019 Fall Symposium, edited by W. Brian Arthur, Eric D. Beihocker, and Allison Stranger. Santa Fe, NM: Santa Fe Institute Press, 2020.

Miller, David W. "The Historical Development of the Oil and Gas Laws of the United States." *California Law Review* 51, no. 3 (1963): 506–34.

Mills, Warner E., Jr. "Martial Law in East Texas." The Inter-University Case Program, no. 53. Tuscaloosa, AL: University of Alabama Press, 1960.

Mitchell, Alfred Cameron. *Market Demand and Proration of Texas Crude Petroleum*. Austin: Bureau of Business Research, University of Texas, 1964.

Mitchell, Melanie. *Complexity: A Guided Tour*. New York: Oxford University Press, 2009.

Montgomery, Carl T., and Michael B. Smith. "Hydraulic Fracturing: History of an Enduring Technology." *Journal of Petroleum Technology* 62 (December 2010): 26.

Montgomery, R. H. *The Brimstone Game*. New York: Vanguard Press, 1940.

Moody, Dan, and Charles B. Wallace. "Texas Antitrust Laws and Their Enforcement—Comparison With Federal Antitrust Laws." *Southwestern Law Journal* 11 (Winter 1957): 1–26.

Morgan, John, and Ken Steiglitz. "Pairwise Competition and the Replicator Equation." *Bulletin of Mathematical Biology* 65 (November 2003): 1163–72.

Morsky, B., and C. T. Bauch. "Truncation Selection and Payoff Distributions Applied to the Replicator Equation." *Journal of Theoretical Biology* 404 (7 September 2016): 383–90.

Muir, W. M. "Group Selection for Adaptation to Multiple-Hen Cages: Selection Program and Direct Responses." *Poultry Science* 75, no. 4 (1995): 447–58.

Müller, B., F. Bohn, G. Dreßler, J. Groeneveld, C. Klassert, R. Martin, M. Schlüter, et al. "Describing Human Decisions in Agent-Based Models—ODD + D, an Extension of the ODD Protocol." *Environmental Modelling and Software* 48 (October 2013): 37–48.

Müller, B., S. Balbi, C. M. Buchmann, L. De Sousa, G. Dressler, J. Groeneveld, C. J. Klassert, et al. "Standardised and Transparent Model Descriptions for Agent-Based Models: Current Status and Prospects." *Environmental Modelling & Software* 55 (May 2014): 156–63.

Myres, Samuel D. *The Permian Basin: Petroleum Basin of the Southwest—Era of Discovery*. El Paso, TX: Permian Basin Press, 1973.

Nelson, Richard R., and Sidney G. Winter. *An Evolutionary Theory of Economic Change*. Cambridge, MA: Harvard University Press, 1982.

———. "Evolutionary Theorizing in Economics." *Journal of Economic Perspectives* 16, no. 2 (2002): 23–46.

Nettle, Daniel. "Selection, Adaptation, Inheritance, and Design in Human Culture: The View from the Price Equation." *Philosophical Transactions of the Royal Society B* 375 (27 April 2020): https://doi.org/10.1098/rstb.2019.0358.

Nicholson, Barclay, and Brian Albrecht. "Hydraulic Fracturing as a Subsurface Trespass: A New Ruling in West Virginia Provides Protection for the Landowner." *The Energy Law Advisor*. Plano, TX: The Center for American and International Law, n.d.

Nicholson, Patrick J. *Mr. Jim: The Biography of James Smither Abercrombie*. Houston: Gulf Publishing Company, 1983.

Nisbett, Richard E., and Dov Cohen. *Culture of Honor: The Psychology of Violence in the South*. Boulder, CO: Westview Press, 1996.

Noailly, J., C. A. Withagen, and J. C. Van den Bergh. "Spatial Evolution of Social Norms in a Common-Pool Resource Game." *Environmental and Resource Economics* 36, no. 1 (2007): 113–41.

Noailly, J., J. C. Van den Bergh, and C. A. Withagen. "Local and Global Interactions in an Evolutionary Resource Game." *Computational Economics* 33, no. 2 (2009): 155–73.

Nordhauser, Norman E. *The Quest for Stability: Domestic Oil Regulation, 1917–1935*. New York: Garland, 1979.

Nosil, Patrik, Romain Villoutreix, Clarissa F. de Carvalho, Timothy E. Farkas, Víctor Soria-Carrasco, Jeffrey L. Feder, Bernard J. Crespi, and Zach Gompert. "Natural Selection and the Predictability of Evolution in *Timema* Stick Insects." *Science* 359, no. 6377 (16 February 2018): 765–70.

Nowak, Martin A. "Five Rules for the Evolution of Cooperation." *Science* 314 (8 December 2006): 1560–63.

———. *Evolutionary Dynamics: Exploring the Equations of Life*. Cambridge: Harvard University Press, 2006.

———, ed. "Evolution of Cooperation." *Journal of Theoretical Biology* 299 (12 April 2012): 1–188.

Nowak, Martin A., and Robert M. May. "Evolutionary Games and Spatial Chaos." *Nature* 359 (29 October 1992): 826–29.

Ohtsuki, H., Christoph Hauert, E. Lieberman, and Martin A. Nowak. "A Simple Rule for the Evolution of Cooperation on Graphs and Social Networks." *Nature* 441 (25 May 2006): 502–5.

Okasha, Samir. "Biological Altruism." *The Stanford Encyclopedia of Philosophy* (Fall 2013 edition), edited by Edward N. Zalta and Uri Nodelman. http://plato.stanford.edu/archives/fall2013/entries/altruism-biological/.

———. *Evolution and the Levels of Selection*. Oxford, UK: Clarendon/Oxford, 2006.

———. "Multilevel Selection and the Partitioning of Covariance: A Comparison of Three Approaches." *Evolution* 58, no. 3 (2004): 486–94.

Okasha, Samir, and Jun Otsuka. "The Price Equation and the Causal Analysis of Evolutionary Change." *Philosophical Transactions of the Royal Society B* 375 (27 April 2020): https://doi.org/10.1098/rstb.2019.0365.

Olien, Diana Davids, and Roger M. Olien. *Oil in Texas: The Gusher Age, 1895–1945*. Austin: University of Texas Press, 2002.

Olien, Diana Davids, and Roger M. Olien. "Running Out of Oil: Shaping Petroleum Conservation Discourse in the Public Forum," manuscript, n.d.

Olien, Roger M., and Diana Davids Olien. *Oil and Ideology: The Cultural Creation of the American Petroleum Industry*. Chapel Hill: University of North Carolina Press, 2000.

———. *Wildcatters: Texas Independent Oilmen*. Austin: Texas Monthly Press, 1984.

Olmstead, Alan L., and Paul Rhode. "Rationing without Government: The West Coast Gas Famine of 1920." *American Economic Review* 15 (December 1985): 1044–55.

Olson, Mancur. *The Logic of Collective Action*. Cambridge, MA: Harvard University Press, 1965.

Orbell, John M., Alphons van de Kragt, and Robyn M. Dawes. "Explaining Discussion-Induced Cooperation." *Journal of Personality and Social Psychology* 54, no. 5 (1988): 811–19.

Ostrom, Elinor. "A Behavioral Approach to the Rational Choice Theory of Collective Action." *American Political Science Review* 92 (March 1998): 1–22.

———. "Beyond Markets and States: Polycentric Governance of Complex Economic Systems." *American Economic Review* 100, no. 3 (2010): 641–72.

———. *Governing the Commons: The Evolution of Institutions for Collective Action*. Cambridge, UK: Cambridge University Press, 1990.

Ostrom, Elinor, Thomas Dietz, Nives Dolšak, Paul C. Stern, Susan Stonich, and Elke U. Weber, eds. *The Drama of the Commons*. Washington, DC: National Academy Press, 2002.

Ostrom, Elinor, R. Gardner, and J. Walker. *Rules, Games, and Common-Pool Resources*. Ann Arbor: University of Michigan Press, 1994.

Ostrom, Elinor, James Walker, and Roy Gardner. "Covenants with and without the Sword: Self-Governance Is Possible." *American Political Science Review* 86 (June 1992): 404–17.

Oudshoorn, Nelly, and Trevor Pinch, eds. *How Users Matter: The Co-Construction of Users and Technology*. Cambridge, MA: MIT Press, 2003.

Page, Karen M., and Martin A. Nowak. "Unifying Evolutionary Dynamics." *Journal of Theoretical Biology* 219 (7 November 2002): 93–98.

Palmer, Bruce. *"Man Over Money:" The Southern Populist Critique of American Capitalism*. Chapel Hill: University of North Carolina Press, 1980.

Pan, L., D. Hao, Z. Rong, and T. Zhou. "Zero-Determinant Strategies in the Iterated Public Goods Game." *Scientific Reports* 5 (21 August 2015), https://doi.org/10.1038/srep13096.

Panchanathan, Karthik. "George Price, the Price Equation, and Cultural Group Selection." *Evolution and Human*

*Behavior* 32, no. 5 (2011): 368–71.

Parkes, Oscar. *British Battleships: "Warrior" 1860 to "Vanguard" 1950; A History of Design, Construction and Armament.* Hamden, CT: Archon Books, Shoe String Press, 1970.

Parkinson, C. Northcote. *Parkinson's Law: The Pursuit of Progress.* London: John Murray, 1958.

Parr, Joy. "What Makes Washday Less Blue? Gender, Nation, and Technology Choice in Postwar Canada." *Technology and Culture* 38 (January 1997): 153–86.

Peck, Joel R. "The Evolution of Outsider Exclusion." *Journal of Theoretical Biology* 142 (22 February 1990): 565–71.

———. "Friendship and the Evolution of Cooperation." *Journal of Theoretical Biology* 162 (21 May 1993): 195–228.

Peck, Joel R., and M. Feldman. "The Evolution of Helping Behavior in Large, Randomly Mixed Populations." *American Naturalist* 127 (February 1986): 209–21.

Peebles, Malcolm W. H. *Evolution of the Gas Industry.* London: New York University Press, 1980.

Peña, Jorge, Georg Nöldeke, and Laurent Lehmann. "Evolutionary Dynamics of Collective Action in Spatially Structured Populations." *Journal of Theoretical Biology* 382 (7 October 2015): 122–36.

Penn State College of Earth and Mineral Sciences, John A. Dutton Institute for Teaching and Learning Excellence, PNG 301 Introduction to Petroleum and Natural Gas Engineering; licensed by Creative Commons CC BY-NC-SA 4.0, accessed April 27, 2023.

Pepper, John W. "Simple Models of Assortment Through Environmental Feedback." *Artificial Life* 13, no. 1 (2007): 1–9.

Pepper, John W., and Barbara B. Smuts. "Agent-based Modeling of Multilevel Selection: The Evolution of Feeding Restraint as a Case Study." *Natural Resources and Environmental Issues* 8, no. 1 (2001): Article 10.

Perreault, Charles, Cristina Moya, and Robert Boyd. "A Bayesian Approach to the Evolution of Social Learning." *Evolution and Human Behavior* 33, no. 5 (September 2012): 449–59.

Persky, Joseph J. *The Burden of Dependency: Colonial Themes in Southern Economic Thought.* Baltimore, MD: The Johns Hopkins University Press, 1992.

Pfaffenberger, Bryan. "The Harsh Facts of Hydraulics: Technology and Society in Sri Lanka's Colonization Schemes." *Technology and Culture* 31 (July 1990): 361–97.

Phadnis, Nitin. "Poisons, Antidotes, and Selfish Genes: Genes Masquerade as Essential to Development to Ensure Their Transmission." *Science* 366 (9 June 2017): 1013.

Pichugin, Yuriy, Chaitanya S. Gokhale, Julián García, Arne Traulsen, and Paul B. Rainey. "Modes of Migration and Multilevel Selection in Evolutionary Multiplayer Games." *Journal of Theoretical Biology* 387 (21 December 2015): 144–53.

Pinker, Steven, with responses from the Reality Group. "The False Allure of Group Selection." *Edge,* June 21, 2012. URL: http://edge.org/conversation/the-false-allure-of-group-selection.

Pittman, Blair. *The Stories of I. C. Eason, King of the Dog People.* Denton: University of North Texas Press, 1996.

Plaiss, Adam. "From Natural Monopoly to Public Utility: Technological Determinism and the Political Economy of Infrastructure in Progressive-Era America." *Technology and Culture* 57, no. 4 (October 2016): 806–30.

Plotkin, Samuel L. *The Rational Peasant: The Political Economy of Rural Society in Vietnam.* Berkeley: University of California Press, 1979.

Plummer, F. B. "Petroleum Engineering Education—Present Curricula and Future Possibilities." *Mining and Metallurgy* 17 (October 1936): 485.

Poincaré, Henri. *Science and Method.* "Chapter III: Mathematical Creation." 1908. Reprinted in *The Foundations of Science: Science and Hypothesis; The Value of Science; Science and Method.* Translated by George Bruce Halsted. New York: The Science Press, 1913.

Pollock, Norman. *The Just Polity: Populism, Law, and Human Welfare.* Urbana: University of Illinois Press, 1987.

Porter, Theodore M. *Trust in Numbers: The Pursuit of Objectivity in Science and Public Life.* Princeton, NJ: Princeton University Press, 1995.

Porterfield, Bill. "H. L. Hunt's Long Goodbye." *Texas Monthly,* no. 3, March 1975, 63–69ff.

Poteete, Amy R., Marco A. Janssen, and Elinor Ostrom. *Working Together: Collective Action, the Commons, and Multiple Methods in Practice.* Princeton, NJ: Princeton University Press, 2010.

Potter, Harry Grant, III. "History and Evolution of the Rule of Capture." URL: https://www.twdb.texas.gov/publications/reports, accessed March 12, 2019.

Powers, Simon T., Daniel J. Taylor, and Joanna J. Bryson. "Punishment Can Promote Defection in Group-Structured Populations." *Journal of Theoretical Biology* 311 (21 October 2012): 107–16.

Pratt, Joseph A. "The Petroleum Industry in Transition: Antitrust and the Decline of Monopoly Control in Oil." *Journal of Economic History* 40 (December 1980): 815–37.

Pratt, Joseph A., Tyler Priest, and Christopher J. Castaneda. *Offshore Pioneers: Brown & Root and the History of Offshore Oil and Gas.* Houston: Gulf Publishing, 1997.

Pred, Allan R. *The Spatial Dynamics of U.S. Urban-Industrial Growth, 1880–1914: Interpretive and Theoretical*

*Essays.* Cambridge, MA: MIT Press, 1966.

Press, William H., and Freeman J. Dyson. "Iterated Prisoner's Dilemma Contains Strategies That Dominate Any Evolutionary Opponent." *Proceedings of the National Academy of Sciences* 109 (26 June 2012): 10409–13.

Price, Derek J. de Solla. "Gears from the Greeks. The Antikythera Mechanism: A Calendar Computer from ca. 80 BC." *Transactions of the American Philosophical Society* 64, no. 7 (1974): 1–70.

———. *Science since Babylon.* Enlarged edition. New Haven, CT: Yale University Press, 1975.

Price, George R. "Extensions of Covariance Selection Mathematics." *Annals of Human Genetics* 35 (1972): 485–90.

———. "Fisher's 'Fundamental Theorem' Made Clear." *Annals of Human Genetics* 36 (1972): 129–40.

———. "Selection and Covariance." *Nature* 277, no. 5257 (1970): 520–21.

Prindle, David F. *Petroleum Politics and the Texas Railroad Commission.* Austin: University of Texas Press, 1981.

Pruitt, Jonathan N., and Charles J. Goodnight. "Site-Specific Group Selection Drives Locally Adapted Group Compositions." *Nature* 514, no. 7522 (2014): 359–62.

Queller, David C. "The Gene's Eye View, the Gouldian Knot, Fisherian Swords and the Causes of Selection." *Philosophical Transactions of the Royal Society B* 375 (27 April 2020): https://doi.org/10.1098/rstb.2019.0354.

———. "Quantitative Genetics, Inclusive Fitness, and Group Selection." *The American Naturalist* 139, no. 3 (March 1992): 540–58.

Rajkumar, Karthik, Guillaume Saint-Jacques, Iavor Bojinov, Erik Brynjolfsson, and Sinan Ara. "A Causal Test of the Strength of Weak Ties." *Science* 377, no. 6612 (September 15, 2022): 1304–10.

Ramírez, Juan Camillo, and James A. R. Marshall. "Can Natural Selection Encode Bayesian Priors?" *Journal of Theoretical Biology* 426 (7 August 2017): 57–66.

Rao, Vic. "1984 and Beyond: The Advent of Horizontal Wells." *Journal of Petroleum Technology* 59, no. 10 (October 2007): 118–20.

Reistle, Carl E., Jr. "Reservoir Engineering." In American Petroleum Institute, *History of Petroleum Engineering*, Executive Committee on Drilling and Production Practice, Division of Production, 811–46. Dallas: American Petroleum Institute, 1961.

Reynolds, Terry S. "Defining Professional Boundaries: Chemical Engineering in the Early 20th Century." *Technology and Culture* 27, no. 4 (October 1986): 694–716.

Reznick, David, and Joseph Travis. "Is Evolution Predictable?" *Science* 359, no. 6377 (16 February 2018): 738–39.

Rice, Sean H. "Universal Rules for the Interaction of Selection and Transmission in Evolution." *Philosophical Transactions of the Royal Society B* 375 (27 April 2020): https://doi.org/10.1098/rstb.2019.0353.

Richards, Robert J., and Lorraine Datson, eds. *Kuhn's Structure of Scientific Revolutions at Fifty: Reflections on a Science Classic.* Chicago: University of Chicago Press, 2016.

Richerson, Peter J. "Group Selection Among Alternatively Evolutionarily Stable Strategies." *Journal of Theoretical Biology* 145 (9 August 1990): 331–42.

———. "An Integrated Bayesian Theory of Phenotypic Flexibility." *Behavioral Processes* 161 (April 2019): 54–64.

Richerson, Peter J., R. Baldini, A. Bell, K. Demps, K. Frost, V. Hills, S. Mathew, et al. "Cultural Group Selection Plays an Essential Role in Explaining Human Cooperation: A Sketch of the Evidence." *Behavioral and Brain Sciences* 39 (January 2016): https://doi.org/10.1017/s0140525x1400106x.

Richerson, Peter J., and Robert Boyd. "Simple Models of Complex Phenomena: The Case of Cultural Evolution." In *The Latest on the Best: Essays on Evolution and Optimality*, edited by John Dupré, 27–52. Cambridge, MA: MIT Press, 1987, reproduced in Boyd and Richerson, *Origin and Evolution of Culture*, chapter 19, 397–419.

———. "Complex Societies: The Evolutionary Origins of a Crude Superorganism." *Human Nature: International Biosocial Perspectives* 10, no. 3 (1999): 195–219.

Richerson, Peter J., Robert T. Boyd, and Joseph Henrich. "Cultural Evolution of Human Cooperation." In *The Genetic and Cultural Evolution of Cooperation*, edited by Peter Hammerstein, 357–88. Cambridge, MA: MIT Press, 2003.

Richerson, Peter J., Robert Boyd, and Brian Paciotti. "An Evolutionary Theory of Commons Management." In *The Drama of the Commons*, edited by Elinor Ostrom, Thomas Dietz, Nives Dolšak, Paul C. Stern, Susan Stonich, and Elke U. Weber, 403–42. Washington, DC: National Academy Press, 2002.

Robertson, Alan. "A Mathematical Model of the Culling Process in Dairy Cattle." *Animal Production* 8, no. 1 (February 1966): 95–108.

Rogers, Everett M., with F. Floyd Shoemaker. *Communication of Innovations: A Cross Cultural Approach.* New York: The Free Press, 1971.

Rose, Carol M. "Common Property, Regulatory Property, and Environmental Protection: Comparing Community-Based Management to Tradable Environmental Allowances." In *The Drama of the Commons*, edited by Elinor Ostrom, Thomas Dietz, Nives Dolšak, Paul C. Stern, Susan Stonich, and Elke U. Weber, 233–57. Washington, DC: National Academy Press, 2002.

Rosenberg, Charles E. "Rationalization and Reality in Shaping American Agricultural Research, 1875–1914." *Social*

*Studies of Science* 7 (November 1977): 401–22.

Rosenberg, Nathan. "Technological Change in the Machine Tool Industry, 1840–1910." *Journal of Economic History* 23, no. 4 (December 1963): 414–46.

———. *Technology and American Economic Growth*. New York: Harper Torchbooks/Harper & Row, 1972; 7th printing, Armonk, NY: M. E. Sharpe, Inc.

Rostow, Eugene V. *A National Policy for the Oil Industry*. New Haven CT: Yale University Press, 1948.

Ryan, Bryce Finley. "Boston High School Graduates in Periods of Prosperity and Depression." PhD diss., Harvard University, 1940.

———. "A Sociological Study of the Mail Order Oil Promoter and His Methods." MA thesis, University of Texas at Austin, 1933.

Ryan, Bryce, and Neal C. Gross. "The Diffusion of Hybrid Corn Seed in Two Iowa Communities." *Rural Sociology* 8 (March 1943): 15–24.

Salali, Gul Deniz, Myriam Juda, and Joseph Henrich. "Transmission and Development of Costly Punishment in Children." *Evolution and Human Behavior* 36 (March 2015): 86–94.

Sampson, Anthony. *The Seven Sisters: The Great Oil Companies and the World They Shaped*. New York: Viking, 1975.

Samuelson, Larry. "Evolution and Game Theory." *Journal of Economic Perspectives* 16, no. 2 (2002): 47–66.

Samuelson, Paul A. "Altruism as a Problem Involving Group versus Individual Selection in Economics and Biology." *American Economic Review* 83, no. 2 (May 1993): 143–48.

Sanders, Karin. "Cooperative Behaviors in Organizations." In *Friends and Enemies in Organizations*, edited by Rachel L. Morrison and Sarah L. Wright, 101–21. London, UK: Palgrave Macmillan, 2009.

Sanders, M. Elizabeth. *The Regulation of Natural Gas: Policy and Politics, 1938–1978*. Philadelphia, PA: Temple University Press, 1981.

Sangren, P. Steven. "Dialectics of Alienation: Individuals and Collectivities in Chinese Religion." *Man* 26 (March 1991): 68–69.

———. *History and Magical Power in a Chinese Community*. Stanford, CA: Stanford University Press, 1987.

———. "History and the Rhetoric of Legitimacy: The Ma Tsu Cult of Taiwan." *Comparative Studies of Society and History* 30 (October1988): 674–97.

———. "Ma Tsu's Black Face: Individuals and Collectivities in Chinese Religion." Manuscript, n.d.

———. "Multilectics of Alienation: Worship and Testimony in the Ma Tsu Pilgrimages of Taiwan." Manuscript, n.d.

———. "Ritual as Production, Reproduction, and Transformation: The Ma Tsu Cult of Taiwan." University of Pittsburgh, National Resource Center for East Asian Studies, March 16, 1989.

Santos, Francisco C., Jorge M. Pacheco, Max O. Souza, and Brian Skyrms. "Evolutionary Dynamics of Collective Action in *N*-Person Stag-Hunt Dilemmas." *Proceedings of the Royal Society B* 276, no. 1655 (January 2009): 315–21.

Sasaki, T., and S. Uchida. "The Evolution of Cooperation by Social Exclusion." *Proceedings of the Royal Society B: Biological Sciences* 280 (7 February 2013): https://doi.org/10.1098/rspb.2012.2498.

Saviotti, P. P., and G. S. Mani. "Competition, Variety and Technological Evolution: A Replicator Dynamics Model." *Journal of Evolutionary Economics* 5, no. 4 (1995): 369–92.

Schank, Jeffrey C., Paul E. Smaldino, and Matt L. Miller. "Evolution of Fairness in the Dictator Game by Multilevel Selection." *Journal of Theoretical Biology* 382 (7 October 2015): 64–73.

Schlüter, Maja, Alessandro Tavoni, and Simon Levin. "Robustness of Norm-Driven Cooperation in the Commons." *Proceedings of the Royal Society B: Biological Sciences* 283 (13 January 2016): 1–22.

Schoenmakers, Sarah, Christian Hilbe, Bernd Blasius, and Arne Traulsen. "Sanctions as Honest Signals—The Evolution of Pool Punishment by Public Sanctioning Institutions." *Journal of Theoretical Biology* 356 (7 September 2014): 36–46.

Schulz, Jonathan F., Duman Bahrami-Rad, Jonathan P. Beauchamp, Joseph Henrich. "The Church, Intensive Kinship, and Global Psychological Variation." *Science* 366, no. 6466 (November 8, 2019): 707.

Schumpeter, Joseph A. *Business Cycles: A Theoretical, Historical, and Statistical Analysis of the Capitalist Process*. New York: McGraw-Hill, 1939.

Scott, James C. "Hegemony and the Peasantry." *Politics and Society* 7, no. 3 (1977): 267–96.

———. *The Moral Economy of the Peasant: Rebellion and Subsistence in Southeast Asia*. New Haven, CT: Yale University Press, 1976.

———. *Seeing Like a State: How Certain Schemes to Improve the Human Condition Have Failed*. New Haven, CT: Yale University Press, 1998.

Scranton, Philip. "Manufacturing Diversity: Production Systems, Markets, and an American Consumer Society, 1870–1930." *Technology and Culture* 35 (July 1994): 476–505.

Seftel, Howard. "Government Regulation and the Rise of the California Fruit Industry: The Entrepreneurial Attack on Fruit Pests, 1880–1920." *Business History Review* 59 (Autumn 1985): 369–402.

Seki, Motohide, and Mayuko Nakamaru. "A Model for Gossip-Mediated Evolution of Altruism with Various Types of False Information by Speakers and Assessment by Listeners." *Journal of Theoretical Biology* 407 (21 October 2016): 90–105.

Sen, Amartya K. "Rational Fools: A Critique of the Behavioral Foundations of Economic Theory." *Philosophy and Public Affairs* 6, no. 4 (Summer 1977): 317–44.

Servos, John W. "The Industrial Relations of Science: Chemical Engineering at MIT, 1900–1939." *Isis* 71, no. 4 (December 1980): 530–49.

Sethi, Rajiv, and E. Somanathan. "The Evolution of Social Norms in Common Property Resource Use." *American Economic Review* 86, no. 4 (September 1996): 766–88.

Shakespeare, William S. *Henry IV, Part I: First Folio*, Act 5, Scene 3.

Shapin, Steven. "The House of Experiment in Seventeenth Century England." *Isis* 79 (September 1988): 373–404.

———. *A Social History of Truth: Civility and Science in Seventeenth Century England*. Chicago: University of Chicago Press, 1994.

Shapin, Steven, and Simon Schaffer. *Leviathan and the Air-Pump: Hobbes, Boyle, and the Experimental Life*. Princeton, NJ: Princeton University Press, 1985.

Shelton, Deborah E., and Richard E. Michod. "Group and Individual Selection During Evolutionary Transitions in Individuality: Meanings and Partitions." *Philosophical Transactions of the Royal Society B* 375 (27 April 2020): https://doi.org/10.1098/rstb.2019.0364.

Shennan, Stephen J., Enrico R. Crema, and Tim Kerig. "Isolation-by-Distance, Homophily, and 'Core' vs. 'Package' Cultural Evolution Models in Neolithic Europe." *Evolution and Human Behavior* 3, no. 2 (2015): 103–19.

Shillings, Erica. "SPE: A 1957 Snapshot." *Journal of Petroleum Technology* 59, no. 10 (October 2007): 49–51.

Sigmund, Karl. *The Calculus of Selfishness*. Princeton, NJ: Princeton University Press, 2010.

———. "Moral Assessment in Indirect Reciprocity." *Journal of Theoretical Biology* 299 (21 April 2012): 25–30.

Sigmund, Karl, Hannelore De Silva, Arne Traulsen, and Christoph Hauert. "Social Learning Promotes Institutions for Governing the Commons." *Nature* 466, no. 7308 (2010): 861–63.

Simon, Burton. "Continuous-Time Models of Group Selection, and the Dynamical Insufficiency of Kin Selection Models." *Journal of Theoretical Biology* 349 (21 May 2014): 22–31.

Simon, Herbert A. "Altruism and Economics." *American Economic Review* 83, no. 2 (May 1993): 156–61.

———. "Altruism and Economics." *Eastern Economic Journal* 18 (Winter 1992): 73–83.

———. "A Behavioral Model of Rational Choice." *Quarterly Journal of Economics* 61, no. 1 (1955): 99–118.

———. "Complex Systems: The Interplay of Organizations and Markets in Contemporary Society." *Computational and Mathematical Organization Theory* 7, no. 2 (August 2001): 79.

———. "Near Decomposability and the Speed of Evolution." *Industrial and Corporate Change* 11, no. 3 (1 June 2002): 587–99.

———. "Rational Choice and the Structure of the Environment." *Psychological Review* 63 (1956): 129–38.

———. "Rationality as Process and as Product of Thought." *American Economic Review* 68, no. 2 (1978): 1–16.

———. *Sciences of the Artificial*. Cambridge MA: MIT Press, 1969.

———. "We and They: The Human Urge to Identify with Groups." *Industrial and Corporate Change* 11, no. 3 (June 2002): 607–10.

Skyrms, Brian. *The Stag Hunt and the Evolution of Social Structure*. Cambridge, UK: Cambridge University Press, 2004.

Skyrms, Brian, and Robin Pemantle. "A Dynamic Model of Social Network Formation." *Proceedings of the National Academy of Sciences* 97 (August 2000): 9340–46.

Skytt, Christine Benna, and Lars Winther. "Trust and Local Knowledge Production: Inter-Organizational Collaborations in the Sønderborg Region, Denmark." *Geografisk Tidsskrift-Danish Journal of Geography* 111, no. 1 (2011): 27–41.

Smaldino, Paul E. "The Cultural Evolution of Emergent Group-Level Traits." *Behavioral and Brain Sciences* 37, no. 3 (June 2014): 243–95.

———. *Modeling Social Behavior: Mathematical and Agent-Based Models of Social Dynamics and Cultural Evolution*. Princeton: Princeton University Press, 2023.

Snyder-Mackler, Noah, et al. "Social Status Alters Immune Regulation and Response to Infection in Macaques." *Science* 354, no. 6315 (2016): 1041–45.

Sober, Elliott, and David Sloan Wilson. *Unto Others: The Evolution and Psychology of Unselfish Behavior*. Cambridge, MA: Harvard University Press, 1998.

Society of Petroleum Engineers of American Institute of Mining Engineers, Education Committee. *Petroleum Engineering Schools of the United States—1965*. Dallas, TX: Office of the Executive Secretary, Society of Petroleum Engineers, 1965.

Society of Petroleum Engineers. *Journal of Petroleum Technology, Special Commemorative Issue: Celebrating 50 Years of the Society of Petroleum Engineers* 59, no. 10 (October 2007).

Souza, Max O., Jorge M. Pacheco, and Francisco C. Santos. "Evolution of Cooperation Under N-Person Snowdrift Games." *Journal of Theoretical Biology* 260 (21 October 2009): 581–88.

Spolaore, Enrico, and Romain Wacziarg. "How Deep Are the Roots of Economic Development?" *Journal of Economic Literature* 51, no. 2 (June 2013): 325–69.

Spratt, John Stricklin. *The Road to Spindletop: Economic Change in Texas, 1875–1901.* Dallas: Southern Methodist University Press, 1955. Reprint, Austin: University of Texas Press, 1983.

Stark, Oded, Doris A. Behrens, and Yong Wang. "On the Evolutionary Edge of Migration as an Assortative Mating Device." *Journal of Evolutionary Economics* 19, no. 1 (February 2009): 95–109.

Stason, E. Blythe. "'Substantial Evidence' in Administrative Law." *University of Pennsylvania Law Review* 89 (June 1941): 1026–51.

Steffy, Loren. *George P. Mitchell: Fracking, Sustainability, and an Unorthodox Quest to Save the Planet.* College Station: Texas A&M University Press, 2019.

———. "The First Barnett Shale Gas Well, the One That Ignited the Revolution, Remains Standing as the Suburbs Encroach." *Dallas Morning News,* October 6, 2019.

Steinbeck, John. *The Grapes of Wrath.* N.p: n.p, 1939.

Sterling, Ross S., with Ed Kilman and Don Carleton. *Ross Sterling, A Texan. A Memoir by the Founder of the Humble Oil and Refining Company.* Austin: University of Texas Press, 2007.

Steward, Dan. *The Barnett Shale Play: Phoenix of the Fort Worth Basin, A History.* Fort Worth: Fort Worth Geological Society, 2007.

Stewart, Alexander, and Joshua Plotkin. "Extortion and Cooperation in the Prisoner's Dilemma." *Proceedings of the National Academy of Sciences* 109, no. 26 (June 26, 2012): 10134–35.

Stewart, James I. "Cooperation When N Is Large: Evidence from the Mining Camps of the American West." *Journal of Economic Behavior and Organization* 69, no. 3 (2009): 213–25.

Stigler, George J., and Gary S. Becker. "De Gustibus Non Est Disputandum." *American Economic Review* 67, no. 2 (March 1977): 76–90.

Stocking, George Ward. *The Oil Industry and the Competitive System: A Study in Waste.* Boston, MA: Houghton Mifflin, 1925.

———. "Stabilization of the Oil Industry: Its Economic and Legal Aspects." *American Economic Review* 23, no. 1, Supplement, Papers and Proceedings of the Forty-fifth Annual Meeting of the American Economic Association (March 1933): 55–70.

Stockton, John R., Richard C. Henshaw, and Richard W. Graves. *Economics of Natural Gas in Texas.* Austin: Bureau of Business Research, University of Texas, 1952.

Strassmann, J. E., and David C. Queller. "The Social Organism: Congresses, Parties, and Committees." *Evolution* 64 (March 2010): 609.

Suárez-Menéndez, Marco, Martine Bérubé, Fabrício Furni, Vania E Rivera-León, Mads-Peter Heide-Jørgensen, Finn Larsen, Richard Sears, et al. "Wild Pedigrees Inform Mutation Rates and Historic Abundance in Baleen Whales." *Science* 381 no. 6661 (1 September 2023): 990–95.

Sui, Xiukai, Rui Cong, Kun Li, Long Wang. "Evolutionary Dynamics of N-Person Snowdrift Game." *Physics Letters A* 379 nos. 45–46 (December 2015): 2922–34.

Sullivan, Robert E. "Texas," from "Reports Prepared by Robert E. Hardwicke since 1949." In *Conservation of Oil and Gas, A Legal History, 1958,* edited by Robert E. Sullivan, 218–46. Chicago: American Bar Association, Section of Mineral and Natural Resources Law, 1960.

Suman, John R. "Evolution by Companies." In *History of Petroleum Engineering.* Executive Committee on Drilling and Production Practice, Division of Production, 63–132. Dallas: American Petroleum Institute, 1961.

Summers, L. H. "The Scientific Illusion in Empirical Macroeconomics." *Scandinavian Journal of Economics* 93, no. 2 (1991): 129–48.

Summers, Suzanne L. "The Geographic and Social Origins of Antebellum Merchants in Houston and Galveston, Texas, 1836–1860." *Essays in Economic and Business History* 15 (1997): 95–107.

Swift, Joseph. "Classics Revisited. Arbiters of Truth, Then and Now: A 40-Year-Old Tome's Prescient Observations about Scientific Fact-Making Resonate Today." *Science* 366, no. 6469 (29 November 2019): 1081.

Swigart, Theodore E. "Handling Oil and Gas in the Field." In *History of Petroleum Engineering,* Executive Committee on Drilling and Production Practice, Division of Production, 907–97. Dallas: American Petroleum Institute, 1961.

Tarr, Joel A. "Transforming an Energy System: The Evolution of the Manufactured Gas Industry and the Transition to Natural Gas in the United States (1807–1954)." In *The Governance of Large Technical Systems,* edited by Olivier Coutard. New York: Routledge, 1999.

Tarr, Joel A., and Bill C. Lamperes. "Changing Fuel Use Behavior and Energy Transitions: The Pittsburgh Smoke Control Movement, 1940–1950." *Journal of Social History* 14 (Summer 1981): 561–88.

Tavoni, Alessandro, Maja Schlüter, and Simon Levin. "The Survival of the Conformist: Social Pressure and Renewable Resource Management." *Journal of Theoretical Biology* 299 (21 April 2012): 152–61.

Taylor, C., and Martin A. Nowak. "Transforming the Dilemma." *Evolution* 61 (17 August 2007): 2281–92.

Taylor, Peter D., and Leo B. Jonker. "Evolutionary Stable Strategies and Game Dynamics." *Mathematical Biosciences* 40 (July 1978): 145–56.

Teece, David J. "The Dynamics of Industrial Capitalism: Perspectives on Alfred Chandler's *Scale and Scope*." *Journal of Economic Literature* 31 (March 1993): 119–225.

Texaco, Inc. "Texaco Canada Announces Promising Tertiary Recovery Project in Alberta." Texaco Inc. Second Quarter Report, 1983. White Plains, NY: Texaco Inc., 1983.

*The Texas Almanac, 1941–1942*. Dallas, Dallas Morning News, 1943.

*The Texas Almanac, 1958–1959*. Dallas: Dallas Morning News, 1960.

Texas Railroad Commission. "Crude Oil Production and Well Counts (since 1935): History of Texas Initial Crude Oil, Annual Production, and Producing Wells," includes onshore and offshore wells within Texas's tidelands (up to ten miles out), accessed January 18, 2021.

Thackeray, William Makepeace. *Vanity Fair: Pen and Pencil Sketches of English Society or Vanity Fair: A Novel without a Hero*. London: *Punch* (serial); Bradbury and Evans (novel), 1847–1848.

Thalhelm, Thomas, Xuemin Zhang, and Shigehiro Oishi. "Moving Chairs in Starbucks: Observational Studies Find Rice-Wheat Cultural Differences in Daily Life in China." *Science Advances* 4 (25 April 2018): 8469.

Thomas, James, and Simon Kirby. "Self-Domestication and the Evolution of Language." *Biology and Philosophy* 33, no. 9 (2018): 1–30.

Thompson, Edward Palmer. "The Moral Economy Reviewed." In *Customs in Common: Studies in Traditional Popular Culture*, 259–351. London: Merlin Press, 1991.

———. "The Moral Economy of the English Crowd in the Eighteenth Century." *Past & Present* 50, no. 1 (February 1971): 76–136.

Thompson, Ernest O. "An Administrator's Views on Proration," 1939.

———. "The Application of the Market Demand Statute to Oil Production in Texas." *Journal of Petroleum Technology* 2 (March 1950): Section 1, 10.

Thöni, Christian. "Trust and Cooperation: Survey Evidence and Behavioral Experiments." In *Social Dilemmas: New Perspective on Trust*, edited by P. A. M. Lange, Bettina Rockenbach, and T. Yamagishi. Oxford, UK: Oxford University Press, 2017.

Tilman, A. R., S. Levin, and J. R. Watson. "Revenue-Sharing Clubs Provide Economic Insurance and Incentives for Sustainability in Common-Pool Resource Systems." *Journal of Theoretical Biology* 454 (7 October 2018): 205–14.

Tinkle, Lon. *Mr. De: A Biography of Everette Lee DeGolyer*. Boston, MA: Little, Brown and Company, 1970.

Tomassini, Marco, and Alberto Antonioni. "Lévy Flights and Cooperation among Mobile Individuals." *Journal of Theoretical Biology* 364 (7 January 2015): 154–61.

Traulsen, Arne, and Martin A. Nowak. "Evolution of Cooperation by Multilevel Selection." *Proceedings of the National Academy of Sciences USA* 103 (18 July 2006): 10952–55.

Trivers, Robert L. "The Evolution of Reciprocal Altruism." *Quarterly Review of Biology* 46, no. 4 (1971): 35–57.

———. "The Evolution of Cooperation." In *The Nature of Prosocial Development: Interdisciplinary Theories and Strategies*, edited by Diane Bridgeman, 43–60. Waltham, MA: Academic Press, 1983.

Trotter, Joe W., Jr. *Coal, Class, and Color: Blacks in Southern West Virginia, 1915–1932*. Urbana: University of Illinois Press, 1990.

Tudor, Tasha. *The Tasha Tudor Book of Fairy Tales*. New York: Platt & Munk, 1961.

Trut, L. "Early Canid Domestication: The Farm-Fox Experiment." *American Scientist* 87 (1999): 160–69.

Turner, Frederick Jackson. "The Significance of the Frontier in American History" (1910). In *The Frontier in American History*, 1–38. New York: Henry Holt and Company, 1921.

US Energy Information Administration. *Monthly Energy Review*, Table 3.1, March 2020, accessed January 21, 2021.

———. "U.S. Field Production of Crude Oil, 1870–2020."

US Treasury Department, Internal Revenue Service. *Manual for the Oil and Gas Industry Under the Revenue Act of 1918*. Washington, DC: Internal Revenue Service, 1919.

Vanberg, Viktor J. "The Rationality Postulate in Economics: Its Ambiguity, Its Deficiency, and Its Evolutionary Alternative." *Journal of Economic Methodology* 11, no. 1 (2004): 1–29.

Vance, Harold. "Bank Loans on Oil and Gas Production." *Petroleum Engineer* 26, no. 11 (October 1954): E4-10.

———. "Completion Methods." In *History of Petroleum Engineering*, Executive Committee on Drilling and Production Practice, Division of Production, 579–616. Dallas: American Petroleum Institute, 1961.

———. "Evaluation." In *History of Petroleum Engineering*, Executive Committee on Drilling and Production Practice, Division of Production, 999–1080. Dallas: American Petroleum Institute, 1961.

van den Bergh, Jeroen C. J. M., and John M. Gowdy. "A Group Selection Perspective on Economic Behavior

Institutions and Organizations." *Journal of Economic Behavior & Organization* 72, no. 1 (2009): 1–20.

Vandiver, Frank E. *Their Tattered Flags: The Epic of the Confederacy.* New York: Harper Collins, 1970.

van Fraassen, Bas C. *The Scientific Image.* Oxford, UK: Oxford University Press, 1980.

van Veelen, Matthijs. "On the Use of the Price Equation." *Journal of Theoretical Biology* 237 (21 December 2005): 412–26.

———. "The Problem with the Price Equation." *Philosophical Transactions of the Royal Society B* 375 (27 April 2020): https://doi.org/10.1098/rstb.2019.0355.

Veasey, James A. "The Law of Oil and Gas." *Michigan Law Review* 18, no. 6 (April 1920): 445–69.

Veblen, Thorstein. "Why Is Economics Not an Evolutionary Science?" *Quarterly Journal of Economics* 12, no. 4 (1898): 373–97.

Vignieri, Sacha. "Social Evolution of Mammal Societies." *Science* 373, no. 6561 (17 September 2021): 1322.

Villani, Dario. "From Theory to Application." In *Complexity Economics: Dialogues of the Applied Complexity Network*, Proceedings of the Santa Fe Institute's 2019 Fall Symposium, edited by W. Brian Arthur, Eric D. Beihocker, and Allison Stranger. Santa Fe, NM: Santa Fe Institute Press, 2020.

Villena, M. G., and M. J. Villena. "Evolutionary Game Theory and Thorstein Veblen's Evolutionary Economics: Is EGT Veblenian?" *Journal of Economic Issues* 38, no. 3 (2004): 585–610.

Vincenti, Walter G. "The Air Propeller Tests of W. F. Durand and E. P. Lesley: A Case Study in Technological Methodology." *Technology and Culture* 20 (October 1979): 712–51.

———. "Control Volume Analysis: A Difference in Thinking Between Engineering and Physics." *Technology and Culture* 23 (April 1982): 145–74.

———. "Engineering Knowledge, Type of Design, and Level of Hierarchy: Further Thoughts about *What Engineers Know . . . ,*" In *Technological Development and Science in the Industrial Age*, edited by P. Droes and M. Bakker, 17–34. Amsterdam: Kluwer Academic Publishers, 1992.

———. "The Retractable Airplane Landing Gear and The Northrop 'Anomaly': Variation-Selection and the Shaping of Technology." *Technology and Culture* 35 (January 1994): 1–33.

———. "The Scope for Social Impact in Engineering Outcomes: A Diagrammatic Aid to Analysis." *Social Studies of Science* 21 (1991): 761–67.

———. *What Engineers Know and How They Know It: Analytical Studies from Aeronautical History.* Baltimore, MD: Johns Hopkins University Press, 1990.

Vogt, Evon Z. "American Subcultural Continua as Exemplified by the Mormons and Texans." *American Anthropologist* 57 (1955): 1163–72.

von Hippel, Eric. "Cooperation Between Rivals: Informal Know-How Trading." *Research Policy* 16, no. 6 (December 1987): 291–302.

von Neumann, John, and Otto Morgenstern. *Theory of Games and Economic Behavior.* Princeton, NJ: Princeton University Press, 1944.

Voosen, Paul. "Climate Scientists Open up Their Black Boxes to Scrutiny: Modelers Become Less Hush-Hush about Tuning, the 'Secret Sauce' That Controls Fine-Scale Processes." *Science* 354, no. 6311 (2016): 401–2.

Wade, Michael J. "An Experimental Study of Group Selection." *Evolution* 31 (1 March 1977): 134–53.

Wade, Michael J., Piter Bijma, Esther D. Ellen, and William Muir. "Group Selection and Social Evolution in Domesticated Animals." *Evolutionary Applications* 3 nos. 5–6 (2010): 453–65.

Wade, Michael J., David Sloan Wilson, Charles Goodnight, Doug Taylor, Yaneer Bar-Yam, Marcus A. M. de Aguiar, Blake Stacey, et al., "Multilevel and Kin Selection in a Connected World." *Nature* 463 (2010): E8–E9.

Walker, Caroline D. "Using Computer-Generated Animation and Simulation Evidence at Trial: What You Should Know." American Bar Association, PRACTICE POINTS, January 11, 2018; accessed April 20, 2023.

Wang, Dashun, and Brian Uzzi. "Weak Ties, Failed Tries, and Success." *Science* 377, no. 6612 (15 September 2022): 1256–58.

Wang, Zhen, Marko Jusup, Rui-Wu Wang, Lei Shi, Yoh Iwasa, Yamir Moreno, and Jürgen Kurths. "Onymity Promotes Cooperation in Social Dilemma Experiments." *Science Advances* 3 (29 March 2017): https://doi.org/10.1126/sciadv.1601444.

Waring, Timothy M., Sandra H. Goff, and Paul E. Smaldino. "The Coevolution of Economic Institutions and Sustainable Consumption via Cultural Group Selection." *Ecological Economics* 131 (January 2017): 524–32.

Warner, Charles A. "Sources of Men." In *History of Petroleum Engineering*, Executive Committee on Drilling and Production Practice, Division of Production, 35–61. Dallas: American Petroleum Institute, 1961.

Watts, Duncan J. *Small Worlds: The Dynamics of Networks between Order and Randomness.* Princeton, NJ: Princeton Studies in Complexity, Princeton University Press, 1999.

Watts, Duncan J., and Steven H. Strogatz. "Collective Dynamics of 'Small-World' Networks." *Nature* 393 (14 June 1998): 440–42.

Weaver, Bobby D. "Cities Service Company." *The Encyclopedia of Oklahoma History and Culture.* www.okhistory.

org, accessed September 23, 2017.

———. "Doherty, Henry Latham." *The Encyclopedia of Oklahoma History and Culture*. www.okhistory.org, accessed September 23, 2017.

Weaver, Jacqueline Lang. *Unitization of Oil and Gas Fields in Texas: A Study of Legislative, Administrative, and Judicial Politics*. Baltimore, MD: Johns Hopkins University Press, 1986.

Webb, Walter Prescott. *The Great Plains*. Lincoln: University of Nebraska Press, 1981 (1931).

Webber, Melvin M. "Production of Crude Petroleum in Texas: A Statistical Analysis." MA thesis, University of Texas at Austin 1948.

———. "Order in Diversity: Community without Propinquity." In *Cities and Space*, edited by L. Wirigo. Baltimore, MD: Johns Hopkins University Press, 1963.

Weber, Max. "Objectivity in Social Science and Social Policy." In *The Methodology of the Social Sciences*, edited and translated by E. A. Shils and H. A. Finch. New York: Free Press, 1949 (1904).

———. *The Protestant Ethic and the Spirit of Capitalism*. Translated by Talcott Parsons. London: George Allen & Unwin, 1930.

Weinberger, Edward D., and Peter F. Stadler. "Why *Some* Fitness Landscapes are Fractal." *Journal of Theoretical Biology* 163, no. 2 (21 July 1993): 255–75.

Welch, June Rayfield, and J. Larry Nance. *The Texas Courthouse*. Dallas: G.L.A. Press, 1971.

West, Stuart A., and Andy Gardner. "Altruism, Spite, and Greenbeards." *Science* 327 (12 March 2010): 1341–44.

West, Stuart A., A. S., Griffin, and A. Gardner. "Social Semantics: Altruism, Cooperation, Mutualism, Strong Reciprocity and Group Selection." *Journal of Evolutionary Biology* 20 (16 October 2007): 415–32.

White, T. H. *The Book of Merlyn*. Austin: University of Texas Press, 1977.

Wiggins, Steven N., and Gary D. Libecap. "Firm Heterogeneities and Cartelization Efforts in Domestic Crude Oil." *Journal of Law, Economics, and Organization* 3 (Spring 1987): 1–25.

Wikipedia. "CGG (company)." Accessed January 28, 1921.

Wilczek, Frank. "Freeman Dyson (1923–2020): Brilliant Polymath Who Reshaped Quantum Physics." *Science* 368, no. 6492 (15 May 2020): 715.

Wilgus, Henry LaFayette. "State Regulations Affecting Interstate Commerce." *Michigan Law Review* 8 (1910): 656–60.

Willbern, York Young. "Administrative Control of Petroleum Production in Texas." PhD diss., University of Texas at Austin, 1949.

———. "Administrative Control of Petroleum Production in Texas." In *Public Administration and Policy Formation: Studies in Oil, Gas, Banking, River Development, and Corporate Investigations*, edited by Emmette S. Redford, 3–50. Austin: University of Texas Press, 1956.

Williams, George C. *Natural Selection: Domains, Levels and Challenges*. Oxford, UK: Oxford University Press, 1992.

Williamson, Oliver E. "Calculativeness, Trust, and Economic Organization." *Journal of Law & Economics* 36 (April 1993): 453–86.

———. "Contested Exchange Versus the Governance of Contractual Relations." *Journal of Economic Perspectives* 7, no. 1 (Winter 1993): 103–8.

———. "Economic Organization: The Case for Candor." *Academy of Management Review* 21, no. 1 (1996): 48–57.

——— "The Economics of Organization: The Transaction Cost Approach." *American Journal of Sociology* 87 (November 1981): 548–77.

———. "The Fading Boundaries of the Firm: Comment." *Journal of Institutional and Theoretical Economics* 152 (March 1996): 85–88.

———. "Hierarchies, Markets and Power in the Economy: An Economic Perspective." *Industrial and Corporate Change* 4, no. 1 (1995): 21–49.

———. *The Mechanisms of Governance*. New York: Oxford University Press, 1996.

———. "Opportunism and Its Critics." *Managerial and Decision Economics* 14 (March–April 1993): 97–107.

———. "Transaction Cost Economics: The Governance of Contractual Relations." *Journal of Law and Economics* 22 (October 1979): 233–61.

———. "Visible and Invisible Governance." *American Economic Review* 84 (May 1994): 323–26.

Wilson, David Sloan. "A Theory of Group Selection." *Proceedings of the National Academy of Sciences* 72 (January 1975): 143–46.

———. "Structured Demes and the Evolution of Group-Advantageous Traits." *American Naturalist* 111 (January–February 1977): 157–85.

———. *The Natural Selection of Populations and Communities*. Menlo Park, CA: Benjamin Cummings, 1980.

———. "Species of Thought: A Comment on Evolutionary Epistemology." *Biology and Philosophy* 5 (January 1990): 37–62.

——— "Altruism and Organization: Disentangling the Themes of Multilevel Selection Theory." *American Naturalist*

150, no. S1 (July 1997): S122–S134.

Wilson, David Sloan, and Elliott Sober. "Reintroducing Group Selection to the Human Behavioral Sciences," with Open Peer Commentary. *Behavioral and Brain Sciences* 17 (December 1994): 585–654.

Wilson, David Sloan, and Edward O. Wilson. "Rethinking the Theoretical Foundation of Sociobiology." *Quarterly Review of Biology* 82, no. 4 (December 2007): 327–48.

Wilson, David Sloan, Steven C Hayes, Anthony Biglan, and Dennis D Embry. "Evolving the Future: Toward a Science of Intentional Change." *Behavioral and Brain Sciences* 37, no. 4 (2014): 395–416.

Wilson, Edward O. *Sociobiology: The New Synthesis.* Cambridge, MA: Belknap Press, 1975.

Wilson, James, Liying Yan, and Carl Wilson. "The Precursors of Governance in the Maine Lobster Fishery." *Proceedings of the National Academy of Sciences U.S.A.* 104, no. 39 (25 September 2007): 15212–17.

Wilson, Thomas Woodrow. *A History of the American People,* 5 vols. New York: Harper and Brothers, 1902.

Wilson, Wallace. *Bank Financing of Oil and Gas Production Payments.* Dallas: Southwest School of Banking, 1962.

Wimsatt, William C. "The Role of Generative Entrenchment and Robustness in the Evolution of Complexity." In *Complexity and the Arrow of Time,* edited by Charles H. Lineweaver, Paul C. Davies, and Michael Ruse, 308–31. New York: Cambridge University Press, 2013.

Wineburgh, H. Harold. *The Texas Banker: The Life and Times of Fred Farrel Florence.* Dallas: H. Harold Wineburgh, 1981.

Winter, Sidney G., Gino Cattani, and Alex Dorsch. "The Value of Moderate Obsession: Insights from a New Model of Organizational Search." *Organization Science* 18, no. 3 (May–June 2007): 403–19.

Wolfson, Albert. "Origin of the North American Bird Fauna: Critique and Reinterpretation from the Standpoint of Continental Drift." *American Midland Naturalist* 53, no. 2 (1955): 353–80.

Wolpert, David, and James W. Bono. "Predicting Behavior in Unstructured Bargaining with a Probability Distribution." *Journal of Artificial Intelligence Research* 46 (2013): 579–605.

Woolrich, W. R., et al. *Men of Ingenuity: From Beneath the Orange Tower, 1884–1964.* Austin: The College of Engineering of the University of Texas, 1969.

Wyatt-Brown, Bertram. *Southern Honor: Ethics and Behavior in the Old South.* New York: Oxford University Press, 1982.

Young, H. P., and M. A. Burke. "Competition and Custom in Economic Contracts: A Case Study of Illinois Agriculture." *American Economic Review* 93, no. 3 (1998): 559–73.

Zaman, Mohammad Qamar. "Efficient Rates of Oil and Gas Production through Regulation." MS thesis, University of Texas at Austin, 1953.

Zelden, Charles. "Regional Growth and the Federal District Courts: The Impact of Judge Joseph C. Hutcheson Jr. on Southeast Texas, 1918–1931." *Houston Review* 11, no. 2 (1989): 66–94.

Ziman, John M., ed. *Technological Innovation as an Evolutionary Process.* New York: Cambridge University Press, 2000.

Zimmermann, Erich W. *Conservation in the Production of Petroleum: A Study in Industrial Control.* New Haven, CT: Yale University Press, 1957.

Zuckerman, Gregory. "Breakthrough: The Accidental Discovery That Revolutionized American Energy." *The Atlantic,* November 6, 2013.

———. *The Frackers: The Outrageous Inside Story of the New Billionaire Wildcatters.* New York: Portfolio/Penguin, 2013.

# TECHNICAL ARTICLES, PAPERS, AND PUBLICATIONS

"Acidizing Treatment in Oil and Gas Operators." API Digital Media (DM2014-113 | 06.14), 2014, accessed February 15, 2021.

Alhakeem, Aamer Ali. "3D Seismic Data Interpretation of Boonsville Field, Texas." MS thesis, Missouri University of Science and Technology, 2013.

Allen, R. E. (Assistant Oil Umpire of California). "Theory and Practice of Directed Drilling." *Transactions of the American Institute of Mining Engineers* 107, no. 1 (December 1934). https://doi.org/10.2118/934034-G.

Archie, Gustave E. "Gus." "The Electrical Resistivity Log as an Aid in Determining Some Reservoir Characteristics." *Transactions of the American Institute of Mining Engineers* 146, no. 1 (December 1942): 54–62.

Aronofsky, J. S., and R. Jenkins (Magnolia Petroleum, Dallas). "A Simplified Analysis of Unsteady Radial Gas Flow." *Journal of Petroleum Technology* 6, no. 7 (July 1954): 23–28.

Aronofsky, J. S., and R. Jenkins (Magnolia Petroleum, Dallas). "Unsteady Flow of Gas Through Porous Media—One Dimensional Case." *Proceedings of the First U.S. National Congress of Applied Mechanics.* Ann Arbor, MI: J. Edwards, 1952.

Arps, J. J. "Profitability of Capital Expenditures for Development Drilling and Appraisal of Producing Properties." *AIME Journal of Petroleum Technology* 13 (July 1958).

Arps, J. J., and J. L. Arps. "The Subsurface Telemetry Problem—A Practical Solution." *Journal of Petroleum Technology* 16, no. 5 (May 1964): 487–93.

Beal, Carl H. "The Decline and Ultimate Production of Oil Wells." *U.S. Bureau of Mines Bulletin 177* (1919).

Beecher, C. E., and I. P. Parkhurst. "Effect of Dissolved Gas upon the Viscosity and Surface Tension of Crude Oil." *Petroleum Development and Technology in 1926, Transactions AIME*, 51–69. American Institute of Mining, Metallurgical, and Petroleum Engineers, 1926.

Bell, C. E., B. W. Holmes, and A. R. Rickards. "The Effective Diverting on Horizontal Wells in the Austin Chalk." Paper presented at Annual Technical Conference and Exhibition, Houston, October 3–6, 1993.

Bell, John S., and J. M. Shepherd. "Pressure Behavior in the Woodbine Sand." *Transactions of the American Institute of Mining Engineers* 192 (1951): 19. *Petroleum Transactions, AIME* 192 (1951); *Journal of Petroleum Technology* 3, no. 1 (January 1951): SPE-951019-G.

Botset, Holbrook G. "Flow of Gas-Liquid Mixtures through Consolidated Sand." *Transactions of the American Institute of Mining Engineers* 136, no. 1 (December 1940): 91–105.

Bridgman, Percy W. "Thermodynamic Properties of Liquid Water to 80° (C) and 12,000 KG N (Kg/Cm²)." *Proceedings of the American Academy of Arts and Sciences* 48, no. 9 (September 1912).

Briggs, Lyman J. *The Mechanics of Soil Moisture*. Washington, DC: Department of Agriculture, Division of Soils, Government Printing Office, 1897.

Briggs, Robert (?). *Proceedings, American Philosophical Society* 10 (1865): 136, cited in Everette Lee DeGolyer Sr., "Waste in Oil and Gas Production" (typescript), 1941, Mss 60, Box 25, Folder 2458, DeGolyer Papers, 6.

Bruce, G. H., D. W. Paceman, H. H. Rachford Jr., and J. D. Rice. "Calculations of Unsteady-State Gas Flow through Porous Media." *Journal of Petroleum Technology* 5, no. 3 (March 1953): 79–92.

Bruce, W. A. "An Electrical Device for Analyzing Oil-reservoir Behavior." *Transactions of the American Institute of Mining and Metallurgical Engineers* 151, no. 1 (December 1943): 112–24.

Buckley, Stuart E. "The Pressure-Production Relationship in the East Texas Field." *API Drilling and Production Practice* (1938), conference paper.

Buckley, Stuart E., and M. C. Leverett. "Mechanism of Fluid Displacement in Sands." *Transactions of the American Institute of Mining and Metallurgical Engineers: Petroleum Development and Technology* 146, no. 1 (December 1942): 107–16.

Buckley, Stuart E., and J. H. Lightfoot. "Effects of Pressure and Temperature on Condensation of Distillate from Natural Gas." *Transactions of the American Institute of Mining Engineers* 142, no. 1 (December 1941): 232–45.

Burns, Alan D., Thomas Frank, Ian Hamill, and Jun-Mei Shi. "The Favro Averaged Drag Model for Turbulent Dispersion in Eulerian Multi-Phase Flows." Paper presented at 5th International Conference on Multiphase Flow, Yokohama, Japan, May 30–June 4, 2004.

Carll, J. F. *The Geology of the Oil Regions of Warren, Venango, Clarion and Butler Counties*. Harrisburg, PA: Board of Commissioners for the Second Geological Survey, 1880.

Carpenter, C. B., and G. B. Spencer. "Measurement of Compressibility of Consolidated Oil-Bearing Sandstones." *Bureau of Mines RI 3540* (October 1940).

Cattell, R. A., and H. C. Fowler. "Fluid-Energy Relations in Production of Petroleum and Natural Gas." In *Minerals Yearbook, 1934*, edited by O. E. Kiessling, 707–21. Washington, DC: Government Printing Office, 1935.

Clark, J. B. "A Hydraulic Process for Increasing the Productivity of Wells." *Journal of Petroleum Technology* 1, no. 1 (January 1, 1949): 1–8. T.P. 2510.

Coleman, Stewart, H. D. Wilde Jr., and Thomas W. Moore. "Quantitative Effects of Gas-Oil Ratios on Decline of Average Rock Pressure." *Transactions of the American Institute of Mining and Metallurgical Engineers* 86, no. 1 (December 1930): 174–84.

Collom, R. E. "Prospecting and Testing for Oil and Gas." *U.S. Bureau of Mines Bulletin 201* (1922).

Craze, Rupert C., and Stuart E. Buckley. "A Factual Analysis of the Effect of Well Spacing on Oil Recovery." *API Drilling and Production Practice* (1945): 144.

Cutler, Willard W., Jr. *Estimation of Underground Oil Reserves by Oil-Well Production Curves*. U.S. Bureau of Mines Bulletin 228 (1924).

Darcy, Henry. *Les Fontaines Publiques de la Ville de Dijon*. Paris: Dalmont, 1856.

Day, David T. "The Petroleum Resources of the U.S." *U.S. Geological Survey Bulletin 394* (February 1909).

Deussen, Alexander, and Eugene C. Leonardon. "Electrical Exploration of Drill Holes." *Drilling and Production Practice 1935*, Central Committee on Drilling and Production Practice, Division of Production, American Petroleum Institute, 47–59. New York: API, 1936.

Donoghue, David. "Elasticity of Reservoir Rocks and Fluids, with Special Reference to East Texas Oil Field." *Bulletin of the American Association of Petroleum Geologists* 28, no. 7 (July 1944): 1032–56.

Dow, D. B., and L. P. Calkin. "Solubility and Effects of Natural Gas and Air in Crude Oils." *U.S. Bureau of Mines RI 2732* (February 1926).

Dow, D. B., and Carl E. Reistle Jr. "Absorption of Natural Gas and Air in Crude Petroleum." *Mining and Metallurgy* 5, no. 5 (July 1924): 336–37.

Dowell, Iain, and Andrew A. Mills. "Chapter 4—Measurement-While-Drilling [MWD], Logging-While-Drilling [LWD] and Geosteering." In *Petroleum Well Construction*, edited by Michael J. Economides, Larry T. Watters, and Shari Dunn-Norman, 155–211. New York: John Wiley and Sons, 1998.

"Drilling Mud Motor Components." URL: www.drillingmanual.com, accessed January 31, 2021.

Dumble, E. T., and G. D. Harris. "The Galveston Deep Well." *American Journal of Science* (1893): 46.

Dunlap, H. F., H. L. Bilhartz, Ellis Shuler, and C. R. Bailey. "The Relation Between Electrical Resistivity and Brine Saturation in Reservoir Rocks." *Journal of Petroleum Technology* 1, no. 10 (October 1949).

Engineering Laboratories, Inc. *Instruction Manual: The Maintenance and Operation of the Humble Type Subsurface Sampler and the Technique of Subsurface Sample Evaluation.* Tulsa, OK: Engineering Laboratories, n.d.

English, Walter A., Willard H. Tracy, Arthur Normann, Frank Ittner, and P. C. Kelly. "Seismograph Prospecting for Oil." *Petroleum Technology* 2, no. 2 (May 1939).

Evinger, H. H., and Morris Muskat. "Calculation of Productivity Factors for Oil-Gas-Water Systems in the Steady State." *Transactions of the American Institute of Mining and Metallurgical Engineers: Petroleum Development and Technology* 146, no. 1 (December 1942): 194–203.

———. "Calculation of Theoretical Productivity Factor." *Transactions of the American Institute of Mining and Metallurgical Engineers: Petroleum Development and Technology* 146 (December 1942): 126–27.

Fan, Li, Billy W. Harris, A. (Jamal) Jamaluddin, Jairam Kamath, Robert Mott, Gary A. Pope, Alexander Shandrygin, Curtis Hays Whitson. "Understanding Gas-Condensate Reservoirs." *Oil Field Review* 17, no. 4 (Winter 2005–2006): 14, no. 3 (published by Schlumberger Technology Corporation).

Fancher, G. H., J. A. Lewis, and K. B. Barnes. "Some Physical Characteristics of Oil Sands." In *Proceedings of the Third Pennsylvania Mineral Industries Conference, Petroleum and Natural Gas Section held at the Pennsylvania State College, May 5–6, 1933. The Pennsylvania State College Bulletin. Mineral Industries Experiment Section, Bulletin* 12 (1933): 65–167.

Fettke, C. R. "Core Studies of the Second Sand of the Venango Group from Oil City, Pennsylvania." *Transactions of the American Institute of Mining and Metallurgical Engineers: Petroleum Development and Technology* G26, no. 1 (December 1926): 219.

Foran, E. V. "Development and Production Problems in High-pressure Distillate Pools." *Transactions of the American Institute of Mining Engineers* 132 (December 1939): 22–32.

———. "Interpretation of Bottom-Hole Pressure in East Texas Oil Field." *American Association of Petroleum Geologists Bulletin* 16 (1932): 907.

Gardner, Robert A., Jr. "The Boonsville (Bend Conglomerate Gas) Field, Wise County, Texas." *Abilene Geological Society: Geological Contributions 1960* (1960): 7–18.

Garrison, Allen D. "Selective Wetting of Reservoir Rocks and its Relation to Oil Production." *API Drilling and Production Practice* (January 1, 1935): 330–41.

Gaydos, J. S., and P. C. Harris. "Foam Fracturing: Theories, Procedures and Results." Paper presented at the SPE Unconventional Gas Recovery Symposium, Pittsburgh, PA, May 1980; Paper Number: SPE-8961-1-MS, published May 18, 1980.

Gearhart, M., L. M. Moseley, and M. Foste. "Current State of the Art of MWD and Its Application in Exploration and Development Drilling." Paper presented at the International Meeting on Petroleum Engineering, Beijing, China, March 1986. Paper Number: SPE-14071-MS, published March 17, 1986.

Gearhart, Marvin, Kelly A. Ziemer, and Orien Knight. "Mud Pulse MWD Systems Report." *Journal of Petroleum Technology* 3, no. 12 (December 1981): 2301–6.

Geffen, T. M., D. R. Parrish, G. W. Haynes, and R. A. Morse. "Efficiency of Gas Displacement from Porous Media by Liquid Flooding." *Journal of Petroleum Technology* 4, no. 2 (February 1952).

Goldston, W. L., Jr., and George D. Stevens. "Esperson Dome, Liberty County, Texas." *Bulletin, AAPG* (1934): 857–58.

Gravesen, Jens. "The Geometry of the Moineau Pump." Technical University of Denmark, Department of Mathematics, March 26, 2008. Accessed February 4, 2021.

Green, W. Heber, and G. A. Ampt. "Studies on Soil Physics." *Journal of Agricultural Science* 4 Part I (May 1911): 1–24.

Griswold, E. H. "Recoverable Oil and Gas Content of Land as Suitable Standard of Each Owner's Rights in Pool." *Transactions of the American Institute of Mining Engineers* 103, no. 1 (December 1933): 33–44.

Harris, G. D. "Preliminary Report on the Organic Remains Obtained from the Deep Well at Galveston, Together with Conclusions Respecting the Age of the Various Formations Penetrated." *Texas Geological Survey Annual Report No. 4* (1893).

Hassler, G. L., R. R. Rice, and E. H. Leeman. "Investigations on the Recovery of Oil from Sandstones by Gas Drive." *Transactions of the American Institute of Mining and Metallurgical Engineers: Petroleum Development and Technology* 118, no. 1 (December 1936): 116–37.

Hawkins, Murray F. "Phase Diagrams." In *Petroleum Production Handbook*, edited by Thomas C. Frick and R. William Taylor, chapter 20, 1–15. 2 vols. New York: McGraw-Hill, 1962.

Hawthorn, D. G. "Subsurface Pressures in Oil Wells and Their Field of Application." *Transactions of the American Institute of Mining Engineers* 103, no. 1 (December 1933): 148–69.

Hayward, J. T. "Probabilities and Wildcats." *Drilling and Production Practice* (1 January 1934). Conference Paper.

Helms, Lynn. "Horizontal Drilling." *DMR Newsletter* 35, no. 1 (n.d.): 1–3.

Herold, Stanley Carrollton. *Analytical Principles of the Production of Oil, Gas, and Water from Wells: A Treatise Based upon a System of Fluid Mechanics Particularly Adapted to the Study of the Performance of Natural Reservoirs*. Stanford, CA: Stanford University Press, 1928.

———. *Analytic Principles of the Production of Oil, Gas, and Water from Wells: Part II, Reservoirs in Hydraulic Control*. Stanford, CA: Stanford University Press, 1928.

Hiltz, R. G., J. V. Huzarevich, and Robert M. Leibrock. "Performance Characteristics of the Slaughter Field Reservoir." In *Proceedings for Secondary Oil Recovery Conference: A Symposium on Carbonate Reservoirs*, 146–53. Austin: Texas Petroleum Research Committee, 1951.

Hubbert, M. King, and D. G. Willis. "Mechanics of Hydraulic Fracturing." *Petroleum Transactions, American Institute of Mining Engineers* 210, no. 1 (January 1957): 153–68. T. P. 4597.

Hughes, James D. "Value of Oil-Well Surveying and Applications of Controlled Directional Drilling." *Drilling and Production Practice 1935*, Central Committee on Drilling and Production Practice, Division of Production, American Petroleum Institute, 60–106. New York: API, 1936.

Huntley, L. G. "Possible Causes of the Decline of Oil Wells and Suggested Methods of Prolonging Yield." *U.S. Bureau of Mines Technical Paper 51* (1913).

Hurst, William. "Unsteady Flow of Fluids in Oil Reservoirs." *Physics* 5, no. 1 (January 1934): 20–30.

———. "Water Influx into a Reservoir and Its Application to the Equation of Volumetric Balance." *Transactions of the American Institute of Mining Engineers* 151, no. 1 (December 1943): SPE-943057-G.

Hurst, William, and A. F. van Everdingen. "The Application of the La Place Transformation to Flow Problems in Reservoirs." *Transactions of the American Institute of Mining and Metallurgical Engineers: Petroleum Development and Technology* 186 (1949): 305.

Jamin, Jules Célestin. "Leçons sur les Lois de l'Équilibre et du Mouvement des Liquides dans les Corps Poreux." Professées les Février et Mars, 1861, Société Chimique de Paris.

———. "Mémoire sur l'Équilibre et le Mouvement des Liquides dans les Corps Poreux." *Comptes Rendus* 50 (January–June 1860), 3 parts, at 172, 311, 385.

Jenkins, R., and J. S. Aronofsky. "Unsteady Radial Flow of Gas Through Porous Media." *Journal of Applied Mechanics* 20, no. 2 (June 1953).

Jennings, Alfred R., Jr. "Fracturing Fluids—Then and Now." *Journal of Petroleum Technology* 48, no. 7 (1 July 1996): 604–10.

Johnson, N., and N. van Wingen. "Reservoir Fluid Flow Research." *Petroleum Engineering* 16 (July 1945): 108.

Johnston, Norris, and Carrol M. Beeson. "Water Permeability of Reservoir Sands." *American Institute of Mining Engineers, Petroleum Transactions* 160 (1945): 43–55.

Joshi, S. D. "Cost/Benefits of Horizontal Wells." Paper presented at the SPE Regional/AAPG Pacific Section Joint Meeting, Long Beach, CA, May 19–24, 2003.

Jousset, Philippe. "Illuminating Earth's Faults." *Science* 366, no. 6469 (29 November 2019): 1076–77.

Judson, Sidney A., H. D. Easton Jr., and W. A. Schaeffer Jr. "Estimation of Petroleum Reserves in Prorated Fields." *Transactions of the American Institute of Mining Engineers* 114, no. 1 (December 1935): 11–24.

Katz, D. L. "A Method of Estimating Oil and Gas Reserves." *Transactions of the American Institute of Mining and Metallurgical Engineers: Petroleum Development and Technology* 118, no. 1 (December 1936): 18–32.

Katz, Donald L. "Overview of Phase Behavior in Oil and Gas Production." *Journal of Petroleum Technology* 35, no. 6 (June 1983): 1205.

Katz, D. L., and W. H. Barlow. "Relation of Bottom-Hole Pressure to Production Control." *Drilling and Production Practice 1935*, Central Committee on Drilling and Production Practice, Division of Production, American Petroleum Institute, 116–19. New York: API, 1936.

Katz, Donald L., and Frederick Kurata. "Retrograde Condensation." *Industrial and Engineering Chemistry* 32, no. 6 (1 June 1940): 817.

Kelly, R. B. "The Potential or Productivity Factor in Allocation Formulas." *Drilling and Production Practice* (1 January 1934). Conference Paper.

King, F. H. "Conditions and Movements of Underground Waters." In *Nineteenth Annual Report of the United States: Geological Survey 19th Annual Report, 1897–1898, Part II—Papers Chiefly of a Theoretic Nature* 59–294. Washington, DC: Government Printing Office, 1899.

Kuenen, J. P. "On Retrograde Condensation and the Critical Phenomena of Mixtures of Two Substances."

Communication, Physics Laboratory, University of Leiden, The Netherlands (1892) no. 4B, 7–14.

Leverett, M. C. "Capillary Behavior in Porous Solids." *Transactions of the American Institute of Mining and Metallurgical Engineers* 142 (December 1941): 152–74.

———. "Flow of Oil-Water Mixtures through Unconsolidated Sands." *Transactions of the American Institute of Mining and Metallurgical Engineers* 132, no. 1 (December 1939): 149–71.

Leverett, M. C., and W. B. Lewis. "Steady Flow of Gas-Oil-Water Mixtures through Unconsolidated Sands." *Transactions of the American Institute of Mining and Metallurgical Engineers* 142, no. 1 (December 1941): 113–14.

Lewis, James A., William L. Horner, and Marion Stekoll. "Productivity Index and Measurable Reservoir Characteristics." American Institute of Mining and Metallurgical Engineers, Technical Publication No. 1467 (Class G. Petroleum Division, No. 160), 1942.

Lewis, J. O. "Methods for Increasing the Recovery from Oil Sands." *Bureau of Mines Bulletin 148* (1917).

———. "Rock Pressure." *Petroleum Engineer* (May 1934).

Lewis, J. O., and Carl H. Beal. "Some New Methods for Estimating the Future Production of Oil Wells." *Transactions of the American Institute of Mining and Metallurgical Engineers* 59 (1918): 492.

Liapidevskii, V. Yu, and V. Tikhonov. "Lagrangian Approach to Modeling Unsteady Gas-Liquid Flow in a Well." *Journal of Physics: Conference Series* 722 (2016): 012026.

Lindsey, Nathaniel J., T. Craig Dawe, and Jonathan B. Ajo-Franklin. "Illuminating Seafloor Faults and Ocean Dynamics with Dark Fiber Distributed Acoustic Sensing." *Science* 366, no. 6469 (29 November 2019): 1103–7.

Lindsly, Ben E. "Effect of Gas Withdrawal upon Reservoir Fluids." *Transactions of the American Institute of Mining and Metallurgical Engineers: Petroleum Development and Technology* 107, no. 1 (December 1934): 94–97.

———. "Solubility and Liberation of Gas from Natural Oil-Gas Solutions." *U.S. Bureau of Mines Technical Paper 554* (1933).

———. "A Study of Bottom-Hole Samples of East Texas Crude Oil." *United States Bureau of Mines RI* 3212 (May 1933). Paper presented to the American Petroleum Institute, Division of Production, Tulsa, Oklahoma, May 19, 1933.

Lindtrop, N. T., and V. M. Nicolaef. "Oil and Water Content of Oil Sands, Grozny, Russia." *American Association of Petroleum Geologists Bulletin* 13, no. 7 (July 1929): 811–22.

Mathews, C. S., E. Brons, and P. Hazebroek. "A Method for Determination of Average Pressure in a Bounded Reservoir." *Petroleum Transaction, American Institute of Mining Engineers* 201, no. 1 (December 1954): 182–91.

May, Cecil J. "Efficiency of Flowing Wells." *Transactions of the American Institute of Mining Engineers* 114, no. 1 (December 1935): 99–115.

Melcher, A. F. "Determination of Pore Space of Oil and Gas Sands." *Transactions of the American Institute of Mining and Metallurgical Engineers* 65 (1921): 469.

Michael, Andreas. "Orientation of Hydraulic Fracture Initiation from Perforated Horizontal Wellbores." Paper presented at the SPE Annual Technical Conference and Exhibition, Calgary, Alberta, Canada, September 2019.

Miller, F. G. *Multiphase-Flow Theory and the Problem of Spacing Wells, U.S. Bureau of Mines Bulletin 529* (1954).

Miller, Harold Carl. *Function of Natural Gas in the Production of Oil.* New York: American Petroleum Institute, 1929.

Miller, Harold Carl, and R. V. Higgins. "Review of Cutler's Rule of Well Spacing." *U.S. Bureau of Mines RI 3479* (1939).

Millikan, C. V. "Reservoir and Bottom-Hole Producing Pressures as a Basis for Proration." *Transactions of the American Institute of Mining Engineers* 103 (December 1933): 137–41.

———. "Geological Application of Bottom-Hole Pressures." *American Association of Petroleum Geologists Bulletin* 16, no. 9 (September 1, 1932): 891–906.

Millikan, C. V., and C. V. Sidwell. "Bottom-Hole Pressure in Oil Wells." *Transactions of the American Institute of Mining and Metallurgical Engineers: Petroleum Development and Technology* 92, no. 1 (December 1931): 194–205.

Mills, R. Van A. "Experimental Studies of Subsurface Relationships of Oil and Gas Fields." *Economic Geology* 15, no. 5 (1920): 398.

———. "Relations of Texture and Bedding to the Movements of Oil and Water Through Sand." *Economic Geology* 16, no. 2 (1 March 1921): 124–41.

Monaghan, J. J., and A. Kocharyan. "SPH Simulation of Multi-phase Flow." *Computer Physics Communications* 87, no. 1–2 (2 May 1995): 225–35.

Moore, T. V. "Determination of Potential Production of Wells Without Open-Flow Test." *Proceedings of the American Petroleum Institute, Production Bulletin* 206 (1930): 27–29.

——— "The Effect of Curtailment on Ultimate Recovery." *Drilling and Production Practice* (1 January 1934). Conference Paper.

Moore, T. V., and Ralph J. Schilthuis. "Calculation of Pressure Drops in Flowing Wells." *Transactions of the American*

*Institute of Mining Engineers* 103, no. 1 (December 1933): 170–90.

Moore, T. V., Ralph J. Schilthius, and William Hurst. "The Determination of Permeability from Field Data." *Proceedings of the American Petroleum Institute, Production Bulletin 211* (1933); also published in *Oil Weekly*, May 21, 1933.

Murray, William F., and James O. Lewis. "Underground Wastes in Oil and Gas Fields and Methods of Prevention." *U.S. Bureau of Mines Technical Paper 130* (1916).

Muskat, Morris. "The Flow of Compressible Fluids Through Porous Media and Some Problems in Heat Conduction." *Physics* 5, no. 1 (January 1934): 71.

———. *The Flow of Homogeneous Fluids Through Porous Media*. New York: McGraw-Hill, 1937 (second printing, Ann Arbor, MI: J. W. Edwards, 1946).

———. *Physical Principles of Oil Production*. New York: McGraw-Hill, 1949.

———. "Potential Distributions in Large Cylindrical Disks with Partially Penetrating Electrodes." *Physics* 2, no. 5 (May 1932): 329.

———. "Principles of Well Spacing." *Transactions of the American Institute of Mining Engineers* 136, no. 1 (December 1940): 37–56.

Muskat, Morris, and H. G. Botset. "Flow of Gases Through Porous Media." *Physics* 1, no. 1 (July 1931): 27–47.

Muskat, Morris, and M. W. Meres. "The Flow of Heterogeneous Fluids Through Porous Media." *Journal of Applied Physics* 7, no. 9 (September 1936): 346–63.

Muskat, Morris, and R. D. Wyckoff. "An Approximate Theory of Water-Coning in Oil Production." *Transactions of the American Institute of Mining Engineers* 114, no. 1 (December 1935): 144–63.

Muskat, Morris, R. D. Wyckoff, H. G. Botset, and M.W. Meres. "Flow of Gas-Liquid Mixtures Through Sands." *Transactions of the American Institute of Mining and Metallurgical Engineers* 123 (1937): 69–96.

Nye, G. L., and Carl E. Reistle Jr. "Recent Changes in Reservoir Pressure Conditions in the East Texas Field." *Transactions of the American Institute of Mining Engineers* 107, no. 1 (December 1934): 77–83.

Old, R. E., Jr. "Analysis of Reservoir Performance." *Transactions of the American Institute of Mining and Metallurgical Engineers* 151, no. 1 (December 1943): 86–98.

Oliver, Earl. "Stabilizing Influences for the Petroleum Industry." *Transactions of the American Institute of Mining and Metallurgical Engineers, Petroleum Division* (1932).

Osiptsov, Andrei A. "Multiphase Flow Models for Hydraulic Fracturing Technology." *Journal of Physics: Conference Series* 894 (2017): 012068.

PERM Inc., TIPM Laboratory. "Fundamentals of Fluid Flow in Porous Media: Chapter 2. Porosity: Laboratory Porosity Measurement." Downloaded April 24, 2020.

Pierce, H. R., and E. L. Rawlins. "The Study of a Fundamental Basis for Controlling and Gauging Natural Gas Wells: Part 1. Computing the Pressure at the Sand in a Gas Well." *Bureau of Mines RI 2929* (1929).

———. "The Study of a Fundamental Basis for Controlling and Gauging Natural-Gas Wells: Part 2. A Fundamental Relation for Gauging Gas-Well Capacities." *U.S. Bureau of Mines Report of Investigations 2930* (1929).

Pyle, Howard C., and John E. Sherborne. "Core Analysis." *Transactions of the American Institute of Mining and Metallurgical Engineers: Petroleum Development and Technology* 132, no. 1 (December 1939): 33–61.

Pym, L. A. "The Measurement of Gas-Oil Ratios and Saturation Pressures and Their Interpretation." *Proceedings of the World Petroleum Congress* 1 (1934): 452.

Reid, L. S., and R. L. Huntington. "Flow of Oil-gas Mixtures Through Unconsolidated Sands." *Transactions of the American Institute of Mining and Metallurgical Engineers* 127, no. 1 (December 1938): 232–33.

Reistle, Carl E., Jr. "East Texas Production." *API Drilling and Production Practice* (1 January 1934). Conference paper.

Reistle, C. E., Jr., and E. P. Hayes. "A Study of Subsurface Pressures and Temperatures in Flowing Wells in the East Texas Field and the Application of These Data to Reservoir and Vertical Flow Problems." *United States Bureau of Mines RI* 3211 (May 1933).

Roberts, Andrew, Robert Newton, and Frederick Stone. "MWD Field Use and Results in the Gulf of Mexico." Paper presented at the SPE Annual Technical Conference and Exhibition, New Orleans, Louisiana, September 1982; SPE-11226-MS, published September 26, 1982.

Rumble, R. C., H. H. Spain, and H. E. Stamm III. "A Reservoir Analyzer Study of the Woodbine Basin." T.P. 3219, *Petroleum Transactions, AIME* 192 (1951): 331–40. Alternative citation: *Journal of Petroleum Technology* 3, no. 12 (December 1951): SPE-122-G. George W. Barineau, "Supplement, 1962: The East Texas Field."

Russell, W. L. "A Quick Method for Determining Porosity." *American Association of Petroleum Geologists Bulletin* 10 (October 1926): 931–38.

Sage, Bruce H., and W. N. Lacey. "Energy Relations in a Flowing Well." In American Petroleum Institute, *Drilling and Production Practice 1935*, 107–15. New York: API, 1936.

———. "Phase Equilibria in Hydrocarbon Systems." *Industrial and Engineering Chemistry* 31, no. 12 (1 December 1939): 1497–1509.

————. *Volumetric and Phase Behavior of Hydrocarbons.* Stanford, CA: Stanford University Press, 1939.

Sage, Bruce H., William N. Lacey, and Jan G. Schaafsma. "Phase Equilibria in Hydrocarbon Systems II. Methane-Propane System." *Industrial and Engineering Chemistry* 26, no. 2 (1 February 1934): 214–17.

Sage, Bruce H., Jan G. Schaafsma, and William N. Lacey. *American Petroleum Institute Bulletin 212* (1933).

Sage, Bruce H., Sherborne Davies, and W. N. Lacey. "Phase Equilibria in Hydro-Carbon Systems XVII. Ethane-Crystal Oil Systems." *Industrial and Engineering Chemistry* 28, no. 11 (1 November1936): 1328–33.

Schilthuis, R. J. "Active Oil and Reservoir Energy." *Transactions of the American Institute of Mining and Metallurgical Engineers: Petroleum Development and Technology* 118, no. 1 (December 1936): 33–52.

————. "Connate Water in Oil and Gas Sands." *Transactions of the American Institute of Mining and Metallurgical Engineers: Petroleum Development and Technology* 127, no. 1 (December 1938): 199–214.

————. "Technique for Securing and Examining Sub-Surface Sample of Oil and Gas." In American Petroleum Institute, *Drilling and Production Practice 1935*, 120–29. New York: API, 1936.

Schilthuis, R. J., and William Hurst. "Variations in Reservoir Pressure in the East Texas Field." *Transactions of the American Institute of Mining and Metallurgical Engineers: Petroleum Development and Technology* 114 (1935): 164–76.

Sclater, Kenneth C., and B. R. Stephenson. "Measurements of Original Pressure, Temperature, and Gas-Oil Ratio in Oil Sands." *Transactions of the American Institute of Mining and Metallurgical Engineers* 82 (1928–1929): 119–36.

Sewell, Ben W. "The Carter Pressure Core Barrel." *Drilling and Production Practice 1940*, Central Committee on Drilling and Production Practice, Division of Production, American Petroleum Institute, 69–78. New York: API, 1940.

Sherrod, Gerald E. "What Makes Those Bank Engineers So Conservative?" *Journal of the Society of Petroleum Evaluation Engineers* 1, no. 1 (January 1968): 18–22.

Slichter, C. S. "Theoretical Investigation of the Motion of Ground Waters." In *Nineteenth Annual Report of the United States: Papers Chiefly of a Theoretic Nature*: 295–384. Washington, DC: Government Printing Office, 1899.

Snow, Dale R. "Water Encroachment in Bartlesville Sand Pools of Northeastern Oklahoma and Its Bearing on East Texas Recovery Problem." *American Association of Petroleum Geologists Bulletin* 16, no. 9 (1 September 1932): 881–90.

Suman, John R. "The Well-Spacing Problem—Low Well Density Increases Ultimate Recovery." *API Drilling and Production Practice* 158 (1 January 1934). Conference Paper.

Swigart, Theodore E. *Underground Problems in the Comanche Oil and Gas Field, Stephens County, Oklahoma.* Washington, DC: Department of the Interior, Bureau of Mines, in cooperation with the State of Oklahoma, 1911.

Swigart, Theodore E., and C. R. Bopp. "Experiments in the Use of Backpressure on Oil Wells." *U.S. Bureau of Mines TP 322* (1924).

Taliaferro, D. B., and R. E. Heithecke. "Bureau of Mines-American Petroleum Institute Pressure Core Barrel." *Drilling and Production Practice 1940*, Central Committee on Drilling and Production Practice, Division of Production, American Petroleum Institute, 53–68. New York: API, 1940.

Tanakov, M. Y., and M. Kelkar. "Integrated Reservoir Description for Boonsville, Texas Field Using 3D Seismic Well and Production Data." Paper presented at the SPE/DOE Improved Oil Recovery Symposium, Tulsa, Oklahoma, April 2000; published April 3, 2000: SPE-59349-MS.

Taylor, Hugh S. *The ABC's of Science in Oil Recovery.* New York: American Petroleum Institute, 1927.

Texas Water Development Board. "Texas Water Use Estimates, 2018 Summary, June 15, 2020."

Thorogood, John. "Drilling in 2032—Back to the Future." *Journal of Petroleum Technology* 59, no. 10 (October 2007): 108–10.

Umpleby, Joseph Bertram. "A.I.M.E. Stabilization Forum, Article 8; Reservoir Energy: Source, Ownership, Use." *Oil & Gas Journal*, July 20, 1933.

————. "Changing Concepts in the Petroleum Industry." *Transactions of the American Institute of Mining and Metallurgical Engineers: Petroleum Division*, 98 (December 1932): 38–49.

————. "Production Engineering in 1927." *Petroleum Development and Technology in 1926*. In *Transactions of the American Institute of Mining and Metallurgical Engineers: Petroleum Division*, 12–13, 193–94.

————. "Reservoir Energy: Its Source, Ownership and Utilization in the Production of Petroleum." *Transactions of the American Institute of Mining and Metallurgical Engineers* 103, no. 1 (December 1933): 22–32.

Uren, Lester C. "The 'Gas Factor' as a Measure of Oil-Production Efficiency." *Transactions of the American Institute of Mining and Metallurgical Engineers* 77, no. 1 (December 1927): 146–57.

Uren, Lester C., and E. J. Bradshaw. "Experimental Study of Pressure Conditions within the Oil Reservoir Rock in the Vicinity of a High-Pressure Producing Well." *Transactions of the American Institute of Mining Engineers: Petroleum Development and Technology, 1932*: 438–60.

Uren, Lester C., and E. H. Fahmy. "Factors Influencing the Recovery of Petroleum from Unconsolidated Sands by Water-flooding." *Transactions of the American Institute of Mining and Metallurgical Engineers* 77, no. 1 (December 1927): 318.

US Energy Information Administration, Department of Energy. "Review of Emerging Resources: U.S. Shale Gas and Shale Oil Plays," July 2011.

U.S. Treasury Department, Internal Revenue Service. *Manual for the Oil and Gas Industry Under the Revenue Act of 1918.* Washington, DC: US Internal Revenue Service, 1919.

van Everdingen, A. F. and William Hurst. "The Application of the La Place Transformation to Flow Problems in Reservoirs." *Journal of Petroleum Technology* 1, no. 12 (December 1949): 305–24.

Versluys, Jan. "Applied Geology: The Equation of Flow of Oil and Gas to a Well After Dynamic Equilibrium Has Been Established." *Akademie van Weenschappen, Amsterdam: Proceedings* 33-3 nos. 6–10 (1930): 578–86.

———. "Can Absence of Edge-Water Encroachment in Certain Oil Fields Be Ascribed to Capillarity?" *American Association of Petroleum Geologists Bulletin* 15 (1931): 197.

——— "De Capillaire in den Bodem" (Capillarity in Soils). Thesis (in Dutch), Amsterdam, 1916.

———. "Factors Involved in Segregation of Oil and Gas from Subterranean Water." *American Association of Petroleum Geologists Bulletin* 16 (1932): 924.

———. "Hydraulics in Flowing Wells: Mathematical Development of the Theory of Flowing Oil Wells." Technical Publication 213 (1929) and *Transactions of the American Institute of Mining Engineers, Petroleum Division* (1930): 192.

———. "Die Kapillarität im Boden." *Internationale Mitteilungen für Bodenkunde* (1917): 117–40.

———. "The Origin of Artesian Pressure." *Proceedings of the Royal Netherlands Academy of Arts and Sciences* 33, no. 3 (1930): 217.

Wanniarachichi, W. A. M., P. G. Ranjith, M. S. A. Perera, A. Lashin, N. Al Arifi, and J. C. Li. "Current Opinions on Foam-Based Hydro-Fracturing in Deep Geological Reservoirs." *Geomechanics and Geophysics for Geo-Energy and Geo-Resources* 1 (12 November 2015): 121–34.

Weatherby, B. B. "The History and Development of Seismic Prospecting." *Geophysical Case Histories*, vol. I, edited by L. L. Nettleton, 13–17. n.p.: Society of Exploration Geophysicists, 1949.

Weaver, D. K. "Practical Aspects of Directional Drilling." *API Drilling and Production Practice* 1 (1 January 1946).

Weijers, Leen, Chris Wright, Mike Mayerhofer, Mark Pearson, Larry Griffin, and Paul Weddle. "Trends in the North American Frac Industry: Invention through the Shale Revolution." Paper presented at SPE Hydraulic Fracturing Technology Conference and Exhibition, February 5–7, 2019, The Woodlands, Texas.

Welge, Henry J. "A Simplified Method for Computing Oil Recovery by Gas or Water Drive." *Journal of Petroleum Technology* 4, no. 14 (April 1952): 91–98.

Wendlandt, Edward Allen, Thomas H. Shelby, and John Smith Bell. "Hawkins Field, Wood County, Texas." *Bulletin of the American Association of Petroleum Geologists* 30, no. 11 (November 1946): 1830–56.

Wilde, H. D., Jr. "Why Measure Bottom Hole Pressures?" Paper presented at API Midyear Meeting, Tulsa, Oklahoma, June 3, 1932.

Wilde, H. D., Jr., and F. H. Lahee. "Simple Principles of Efficient Oil Field Development." *Bulletin of the American Association of Petroleum Geologists* 17, no. 8 (August 1933): 981–1002.

Wilde, H. D., Jr., and T. V. Moore. "Hydrodynamics of Reservoir Drainage and Its Relation to Well Spacing." *Proceedings of the American Petroleum Institute, Production Bulletin 210* (1932): 83–90.

Woods, Rex W., and Morris Muskat. "An Analysis of Material-Balance Calculations." *Transactions of the Institute of Mining Engineers* 160, no. 1 (December 1945): 124–39.

Wyckoff, R. D. "The Relation of Well Potentials, Sand Permeability, and Well Pressures to Allocation of Production." *Drilling and Production Practice* 1 (1 January 1934).

Wyckoff, Ralph D. "Discussion." Ralph J. Schilthuis and William Hurst, "Variations in Reservoir Pressure in the East Texas Field." *Transactions of the American Institute of Mining and Metallurgical Engineers: Petroleum Development and Technology* 114, no. 1 (December 1935): 173–74.

Wyckoff, R. D., and H. G. Botset. "The Flow of Gas-Liquid Mixtures Through Unconsolidated Sands." *Physics* 7, no. 9 (September 1936): 325–45.

Wyckoff, R. D., H. G. Botset, Morris Muskat, and D. W. Reed. "Measurement of Permeability of Porous Media." *Bulletin of the American Association of Petroleum Geologists* 18, no. 2 (February 1934): 161–90.

## CASES CITED

*Aluminum Company of America v. Railroad Commission of Texas*, 368 S.W.2d 818 (Court of Civil Appeals of Texas, 1963).

*Amazon Petroleum Corporation, et al. v. Railroad Commission of Texas, et al.*, 5 F. Supp. 633 (Federal District Court for the Eastern District of Texas, Tyler, 1934).

*Amazon Petroleum Corporation, et al. v. Railroad Commission of Texas, et al.*, 71 F.2d 1 (Federal Fifth Circuit Court of Appeals, 1934).

*Amerada Petroleum Corp. v. Railroad Commission*, 395 S.W.2d 403 (Court of Civil Appeals of Texas, 1965).

*Amoco Production Co. v. Alexander*, 622 S.W.2d 563 (Supreme Court of Texas, 1981).

*Atlantic Oil Production Company v. Railroad Commission*, 85 S.W. 2d 655 (Court of Civil Appeals of Texas).

*Atlantic Refining Company et al. v. Bright & Schiff*, 321 S.W. 2d 167 (Court of Civil Appeals of Texas).

*Atlantic Refining Company v. Railroad Commission of Texas*, 330 S.W. 2d 494 (1960).

*Atlantic Refining Company et al., Appellants v. Railroad Commission of Texas et al., Appellees*, 346 S.W. 2d 801 (Supreme Court of Texas, No. A-7355, 1961). AKA "Normanna."

*Bandini Petroleum Company v. Superior Court of California ex rel. Los Angeles County*, 284 U.S. 8 (U.S. Supreme Court, 1931).

*Barnard v. Monongahela Gas Company*, 216 Pa. 362, Atl. 801 (Supreme Court of Pennsylvania,1907).

*Alphonza E. Bell Corporation and Union Oil of California v. Bell View Oil Syndicate, et al.*, 76 2d Pacific Reporter 166, 24 Cal. App. 2d 747 (District Court of Appeal, Third District, California, 1938).

*Bender v. Brooks*, 103 Tex. 329 (Supreme Court of Texas, 1910).

*Boxrollium v. Smith*, 4 F. Supp. 624 (Federal District Court, Southern District of Texas, 1933).

*Briggs v. Southwestern Energy Production Company*, 184 A. 3d 153 (Pennsylvania Superior Court, 2018).

*Broussard v. Texaco, Inc.*, 479 S.W.2d 270 (Supreme Court of Texas, 1972).

*Brown v. Vandergrift, et al.*, 80 Penn. St. Reports 142 (Pennsylvania Supreme Court, 1875).

*Brown, et al. v. Humble Oil and Refining Co.*, 68 S.W.2d 622 (Court of Civil Appeals of Texas, 1934).

*C. H. Brown, et al. v. Humble Oil and Refining Co.*, 126 Tex. 296 and 314; 83 S.W. (2d) 935 (Supreme Court of Texas, 1935).

*C. H. Brown et al. v. Humble Oil & Refining Co.*, 87 S.W 2d 1069 (Supreme Court of Texas, 1935); motion for rehearing denied.

*Buford, et al. v. Sun Oil Co. et al., Sun Oil Co. et al. v. Buford, et al.*, 319 U.S. 315, 63 S. Ct. 1098 (U.S. Supreme Court, 1943).

*Cambria Natural Gas v. De Witt*, U.S. 284 (U.S. Supreme Court, 1899).

*Canadian River Gas Co. v. Terrell*, 4 F. Supp. 222 (Federal District Court, Western District of Texas, 1933).

*Champlin Refining Co. v. Corporation Commission of Oklahoma, et al.*, No. 1156 in Equity (U.S. District Court for the Western District of Oklahoma, R75-B, 1931).

*Champlin Refining Co. v. Corporation Commission of Oklahoma, et al.*, 286 U.S. 210 (U.S. Supreme Court, 1932).

*Chevron U.S.A., Inc. v. Natural Resources Defense Council*, 467 U.S. 837 (U.S. Supreme Court, 1984).

*Chicago, Burlington & Quincy Railway Co. v. Illinois*, 200 U.S. 561 (U.S. Supreme Court, 1906).

*Clymore Production Company v. Thompson*, 11 F. Supp. 791 (Federal District Court, Western District of Texas, 1935 [prelim. inj.]); 13 F. Supp. 469 (Federal District Court, Western District of Texas, 1936 [final hearing]).

*Coastal Oil & Gas Corp. v. Garza Energy Trust*, 268 S.W.3d 1 (Supreme Court of Texas, 2008).

*George H. Coates, et al. v. Genoveva O. De Garcia, et al.*, 286 S. W. 2d 691 (Court of Civil Appeals of Texas, 1956).

*Coloma Oil & Gas Corp. v. Railroad Commission*, 348 S.W.2d 390 (Court of Civil Appeals of Texas, 1961).

*Coloma Oil & Gas Corp. v. Railroad Commission*, 358 S.W.2d 566 (Supreme Court of Texas, 1962).

*Colorado Interstate Gas Co. v. Sears*, 362 S.W.2d 396 (Court of Civil Appeals of Texas, 1962).

*Consolidated Edison Co. v. National Labor Relations Board*, 305 U.S. 197 (U.S. Supreme Court, 1938).

*Consolidated Gas Utilities Corporation v. Thompson, et al.*, 14 F. Supp. 318 (Federal District Court, Western District of Texas, Austin, 1936).

*Constantin, et al. v. Smith, et al.*, 67 F. 2d 227 (Federal District Court, Eastern District of Texas, Tyler, 1932).

*Corzelius, et al. v. Harrell*, 186 S. W. 2d 961 (Supreme Court of Texas, 1945).

*Corzelius v. Railroad Commission*, 182 S.W.2d 412 (Texas Court of Civil Appeals, Austin, 1944).

*Cox v. Robison (Commissioner of General Land Office)*, 150 S.W. 1149 (Supreme Court of Texas, 1912).

*Danciger Oil & Refining Co. v. Railroad Commission of Texas, et al.*, 49 S. W. 2d 837 (Court of Civil Appeals of Texas, 1932).

*Danciger Oil & Refining Co. v. Railroad Commission of Texas, et al.*, 56 S. W. 2d 1075 (Supreme Court of Texas, 1933).

*Danciger Oil & Refining Co., et al. v. Smith, et al.*, 4 F. Supp. 236 (U.S. Federal District Court, Northern District of Texas, Amarillo, 1933).

*Daubert v. Merrell Dow Pharmaceuticals, Inc.*, 509 U.S. 579, 113 S. Ct. 2786 (U.S. Supreme Court, 1992).

*Delhi-Taylor Oil Corporation v. Holmes*, 344 S.W. 2d 317 (Supreme Court of Texas, 1961).

*Erie R. R. Co. v. Tompkins*, 504 U.S. 65 (U.S. Supreme Court, 1938).

*Exxon Corp. v. First National Bank*, 529 S.W.2d 110 (Court of Civil Appeals of Texas, 1975).

*Exxon Corp. v. Railroad Commission*, 471 S.W.2d 497 (Supreme Court of Texas, 1978).

*Frye v. United States*, 293 F. 1013 (U.S. Circuit Court of Appeals, District of Columbia, 1923).

*Funk v. Haldeman*, 53 Pa. St. Reports 229 (Pennsylvania Supreme Court, 1866).

*Geo Viking, Inc. v. Tex-Lee Operating Company*, 839 S.W. 2d 797 (Supreme Court of Texas, 1992).

*Gifford Operating Company v. Indrex, Inc.*, No. 2:89-CV-0189, 1992 U.S. Dist. LEXIS 22505 (Federal Northern District of Texas, 1992).

*Gregg v. Delhi-Taylor Oil Corp.*, 337 S.W.2d 216 (Court of Civil Appeals of Texas, 1960).

*Gregg v. Delhi-Taylor Oil Corp.*, 344 S.W.2d 411 (Supreme Court of Texas, 1961).

*Grubb v. McAfee*, 109 Tex. 527, 212 S.W. 464 (Supreme Court of Texas, 1919).

*J. M. Guffey Petroleum Company v. Chaison Townsite Company*, 48 Texas Civil Appeals 555, 107 S.W. 609, at 612 (Texas Court of Civil Appeals, 1908).

*Gulf Land Co. v. Atlantic Refining Co.*, 134 Tex. 74, 131 S. W. (2d) 73 (Supreme Court of Texas, 1939).

*Gulf Land Co. v. Atlantic Refining Co.* 113 F. 2d 902 (Federal Circuit Court of Appeals, Fifth Circuit, 1940).

*Hail, Etc. v. Reed, Etc.*, 15 Ben Monroe's Reports (Kentucky) 479 (Kentucky Supreme Court, 1854).

*Halbouty v. Darsey*, 326 S.W.2d 528 (Court of Civil Appeals of Texas, 1959).

*Michel T. Halbouty et al., Appellants v. Railroad Commission of Texas et al., Appellees*, 347 S.W. 2d 364 (Supreme Court of Texas, No. A-8200, 1962). AKA "Port Acres."

*H. M. Harrington Jr., Appellant, v. Railroad Commission of Texas et al., Appellees*, 375 S.W.2d 892 (Supreme Court of Texas, Nos. A-9702 to A-9704, 1964).

*Hastings Oil Company v. Texas Company*, 149 Tex. 416, 234 S.W. 2d 389 (Supreme Court of Texas, 1950).

*Hawkins v. Texas Company*, 203 S.W.2d 1003 (Court of Civil Appeals of Texas, 1947).

*Hawkins v. Texas Company*, 209 S.W.2d 338 (Supreme Court of Texas, 1948).

*F. C. Henderson, Inc. v. Railroad Commission*, 56 F. (2d) 218 (U.S. District Court for the Western District of Texas, 1932).

*Henderson Co. v. Thompson, et al.*, 300 U.S. 258 (U.S. Supreme Court, 1937).

*Herman v. Thomas*, 143 S.W. 195 (Texas Court of Civil Appeals, 1911).

*Houston &Texas Central Railroad v. East*, 98 Tex. 146, 81 S.W. 279 (Supreme Court of Texas, 1904).

*Hufo Oils et al. v. Railroad Commission of Texas*, No. 382,447 (District Court of Travis County, 250th Judicial District of Texas, 1985).

*Hufo Oils, et al. v. Railroad Commission of Texas*, No. 14,603 (Court of Civil Appeals of Texas, 1986).

*Humble Oil & Refining Co. v. L. & G. Oil Co. et al.*, 259 S.W.2d 933 (Court of Civil Appeals of Texas, Austin, 1953).

*Humble Oil & Refining Company v. Lasseter*, 120 S.W. 2d 541 (Court of Civil Appeals of Texas, 1938).

*Humble Oil and Refining Company v. Railroad Commission of Texas et al.*, 68 S.W. 2d 622 (Court of Civil Appeals of Texas, Austin, 1934).

*Humble Oil & Refining Co. v. Railroad Commission*, 112 S.W. 2d 222 (Court of Civil Appeals of Texas, 1937).

*Interstate Commerce Commission v. Illinois Central*, 215 U.S. 452 (U.S. Supreme Court, 1910).

*Interstate Commerce Commission v. Union Pacific Railroad*, 222 U.S. 541 (U.S. Supreme Court, 1911).

*Julian Oil and Royalties Company v. Capshaw, et al.*, 145 Okla. 237 (Supreme Court of Oklahoma, 1930).

*Kelly v. Ohio Oil Company*, 57 Ohio St. 317, 49 N.E. 399 (Ohio Supreme Court, 1897).

*Killingsworth v. Jones*, 403 S.W.2d 325 (Supreme Court of Texas, 1965).

*Kraker v. Railroad Commission*, 188 S.W.2d 912 (Court of Civil Appeals of Texas, 1946).

*L&G Oil Co. v. Railroad Commission*, 368 S.W.2d 187 (Supreme Court of Texas, 1963).

*Leonard v. Prater*, 18 S.W. 2d 681 (Court of Civil Appeals of Texas, 1929).

*Livingston Oil Corp. v. Waggoner*, 273 S.W. 903 (Texas Court of Civil Appeals, 1925).

*Lone Star Gas Co. v. State*, 153 S.W.2d 681 (Supreme Court of Texas, 1941).

*Loper Bright Enterprises v. Raimondo* (U.S. Supreme Court, pending 6/5/23).

*MacMillan et al., v. Railroad Commission of Texas et al.*, 51 F. 2d 400 (Federal District Court, Western District of Texas, Austin, 1931).

*Magnolia Petroleum Co., et al. v. Railroad Commission of Texas, et al.*, 90 S. W. 2d 659 (Court of Civil Appeals of Texas, 1936).

*Marbury v. Madison*, 5 U.S. 137, 2 L. Ed. 60, 1803 U.S. LEXIS 352 (U.S. Supreme Court, 1803).

*Marrs v. Railroad Commission*, 142 Tex. 293, 177 S. W. 2d 941 (Supreme Court of Texas, 1944).

*Medina Oil Development Co. v. Murphy, et al.*, 233 S. W. 333 (Court of Civil Appeals of Texas, 1921).

*Munn v. Illinois*, 94 US 113, 24 L. Ed. 77, 1876 U.S. LEXIS 1842 (U.S. Supreme Court, 1877).

*Murphy Oil Company v. Burnet, Commissioner of Internal Revenue*, 287 U.S. 299 (U.S. Supreme Court, 1932).

*Nash v. Shell Petroleum Corp.*, 120 S. W. 2d 522 (Court of Civil Appeals of Texas, Austin, 1938).

*Ohio Oil Company v. Indiana*, 177 U.S. 190 (U.S. Supreme Court, 1900).

*Oxford Oil Co., et al. v. Atlantic Oil & Producing Co., et al.*, 16 F. 2d 639 (Federal District Court, Northern District of Texas, Dallas, 1926).

*Oxford Oil Co. v. Atlantic Oil & Producing Co.*, 22 F.2d 597 (Federal Fifth Circuit Court of Appeals, 1927).

*Panama Refining Co. v. Ryan, et al.*, 5 F. Supp. 639 (Federal District Court, Eastern District of Texas, Tyler, 1934).

*Panama Refining Co. v. Ryan, et al.*, 293 U.S. 388, 55 S. Ct. 241 (U.S. Supreme Court, 1935).

*People v. Associated Oil Company et al.*, 211 Cal. 93, 294 Pacific 717 (Supreme Court of California, 1930).

*People's Petroleum Producers, Inc., et al. v. Smith et al., and six other cases*, 1 F. Supp. 361 (Federal District Court, Eastern District of Texas, Tyler, 1932).

*People's Petroleum Producers, Inc., et al. v. Sterling, et al.*, 60 F. 2d 1041 (Federal District Court, Eastern District of Texas, Tyler, 1932).

*Peterson, et al. v. Grayce Oil Co., et al.*, 37 S. W. 2d 367 (Court of Civil Appeals of Texas, 1931).

*Pickens v. Railroad Commission*, 387 S.W.2d 35 (Supreme Court of Texas, 1965).

*Pierson v. Post*, 3 Cal. R. 175, Am. Dec. 264 (Supreme Court of Judicature of New York, 1805).

*Rabbit Creek Oil Co. v. Shell Petroleum Corporation*, 66 S. W. 2d 737 (Court of Civil Appeals of Texas, 1933).

*Railroad Commission v. Aluminum Co. of America*, 380 S.W.2d 599 (Supreme Court of Texas, 1964).

*Railroad Commission of Texas v. Bass*, 10 S. W. 2d 586 (Court of Civil Appeals of Texas, 1928).

*Railroad Commission v. City of Austin*, 524 S.W.2d 262 (Supreme Court of Texas, 1975).

*Railroad Commission v. Coleman*, 445 S.W.2d 790 (Court of Civil Appeals of Texas, 1970).

*Railroad Commission v. Coleman*, 460 S.W.2d 404 (Supreme Court of Texas, 1970).

*Railroad Commission v. Continental Oil Co.*, 157 S.W.2d 695 (Court of Civil Appeals of Texas, 1941).

*Railroad Commission v. Flour Bluff Oil Corp.*, 219 S.W.2d 506 (Court of Civil Appeals of Texas, 1949).

*Railroad Commission v. Graford Oil Corp.*, 557 S.W.2d 946 (Supreme Court of Texas, 1977).

*Railroad Commission, et al. v. Humble Oil & Refining Co., et al.*, 193 S. W. 2d 824 (Court of Civil Appeals of Texas, 1946). AKA The "Hawkins Case" (not to be confused with *Hawkins v. Texas Company*).

*Railroad Commission of Texas v. Konowa Operating Co., et al.*, 174 S. W. 2d 605 (Court of Civil Appeals of Texas, 1943).

*Railroad Commission v. Magnolia Petroleum Co.*, 109 S.W. 2d 967 (Supreme Court of Texas, 1937).

*Railroad Commission v. Manziel*, 361 S.W. 2d 560 (Supreme Court of Texas, 1962).

*Railroad Commission of Texas and the Atlantic Refining Company v. Permian Basin Pipeline Company,* and *Railroad Commission of Texas* and *the Atlantic Refining Company v. Phillips Petroleum Company* (combined opinion), 302 S.W. 2d 238 (Court of Civil Appeals of Texas, 1957).

*Railroad Commission v. Rio Grande Valley Gas Co.*, 405 S.W. 2d 304 (Supreme Court of Texas, 1966).

*Railroad Commission v. Rowan Oil Co.*, 259 S.W. 2d 173 (Supreme Court of Texas, 1953). AKA The "Spraberry Case."

*Railroad Commission of Texas et al. v. Rowan & Nichols Oil Co.*, 107 F. 2d 70 (United States Court of Appeals, Fifth Circuit, 1939).

*Railroad Commission of Texas, et al. v. Rowan & Nichols Oil Co.*, 310 U.S. 573, 60 S. Ct. 1021 (U.S. Supreme Court, 1940).

*Railroad Commission of Texas, et al. v. Rowan & Nichols Oil Co.* 311 U.S. 570 (U.S. Supreme Court, 1941).

*Railroad Commission, et al. v. Shell Oil Co., Inc., et al.*, 154 S. W. 2d 507 (Court of Civil Appeals of Texas, 1941). AKA "Trem Carr."

*Railroad Commission, et al. v. Shell Oil Co., Inc., et al.*, 161 S. W. 2d 1022 (Supreme Court of Texas, 1942). AKA "Trem Carr."

*Railroad Commission et al. v. Shell Oil Co., Inc. et al.*, 206 S.W. 2d 235 (Supreme Court of Texas, 1947). AKA "Seeligson."

*Railroad Commission of Texas et al., Petitioners, v. Shell Oil Company et al., Respondents,* 380 S. W. 2d 556 (Supreme Court of Texas, No. A-9759, 1964). AKA "Quitman."

*Railroad Commission v. Sterling Oil and Refining*, 147 Tex. 547, 218 S.W. 2d 415 (Supreme Court of Texas, 1949).

*Railroad Commission v. Williams*, 356 S.W.2d 131 (Supreme Court of Texas, 1961).

*Reed Roller Bit Co. v. Hughes Tool Co.*, 12 F.2d 207 (Federal Fifth Circuit Court of Appeals, 1926).

*Rowan & Nichols Oil Co. v. Railroad Commission of Texas*, 28 F. Supp. 131 (Western District of Texas, 1939).

*Rudman v. Railroad Commission*, 349 S.W. 2d 717 (Supreme Court of Texas, 1961).

*Ryan v. Amazon Petroleum Corporation*, 71 F 2d 1 (Federal Fifth Circuit Court of Appeals, 1934).

*Ryan Consolidated Petroleum Corporation v. W. L. Pickens, et al.*, 285 S. W. 2d 201 (Supreme Court of Texas, 1955).

*Schechter Poultry Corp. v. United States* 295 U.S. 495 (U.S. Supreme Court, 1935).

*Smith v. Stewart*, 68 S.W. 2d 627 (Court of Civil Appeals of Texas, 1934).

*Smith v. Stewart*, 83 S.W. 2d 945 (Supreme Court of Texas, 1935).

*Standard Oil Company v. Railroad Commission*, 215 S.W. 2d 633 (Court of Civil Appeals of Texas, 1948).

*Stanolind Oil & Gas Co. v. Midas Oil Co.* 123 S.W. 2d 911 (Court of Civil Appeals of Texas, 1938).

*Stanolind Oll & Gas Co., v. Railroad Commission*, 198 S.W.2d 420 (Supreme Court of Texas, 1946).

*State v. Jarmon*, 25 S. W. 2d 934 (Court of Civil Appeals of Texas, 1930).

*Stephens County v. Mid-Kansas Oil & Gas Co.*, 254 S. W. 290 (Supreme Court of Texas, 1923).

*Sterling, Governor of Texas, et al. v. Constantin, et al.*, 287 U.S. 378 (U.S. Supreme Court, 1932).

*Stewart v. Humble Oil & Refining Co.*, 377 S.W.2d 830 (Supreme Court of Texas, No. A-9902, 1964).

*Marion Stone and Brian Corwin v. Chesapeake Appalachia, LLC, Statoil USA Onshore Properties, Inc., and Jamestown Resources, Inc.*, No. 5: 12-CV-102, 2013 U.S. Dist. Lexis 71121 (Federal District Court, Northern District of West Virginia, Wheeling, WV, 2013).

*Sun Oil Co. v. Railroad Commission, et al.*, 68 S. W. 2d 609 (Court of Civil Appeals of Texas, 1933).

*Superior v. Railroad Commission*, 519 S.W.2d 479 (Court of Civil Appeals of Texas, 1975).

*Superior Oil Co. v. Railroad Commission*, 546 S.W.2d 121 (Court of Civil Appeals of Texas, 1977).

*Swift v. Tyson*, 16 Peters 1 (U.S. Supreme Court, 1842).

*Texaco v. Railroad Commission*, 583 S.W.2d 307 (Supreme Court of Texas, 1979).

*Texas Company v. Daugherty*, 107 Tex. 226, 176 S. W. 717 (Supreme Court of Texas, 1915).

*Texas Company v. Davis*, 254 S.W. 304 (Supreme Court of Texas, 1923).

*Texoma Natural Gas Co. v. Railroad Commission of Texas et al. and three other cases*, 59 F. 2d 750 (Federal District Court, Western District of Texas, 1932).

*Texoma Natural Gas Company v. Terrell*, 2 F. Supp. 168 (Western District of Texas, December 23, 1932).

*Thomas v. Stanolind Oil and Gas Company*, 145 Tex. 270, 198 S.W. (2d) 420 (Supreme Court of Texas, 1946).

*Thompson v. Consolidated Gas Utilities Corporation et al.*, 57 S. Ct. 364, 300 U.S. 55 (U.S. Supreme Court, 1937).

*Tidewater Associated Oil Co. v. Stott*, 159 F.2d 174 (Federal Fifth Circuit of Appeals, 1947).

*Trapp v. Shell Oil Co.*, 145 Tex. 323, 198 S.W. 2d 424 (Supreme Court of Texas, 1946).

*United States v. Causby*, 328 U.S. 256, at 261, 66 S. Ct. 1062, 90 L. Ed. 1206 (United States Supreme Court, 1946).

*United States v. Standard Oil of New Jersey et al.*, 221 U.S. I, 31 S. Ct. 502, 55 L. Ed. 619, 1911 U.S. LEXIS 1725 (U.S. Supreme Court, 1911).

*United States v. Trans-Missouri Freight Association* 166 U.S. 290, 17 S. Ct. 540, 41 L. Ed. 1007, 1897 U.S. LEXIS 2025 (U.S. Supreme Court, 1897).

*Utah Power and Light Company v. Pfost*, 286 U.S. 299 (U.S. Supreme Court, 1932).

*Westmoreland Natural Gas Company v. DeWitt, et al.*, 130 Pa. St. 235, 18 Atl. 724 (Pennsylvania Supreme Court, 1889).

*H. F. Wilcox Oil & Gas Co. v. State, et al.*, 162 Okla. 89, Okl. 19 Pacific 2d 347 (Supreme Court of Oklahoma, 1933).

*Windsor Gas Corporation v. Railroad Commission*, 529 S.W.2d 834 (Court of Civil Appeals of Texas, 1975).

# INDEX

*Note: Page numbers in italics refer to images.*

AAPG. *See* American Association of Petroleum Geologists

ABC financing, 209–10

Abercrombie, James Smither, *200*

Abercrombie & Harrison, 215

acidizing, 249

agencies, quasi-judicial, 200–201

agent-based models, 77–78, 277, 280n99

agrarian discontent, 9

AIME (American Institute of Mining Engineers), 99, 104, 106, 108, 160, 193

alienated representations, collective, 58, 79–82, 84, 244, 258

altruism, paradox of, 58, 61, 66

altruistic cooperation, 3–4, 33, 36, 57–60, 66, 71, 73, 75–78, 208, 268n7, 272n14

evolution of, 76, 82

altruistic cooperators, 69–70, 75, 88, 90, 138–39

*Amazon Petroleum Corporation v. Railroad Commission of Texas*, 143, 145–148, 152, 157, 165–66, 168–69, 171, 201–2, 236, 289nn80/90, 303n9

Amerada Petroleum Corp., 230, 292–93n29

American Association of Petroleum Geologists (AAPG), 160, 164, 193

American Bar Association, 140, 214

American Institute of Mining and Metallurgical Engineers (AIME), 156–57, 173

American Petroleum Institute (API), 98–99, 104–8, 192–93

Amoco, 209

Anahuac and Tomball oil fields, 186

analyses, volumetric, 225

ancestor-descendant lineages, 17, 68, 81

Anderson, Dillon, 265n52

Antikythera mechanism, 190

anti-social punishment, 63–64

antitrust laws, 43, 108, 111

API. *See* American Petroleum Institute

API gravities, 101, 156, 283n25

Aquinas, Saint Thomas, 296

Archbold, John D., 11

Archie, Gustave E. "Gus," 222–23

Archimedes, 172, 254

Arkansas Fuel Oil Company, 166

Arkansas Natural Gas Company, 166

Aronofsky, J. S., 182

Ashland oil field, 41

Associated Oil Co., 141, 147

associative state, 114

associations

    cattlemen's, 16, 23

    stockmen's, 21, 26, 46

assortative group formation, 70–72, 90, 278n77

Atlantic Refining and Tidewater Oil Company, 224

Austin Chalk, 51, 250, 253–54

Autry, James Lockhart, 44

Avery Island, 11

backpressure, on oil wells, 102

backward induction, 60, 272n10

Baker, Hines, 218, 222, 227

Baker, Rex, 110

Baker Hughes, 246

Bakken (formation), 254

Ballanfonte, Joseph, 225–26

Bandini Petroleum Company, 147

Barnett Shale, 51, 252–54

Barnsdall Oil Co., 153

Bartlesville, OK, 98, 100, 103, 133, 152, 160, 193

bass, large, 196

Bass, Sam, 236, 239

Bataafsche Petroleum Maatschappij, 154

Batson's Prairie, 7, 28

battleship, big gun, 245

Baumel, Jack K., 222

bean job, 40

Beecher, C. E., 101–3, 106, 129, 147, 149, 157, 193, 250

    and Parkhurst, Ivan P., paradigm, 106, 120, 122, 131, 137, 146, 149, 152

Beecher and Parkhurst, 101–5, 107, 125, 138, 147, 150, 156, 162, 231

    experiments, 104

    findings and analysis, 105–6, 108

    gas-drive hypothesis, 130

gas-drive paradigm, 127, 153, 156–57, 165

*bellum omnium contra omnes*, 66

Benedum-Trees, 38

Big Lake Field in West Texas, 184

Boatright, Mody, 22–23,

Boonsville Bend Field, 251–52

Botset, Holbrook G., 159–61, 173, 175–78

bottom hole assembly (BHA), 247–48

bottom-hole measurements, 149–51

bottom-hole money, 40, 207

bottom-hole pressure tests in East Texas, 152

bottom-hole sampler, 152–53

Brandeis, Louis D., 213

Bridgman, Percy W., 163

Bright & Schiff, 224–28

Brown, George, 204

Brown, Herman, *200*

Brown and Root, 200, 204

Bryan, William Jennings, 28, 47

Bryant, Randolph J., 121, 132, 148, 165

*Buford et al. v. Sun Oil Co.*, 144

Bureau of Mines, 99–100, 103, 105, 107, 124,
   155, 160, 190

Burkburnett field, North Texas, 17, 37, 39, 47–50,
   109, 118–19


Caddo Lake, 44

Calhoun, John C., Jr., 244

California Company, 40

California Institute of Technology, 104, 156,
   159, 183, 192

California Supreme Court, 141, 147

Calkin, L. P., 102–4, 106, 156

Campbell, Donald Thomas, 164, 260, 263

Carter Oil Company, 182, 190

Carter Oil Pressure Core Barrel, 181

casing, 10, 31, 47, 232, 247, 249

casinghead gas, 204–6, 304n34

Catahoula dogs, 266n79

Central Advisory Committee, 113

Central Proration Committee, 77, 112–13

Champlin case, 127–29, 135–36, 138, 140,
   146–47, 167

Chandler, Alfred Dupont, 55

chemical engineering, 103

Chevron, 253, 267n110, 303n9

Christmas trees, 119

Cities Service Co., 98–99, 282n14

City National Bank, 204

Clymore Production Company, 300n64

Coastal Oil & Gas Corporation, 256

coevolution, 5, 57

collateral loan amount, 209

collateralization, large-scale, 203

collective representations, 80, 82

collectivities, social, 17, 35, 48, 53, 58, 79

Collins, Ray M., 128

colonization events, regular, 75

committees
   advisory, 192, 214
   field safety, 16, 47, 203
   vigilance, 16, 23, 46, 155

common-pool resources (CPR), 60–61, 63, 75, 83, 87

commons, 25, 60

communalism, traditional, 21

community without propinquity, 53

comparative advantages, 55

Connally Hot Oil Act, 234

Conoco, 291n26

Conroe, TX, 94, 233, 235

conscientious development, 14

conservation
   acts, 125, 127, 216
   laws, 294n74, 304n39
   statutes, 144, 147

Consolidated Edison Co., 201

Consolidated Gas, 144

Constantin, Eugene, 130, 135, 142

Continental Illinois Bank, 210

Continental Oil Co., 211, 235, 291n26

convergent validity, 164, 231

cooperation, self-interested, 61

Copernicus, 172, 284, 297n145

Core Laboratories Inc., 180

core samples, 51, 148, 159, 185–86, 253

core sampling, 185

cowhunts, 20, 41

crater, 28, 69, 233, 276n58

creative destruction, 221, 258

creekology, 28, 51

crop liens, 9

crude oil production, 268n123

Crutchfield, John W., 224–26

Crutchfield and Pruett, 224–25

Cullen, Roy, 51, 114

Cullinan, Joseph Stephen, 18, 42–44, 46

cultural group selection, 277n74

cultural transmission, 82, 266n81

Cutler, Willard W., Jr., 176, 217


Daisy Bradford No. 3, 37, 39, 117, 119, 126,
   135, 148, 250

Danciger, Joe, 119–20, 137

Danciger cases, 113, 120–21, 125, 137–38, 145–46

Danciger Oil, 142–43

Danciger Oil & Refining Company, 120

Darcy, Henry, 158–59, 161, 171
Darcy's law for water flow in sand-filled
    pipes, 158, 173
*Daubert v. Merrell Dow Pharmaceuticals Inc.*, 257
Davis, Morgan, 55
deal selection, 4, 76
Deane, Early C., 11
debt financing, 205
de facto credit, 207
deference, 24, 49, 232
DeGolyer, Everette Lee, Sr., 105–6, 211, 258, 263
Delhi-Taylor Oil Corporation, 256
democracy, rural participatory, 23
density-dependent bias, 17
depletion allowance, 115, 210
Desdemona field, 7, 9, 29, 37, 39
deviated wells, 232–34
deviation surveys, 232, 235
Dictator Game (DG), 63
differentiation, functional, 54–55
directed drilling, 310
directional drilling, precision, 232–33, 235
dissolved gas, effect of, 101, 104
Doherty, Henry L., 98–101, 103–4, 109–10, 215, 231
Dow and Calkin, 102–3, 106, 156
    experiments, 103
downhole pressure gauges, 149, 225
    deployed, 151
drainage radius, 155
drilling, directional, 231–34
drilling logs, 51, 115, 122, 148
drilling mud, 52, 185–86, 231, 248
    circulating, 248
dry-hole money, 40, 55
Durkheim, Émile, 80

E&P financing, 254–55
Eagle Ford, TX, 254
East Texas Engineering Association, 103, 155, 193
East Texas Oil Field, 37, 117–18, 129, 150, 157,
    162–63, 166, 231
East Texas Oil Museum, 271n72
East Texas oil operators, 236, 311n81
East Texas production, 135, 153, 168
East Texas Salt Water Disposal Company, 203
Eastman, John, 232
economic theory, 270
economics
    behavioral, 4, 59, 61–62, 69
    experimental, 57
effective water drive, 165, 180
egoistic incentive assumptions, 57
egoistic incentives, 4, 57, 60–61, 65–66, 70

Einstein, Albert, 172
El Dorado, 7–8, 12
Electra, 38, 40, 44
electrical log, 186
Elkins, James A., 110, 204
Emerson, Ralph Waldo, 285n71
Empire Gas and Fuel Company, 98, 100
engineering
    disciplines, 192, 194
    education, 193–95, 244
    enrollments, 196
    practice, 99, 106, 129, 157, 176, 191, 195, 199,
        208, 211, 222
    practitioners, 104, 107, 164, 192–93,
        195, 199, 211
    programs, 131–32, 194, 196
engineers
    bank, 210
    consulting, 211
English crowd, eighteenth-century, 3, 34
equation of flow of oil and gas, 159
Esperson Building, 251
Eugene Island, 275n56
evangelical, 19, 27–28
evidence, substantial, 188, 201–2, 224, 230
evolution of altruism, 74, 271, 274, 281n127
evolutionary biology, 35, 58–59, 66, 261, 274, 281–82
evolutionary games, 69
evolutionary game theory and evolutionary econom-
    ics, 276n64
exaptation, 271n4
exapted subpopulations, 72
expectation, 48, 265n49, 278n77
expert witnesses, 122, 127, 136, 166–67, 218, 225–26,
    257, 289n80
Exxon Neftegas, 248

Farish, William S., 46, 109–10
farm outs, 25, 40, 55
Farris, Floyd, 249
Federal Oil Conservation Board (FOCB), 99, 105
Federal Power Commission (FPC), 200, 205
*ferae naturae*, 140, 264n31
Fettke, Charles R., 185
field rules, 16, 214, 216, 223, 230
field unitization, 47, 108
fields, water-drive, 150, 156, 162, 167, 233
First National Bank, 281
Fischer, Ronald A., 278n85
fitness advantages, differential, 67
fitnesses, 60, 70–71, 89
    inclusive, 59,
Florence, Fred, 209

Flour Bluff field, 206
Flour Bluff Oil Corporation, 206
flow of compressible fluids, 163, 295
flow of gas-liquid mixtures, 173, 178
flow of heterogeneous fluids, 173, 181–82
flow of homogeneous fluids, 171
fluid displacement in sands, 180
fluids, single-phase, 155, 183
flush production, 46, 92–94, 98–100, 111–12, 139
Fondren, Walter, 45
Foote, Paul D., 160
Forrest, Nathan Bedford, 275n54
Fowler, Harry, 183
fracking, 208, 245, 249–50, 252–54, 256–57
    slickwater, 253–54
fracking fluids, 249, 252, 254, 256
fractures, induced, 250, 257
fracturing, 249
Frankfurter, Felix, 199, 202
free rider problem, second-order, 272n95
free riders, second-order, 279n94
fruitful empiricism, 144, 199–219
function of natural gas, 105, 219

Gächter, Simon, 61
Galambos, Louis, 261
Galbraith, John Kenneth, 231
Galey, John H., 10, 14
game theory, 59, 66, 84, 261
games
    iterated prisoner's dilemma, 273n40
    normal form, 69, 85, 273n40
Garza Energy Trust, 256–57
Gas Conservation Committee, 105–6
gas drive, dissolved, 102, 150, 152
gas-drive fields, 131, 146, 150–52, 166–67, 178
gas energy, 100, 125, 127, 129, 147, 155, 158
gas factor, 104, 131, 283
gas-oil ratios, 104, 107, 122–23, 129–33, 147,
    149–50, 152, 165, 167, 177
    efficient, 131, 137
gases, available, 205, 207
Gates, Charles G., 44
General American Oil Company of Texas, 209
Ghiselin, Michael, 59
Gill, Stanley, 166
Goodnight, Charles, 24
Goodnight-Loving Trail, 24
Grayce Oil Company, 257
greenbeard scenario, 71
Greenhill, Joe R., 285n94
group-advantageous traits, 277
group-selection, 4, 57, 66–67, 72, 74–77, 274–75n14

group selection models, 58, 67
group size, 70, 86, 88–89
Grozny oil field, 185
GSD model, 66, 70, 72, 74–76, 78, 88, 138, 279n91
Guffey Petroleum Company, 14
Gulf, 19, 45, 52, 54, 109–10, 133, 166,
    204, 275–76n56
Gulf Oil, 100, 133, 159, 178, 180, 292n39
Gulf Oil Research, 158
Gulf Oil Research & Development Company, 159
Gulf Pipeline, 100
Gulf Pipeline Company, 132
Gulf Production Company, 166, 293n62
Gulf Research, 160, 173
gushers, 9, 17, 92, 102, 265n49

Habermas, Jürgen, 80
Hadamard, Jacques, 291n13
Halbouty, Michel T., 29, 40, 51–52, 229, 237–38,
    264–65, 268–70, 276, 285n2
Halliburton, 245–46, 249
Hamill brothers, 11
Hamilton, William B., 41, 50
Hamilton's Rule, 59, 271n3
Hardin, Garrett, 60
Hardwicke, Robert E., 117, 124, 168–69
Hardwicke's claims, 169
Harrison, Dan, 232–33
Harrison & Abercrombie, 216, 232–33
Hawkins case, 188–89, 220–22, 224, 226–27
Hawkins field, 189, 220, 233
Hayes, Willard C., 11
hegemony, 4, 58
Henderson, 117, 239
Hendricks field, 94, 110, 113
Herodotus, 81
heterogeneous flow theory, 172, 175–76, 178,
    180–82, 184–86, 229, 257
heterogeneous fluids, 104, 173, 181
Higgins, Patillo, 11, 250, 252
High Island field, 51–52
Hines Baker's argument, adopted, 219
Hobbesian, 66
Hobsbawm, Eric, 25
Hogg, James S., 12, 26–27, 44, 267n110
Hogg, William C., 44
Hogg-Swayne Syndicate, 12, 43
holes, slant, 235, 237
Holifield, Ray, 250
Homo moralis, 274n42
Homophily, 266, 337
horizontal drainhole wells, 256
horizontal drilling, 208, 245, 247–50, 253–54, 256

hot oil, 129–30, 135
Hot Oil Act, 116
House, Edward M., 12
Houston & Texas Central Railroad, 141
Hubbert, M. King, 249
huff and puff, 185
Hughes, Charles Evans, 201
Hughes, Thomas P., 99
Hull, David L., 17, 68, 80
Humble, 28
Humble Oil & Refining, 45, 50, 55, 100, 103, 109,
    125–26, 143, 151, 161, 176, 178–79, 189–90,
    217–19, 225, 232–33, 235–36, 293n62
Hundall, J. S., 164
Hunt Oil, 236
hunting seasons, 25
Huntington Beach, 98, 232–33
Hurst, William, 158–59, 162–65, 167, 171,
    188–90, 211
Hutcheson, Judge Joseph C., Jr., 121, 124–25,
    130, 132–34, 136–37, 145–46, 148–49, 151,
    166–68, 201–2
    on judicial decision, 146
Hydrafrack, 249
hydrafracking, 249–50, 256
hydraulic fracturing, 185, 245, 249
Hydraulic Fracturing Chemical Disclosure
    Requirements, 256
hydrostatic control hypothesis, 122

Ickes, Harold, 143, 165
ideology, scientific, 100, 195
idiom, collective, 80–81
immigrants, southern, 19, 21
implied covenants, 14, 264n41
indefinitely repeated prisoner's dilemma
    (IRPD), 271n10
independent oil operators, 29, 55, 110, 197, 200, 211,
    232, 241, 251
informal know-how trading, 100
instruments, 51, 53, 149, 223, 232
Insull, Samuel, 98
insurance companies, 26–27, 225
Interstate Commerce Commission (ICC), 36, 200
Interstate Oil Compact Commission (IOCC), 54
invisible colleges, 65, 133, 178
Iterated Prisoner's Dilemma (IPD), 60, 86
Iterated Public Goods (Iterated PG), 60, 86

Jackson, Andrew, 265n43
Jennings, Waylon, 241
Joiner, Columbus Marion "Dad," 40, 117
Jones, Jesse H., 211

Jordan, Terry G., 19–22
judgement intuitive, 145
Julian Oil, 126, 141

Kauffman, Stuart A., 172
kelly joints, 10
Kilgore, 117–18, 155, 234, 238–39, 271n72
kin selection, 59, 274–75
Kuhn, Thomas S., 156, 183–84

Lacey, William N., 104, 192
landowners, small, 215–16, 257
law of capture, 226–27, 231
lending, mezzanine, 255
Leverett, M. C., 178–80, 182, 223
Lewis, James O., 164, 179–80, 224–25
life insurance companies, 204–5
Lindsly, Ben Edwin, 108, 152–54, 156, 163, 166–67,
    171, 183–84, 236
    results, 154
lineages, cultural, 17, 22
Little Big Inch Pipeline, 204
loans, reserve-backed term, 210
logging, 222, 245, 247–48, 263
logging while drilling (LWD), 247–48
logs, electric, 208, 222, 224
Lucas, Captain Anthony F., 9, 11, 13, 16, 258

MacKenzie, Donald, 107
MacMillan case, 117, 120–27, 129–30, 132, 134,
    146–50, 152, 167, 169
Magnolia, 38, 54, 94
majors and independents, 54–55, 155
manager-entrepreneurs, 99, 103
Marcellus Shale, 254, 257
marginal wells, 55, 137, 214, 235, 243
market demand, 31, 112, 116, 125–27, 136, 147, 214
Market Demand Act, 116, 135, 166–67
market demand for oil, 116, 125
marketing, collective, 46–47
markets, capitalist, 65
Marland Production Company, 114, 149, 211
martial law in East Texas, 126–27, 150, 235
Marx, Karl, 80
Mayerhofer, Mike, 252
McCarthy, Glenn H., 114
McMath, Robert C., Jr., 19
measurement while drilling (MWD), 245, 247–48
measurements, downhole pressure, 152, 231
Meres, Milan W., 173, 175–79, 181
Meres's theory, 176–77
Merlyn, 57, 78, 85
Merrell Dow Pharmaceuticals, 257

Mexia-Talco fault zone, 189
Miller, Harold Carl, 105
mine and thine, 199
mineral rights, 13–14, 16
Mitchell, George Phydias, 251–53
Mobil, 38, 54
models
    abstract, 58, 78, 84
    conceptual, 78–79
    multilevel selection, 58, 66
modus vivendi, 107, 228, 230
Moineau pumps, 248
Monongahela Gas Company, 141
moral economies, 3–5, 33–37, 41–59, 65–67, 69,
    75–78, 83–85, 138–39, 191–92, 199, 211–12,
    238–39, 244–45, 254–5
mud, 10–11, 52, 248, 250
mud circulation, 52
mud motors, 245, 247–48
Mudmotor Power Section, 247–48
mud-pulse telemetry, 248
Multilevel Selection Theory, 68, 281n130
multiphase flow in porous media, 163, 231
Murray, Alfalfa Bill, 126–27, 129, 235
Murray, Railroad Commissioner William J., 235
Murray, William J., 204
Muskat, Morris, 104, 158–64, 167, 171–73, 178,
    180–82, 184, 188–90
Muskat and Botset, 159–60
Muskat and Meres, 175–77, 179

Nash equilibrium, 59–60, 84, 86–87
National Industrial Recovery Act. See NIRA
National Recovery Administration, 116
natural gas pipelines, 205
Navier-Stokes equations, 159
Nazro, Underwood, 54, 100, 114
Neff, Governor Pat, 135
negative appropriation externalities, 60
networks, small-world, 75
Newton, Isaac, 5, 172, 182, 254
New York life insurance companies, 204
New York Mercantile Exchange, 255
Nichols Oil Co., 144
NIRA (National Industrial Recovery Act), 137,
    147–48, 165, 282n8
nitroglycerin, 249
Non-Cooperative Game, 60, 63, 87
Normanna case, 222–30
n-person Prisoner's Dilemma, 59–60, 69–70, 76, 86
Nye and Reistle, 156–57

obligations, implicit, 14

Occam's razor, 164
Ohio Oil Company, 127–28, 140
oil, running out of, 97–98
oil booms, 13, 16–17, 52, 69, 75, 92, 119, 208,
    254, 266n68
oil companies, major, 46, 109, 113, 196–97, 234
oil conservation, 126, 135, 168, 285n2, 305n65
oil firms, major, 99, 236
oil-gas mixtures, 131, 152, 156
oil-gas ratios, 129–30, 136
oil payment, 41
oil pools, 51–52, 75, 107–9, 125, 162, 212
oil prices, 134, 142, 242, 246, 255
oil prorationing, 25, 212–13
oil-state preeminence in petroleum engineering
    education, 194
oil viscosity, 100, 104, 127–28, 180
    increased, 100, 111
    reduced, 99, 165
Oklahoma City fields, 126–28
Oklahoma Corporation Commission's
    orders, 127–29
Oklahoma prorationing, 127
Oklahoma Supreme Court, 97, 126–27, 141, 168,
    257, 288n53
Oklahoma's prorationing statute, 287n51
Old Ocean Field, 215, 223
OPEC oil boycott, 250
open-range cattle ranching, western, 19–20
operators, slant-hole, 234–35, 237
oral history, 36, 276n63
orders, commission's proration, 134, 227, 230
Ostrom, Elinor, 63, 84
ownership of oil in place, 14, 222

Pan American Petroleum, 235–36
Panama Refining Company, 143, 147–48, 165
Panhandle field, 113, 120, 137, 142–43, 212
Panhandle Stock Association of Texas, 24
Parkhurst, Ivan P., 101–7, 128–29, 133, 138, 147,
    149–50, 156–57, 162, 165, 231
People's Petroleum case, 132, 136, 146–48,
    150, 165, 169
pools, common, 115, 125, 127–28, 141, 219
population of Texas, 19, 21
pore spaces, 102, 161, 164, 174, 185–87, 190, 222
porosities, defined, 161
porous formations, 102, 251
porous media, consolidated, 172, 178
Port Acres case, 209, 222, 229–30
Pratt, Wallace, 122
preferences, other-regarding, 64
preferential attachment, 75

present value (PV), 83–84, 91–93, 183, 208, 210, 265n42

present value, discounted, 91, 184, 206

Price, George R., 67

Price's Equation, 82, 91

Prisoner's Dilemma (PD), 59–60, 66, 69, 70, 76, 78, 85–86, 271–72n10, 273–74n40, 277n71

privileged access, 3, 33–36, 43, 45, 75–76, 212, 236, 238–39

Proctor, F. C., 45, 54, 100, 109–10, 114, 270n39

Producers Oil Company, 43

production, mass, 55, 214

production engineers, 101, 108, 124, 149

production payment, 208–10

production tax, 115

productivity index, 180

property rights, subsurface, 31, 217

proppant, 249, 252–53, 257

proration, 109–10, 112–13, 120–21, 124–26, 132, 137, 141–42, 165, 215, 233

prorationing, 48, 108–12, 114, 119, 123–24, 126–29, 137–38, 144, 147, 155, 202–3, 226
   market-demand, 147, 213

prorationing orders
   first statewide, 112, 114
   statewide, 126, 138

Public Goods Games, 61–63, 78

public goods experiments, 61, 63

punishment, antisocial, 63

Pure Oil, 54, 100, 109–11


Quitman case, 230

Quivik, Fred, 263n11


Railroad Commission Act, 115

RAND Corporation, 59

Ranger field, 49

reasonable market demand, 112, 115–16, 135, 147, 165

reconstruction, historical, 190, 265n44

recoverable oil, 128, 137, 218, 229

Red Fish Reef, 9

Reistle, Carl E., Jr., 103–4, 152, 154, 156–57, 164, 167, 171, 190, 236

Reistle and Hayes, 152, 154–55, 158

replicator dynamics, 58, 66–69, 70, 74–76, 84, 87, 139, 279n91

representations, 16, 23, 27–28, 36, 78, 80–85

repressurization, 183–84

Republic National Bank, 209, 255

research, cooperative, 152, 191

research groups, 63, 183

reservoir energy, 31, 100, 104, 108, 120, 125, 128, 133, 135, 141–42

reservoir fluids, 149, 152–53, 162, 167

reservoir pressures, 111, 129, 131, 135–36, 149–50, 152–57, 167, 183–84

reservoir temperatures, 149, 151–53, 158

reservoirs, condensate, 182, 206

retrograde condensation, 183–84, 223

revivalism, nineteenth-century, 72

revolving-loan facilities, 210–11, 255

rigs, rotary, 10–11, 185

Rio Bravo Oil Company, 38, 40, 54, 108

risks, operational, 207–8

Rittenhouse Orrery, 190

ritualistic affirmation of community, 26

Roberts, John, 303

Robin Hood, 236

Rockefeller, John Davison, 11, 13, 46, 192

role of dissolved gas in oil production, 129, 231

Rowan & Nichols Oil Company, 136, 144, 151, 199, 201–2, 236

Roxana, 54, 108

Royal, Darrel, 166

Royal Dutch Shell, 154

royalty, 41, 55, 216, 256

Royalty Owners Association, 17, 276n57

Ruth, Babe, 163

Ryan, Archie D., 143, 147–48, 165, 220, 222, 224, 226–27


Sabine Uplift, 118, 188, 234

Sage, Bruce H., 104, 192

salt water reinjection, 191, 203

saltwater disposal, 202–4, 214

samples, bottom-hole fluid, 161, 163

San Jacinto, 219

sanctioning institutions, 61–63, 71, 87

Sangren, Steven, 58, 79

Sangren's model, 80

Santa Rita, 241, 244, 252

Santa Rita No. 1, 37, 109, 241, 250

Saratoga, 28

saturation pressures, 174

scandal, slant-hole, 233, 237

Schilthuis, Ralph J., 154, 158, 162–64, 171, 186

Schilthuis and Hurst, 163, 165, 188

Schilthuis's connate water experiments, 186

Schlumberger, Marcel, 222

Schlumbergers, 222–24, 246

sciences, packed down, 104

scientific consensus, 117, 122, 134, 136

scientific evidence, 127, 136, 166

scientific hypotheses, 168, 297n145

scientific practice, 65

Sclater, Kenneth C., 149
Scott, James C., 34
Scurry reef, 51
secondary recovery projects, 55, 180
Second National Bank of Houston, 211
Seeligson case, 206
seismic surveys, 51
selections, between-group, 73–74, 91
self-productions, 56, 80, 195, 236
    individual, 80–81, 211
Seminole, OK, 94, 119
service firms, 249
shale revolution, 249–50
shales, 10, 52, 251–54
Sharp, Associate Justice John H., 219
Sharp brothers, 43
Shell, 40, 54, 108, 111, 230–31, 248
Shell Company of California, 108
Shell Development Company, 249
Shell Oil Company, 222, 230
Showers & Moncrief, 293n62
Signal Hill, 98
Sinclair Prairie Oil and Gas, 153, 292–93n39
slant holes in East Texas, 235
slickwater, 251, 253
slickwater fracking and horizontal drilling, 254
slickwater solution, 252
small world networks, emergent, 53
Smith, Justice Clyde E., 227
Smith, Lon A., 137
Snowdrift (game), 66
Sober, Elliott, 89
social groups, 17, 79–81
social networks, 52–53
    small-world, 72, 133
Society of Petroleum Engineers (SPE), 17, 193,
    196–97, 244–45
Society of Petroleum Evaluation Engineers
    (SPEE), 211
Socony Vacuum, 38, 54, 100, 182
Sour Lake, 7, 28, 119
South Carolina, 20–23, 27, 49
South Liberty, 7
Southern Pacific Railroad, 38, 54, 108
spacing, 31, 127, 172, 176–77, 182, 215–16,
    218–19, 256
    wide, 176
spacing regulations, 109, 203, 215
spacing rules, 218–19, 222–23
spatiotemporal particulars, 17, 34, 65
SPE. See Society of Petroleum Engineers
Sperry gyroscope, 232
spherical flow, 172, 175

Spindletop, 9, 11–16, 28, 41–42, 49–52, 117–19
Spooner, W. C., 166
Sri Lanka, 5
Stag Hunt, 66
Standard Oil, 11, 13, 15, 54, 76, 100, 182, 236, 255
Stanley, Edwin G., 234
Stanolind Oil & Gas Co., 249
steady-state flow, 159, 161–62, 172, 179
Steinsberger, Nick, 252–53
Sterling, Ross S., 125–26, 129–30, 132, 135, 142, 150
Sterling Oil, 206
Stone v. Chesapeake Appalachia, 257
straight-hole clause, 234
Strake, George, 232
stratigraphy, 51
strong altruism, 75
subjects
    cultural, 58, 79
    experimental, 58, 61, 63
submarines, German, 204
Suman, John R., 38, 40, 107, 109–10, 176, 233
Sun Oil Company, 45, 100, 111, 144, 153, 202,
    217, 230, 232
surface tension, 101–4, 156, 175
surface waste, 101, 112, 121, 127
surveys
    geological, 50
    geophysical, 161
    inclination, 232, 234
    systematic, 135
Sutton, Don, 261
Swigart, Theodore E., 102–3, 107–8, 147
swindlers, mail order, 47
Swindletop, 12, 15, 73, 83
systems
    heterogeneous fluid, 173, 177
    homogeneous fluid, 171, 173

Teapot Dome, 103
technologies, disruptive, 245–46, 254
Terrell, C. V., 35, 136
tests, first field, 52, 152
Texaco Inc., 235–36
Texas A&M, 29, 132, 194, 196, 211, 241, 244, 251
Texas Antitrust Act, 114–15
Texas Company, 41–45, 100, 111, 141, 194,
    232, 267n110
Texas courts, 27, 137, 142, 199–200, 202, 216–17,
    220–21, 256–57
Texas Cross Timbers, 19, 25, 27
Texas Eastern Transmission Corporation, 204
Texas Exchange, 26
Texas Farmers' Alliance, 26

Texas Independent Producers, 17
Texas Rangers, 129, 252
Texoma Company, 37, 41, 50
theory, compressible fluid, 163–64
Thompson, Edward Palmer, 33–36, 238–39,
Thompson, Ernest O., 31, 135–36, *200*, 203,
    216, 232, 258
three-phase heterogeneous flow, 180, 187
Tidewater Oil Company, 224
Tomball field, 186
Townes, Edgar Eggleston, 42, 45–46, 53, 109–11
tracts, segregated, 220
Transaction Cost Economics, 275n55
transaction costs, 68–69, 88–89, 204, 211, 214
Trem Carr, 220
Trivers, Robert L., 59, 271n5
trough of certainty, 107
truncation and selection, 71, 89
truncation selection, 70–71, 74, 91, 277n74
turbulent flow, 159, 161
Turner, Frederick Jackson, 22
two-phase flow, 179

Ultimatum Game (UG), 63
Umpleby, Joseph, 165
Union Oil, 350
Union Pacific Railroad, 207, 252
Union Pacific Resources, 252
unitization, 98, 100, 103, 107, 109, 111, 176,
    214–15, 217
    compulsory, 101, 104, 108, 168–69
    voluntary, 215
unit operations, 99, 103, 105, 107, 109–11,
    184, 212, 214
    voluntary, 113
universities, oil-state, 191–94
University of Texas (UT), 111, 123, 193–96,
    235, 241, 252
Uren, Lester C., 104, 131, 155
US Bureau of Mines, 97, 165, 183, 193, 211
US oil consumption, 242–43
UT. *See* University of Texas

vagrant character, 13
values, expected, 71
Van voluntary agreement, 121
Vance, Harold, 211
Vandiver, Frank E., 220, 261
Versluys, Jan, 154, 158–59, 186
Vincenti, Walter G., 261
viscosities, 15, 100–103, 156, 158–59, 173–74, 249
visible hand, 55

Wade, Michael J., 70
waste, subsurface, 83, 109, 133–34, 147
water compressibility, 164, 190
water coning, 15, 122, 127
water contamination, 185
water drive, 15, 107, 125, 150–54, 162, 164–66, 168,
    177, 180, 186, 190
water-drive hypothesis, 131, 133, 151, 154
water drive in East Texas, 151, 154, 157, 164, 186
water encroachment, 154, 180
water influx, 236
water intrusion, 110–11, 122, 128, 163
WEIRDs, 61, 63
Welge, Henry J., 182
wells, slant, 237–38
West, DuVal, 121, 145
Westfall, Richard, 5
West Hawkins, 189
Westmoreland Natural Gas Company, 140
WG. *See* within-group
whack-a-mole, regulatory, 214
Whiggery, 168–69
White, Leslie, 20
Wiess, Harry, 46
Wilcox, 307
Wilcox Oil & Gas Co., 97
Wilde, H. D., Jr, 151, 154, 157, 161
Wilson, David Sloan, 89
Wilson, Thomas Woodrow, 81
Wineburgh, Harold, 209
wise use, 50, 97, 207, 212–14
within-group (WG), 4, 58, 65, 67, 73–74, 81, 91
Wolters, Jacob, 126
Woodbine, 118–19, 135, 148, 163, 166, 189–
    91, 203, 234
Woodbine Aquifer, 150–51, 163, 189–90
Woodbine Basin, 167, 188–89, 191, 231
    subsurface, 188, 190
Woodbine formation, 122, 148, 163, 189, 236
Woodbine oil reservoirs, 189
Woodbine salt water, 167
Woodbine saltwater compressibility, 190
Woodbine Sand, 122, 163, 166, 190
    productive, 235
Woodbine sandstone, 164
Wyckoff, Ralph, 161–62, 164, 171, 223

xenophobia, 16, 23, 75

yardstick schedules, 305n65
Yates, 37, 109–10, 113
    Ira, 109
Yates agreement, 110

Yates field, 109–11, 149
Yates pool, 109–10
Yount, Frank, 52
Yount-Lee Oil Company, 31, 209, 293n62

[Created with **TExtract** / www.TExtract.com]